Glass, Wax and Metal

Glass, Wax and Metal

Lighting technologies in Late Antique, Byzantine and medieval times

Edited by

Ioannis Motsianos and Karen S. Garnett

ARCHAEOPRESS ARCHAEOLOGY

ARCHAEOPRESS PUBLISHING LTD
Summertown Pavilion
18-24 Middle Way
Summertown
Oxford OX2 7LG

www.archaeopress.com

ISBN 978-1-78969-216-7
ISBN 978-1-78969-217-4 (e-Pdf)

Cover illustrations:

Top: Representation of wood brands, torch and lantern, detail of fresco, 14th century, Church of Hagios Nicholaos Orphanos, Thessaloniki. ©Ephoreia of Antiquities of the City of Thessaloniki.
Bottom, from left:
Glass lamp, Kolchida Kilkis, 5th century, Museum of Byzantine Culture, inventory no. BY 182. ©Museum of Byzantine Culture.
Representation of a candela, south panel mosaic detail, Rotunda of Galerius, Thessaloniki. ©Ephoreia of Antiquities of the City of Thessaloniki.
Representation of a candlestick, south panel mosaic detail, Rotunda of Galerius, Thessaloniki. ©Ephoreia of Antiquities of the City of Thessaloniki.
Copper alloy polycandelon, 6th century, Museum of Byzantine Culture, inventory no. BA 77/1. ©Museum of Byzantine Culture.

This book is available direct from Archaeopress or from our website www.archaeopress.com

Contents

Glass, Wax and Metal: Lighting Technologies in Late Antique, Byzantine and Medieval Times: an Introduction

Karen S. Garnett and Ioannis Motsianos

The International Lychnological Association (ILA),[1] a non-profit association founded in 2003 by Dr Laurent Chrzanovski, promotes the knowledge of pre-modern lighting devices, in part by organizing congresses and symposiums.

The present collective volume is the result of two Round Tables organized by the International Lychnological Association to study the technological development of lighting and lighting devices during Late Antiquity and the Middle Ages[2] in Western Europe[3] and Byzantium.[4] The third International Round Table under the title 'Dark Ages? History and archaeology of lighting devices in Continental Europe, from Late Antiquity to late Medieval Ages' was held in Olten, Switzerland from 24 to 29 September 2007 in cooperation with the Historical Museum of Olten (Figure 1). The fourth International Round Table under the title 'Lighting in Byzantium' was held at the Museum of Byzantine Culture, Thessaloniki, Greece from 10 to 15 October 2011 and was organized in cooperation with the Archaeological Institute of Macedonian and Thracian Studies, the Aristotle University of Thessaloniki and the Museum of Byzantine Culture (Figure 2).

Both Round Tables[5] were part of the series of colloquia promoted by the International Lychnological Association that allowed specialists to develop a specific theme (chronological, geographical, typological, etc.) during an exclusive meeting, with more time available for exchange than would be available during the large open triennial congresses.

When organizing these Round Tables we took into account that, while the symbolic, political and religious importance of artificial light increased dramatically during Late Antiquity and the Middle Ages, scholarly research on this timespan appeared less frequently in the scientific literature than research for the preceding historical periods. Nevertheless, the major economic, geopolitical and social changes during the Middle Ages are reflected by radical modifications of lighting devices, and the investigation of these relationships deserves great attention. Lighting devices in Medieval Western Europe and Byzantium provided illumination for everyday activities, at work, at home, in public spaces and in places of worship. Lighting devices were also an important element of the ritual ceremonies of private, sacred and social life and, at the end, were an accompaniment at each individual's final resting place. Much information can be derived from the lighting artifacts preserved in museums and private collections as well as from those devices with archaeological contexts. The written sources and depictions of lighting devices on mosaics, frescos, icons, textiles and manuscripts can help scientists to complete their notions about lighting in the Middle Ages. Both ILA Round Tables' considered the use of lighting devices in everyday and ecclesiastical life and discussed many aspects of the lighting devices, including their terminology, typology, chronology, manufacturing techniques, and symbolic functions. The great breadth of lighting technologies available to those 'Dark Ages' became apparent through the diversity of the discussions, which reflected the great variety of materials used to create lighting devices in addition to the traditional terracotta used for wheel-made and mold-made lamps. Hence the introductory title to this volume: glass, wax and metal were all important materials for the lighting technologies of these eras.

And above all, these Round Tables offered to the participating scholars a venue for comparing two different areas of lychnological interest. In medieval Western Europe (geographically in central and north Europe) during the depth of winter the daylight lasts but a few hours -not long enough to let people organize and complete all their everyday activities. Also, the production and the evolution of lighting devices in Mediterranean areas was influenced by the use of the olive oil that was used as a fuel. In northerly regions remote from the Mediterranean there was no local production of olive oil and animal fat was the main fuel for the lighting devices of the everyday life in these regions.

In Byzantine territory the use of the olive oil, and from the end of Late Antiquity the use of beeswax, was taken for granted not only by the Christian churches and monasteries but also by rich and prosperous people. However, it is not

[1] http://www.lychnology.org, viewed 10 December 2018.

[2] According to Encyclopaedia Britannica 'The period of European history extending from about 500 to 1400-1500 CE is traditionally known as the Middle Ages. The term was first used by 15th-century scholars to designate the period between their own time and the fall of the Western Roman Empire. The period is often considered to have its own internal divisions: either early and late or early, central or high, and late'. https://www.britannica.com/topic/history-of-Europe/The-Middle-Ages, viewed 15 December 2018.

[3] We use this term conventionally to describe the European regions outside the borders of the Byzantine Empire's territory.

[4] 'Byzantine Empire' or 'Byzantium' is the conventional name of a medieval state that existed for more than one thousand years. The date of its beginning remains a subject of discussion; most scholars prefer the date 324 (or 330) when Constantinople was founded by Constantine the Great, or 395, when the Roman Empire was divided between the sons of Theodosius I. It ceased to exist in 1453 when Constantinople was captured by the Ottomans. See Kazhdan, A. P. (ed.) 1991. *The Oxford Dictionary of Byzantium*, Prepared at Dumbarton Oaks, Vols. 1-3. New York-Oxford: Oxford University Press. Here, vol. 1, s.v. Byzantium, 344-345.

[5] For the chronicle of their organization, see in this Volume, Laurent Chrzanovski, A Highly Important 'Half-volume' as Acta of two Pioneer Congresses: some Explanations.

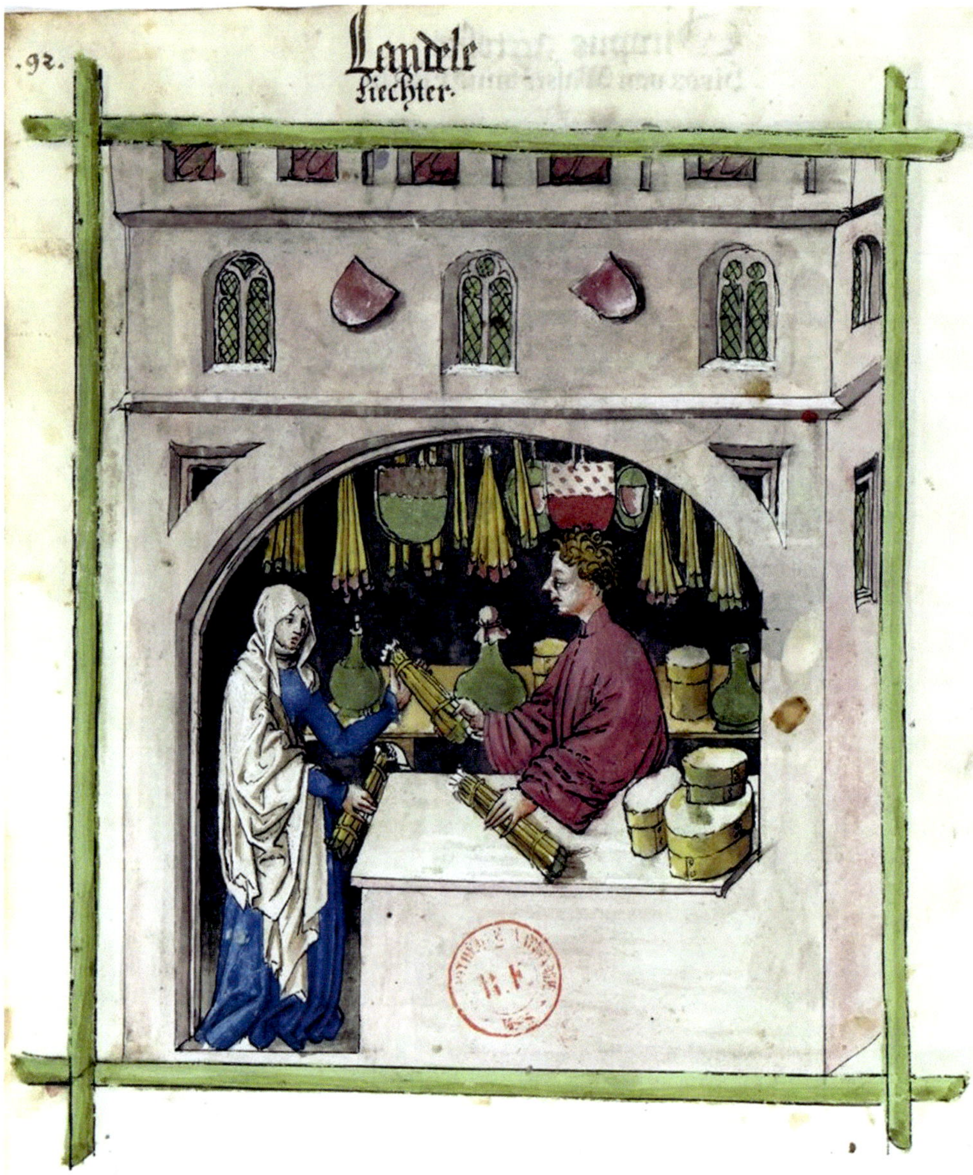

Figure 1. The wax seller, illustration from the 'Tacuinum sanitatis' of Albucasis, bilingual Rhenish manuscript of the 15th century (fol. 92v.), Paris, Bibliothèque nationale de France (Photo from Chrzanovski and Kaiser 2007, color plate 11).[6]

clear if the common folk had the same fiscal ability to make similar choices in their everyday life.[6]

The two ILA Round Tables were accompanied by exhibitions and extensive exhibition catalogues. The *'Dark Ages? Licht im Mittelalter / L'éclairage au moyen-âge'*[7] exhibition opened in the early fall of 2007 and the'Μια ιστορία από φως στο φως/ *Light on light: an illuminating story'*[8] exhibition opened in the fall of 2011[9] in Thessaloniki and subsequently travelled to Athens the following year.

An aftereffect of the exhibition *'Light on light: an illuminating story'* was the idea for the realization of *The Oxford Handbook of Light in Archaeology*[10] published online in June 2017. Since then there have been wonderful lychnological exhibitions

[6] Chrzanovski, L. and P. Kaiser (eds) 2007. *'Dark Ages? Licht im Mittelalter / L'éclairage au moyen-âge'*. Milano: Edizioni Et.

[7] Chrzanovski, L. and P. Kaiser (eds) 2007. *'Dark Ages? Licht im Mittelalter / L'éclairage au moyen-âge'*. Milano: Edizioni Et.

[8] Motsianos, I. and E. Bintsi, (eds) 2011. Μια ιστορία από φως στο φως/ Light on Light: an illuminating story. Thessaloniki: Folklife & Ethnological Museum of Macedonia-Thrace.

[9] For more information see in this Volume, Laurent Chrzanovski, A Highly Important 'Half-volume' as Acta of two Pioneer Congresses: some Explanations.

[10] Papadopoulos, C. and H. Moyes (eds) *The Oxford Handbook of Light in Archaeology*. Online publication: Oxford University Press. http://www.oxfordhandbooks.com/view/10.1093/oxfordhb/9780198788218.001.0001/oxfordhb-9780198788218, viewed 24 January 2019.

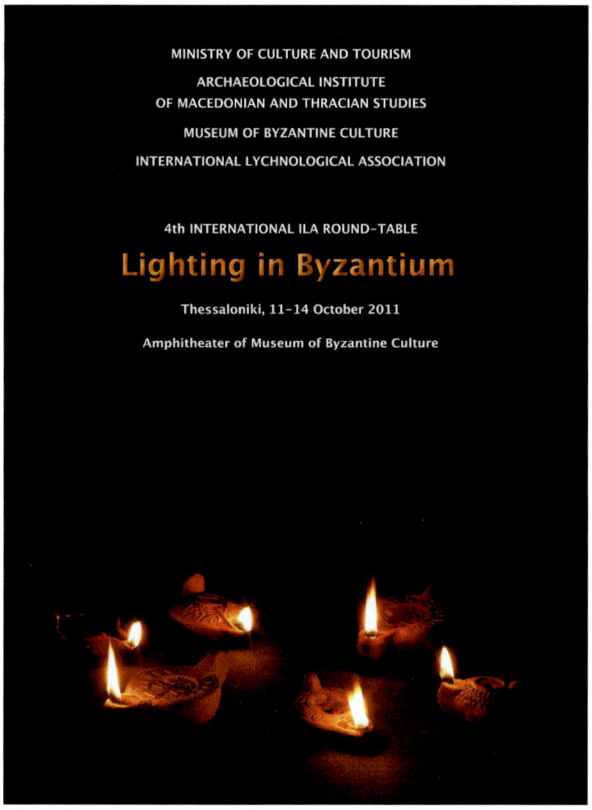

Figure 2. Poster of the 4th International Round Table under the title 'Lighting in Byzantium'.

Figure 3. Logo of the International Year of Light 2015.

in Geneva[11] and in Cluj-Napoca[12] presented in 2012 and 2013 respectively.

Interestingly, according to John Dudley, Chair of the International Year of Light 2015 Steering Committee: 'When discussions of IYL 2015 first began in 2009 within the scientific community, I don't think that anyone could have foreseen the extent of the worldwide enthusiasm around the theme of light, and neither did we anticipate the many new linkages that would appear between science and art and culture'. Continuing this vein, the final report of the International Year of Light 2015 underlines that 'Under the leadership of UNESCO, IYL 2015 brought together hundreds of national and international partners to raise awareness of the importance of light science and technology in areas such as sustainable development, energy, education, climate change, and health.'[13]

Taking into account not only the results of the Round Tables in Olten and Thessaloniki, the exhibitions and the catalogues which accompanied them and the other ones which followed but also the final report of the International Year of Light, we decided that the publication of the present collective volume should include among its goals both to clarify some topics and also to provoke interest in other topics which were not so well-studied to date. Initially the main goal of this volume was to delineate the similarities and the differences concerning lighting and the lighting devices of Western Europe and Byzantium in the Middle Ages and identify relations between them and the social and financial situations of every region.

The majority of the twenty nine papers published in the present volume have been presented in the Round Tables of Olten and Thessaloniki. A few additions provide enhancements to the volume. In many cases the length of each paper reflects how little -or well- studied is the presented topic. Also, we present in this volume discussions on some artifacts dated after 1500 AD because they represent and reflect the technological evolution of lighting related to the Middle Ages.

In the present volume:

Verena Perko with the collaboration of Vesna Tratnik presented '*An Overview of Late Antique Oil Lamps and Lighting Devices from Slovenia*', a short survey of Late Antique lighting devices from the territory of today's Republic of Slovenia.

Thomas Bitterli in his paper '*A Light is on in the Hut. Light and Lighting Equipment in Medieval Everyday Life*' takes into account archaeological evidence as well as literary and iconographic sources which provide vital clues on lighting equipment used in Medieval everyday life.

The paper of Svetlana Avdusina '*Medieval Lighting Devices from the Collection of the State Historical Museum of Russia*' describes the medieval lighting devices from the collections of the State Historical Museum of Russia in Moscow which includes: lamps, candleholders and candles, as well as splinter holders.

Catherine Vincent in the paper '*Lumière et luminaires dans la vie religieuse en Occident au Moyen Âge*' provides a brief panoptic on the elaboration of the lighting, the faith, the symbolism and the use of lighting devices in French regions, not only from ancient Rome but also up to the 13th century AD.

In the paper of Arja Karivieri '*Olaus Magnus the Goth on Fire, Light and Lighting Devices of the Northern People*' a special

[11] Chrzanovski, L. (ed.) 2012. *A la tombée de la nuit... art et histoire de l'éclairage*. Genève-Milano, Musées d'art et histoire de Genève, Genève.
[12] Chrzanovski, L. (ed.) 2013. *LVMEN EST OMEN Arta, istoria şi spiritualitatea iluminatului artificial*. Cluj-Napoca; downloadable from https://laurentchrzanovski.academia.edu
[13] http://www.light2015.org/Home/About/IYL-Final-Report.html, viewed 24 January 2019.

focus is given to the numerous and various instances where Olaus Magnus comments on fire, light and lighting devices in Scandinavia, including the symbolic meaning of fire and light at wedding ceremonies, lighting at guild feasts, light in Christian processions and in the church, lighting in wintertime, light and fire in summertime, fire and lights during warfare, birds and lights, and fishing by means of fire.

Lena Berg Nilsson in her paper *'Mines Illuminated – Reflections upon Lighting in Medieval Mining'* takes into account some written sources and examples from contemporary art in order to present to us some aspects of lighting methods used in the mining industry in medieval Sweden.

Yvonne Seidel studied the Tabula Peutingeriana, a reproduction of an ancient world map, in her paper *'Lighthouses in the Tabula Peutingeriana and their Importance in Late Antiquity,'* greatly expanding our knowledge regarding lighthouses and the use of light signals based on the use of fire in Late Antiquity.

The paper of Ioannis Motsianos *'Lighting Devices in Byzantium: Comparisons in Time and Space'* attempts to address common practices in lighting devices used during the Byzantine era and examines the differences between Byzantine and Roman practices. These differences also help identify the interactivity between corresponding lighting devices used in Central and Northern Europe.

Sophia Akrivopoulou, Paraskevi Leventeli, and Periklis Slampeas in their paper *'Lamps from the Excavation at 45 Basileus Heracleiou Street, Thessaloniki'* added new data to our knowledge about the lamps found in Thessaloniki and new indications for local lamp production, possibly secondary, based on the technique of surmoulage.

The paper *'Lighting Artifacts from the Episcopal Complex at Louloudies Pierias'* written by Evangelia Angelkou and Maria Cheimonopoulou is a panorama of lighting devices used in the Early Christian Episcopal Complex at Louloudies Pierias based on excavational data.

'Bottoms Up! Bases and Handles on Attic KY Shop Lamps from Corinth's Fountain of the Lamps' is the title of the paper in which Karen S. Garnett present new observations and comments concerning the chronological typologies evident when special attention is paid to the bases and handles of lamps produced by a single lampmaker. Her examples were lamps produced by the Athenian lampmaker KY which were part of a single large deposit of over 4000 lamps found in the Fountain of the Lamps in Ancient Corinth.

The paper of Arja Karivieri *'Athenian Lamps in the Early Byzantine period – Export, Import and Imitation'* analyzes in detail what methods the Athenian lamp producers used in their efforts to keep their trade contacts and market towards the end of the 5th century, a time when North African and Asia Minor products became more popular in the Eastern Mediterranean. Another aspect of the article deals with the imitation of imported lamps during this period.

Renate Rosenthal-Heginbottom focused on lamps from domestic and funerary contexts rather than on those in museums and private collections, many of which were acquired on the antiquities market, to support the aim of her article as clearly described by its title *'Aspects of typology, chronology and iconography in the regional lamp production in the area of present-day Israel (4th-7th centuries AD)'*.

Lambrini Koutoussaki in her paper *'Argos and its Lamps from the late 4th to 7th Century AD'* informs us that Argos was, without doubt, one of the main production centers in the region and a lot of the *officines* specialized in the production of copies.

Anastassios Ch. Antonaras in his paper *'A Three-handled, Calyx-shaped glass lamp from Thessaloniki and its Archaeological Context'* proposes a precise chronology for a three-handled calyx-shaped glass lamp, a chronology based closely on its archaeological context.

The article of Anastasia G. Yangaki *'Some Remarks on a Category of Wheel-Made Lamps: A 'Koinè' in the 6th and the 7th Centuries AD?'* explores a category of wheel-made lamp which is characterized by an ovoid body, an oblong nozzle and a cylindrical neck. Based on the diffusion of its six sub-types and the range of the production of each, it is apparent that such wheel-made lamps were common during the 6th and the 7th centuries AD in regions of the eastern Mediterranean.

Christopher S. Lightfoot in his paper *'Lighting Devices found at Byzantine Amorium (Turkey)'* presents us with different types of lighting devices that have been found during the excavations at the Byzantine city of Amorium in central Anatolia. They show that lighting devices of several different types continued to be used throughout the site's existence as a Byzantine city (late 5th–late 11th centuries). In addition to metal and glass lighting devices used in churches and other privileged contexts, small and relatively inexpensive terracotta lamps continued to be produced for use in daily life.

The paper of Stanislav Ryzhov and Tatyana Yashayeva *'Bronze Lightning Devices in Churches of Byzantine Cherson'* presents and interpets a special group of devices composed of bronze church lighting implements from Cherson: icon lamp stands, church chandeliers and candelabra, which played an important practical role in the church interior and also had some symbolical value.

Mara Verykokou focuses on the problems concerning the chronology of Byzantine polykandela that are from museums and private collections and underlines the need for their analysis in her paper *'Problems of Methodology in the Dating of Byzantine Polykandela: The Benaki Museum Examples'*.

Paschalis Androudis and Ioannis Motsianos in their article *'Byzantine metal support fittings for a candlestick from the Monastery of Chilandar on Mount Athos'* propose dates and provide an interpretation for two Byzantine metal support fittings that initially were parts of a candlestick and then re-used as supports for icons.

New research about candelabra is presented in the paper of Mariela Inkova *'Once again on the 'Byzantine-Mediterranean Limoges'?'*. Moreover, Mariela Inkova showed us a possible relation between the Byzantine Candelabra and ones manufactured in Western Europe.

Georgios Velenis and Stavros Zachariadis in their article *'Considerations on the Function and Usage of Pottery Lamps, Inspired by finds from the Forum of Thessaloniki'* takes into account experiments, analytical techniques, tradition and literal sources, supplemented by archaeological evidence, to provide us with a wealth of information concerning the use and function of closed (nozzled) lamps.

Ioannis Motsianos describes the *'Difficulties and Preconditions for the Utilization of Glass in Lighting in Byzantium'* and discusses the technological issues that had to be resolved before glass could become an efficient and practical vessel for lighting.

Naama Sukenik and Yotam Tepper present *'A Linen Wick from the Northern Church at Shivta, Israel'* and give us valuable information about this rare find.

In his article *'Some thoughts about the use and the making of beeswax candles in Byzantium'* Ioannis Motsianos investigates the infrastructure required for the production of beeswax candles: the tools, the utensils, and methods used for this process during the Byzantine and Post Byzantine periods.

Pelli Mastora in her article *'The Virtual Lighting of the Rotunda's Mosaics'* shows us that not only lighting devices but also their depictions can play an important role in our understanding of the lighting of monuments.

The paper of Ioannis G. Iliades *'Light and Lighting Devices in Wall Paintings of Byzantine Churches in Thessaloniki'* presents some of the lighting devices depicted on wall paintings in the monuments of Thessaloniki. In some cases, he is able to make comparative analyses of the lighting devices and their light as depicted in paintings of the same or later period.

The paper of Efterpi Marki *'A beacon from the castle of Kitros, Pieria, Greece'*paper describes a special construction found at the medieval castle of Kitros which could be interpreted as a φρυκτωρία (beacon), i.e., a system of direct communication based on the use of fire which allowed for the rapid transfer of messages over long distances.

The paper of Laurent Chrzanovski *'Lessons of Modernity from the Past: some Amazing Parallels drawn from Antique and Medieval Lychnological Economics'* draws some parallels from the past to the present: rich civilizations without good fuel for lighting, long distance imports, fuel taxes and bureaucracy, fuel poverty, monopolies, monopsonies as well as the use of child labor in industries. All these topics, most of which can be considered as features in contemporary society and among the hottest challenges in global macro-economics, can be found in different civilizations: in ancient and medieval civilizations, in the Classical world and later not only in Medieval Europe but also in the civilizations of today and on other continents.

This volume provides an excellent overview of medieval lighting devices used in medieval Western Europe and Byzantium and coming from a wide range of collections and archaeological research. Scholars have brought forth and coordinated information from sometimes obscure collections or research references to clearly explain and illustrate their ideas.

Additionally, the material covered reaches way beyond basic research into the chronology and typology of the terracotta oil lamp. The articles not only expand our discussions of the scope of materials used as lighting devices, but also expand the timeline for our discussion of ancient lighting devices, going well beyond the Late Roman era when terracotta decreases as the predominant material for lighting vessels. The researchers show us many new ways to look at the physical record presented by the lamps and lighting vessels themselves. And the scholars have also provided information which helps us to recognize the manifestation of these various devices in both pictorial and written records.

In contrast to the lighting common in our electrified world, in Ancient times and indeed right up to the end of the 19th century AD, artificial lighting and fire were one and the same. Whatever gave light was also an agent of combustion, with all the concomitant advantages and disadvantages. That which illuminated, burned; and that which burned, illuminated.[14] Taking into account that one of the five main aims and goals of *The International Year of Light and Light-based Technologies 2015* was 'to reduce light pollution and energy waste'[15] we conclude that maybe the biggest lesson we receive by studying the lighting of the past is that supplies for lamps were the fuels and wicks made from materials that were renewable. It was the nature that could provide people the consumables for their luminaires (tinder, sticks of resinous wood, etc.) or they could acquire them cultivating plants (olive oil, flax, cotton, hemp, etc.) or raising livestock (animal fat). Moreover the people who used them could immediately recognize their abundance or lack –a fact that could prompt them to consume fuels for lighting in moderation, since in most cases the excessive use of lighting devices could also lead to shortages in their diet.[16]

[14] Motsianos, I. 2011. Artificial Lighting during Byzantine and Post-Byzantine Period, in I. Motsianos and E. Bintsi (eds) *Light on light, an illuminating story*, Exhibition Catalogue, 90-125. Thessaloniki: Folklife & Ethnological Museum of Macedonia-Thrace. Here, page 91.

[15] https://www.google.com/url?sa=t&rct=j&q=&esrc=s&source=web&cd=1&ved=2ahUKEwj54rfkxYjgAhWPNOwKHY30CS8QFjAAegQIABAC&url=http%3A%2F%2Fwww.light2015.org%2Fdam%2FAbout%2FResources%2FIYL_EPS_ShortPresentati.pptx&usg=AOvVaw0UqgZblxoQ8q0RiXxr-3Hn viewed 10 January 2019.

[16] Wunderlich, Ch.-H. 2003. Light and Economy an Essay about the Economy of Prehistoric and Ancient Lamps, in L. Chrzanovski (ed.) *Nouveautés Lychnologiques - Lychnological News*, 251-264. Neuchâtel: Chaman Edition.

A Highly Important 'Half-volume' as Acta of two Pioneer Congresses: some Explanations

Laurent Chrzanovski
Doctoral School of History and Humanities, Lucian Blaga State University, Sibiu
Founder and Secretary-General, International Lychnological Association

We remember this moment as if it was yesterday. The scene took place during the 2nd Congress of the International Lychnological Association: held at Zalau and Cluj-Napoca (Romania), 13th to 18th of May 2006, one of the best organized events of our series due to the friendly atmosphere created and managed by the late Dorin Alicu with the staffs of both Zalau Provincial Museum and Cluj-Napoca's National Museum of Transylvanian History.

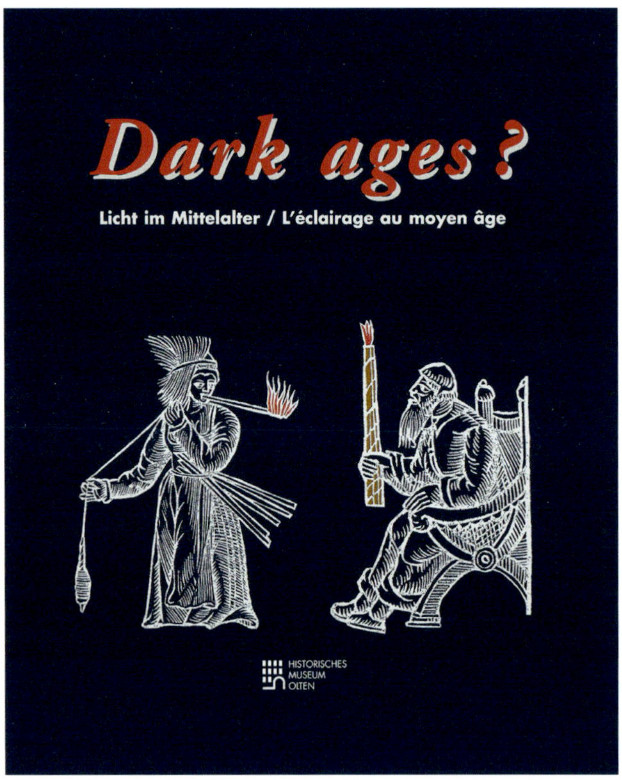

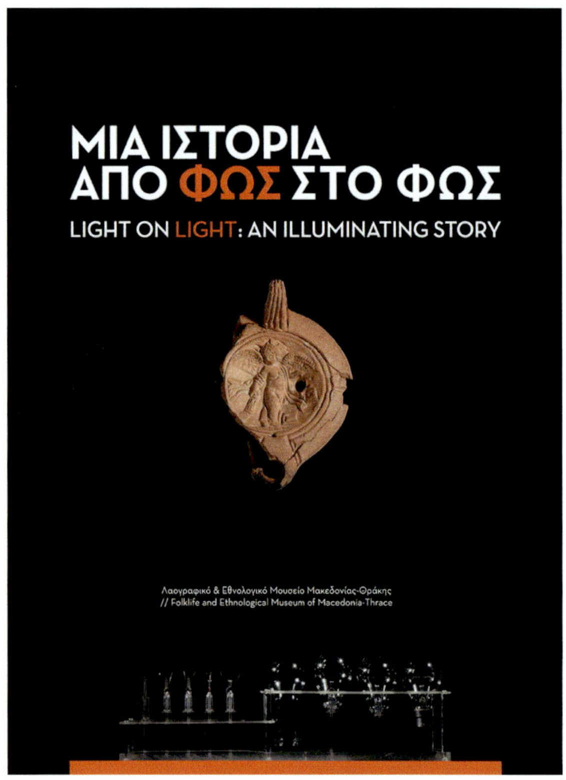

Dorin was one of the few archaeologists we knew who had, no matter the funds at his disposal, hospitality, dialogue and networking set as priorities in his DNA, so it is not a hazard if the conversation mentioned hereunder was held during one of the extremely generous hyper-caloric -how many dishes could have provided fuel for lamps!- and open-bar dinners which lasted long into the night.

During discussions within the board of ILA, we raised the proposal of driving an innovative attempt aimed to foster research on 'what happened next?', meaning how did people light their homes and workplaces after the timeframe beloved and studied by almost all our colleagues -from the Archaic period to the Late Roman era.

Half of the table's guests looked at us with eyes full of skepticism -no words needed to be added. The other half expressed interest, but all responses offered were, in a compulsory way, politely accompanied by all forms of conditional expressions, studded with 'but', 'how', 'if', 'nevertheless', and so on...

'Young and crazy', as we are also now -even if no more so young-, Ioannis Motsianos and myself engaged in a constructive dialogue, shortly ending with 'OK, let's do a medieval times lighting round-table in two sessions: you organize the Continental European part and I take care of the Byzantine one, at Thessaloniki.'

What happened after that was even more 'crazy' if we look at the enthusiasm the idea generated not among lychnologists, but among museum directors as well as among specialists of different periods and topics.

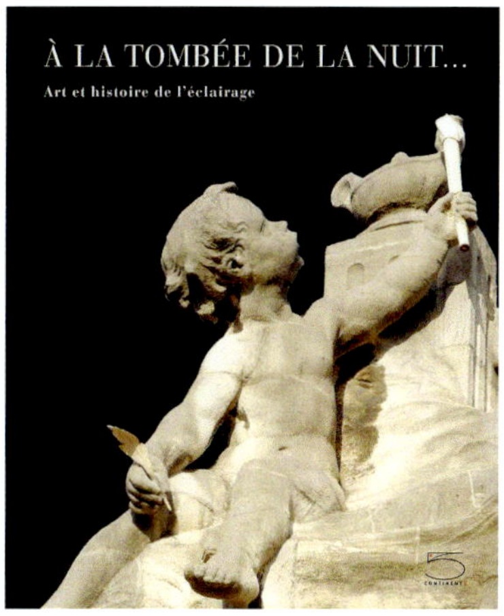

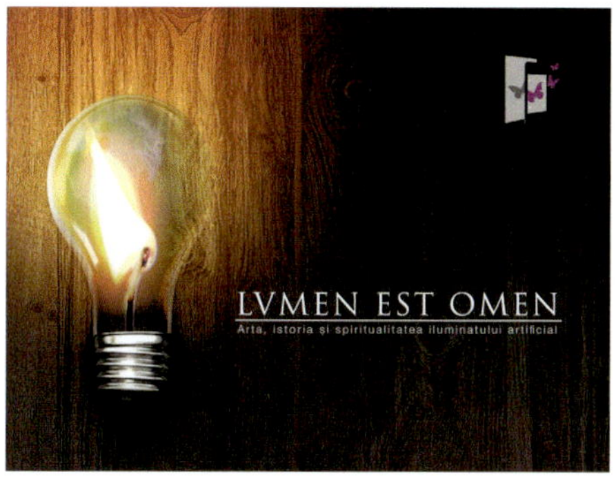

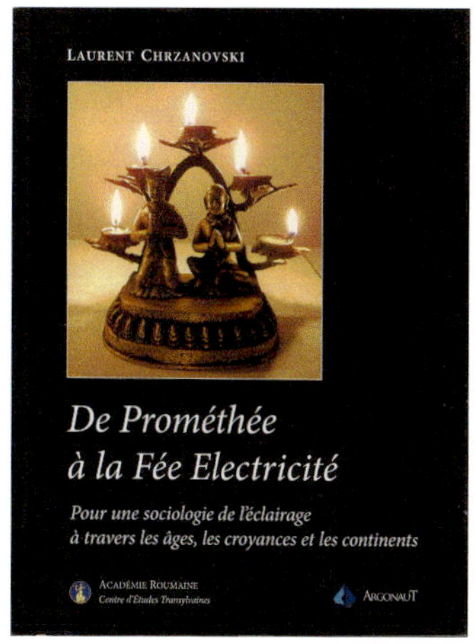

As a result, much more than two small meetings between niche-specialists occurred. Both scholarly events were accompanied by what remains, in our point of view, the two most exemplar exhibitions on the subject ever realized in Europe. Outstanding in terms of their trans-disciplinarity and the innovative ideas they introduced to their visitors, these exhibits achieved a major, yet very rarely reached, objective: to be coherent, attractively arranged and seductive for the general public.

In Switzerland, Peter Kaiser, then Director of the Historishes Museum Olten, took the challenge and we worked hand in hand to gather all the possible Swiss 'exhibition-friendly' materials as well as not a few foreign contributions to help the viewer better understand the objects' contexts by setting the Helvetic artifacts within their broader continental framework.

In Greece, Foteini Oikonomidou-Botsiou, then Director of the Folklife and Ethnological Museum of Macedonia-Thrace in Thessaloniki, gave a green light to Ioannis Motsianos and Eleni Bintsi to build an even more comprehensive and amazing show: a complete history of light and lighting devices.

Furthermore, both exhibitions were opened with huge catalogues (respectively 384 and 462 pages) which are far more than just a visitor's booklet. They offer, along with a crystal-clear and useful introduction to all topics, the presentation of each object shown at the museums and much, much more. Both volumes are bilingual (French and German for the first, Greek and English for the second), in a rare attempt to make the syntheses available to the broadest number of readers possible.

'*Dark Ages? Licht im Mittelalter* / L'éclairage au moyen-âge'[1] exhbition opened in early fall of 2007, and a six day Round-Table of the same title was hosted from the 24th to 29th of September. 'Μια ιστορία από φως στο φως/ *Light on light: an illuminating story*',[2] exhibition opened in the fall of 2011, as the Round-Table 'Lighting in Byzantium' commenced at the Museum of Byzantine Culture from 11th to 14th of October, thanks to its director, Dr Susanna Choulia-Kapeloni.

The provocative title we chose for this introduction, i.e., the '*half-volume*' you have in your hands as *Acta* is in fact the result of all the contributions already published in both volumes, mainly the first, which gathers in its largest chapter a full '*status quaestionis*' as delivered by all the component Swiss cantonal archaeological services which have led excavations and researches to fruitful results on the topic. The 'Continental European' part of the present acts is then considerably diminished in comparison to the speeches presented at Olten and which were already published in the catalogue, which is readable for free on the Association's website and also downloadable on our research website.[3]

Returning to the present work, we have to salute the enormous job made by Ioannis Motsianos, 'constrained' on a voluntary basis, to gather all the texts and edit this volume at the same time we are compiling an analogue work on the *Acta* of the 4th and 5th Lychnological Congresses, held at Ptuj and Sibiu in 2012 and 2015.The financial crisis is still here, and not only in Greece, and is affecting our profession in its most '*politically unattractive*' aspect, its scientific publications.

Thanks to Dr Motsianos efforts, the papers constituting this volume offer to the scientific community a huge contribution to the published scholarship about these poorly known Byzantine lighting devices, especially when compared to the quantity of literature devoted to lamps from preceding periods. This book will be, irony of history, only the second existing 'milestone' on this topic, the first being a huge scholarly effort (2 volumes) published also... in Thessaloniki, in 2010.[4]

As a more personal conclusion, we would like to take a step backward and look again at the volume of the amazing exhibition held in 2011. For us, it was such a revelation in the

[1] Chrzanovski, L. and P. Kaiser (eds) 2007. '*Dark Ages? Licht im Mittelalter / L'éclairage au moyen-âge*'. Milano: Edizioni Et.

[2] Motsianos, I. and E. Bintsi (eds) 2011. *Μια ιστορία από φως στο φως/ Light on Light: an illuminating story*. Thessaloniki: Folklife & Ethnological Museum of Macedonia-Thrace.

[3] For reading online : <http://www.lychnology.org/flip/dark_ages/>, viewed 15 July 2018; for downloading: <https://laurentchrzanovski.academia.edu> , viewed 15 July 2018.

[4] Παπανικόλα-Μπακιρτζή, Δ. and Ν. Κουσουλάκου (eds) 2010. *Κεραμική της ύστερης αρχαιότητας από τον Ελλαδικό Χώρο (3ος-7ος μ.Χ.)*. Επιστημονική συνάντηση, Θεσσαλονίκη, 12-16 Νοεμβρίου 2006. Θεσσαλονίκη (2 volumes), gathering no less than a dozen articles about late Roman and Byzantine lamps.

variety of topics it presented that it has been the source of inspiration not only of the two exhibitions we made shortly thereafter, one in Geneva[5] and another in Cluj-Napoca,[6] the last, being itinerant and gathering more than 500,000 visitors during the two years it travelled, but also for opening our eyes to the religious, cultural and ethnographic aspects of lighting, a subject we could finally research and publish thanks to a generous EU postdoctoral grant.[7]

Dozens of ideas and paths of research would very probably not have captured our interest if it weren't for the exhibition held at the Folklife and Ethnological Museum of Macedonia-Thrace, Thessaloniki, fruit of Ioannis Motsianos' open-mindedness and the collaboration he obtained for conceiving and bringing to fruition the event.

For all this, we have to say an enormous '*thank you, Ioannis*'! We hope the present volume will soon be followed by a resurgence of interest in publications on the illuminating devices of this very eclectic period.

Thessalonki, Motsianos' home, 25th of July 2018.

[5] Chrzanovski, L. (ed.) 2012. *A la tombée de la nuit... art et histoire de l'éclairage.* Genève-Milano, Musées d'art et histoire de Genève, Genève.
[6] Chrzanovski, L. (ed.) 2013. *LVMEN EST OMEN Arta, istoria și spiritualitatea iluminatului artificial.* Cluj-Napoca; downloadable from https://laurentchrzanovski.academia.edu
[7] Chrzanovski, L. 2013. *De Prométhée à la Fée Electricité. Pour une sociologie de l'éclairage à travers les âges, les croyances et les continents.* Cluj-Napoca, Académie Roumaine; downloadable from https://laurentchrzanovski.academia.edu

Acknowledgements

ILA Round Tables form part of a series of conferences promoted by the *International Lychnological Association* (ILA) in order to allow specialists to develop a precise theme in an exclusive meeting. The 3rd ILA colloquium in Olten was organized in 2007 after two preceding meetings devoted to a precise geographical area, the 1st taking place in Amman and Petra (Jordan) in November 2005, on the theme 'Ancient lamps from Bilad esh-Sham, from the Phoenicians to the Ummayads', and the 2nd in Millau (France) in March 2007, on the theme 'Archéologie et histoire de l'éclairage en Gaule, de la Préhistoire au début du Moyen Âge'.

While their symbolic, political and religious importance increased constantly, Late Antique and Medieval lighting devices from Continental Europe have been astonishingly neglected in scientific literature, at least compared to the richness of research concerning the previous periods. Nevertheless, the major economic and geo-political changes of this period caused radical modifications in the means of lighting, which deserve a great scholarly interest. ILA received a strong demand from archaeologists and historians to organize a meeting in order to launch a new discussion, make the research more dynamic and to make a synthesis of the acquired knowledge on lighting devices from the 5th to the 17th century. This was the aim of the Olten colloquium, as well as the organization of a national exhibition on the same theme 'Dark Ages? Licht im Mittelalter', organized by the Museum of Olten in honor of this event.

The 3rd International ILA Round Table 'Dark Ages? History and Archaeology of Lighting Devices in Continental Europe from Late Antiquity to Late Mediaeval Period' was thus organized in Olten, Switzerland, in 24-29 September 2007. The Olten colloquium, organized in the Historical Museum at Olten was placed under the high patronage of Dr Charles Kleiber, the Swiss Secretary of State for Research and Science at that time, and the then mayor of Olten, Mr. Ernst Zingg, and it was supported by the French Centre national de la recherche scientifique (CNRS) and the Swiss National Science Foundation (SNSF). We also wish to thank *Instrumentum* (European Research Group on Crafts in Antiquity), the president of honor, Prof. Catherine Vincent from the University Paris X, the president Dr Peter Kaiser, Curator of the Historisches Museum in Olten, the secretary Dr Laurent Chrzanovski, the Secretary General of the ILA, and the members of the organizing committee, Prof. Massimiliano David, University of Bologna, Dr Michel Feugère, CNRS, Dr Pierre Harb, Director of the Archaeological Service Solothurn, Prof. Jerzy Holubiek, Polish Academy of Sciences, Dr Danièle Foy, CNRS, and Dr Cosmin Rusu, National Museum of History of Transsylvania, for all their support.

The 3rd ILA Round Table is tied into the 4th colloquium, which deals with the same subject but is devoted to the Byzantine and Levantine worlds. The 4th International ILA Round Table 'Lighting in Byzantium' was held four years later, 11-14 October 2011, in the amphitheater of the Museum of Byzantine Culture in Thessaloniki. As was the case with Continental Europe, the research on lighting in Byzantium during Late Antiquity and in the Byzantine era has been astonishingly neglected in the scientific literature, at least in comparison to lighting research for the preceding historical periods. Lighting devices in Byzantium had multiple roles far beyond utilitarian purposes. They provided illumination for everyday activities, at work, at home, in public spaces, and places of worship. Lighting devices were an important element of the significant rituals of Byzantine private, sacred and social lives. Valuable information may be derived from artifacts found in excavations or exhibited in museums and private collections. The Round Table focused on Byzantine lighting devices, including: their typology and chronology, their production and workshops, terminology, the symbolic role of light and lighting devices in Byzantium, and the use of lighting devices in everyday and ecclesiastical life.

The Thessaloniki colloquium about light and lighting devices in the Byzantine world was aimed to cover the Byzantine world, Eastern and Southern Mediterranean, and it was organized with the patronage and the support of the Greek Ministry of Culture and Tourism, Archaeological Institute of Macedonian and Thracian Studies, Museum of Byzantine Culture and International Lychnological Association. During the Round Table the participants had the occasion to visit a special exhibition on artificial lighting from the Byzantine era to the 20th century 'Light on light: an illuminating story', presented in the Folklife and Ethnological Museum of Macedonia-Thrace, as part of the program called 'Thessaloniki: Crossroads of Civilizations', organized by the Greek Ministry of Culture and Tourism.

We wish to thank warmly all the members of the scientific committee, Director, Dr Polyxeni Adam-Veleni, Prof. Georgios Velenis, Aristotle University of Thessaloniki, Prof. Stella Drougou, Aristotle University of Thessaloniki, the then Director of 9th Ephorate of Byzantine Antiquities Dr Despina Makropoulou, the then Director of the Archaeological Institute of Macedonian and Thracian Studies Dr Vasiliki Misailidou-Despotidou, Dr Ioannis Motsianos from the Museum of Byzantine Culture, the Director emerita of the Archaeological Institute of Macedonian and Thracian Studies Dr Demetra Bakirtzi, and the then Director of the Museum of Byzantine Culture Susanna Choulia-Kapeloni, as well as all the members of the organizing committee, Prof. Stella Drougou, Director Dr Vasiliki Misailidou-Despotidou, Dr Ioannis Motsianos, Dr Anastasios Antonaras, Director Susanna Choulia-Kapeloni and Karen Garnett. It would not have been possible to organize the Thessaloniki colloquium without the efficient and important cooperation of the 9th Ephorate of Byzantine Antiquities, 10th Ephorate of Byzantine Antiquities and the Folklife and Ethnological Museum of Macedonia-Thrace. We also wish to thank Mrs Eleni Stoubou-Katsamouri for her help in the production of the information for the colloquium. The colloquium was organized with the support of Malliari Paideia AE.

As the president of the ILA from 2003 to 2015, I am really grateful for all the help provided by the Historical Museum at Olten, as well as the Museum of Byzantine Culture during the 3rd and 4th ILA Round Tables. I am especially thankful to Dr Ioannis Motsianos who as the representative of the ILA for Greece and being responsible for lighting devices for the Byzantine and later periods suggested the organization of a Round Table in Thessaloniki. We owe a great debt of gratitude to him and Karen Garnett for making this volume of the proceedings of the two ILA Round Tables a coherent whole.

For all the ILA members,
in February 2019,
Professor Arja Karivieri
President of the International Lychnological Association (ILA)
from 2003 to 2015

An Overview of Late Antique Oil Lamps and Lighting Devices from Slovenia

Verena Perko
with collaboration of Vesna Tratnik
Regional Museum Kranj, University of Ljubljana, Slovenia

Abstract: The article represents a short survey of Late Antique lighting devices from the territory of today's Republic Slovenia. The western part of the country with the Ljubljana basin had been since the 3rd century included in the Province Venetia et Histria with its center in Aquileia. The defense system was established in the territory of South-Eastern Alps to protect Italy, known as Claustra Alpium Iuliarum. After Diocletian's reforms the territory Noricum Mediterraneum comprised also the town Poetovio. Poetovio, Celeia and Emona became early episcopal seats, which is reflected also in their archaeological heritage. After the mid-5th century the coastal towns were directly included under Byzantine rule. In the hinterland the fortress of Carnium persisted as one of a rare Late Antique lowland strongholds till the beginning of the 7th century. A review of the published oil lamps from Slovenia indicates that in the Western part oil lamps of type Atlante VIII prevailed, while in the coastal regions and Istria oil lamps of type Atlante X amounted to a larger number. In rare cases ecclesiastical equipment was preserved: two bronze candlesticks from Rogoznica near Ptuj and the Christograms from Vipota near Celje and from Emona are known. Glass fragments of lamps were found in Emona and other urban and rural settlements.

Keywords: Lighting device, Emona, Poetovio, Celeia, Carnium, *Claustra Alpium Iuliarum,* peacock-shaped lamp, Christogram, bronze candlestick, Rogoznica.

This paper presents a short survey of Late Antique ceramic oil lamps and other lighting devices, which had already been published in numerous Slovenian site publications and excavation reports. The western part of Slovenia with the Ljubljana basin had been since the emperor Diocletian a part of Province *Venetia et Histria* with its center in Aquileia. In the 3rd and the 4th century a defense system was established in the territory of South-Eastern Alps to protect Italy, known as *Claustra Alpium Iuliarum* (Figure 1). It was protected by fortifications and strongholds of different sizes, connected with roads and enclosed by walls.[1] The strategic center of the defense system was in Aquileia with its military harbor, warehouses and workshops for military equipment.[2] In the interior area of *Claustra* there were two main fortresses, *Castra* and *Ad pirum. Emona,* controlling entrances from Pannonia, represented an important strategic stronghold in the defense system.[3]

The eastern part of today's Slovenia was a part of *Illyricum* in the Early Imperial Period. With the foundation of towns and completion of Romanization processes *Poetovio* became one of the most important towns in the province of *Pannonia Superior*, and *Celeia* one of rich Norican municipia.[4] After Diocletian's reforms the territory *Noricum Mediterraneum* comprised also the town of Poetovio. Poetovio, Celeia and Emona became early episcopal seats, which is reflected in their archaeological heritage.[5]

After the mid-5th century the urban centers broke down. The population moved partly to Istria where some new towns were established on the coast. In the protected areas of the hinterland numerous hilltop settlements started to appear; the larger centers contained complex ecclesiastical buildings, where not only religious services but also administrative and military affairs were performed.[6] Coastal towns survived, were reinforced and continued directly under Byzantine rule. In the hinterland the fortress of Carnium remained as a rare Late Antique lowland stronghold until the beginning of the 7th century. Although Carnium, today's Kranj, used to be a fortress with Germanic military units included as part of the Germanic states, it remained under strong Byzantine influence.[7]

Western Region

Most of the known and studied oil lamps from the western part of today's Slovenia were discovered on Late Roman and Late Antique archaeological sites, sites which are characteristic of their military role or located on important merchant routes, such as the fortresses Hrušica/*Ad pirum,* Ajdovščina/*Castra* (Figure 2).[8] These oil lamps are mainly of African types, the most frequent type being Atlante VIII in different variants.[9] They were discovered in *in Emona-Episcopal Center,* in Ajdovščina/*Castra*, Hrušica/*Ad pirum*, Škocjan, Solkan, Koper/

[1] Ciglenečki 1999, 287-309.
[2] Sotinel 2001; Šašel 1976, 446-461; Christie 1991, 410-430; Ciglenečki 1999, 289.
[3] Ulbert 1981; Plesničar-Gec 1999.
[4] Horvat 2003, 153-189.
[5] Bitenc *et al.* 2001.
[6] Ciglenečki 1999, 287-309.
[7] Vidrih-Perko 2005, 49-77.
[8] Both locations were mentioned in *Tabula Peutingeriana* and *Itinerarium Burdigalense / Hierosolymitanum* (Šašel and Petru 1971).
[9] Atlante 1981, 129.

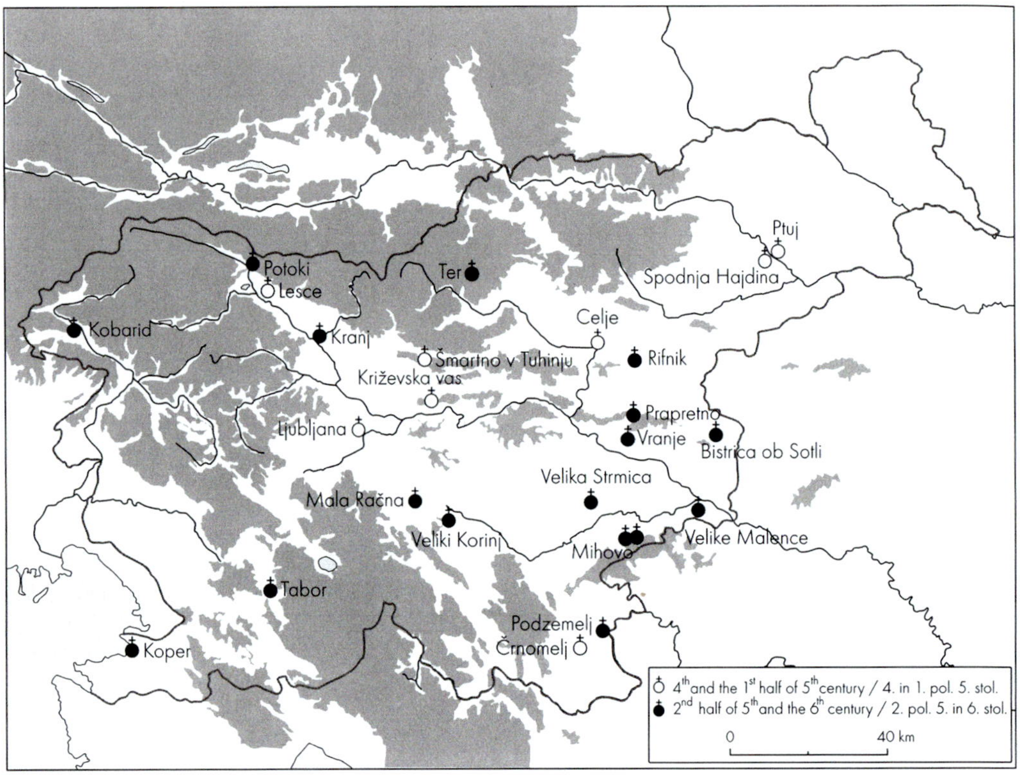

Figure 1. Map of the Slovenian territory in Roman times

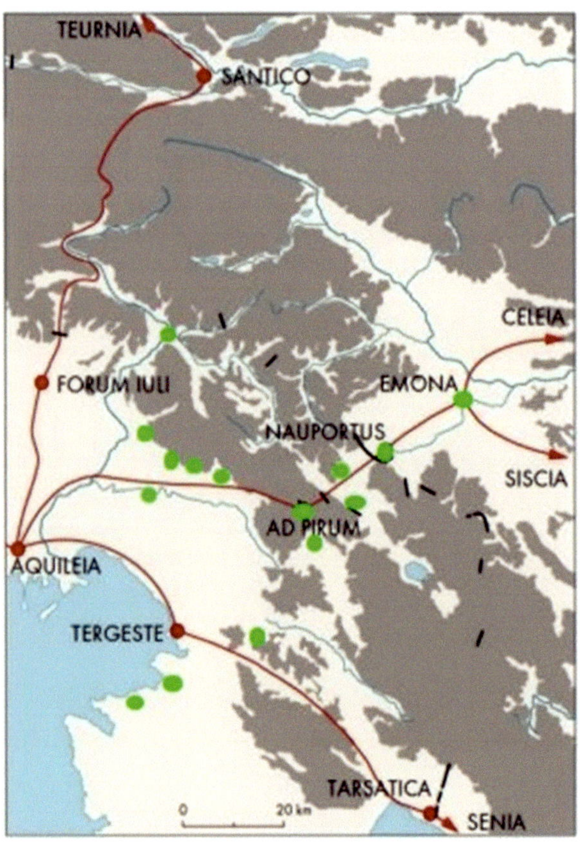

Figure 2. Distribution map of Late Antique oil lamps in western part of Slovenia.

Školarice.[10] The lamps were made of clay fired orange to brick red and have a slip of the same color.[11]

The known workshops where sigillata oil lamps were produced are documented in the Central and Northern Tunisia; among them are *Hencir es Sirna* and *El Mahrine.* The oil lamps of Hayes type I /Atlante VIII, were in use mostly from the second half of the 4th century to the end of the 5th century.[12]

Also the lamps from the discussed sites can be placed exactly in the same chronological time frame. The building of the Episcopal Center in Ljubljana was stratigraphically dated to the end of the 4th century. The destruction of the complex is evidenced by the coins of Honorius dating into the third decade of the 5th century.[13] The fortress at Hrušica/ *Ad pirum,* most probably resisted until the beginning of the 5th century.[14] The characteristic features in the first decades of

[10] For Emona-Episcopal Center: Plesničar-Gec 1983, fig. 37. 3, fig. 42.3; Ajdovščina-*Castra*: Pröttel 1996, 226-228, fig. 64.3-6, 8, 11; Hrušica-*Ad Pirum*: Pröttel 1996, 222, fig. 20.2-5, fig. 5.25; Škocjan: Pröttel 1996, 232, fig. 70.1; Solkan: Pröttel 1996, 231, fig. 96.4,7; Koper-Školarice: Žerjal 2008, 110, fig. 29.

[11] Pröttel 1996, 71.

[12] Oil lamp production was related to the production of African Red Slip Ware, at the beginning to the production C3-C4 (linking up with relief decoration on oil lamps and sigillata vessels' decoration, above all on cups type Hayes 53A), and later to the production D1-D2 (Hayes 1972, 311; Mackensen 1993; Pröttel 1996, 71; Bonifay 2004, 358.

[13] Plesničar-Gec 1983, 29. Some more examples are known from Emona, i.e., Knific and Sagadin, 1991, nos. 47, 48.

[14] Pflaum, 2004, 152.

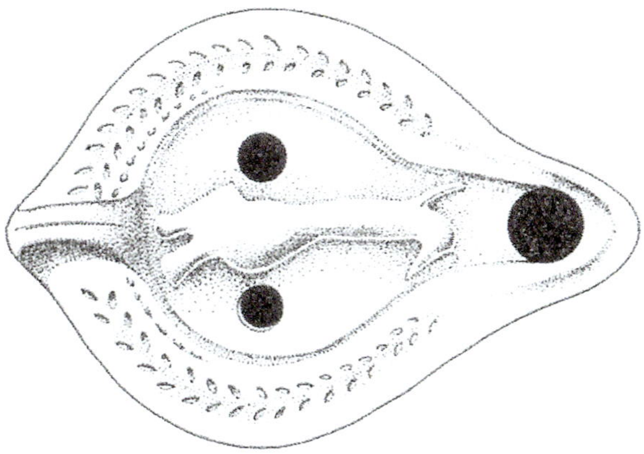

Figure 3. Oil lamp from Bilje (Osmuk 1978, 467. Fig. 1).

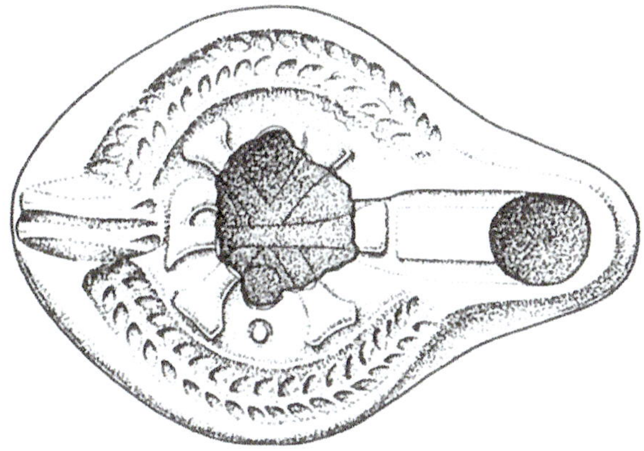

Figure 4. Oil lamp from Solkan (Svoljšak 1985, 278. fig. 107).

the 5th century could be recognized by the materials from the fortification at Ajdovščina/*Castra*.[15]

Also oil lamps of type Atlante VIII from recent excavations of a roman villa in Koper/Školarice are dated by the context to the first half of the 5th century.[16]

Some imitations of oil lamp type Atlante VIII A1 were also identified. The samples were discovered in Ljubljana, Logatec, Hrušica and in the region from Vipava Valley to Most na Soči (Figures 3, 4).[17] Individual lamps were known also from Aquileia, where the local Late Roman production imitating the African types flourished.[18]

The pottery -and brick- workshop was excavated also at site Bilje in Vipava valley. Regarding its brick stamps it had produced them already in the 1st century, and could have operated as a part of a large Aquileian production base, taking into account convenient river transport to Aquileia. The settlement finds bear evidence that workshops in Bilje must have remained in use also in the Late Roman period.[19] However, without analyses, it is difficult to identify the workshop which produced the lamps imitating the oil lamps of type Atlante VIII A1a, which represent frequent finds in that region. Finally, local pottery workshops are confirmed in Emona as well.[20]

In the western part of Slovenia sigillata oil lamps of later type Hayes II/Atlante X were often identified, which is the most typical and widely distributed form of the African types. Those types were discovered in Koper (Figure 5); some also at the site Škocjan and individual items were found at Predjama

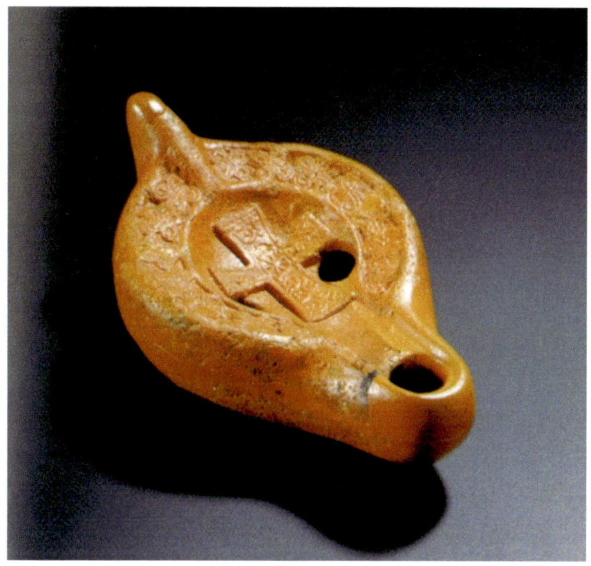

Figure 5. Oil lamp from Koper (Bratož 1999, 295).

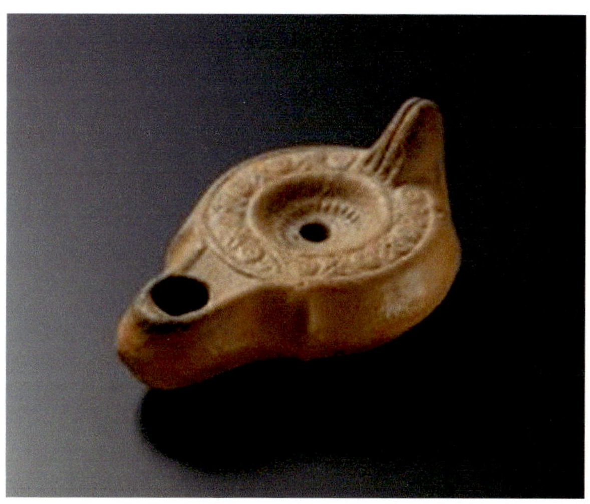

Figure 6. Oil lamp from Predjama (photo by T. Lauko).

[15] Osmuk 1997, 119-130; Vidrih-Perko and Žbona-Trkman 2005, 58.

[16] Žerjal 2008, 112.

[17] Ljubljana: Knific and Sagadin 1991, 66, no. 47; Petru 1972, 134, fig. 98.5; Logatec: Frelih 1988, 24; Hrušica: Pröttel 1996, 222, fig. 5.23,24; Vipava Valley: Osmuk 1978, 464-467; Pröttel 1996, fig. 20.22. fig. 69.1,2,3,7.

[18] An oil lamp model of African type Atlante X is displayed in Museo Nazionale in Aquileia.

[19] Vidrih-Perko and Žbona-Trkman, 2005, 40.

[20] Istenič and Tomanič-Jevremov 2004, 313-341. In the exhibition catalogue *Decline of the Antiquity* (Petru 1976) an oil lamp model from the National Museum of Ljubljana is mentioned, with the depiction of a *Pastor bonus*. Supposedly the model originated from Emona.

(Figure 6), Most na Soči and some more in Emona.[21] The lamps were in use from the beginning of the 5th century to the 7th century. In archaeological contexts from Koper the lamps can be placed in the 5th and 6th century with regard to their stratigraphic position.[22] The Predjama site is dated in the 4th and to the 1st half of the 5th century; considering the defense and strategic position of the site and numerous military finds, Predjama belonged to the context of the *Claustra Alpium Iuliarum* defense system.[23]

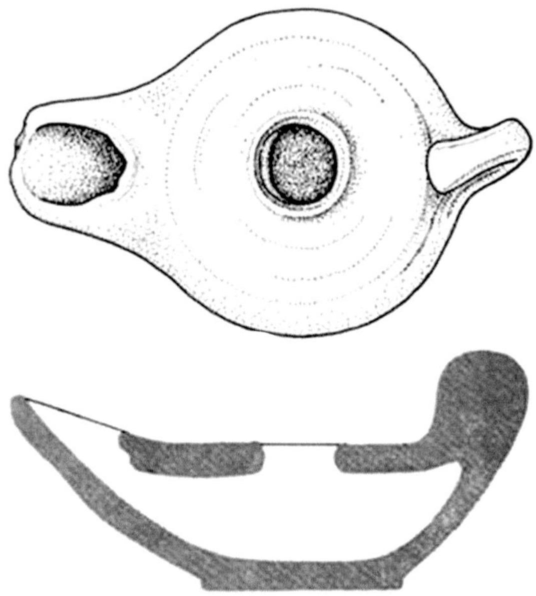

Figure 7. Oil lamp from Javorje near Ljubljana (Guštin and Knific 1973, fig. 5).

The fortress Turnišče near Vrhnika, belongs to the same context and period of time and is where a fragment of an oil lamp type Atlante X A1a was found.[24]

The imitations of oil lamps type Atlante X were discovered in Koper, one item was identified at Vrtovin in Vipava Valley (type Atlante XA1a).

Besides African oil lamps and their imitations some other lamp types appeared in that period. At Ajdovščina a fragment of the oil lamp type Atlante XIII, i.e. Tripoli oil lamp, was identified, which was the northern most find of that type.[25]

There are also some samples of local production. From Emona grave 214 along with a coin of Decius Traianus the so called *Warzenlampe* is known.[26] The decoration was influenced from the type Leibungut 35, where its pear-shaped body and nozzle form are linked to early African types. Some further oil lamps of local production, also influenced by African types, were identified in Emona.[27]

To old Emona grave finds belongs an almond-shaped oil lamp, type Sidon[28] and another one resembling Byzantine forms.[29]

Type Atlante VIII, represented by a mold for a pear-shaped lamp, was found together with a wheel-made lamp of local production (type Iványi XXII) in a grave from Javorje, dated to the second half of the 4th century (Figures 7, 8).[30]

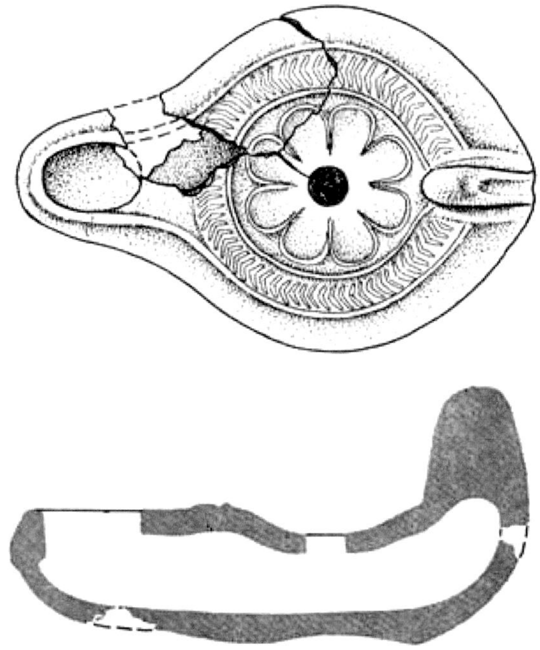

Figure 8. Oil lamp from Javorje near Ljubljana (Guštin and Knific 1973, fig. 5).

Round lamps with a simple turned up handle of type Iványi XXII were characteristic of the Late Roman local production.[31] They appeared in the second half of the 3rd century and together with late samples of type Firmalampen stand for the typical production in numerous local workshops of glazed pottery until the first decades of the 5th century. The oil lamps of type Iványi XXII were identified almost in every local pottery workshop, and are also known from the Emona

[21] Koper: Pröttel 1996, fig. 70.3-8. fig. 71.1-15; Škocjan: Pröttel 1996, fig. 69.10; Emona: Pröttel 1996, fig. 63.8-9. fig. 64.1-2. Some African oil lamps were discovered also in the site Predloka, but the material is still being prepared for publication.
[22] Cunja 1989, 103.
[23] Korošec 1956, 42; Bitenc *et al.* 2001, 27.
[24] Horvat 1990, fig. 19.1.
[25] Pröttel 1996, 87. Contradictory to Pröttel, some researchers suggest these objects as being of South Italian production; (L.Chrzanovsky at the conference in Olten, in September 2007).
[26] Plesničar-Gec 1972, fig. 60.1.
[27] Petru 1972, fig. 98: 5, 8.
[28] Petru 1972, fig. 97.8.
[29] Petru 1972, fig. 97.9. Menzel, fig. 81.2.
[30] Guštin and Knific 1973, 847, fig. 5. 2-69. The oil lamp from Javorje perfectly matches the type 8C 2b, Atlante I, fig. XCVII, with no precise datation (Atlante I, 195).
[31] Iványi 1935, 20, fig. 56.4-11.

pottery production and were found in graves in Emona necropolies.[32]

The drop in production of ceramic oil lamps, which occurred in the Late Roman period, must have resulted from the increased use of glass lamps. Among the numerous fragments of glass beakers, discovered at Emona Episcopal Center, some of them certainly belonged also to glass lamps.[33] The same situation most probably occurred in other Late Roman sites in the western part of today's Slovenian territory, where glass fragments were found (Martinj hrib, Ajdovščina, Hrušica, Predjama, the castle Kalc above Pivka). The latest finds from Koper Cathedral confirm this assumption.[34]

The main traffic route from Aquileia to Transdanubia passes through the Vipava Valley. African oil lamps were brought in as a minor, unimportant part of food cargo, which supplied markets and military posts in the time of agricultural decline. In the 4th, and mostly in the beginning of the 5th century, food supplies together with equipment and clothing implied from time to time a part of military salary, *annona*, which may point to a noticeable difference in the quantity of imported Late Roman vessels in the settlements and posts outside and inside the *Claustra*.[35]

Reviewing the published oil lamps it was determined that in Vipava Valley and Ljubljana basin oil lamps of type Atlante VIII prevailed, while in the coastal regions and Istria oil lamps of the later type Atlante X amounted to a larger number. The later type of oil lamps were also found on important Late Roman military posts, i.e., Most na Soči, Predjama, and Škocjan. The domination of Atlante X type in the coastal towns could be explained by uncertain conditions in the 4th and the 5th centuries, which forced the population to move more often to the safer littoral areas. In the case of the town of Koper it is known that after the collapse of Gothic rule in the mid-6th century, the Byzantine authority seated in Ravenna, restored and reinforced all local ports and protected their territory from the Lombard, Avar and Slav invasions. At the end of the 6th century the town accepted a great number of refugees from the Pannonian region. The town was also mentioned as a episcopal center.[36]

The military posts (Predjama and Most na Soči) undoubtedly continuously existed during the 5th century, which might reflect the developing defenses of the north-eastern Italian borders. Defense gradually changed from the centrally guided defense force into a dispersed countryside structure with numerous minor fortifications and posts along the significant road connecting to Italy.

Eastern part of today's Slovenia – Noricum and Pannonia

Poetovio used to be a rich town. For militarily strategic reasons, it is positioned along a key transportation route and the river Drava. In the Early Imperial Period a legionary camp was established. At the beginning of the 2nd century a status of *colonia* was given to the settlement. Poetovio become the seat of the Illyrian custom system, which accelerated the growth of population and fast economic development with numerous trade workshops.[37] During the Late Roman period strong divisions of military units were stationed in the town. Following the reforms of Diocletian Poetovio become part of *Noricum Mediterraneum*. The town served as the residence of the highest ranking civil and military elite. From the 3rd century onwards it was also the seat of bishop.[38]

The inhabitants' structures were various; especially remarkable is the presence of a population of Eastern origin, as testified by numerous epigraphic monuments and eastern cults spread out over the Poetovio territory. Cultural and ethnic variety is reflected in the rich archaeological heritage of Poetovio.[39]

More than 150 pottery kilns were discovered in Poetovio, confirming numerous pottery workshops.[40] In the Early Imperial period production of ceramic oil lamps is wide spread. Products bearing the names of Italic masters prevail, i.e., Fortis, Cresce, Decimi, Litogene, Aprio, Festi, Exorati, Lucius, Optati and Pulli, while stamps with the names of the local masters Iustinianus and Ursul(i) were active from the second half of the 2nd century to the beginning of the 4th century.[41] Especially thriving were Iustinianus' workshops, producing all kinds of pottery, i.e., *mortaria*. Their products were exported to eastern *Noricum, Pannonia Inferior, Mesia*, and even *Dacia*.[42]

Taking into account the characteristic Roman economy, the phenomenon of large joint-trade exchanges can be presumed, most probably connected with the organization of the custom system and supplying of posts from the seat in Poetovio.

From grave evidence the massive imports of Italic oil lamps, characteristic for the period till the mid-2nd century, were gradually replaced by strong local production.[43]

Among the molds and products from the Poetovio workshops in the Early and Middle Roman Period volute oil lamps and

[32] Petru 1972, fig. 104.6. Among the Emona ceramic finds a lot of glazed pottery and whee lmade oil lamps appeared (unpublished settlement material).

[33] Plesničar-Gec 1983, fig. 23-29; especially the fragments in fig. 23.1, fig. 27.10, 19. fig. 28. 2, 3. 28.4 and 7.

[34] This information was provided by M. Župančič, who is preparing the material for publication.

[35] Vidrih-Perko and Žbona-Trkman, 2005, 61.

[36] Šašel 1989, 10.

[37] Tomanič-Jevremov 2004, 94-99; Horvat 2003, 153-189. Metal, stone, wool, brick and other trades are archaeologically, epigraphically and historically documented.

[38] Horvat 2003, 158.

[39] Besides Mithraism and other oriental cults the presence of a strong Christian community was verified. Poetovio was the Episcopal center already at the end of the 2nd century. The theological opus of bishop Victorinus, one of the most prominent Christian writers, represents an excellent reflexion of the immense cultural background of Poetovio. R. Bratož (2000) one of the most important experts of Early Christianity –considering discoveries of numerous Christian churches and other archaeological materials in connection with Christianity– supposes the existence of a strong Christian community in Poetovio already in the 1st century AD, and the formation of the Poetovian bishop's diocese in the 2nd century. Victorinus' rich philosophical and theological works point to the existence of a comprehensive library, most probably owned by ecclesiastical community in Poetovio (Horvat 2003, 159 with extensive bibliography).

[40] Tomanič-Jevremov 2004, 94-99; Istenič 2004, 108-111.

[41] Istenič 1999, 155, fig. 146. Vomer-Gojkovič 2008.

[42] Istenič 1999, 158.

[43] Istenič 1999; Istenič 2000.

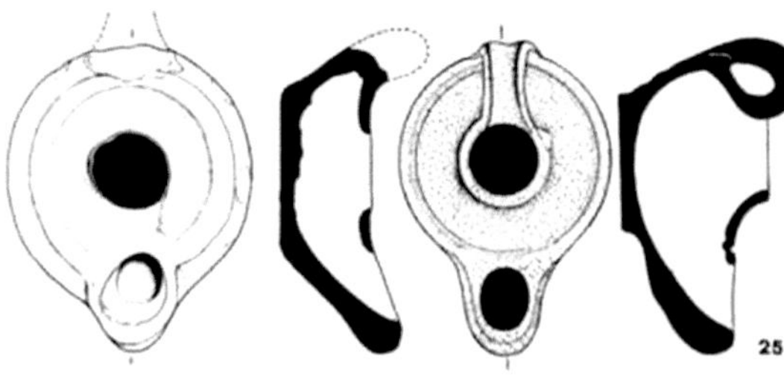

Figure 9. Oil lamps of local production from Ptuj (Šubic 1975, fig. 6.24, 25).

Firmalampen types prevailed. From the pottery kilns are known also waste finds, representing the zoomorphic oil lamps and lamps with face mask decoration.[44]

In the 3rd and 4th centuries the most frequent products of the local workshops were late forms of *Firmalampen*, *Warzenlampen* and round oil lamp of type Ivanyi XXII and XXIII (Figure 9). The oil lamps of type Ivanyi XXIII often appeared among the glazed products and were mostly wheel-made. They were found in abundance in both settlement and grave contexts.[45]

Despite the richness of the town and its important economic and political role in the Late Roman period, the number of African imports in Poetovio, in comparison to Emona, was rather small. This cannot be a result of research status, although it has considerable influence.[46] A rare Poetovian find of a late African import is also an oil lamp of Atlante, type X A 1a, with geometric decoration resembling the form of cross (Figure 10).[47] The oil lamp was found at Hajdina, today's eastern suburb of the modern town of Ptuj, which was the seat of Illyric customs from the Hadrianic Period. In the vicinity many temples had been built, and the remains of an Old Christian basilica were discovered.[48]

The most remarkable finds from Late Roman Poetovian contexts are the bronze peacock-shaped lamp and two bronze candlesticks from Rogoznica.[49]

The peacock-shaped lamp was discovered in the territory of the town area named Zgornji Breg, where a Late Roman cemetery was situated upon the urban remains of the Early Period (Figure 11). The lamp is only partly preserved and the tail of the figure is missing, which at the time of discovery resulted in the identification of the object as a Christian symbol. Because of the site of discovery and incorrect semiotic interpretation, the lamp was dated to the Late Roman period.[50] This datation and symbolic interpretation remained in Slovenian archaeological literature till recently. A similar candlestick in a peacock shape -with a widely

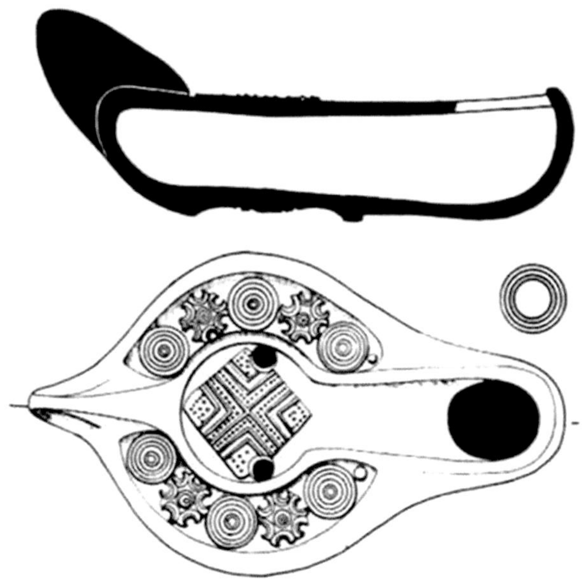

Figure 10. Oil lamp from Ptuj (Šubic 1975, Fig. 5.19).

open tail, but lacking a large opening on the belly- is already known from Menzel's publication.[51] The reliable dating of the Poetovio lamp to the Early Roman period, which corresponds also to the semiotic message of open tail, unfortunately not preserved in the Poetovio example, was confirmed by a recent Swiss discovery. In an intact archaeological context in Gallo-Roman villa from Morat/Combett in Switzerland a completely identical lamp in a peacock shape was found, with a wide open tail and a large opening on the belly. The lamp is dated to the second half of the 2nd century.[52]

Regarding the archaeological context of the Poetovio peacock-shaped oil lamp it can be presumed that the object must have been in use for a long period of time. The same can be stated for the tripod candlestick with folding legs, found in the vicinity of the eastern Late Roman necropolis of Poetovio. It can be placed among earlier finds.[53]

Among the most noteworthy Late Roman finds from Poetovio are two bronze candlesticks, discovered at Rogoznica, today a suburb of Ptuj (Figures 12, 13).

Both objects were discovered in 1858 and because of great cultural and historical value, they were included into the collection of *Kunsthistorisches Museum* in Vienna.[54] The discovery circumstances are not clear. Because the site was in close vicinity to the large Poetovio necropolis along the eastern road, and because the candlesticks were discovered as a pair, it was often mentioned that the candlesticks might have belonged to the graveyard chapel furniture.[55]

[44] Tomanič-Jevremov 2004, 94-99.
[45] Šubic 1975, 83; Mikl-Curk 1976; Istenič 2000, 15: grave 3, fig. 2.1.
[46] Most prevalent of the African imports are those from the production group C, belonging to the period between 230 and 340 AD; later finds are sporadic (Pröttel 1996, 130).
[47] Šubic 1975, fig. 5.19; Knific and Sagadin 1991, 67, fig. 50.
[48] Šubic 1975, 83, fig. 5.19; Mikl-Curk 1976, fig. 18.11.
[49] Knific and Sagadin 1991.
[50] Klemenc 1953.
[51] Menzel, 1969, fig. 90.4.
[52] Agustoni 2005, fig. 1-3.
[53] Vomer-Gojkovič 1996, fig. 20.17.
[54] Kunsthistorisches Museum Wien, inv. no. VI 727.
[55] Korošec 1980, 55-61.

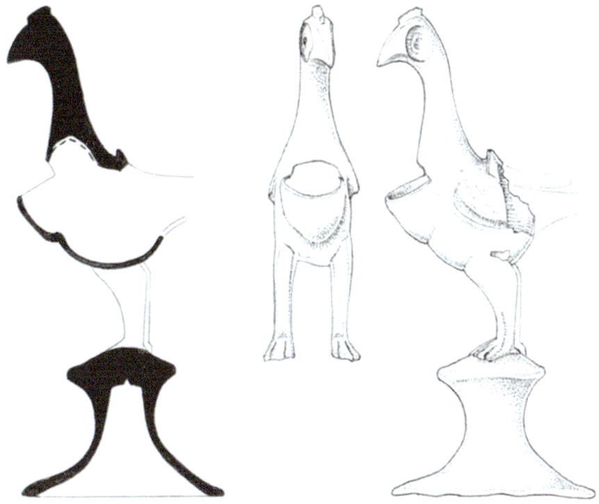

Figure 11. Bronze lamp from Ptuj (Knific and Sagadin 1991, fig. 74).

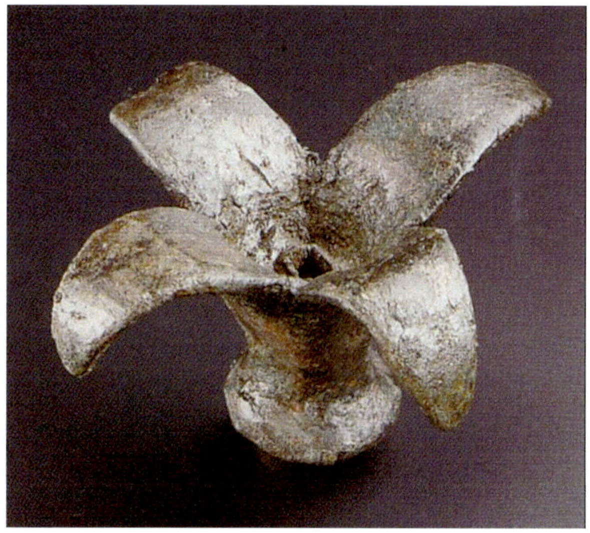

Figure 14. Candleholder from bronze candlestick, found at Črnomelj (Bitenc *et al.* 2001, no. 102).

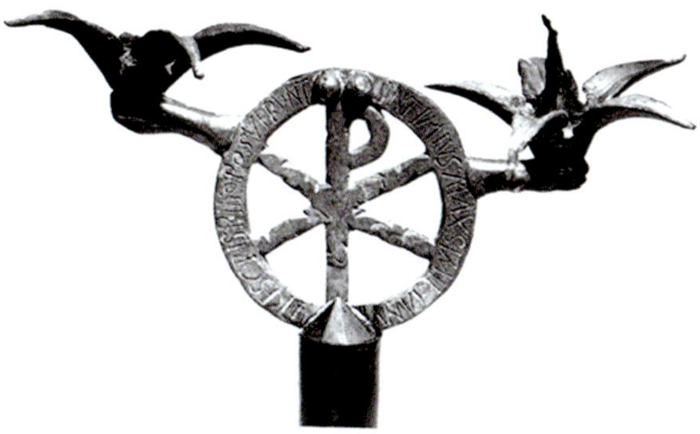

Figure 12. Bronze candlestick from Rogoznica near Ptuj (Knific and Sagadin 1991, fig. 1).

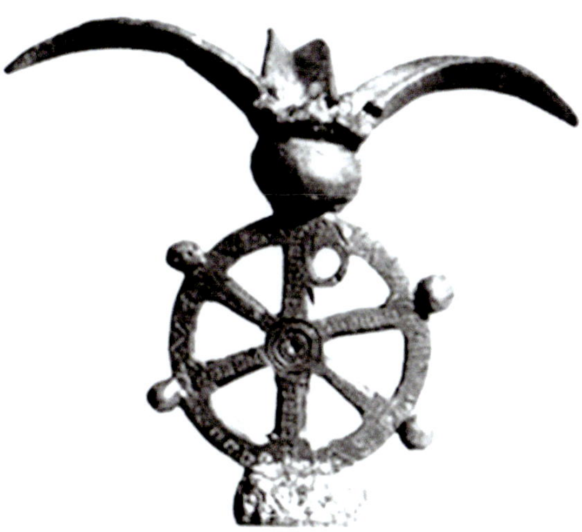

Figure 13. Bronze candlestick from Rogoznica near Ptuj (Knific and Sagadin 1991, fig. 2).

The compact bronze candlesticks are composed of three parts: the lower part for fixing it on the postament, the central part in Christogram form with an incised inscription, and the upper part forming calyces as candleholders. The two objects are almost identical. They only differ in the form of their candleholders. In one case it is shaped as a quinque-foliolate calyx held in two hands, and in the second case as only one four-foliolate flower fixed on the top of the *Christogram*. The inscription on the candlestick with the flower-formed calyx says: *Votum Pusinio posuit*, and on the other: *Intimius, Maximilianus fratres Crispino posuerunt.*[56]

The interpretation of the style, especially the element of the candleholder in a calyx form and hands, reflect in style Coptic spiritual influences. The candlesticks are dated to the second half or end of the 4th century and are perhaps even indicative of the presence of a Coptic church in Poetovio.[57]

The Poetovio candlesticks are not isolated finds as testified by the discovery of a very similar candleholder, a fragment of bronze candlestick. The object was discovered during the extensive archeological excavations in the Late Roman area at the site of Črnomelj (Figure 14), a settlement in the southeast of modern Slovenia.[58]

Celeia had been since its very beginning a Norican town and one of the centers of Celtic indigenous aristocracy.[59] Its rich archaeological heritage comprises many oil lamp finds, and recent archeological work has brought insight into the local ceramic production.[60] Among the published finds of Celeian Late Roman oil lamps is a well known African import motif with the depiction of a menorah candlestick, type Atlante X A 1a (Figure 15).[61]

[56] Knific and Sagadin 1991, 48.
[57] Korošec 1980, 57.
[58] Mason 1998, 285-313; Bitenc *et al.* 2001, 37, no. 102.
[59] Lazar 2002, 71-101.
[60] Unpublished.
[61] Atlante I 1981, 200, fig. 160.3; Lazar 2001, fig. 76.

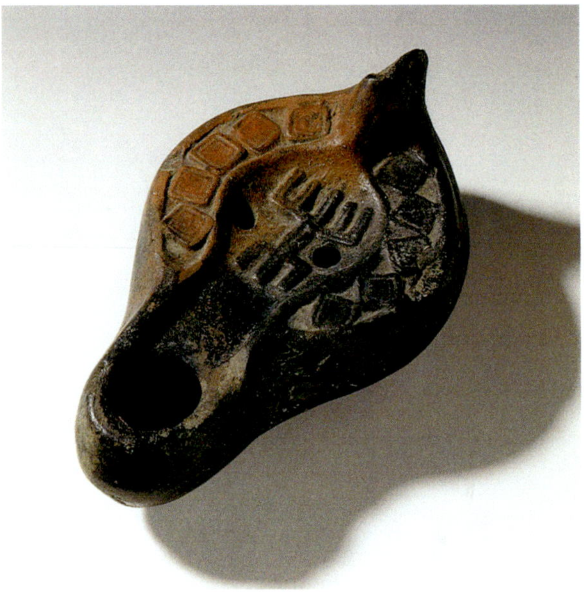

Figure 15. Oil lamp from Celeia (Lazar 2001, fig. 76).

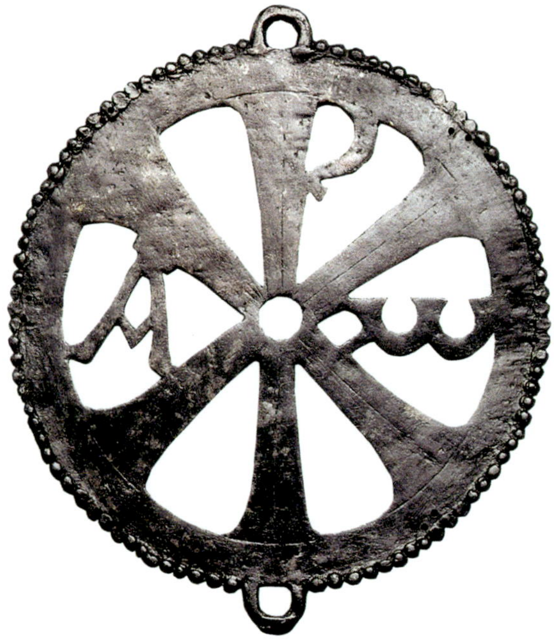

Figure 16. Bronze *christogram* discovered in the hilltop settlement of Vipota near Celje (Lazar 2002, Fig. 36).

Many Late Roman finds were discovered on the neighboring hilltop settlements, where the Celeian population sheltered at the end of the 5th and in the 6th century.

Two bronze Christograms were discovered in the hilltop settlement of Vipota (Figures 16, 17).[62] Both objects have preserved hooks for fixing chains on what may have been hanging glass lamps. The *Christograms* are dated to the end of the 4th century and most probably represent a part of the church equipment from Celeian's Old Christian basilica.[63]

In close vicinity to Celeia another strong Late Antique hilltop settlement of Rifnik with its so called 'group of churches' was discovered.[64] In the fields below the hilltop settlement two exceptional silver oil lamps were found, each with a carinated bowl-body and two flat-topped nozzles with rounded tips. The lamps are partly demolished and slightly flattened (Figure 18). The first one has a raised edge surrounding the top of it and the nozzles with an open canal. The second one has a raised central part without canals and six small holes are placed on the passage towards the nozzle for fixing the hanging chain. A similar braze and bronze objects from the British Museum are dated between AD 200 and 400.[65] The lamps were unearthed during the ploughing of land and are usually considered as a part of treasure rather than as grave goods. It remains unknown whether or not the oil lamps formerly belonged to the sanctuary's furniture of the local deity *Aquonius* and later probably passed over to the church equipment of the Early Christian basilica, erected in the same place. The oil lamps could be also have been brought to the hilltop settlement of

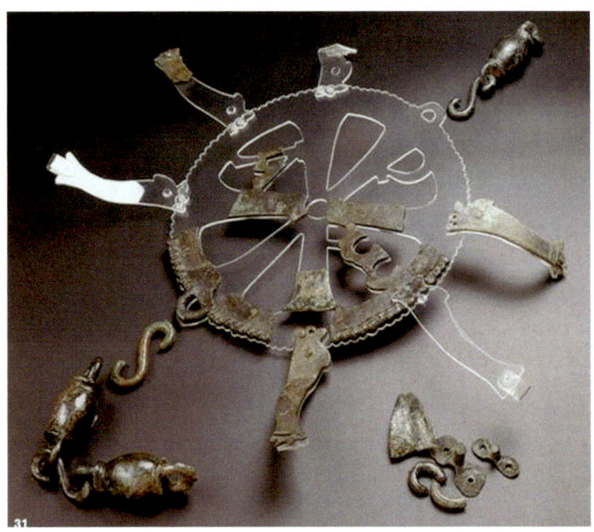

Figure 17. Bronze *christogram* discovered in the hilltop settlement of Vipota near Celje (Lazar 2002, fig. 36).

Rifnik from the near *municipium Celeia* and could have been buried or lost upon the destruction of the hilltop post.[66]

At Vranje near Sevnica, on another hilltop settlement of Late Antiquity, in the southast of modern Slovenia, a large residential area was excavated and another 'group of

[62] Bitenc *et al.* 2001, 18, nos. 30 and 31; Lazar 2002, fig. 36.
[63] Lazar 2002, 97. The town is linked with the Christian tradition by the legend of Saint Maximillian's martyrdom, which together with archaeological finds testifies to the presence of Christianity in Celeia at the end of the 3rd century. The town became the bishop's seat with the Early Christian basilica and baptistry (Lazar 2003, 96).
[64] Pirkmajer 1994, fig. 45.
[65] Bailey 1996, 64, fig. 74, Q 3779.
[66] Pirkmajer 1994, fig. 45.

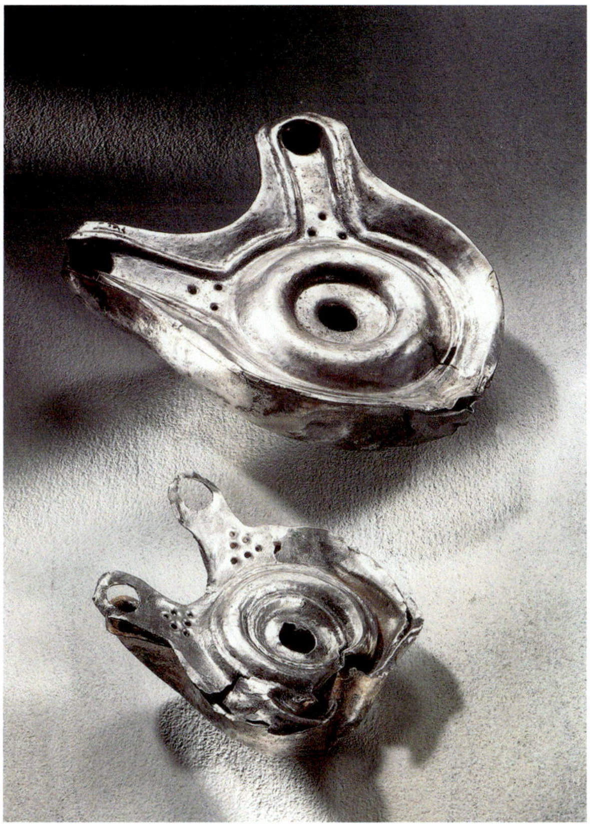

Figure 18. Silver oil lamps from Šentjur near Celje (Pirkmajer 1994, Fig. 45).

churches' was unearthed.[67] No ceramic oil lamps were found, only some glass lamp fragments were excavated in the church presbytery.[68]

On Ajdna, the hilltop settlements above Jesenice (northwest of modern Slovenian territory) glass fragments were discovered in the church altar area and dated to the mid-6th century (Figure 19). From the same site Ajdna, an imitation of an African oil lamp was found within the building layers next to the church.[69] The lamp strongly resembles the example of African production found at Celeia.

Carnium

Ancient *Carnium*, the modern town Kranj in the north of Slovenia, became the most important lowland post after the destruction of the towns *Emona*, *Celeia* and *Poetovio* in the mid-5th century.[70] *Carnium* is known for its rich grave finds belonging to a necropolis of the Germanic and indigenous populations.[71] With regard to the material culture, the persistence of the settlement can be verified up to the first decades of the 7th century.

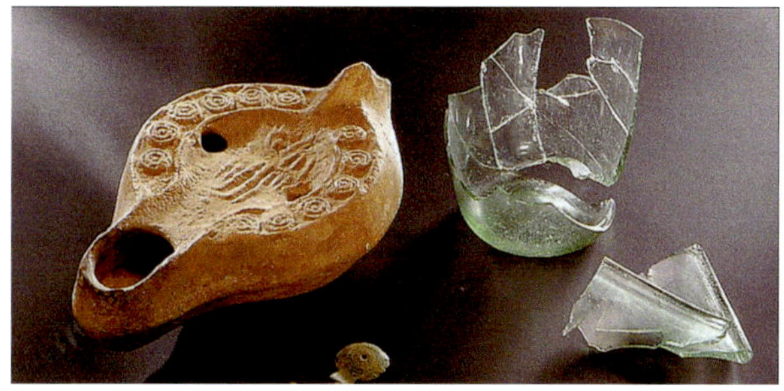

Figure 19. Glass and oil lamp from the hilltop settlement of Ajdna (photo by T. Lauko).

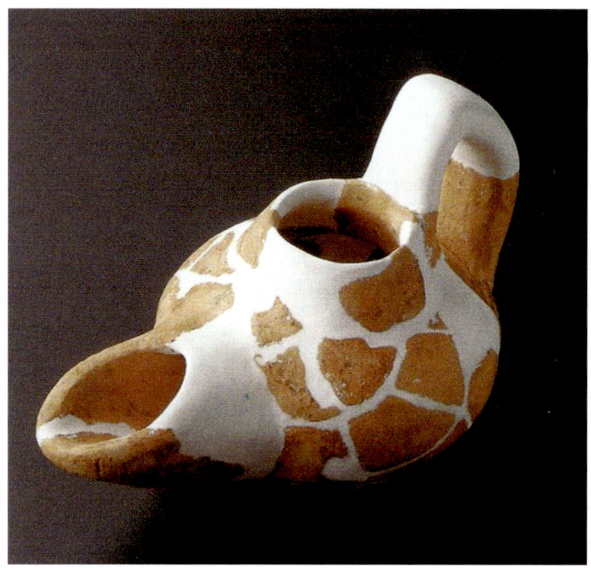

Figure 20. Oil lamps from the Late Antique necropolis at Kranj (Bitenc *et al.* 2001, 63, no. 190).

Two late ceramic oil lamps are known from Kranj –one having been found in grave 230[72] and fragments of another unearthed in settlement layers (Figure 20). In both contexts the oil lamps date to the 6th century. They belong to Late Antiquity wheel made oil lamp types without slip and with a surmounted handle. Similar oil lamps are known from the contexts of the 5th and the 6th century at the site El Ashmunein in Egypt and elsewhere.[73] In the Late Roman period ceramic oil lamps of local production prevailed. The African imports are, in contrast to the coastal zone and Ljubljana basin, not so frequent. A remarkable difference occurred during the late 5th and 6th century. In the coastal zone under the direct Byzantine rule African imports still remained numerous, while in the hinterland they appeared only sporadically.[74]

In rare cases ecclesiastical equipment was preserved, such as the two bronze candlesticks, discovered at Rogoznica and the Christogram from Vipota. The glass fragments of beakers from the Vranje and Ajdna church interiors testify to a frequent use

[67] Petru and Ulbert 1975; Knific 1979.
[68] Knific 1979, fig. 35-41.
[69] Bitenc *et al.* 2001, 44, fig. 120.
[70] Vidrih-Perko 2005, 77.
[71] Stare 1980.
[72] Stare 1980, 67, fig. 72.5-7; Bitenc *et al.* 2001, 63, no. 190.
[73] Bailey 1998, 149, fig. 90.167.
[74] Vidrih-Perko 2005, 77.

of glass lamps. Due to the practice of recycling fragmented glass vessels, only a few of them have been preserved. In the Late Roman settlement of Poetovio glass production is strongly represented; and therefore glass lamps are to be identified in quantity among the glass fragments.[75] A glass workshop was discovered in Kranj in the settlement layers of the 6th century (Figure 21). Among numerous fragments of glass beakers must have been some lamp fragments, too.[76]

It can be also be stated that in the 5th and the 6th centuries, beside ceramic and glass oil lamps, wax and tallow candles were also generally in use.

Addendum

The study of Roman lamps always raises two fundamental questions; the first being why was the Early Imperial period associated with the massive production of mold-made oil lamps which were intensively exported to all parts of the Empire; and the second, why and when did this massive production cease.

The production of ceramic oil lamps was closely associated with olive growing and olive oil production, which occurred after the spread of the Phoenicians across the Mediterranean region. Greek and Hellenistic production of wheel-made oil lamps is remarkable and occurred at many Mediterranean sites. But the massive, almost industrial production of oil lamps was first noticed in the Early Imperial Roman period. That extensive production was linked to mold-made types, which had originated from the Hellenistic Asia Minor workshops and later continued in the Roman centers. The increasing need for oil lamps was fundamentally linked to the immense production of olive oil, which took place in Italy only after the Second Punic War. With the expansion of villas and slave labor the production of wine and oil exploded.[77] Through the foundation of provinces and towns, which represented focuses of Romanization, extensive new markets appeared. The need for food products, above all wine, oil, and grains, increased and so did the need for ceramics and lamps. The use of lamps culminated with the appearance of relief-decorated types of volute lamps and Firmalampen. But the massive production of oil lamps obviously did not result only from the market needs. Oil amphora studies recognized oil exportation into Northern provinces as an important instrument of Romanization, which affected trade, and the systematic army provisioning of all supplies for provincial and municipal bureaucracy. Not only the Latin language, legislation, religion and clothing, but also the way of living and dieting reflected the *romanitas* of the political and military elite, in contrast to the peregrine, indigenous and other lawless populations. In the Early Imperial period also urbane organizations and arts represented the instruments of power, which spread into all parts of the society. It was reflected even on such a bizarre field as in the decoration of vessels.

Therefore the influences of the political agenda -the images of power- can be found also in the decoration of oil

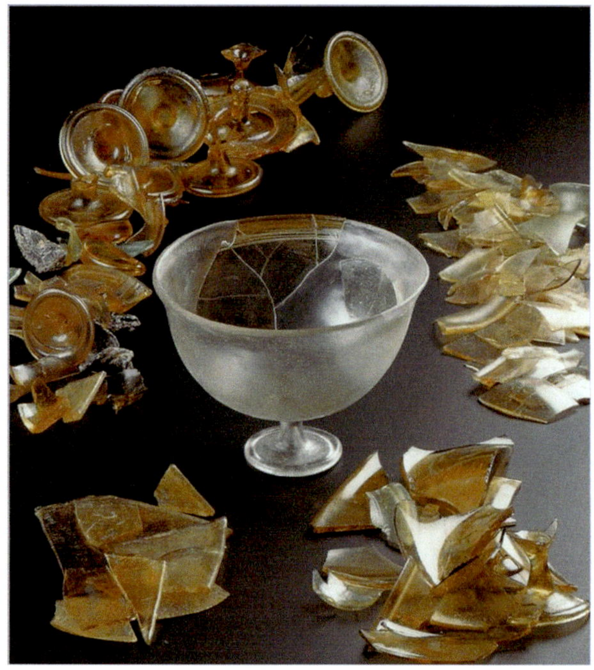

Figure 21. Fragments of glass beakers and lamps from a glass workshop discovered in Kranj (photo by T. Lauko).

lamps. On the disks of volute oil lamps there was enough space for relief decoration, which symbolically depicted heroes from Mediterranean myths and their deeds, Greco-Roman deities with their attributes, or merely to everyone intelligible allegories with the symbols of fortune, fertility and abundance. The image of a long-legged ibis -not only on the frescos in imperial palaces or in urban and suburban villas in southern Italy, but also in the beam of light on a tiny oil lamp reminds one of Augustan victories in Egypt; and Fortuna with a horn of abundance in her hands announces *Pax Romana* welfare. Perhaps the round acorns' caps were mighty enough to recall the image of fertile oak forests of *Cisalpina* into the consciousness of the Italic population even in the farther provinces. Nevertheless, the tiny lamp with its relief decoration and simple Latin inscription was powerful enough to distinguish between the Italic and the indigenous population -between the *civis Romanus* with the lamp filled with olive oil, and the other population who lit wood splinters and used stinking tallow candles. The former one was literate or at least understood Latin, while the latter one was illiterate and spoke a Barbarian language. *Cives Romani* illuminated their dwellings by oil lamps, with respect to inscriptions and relief decoration, which might have even unconsciously influenced them. The language of splinters and tallow candles remained the expression of poverty and the peregrines' deprivation of rights. Oil lamps can be acknowledged as objects of identification, testified by the producers' careful attention to the decoration and inscriptions. The latter had been prominent on oil lamps neither before nor later, when in the 60s of the 1st century *Firmalampen* appeared as the technically most advanced type of ceramic lamps,[78] and disk decoration gradually lost its meaning. The lamp appeared with the producers' names, confirming their quality.

[75] Glass beakers are frequently found in Poetovian graves of the late 4th and the first half of the 5th century (Vomer-Gojkovič 1996. Vomer-Gojkovič 1996a).

[76] Sagadin 2000, 13-22; Vidrih-Perko 2005, 77, figs 16-17.

[77] Carandini 1999, 775-804.

[78] Wunderlich 2006, 40-45.

It could be stated with no exaggeration that the symbolic meaning of the object had surmounted a complex, generally known semiotic message, embodied in the object itself. The message brought to light only the most important data and some rich symbolic elements of art, while the contexts and archetypical support became unnecessary and become neglected. A technically perfect object itself became the essence of the semiotic identity message. Its genuineness was confirmed by the name of its producer.

It is still questionable why, despite numerous provincial workshops in the first two centuries AD, the majority of lamps discovered in provinces were usually of Italic production. Among Early Roman oil lamps volute forms prevailed, mostly of types Loeschcke 1, 4 and 8, less often 2, 3, or 5. There followed *Firmalampen* types Loschcke 9 and 10. In spite of massive imports, all these forms were produced also in local workshops.

From the Flavian period on, oil lamps were generally spread to almost all rural areas of the provinces. They indicated an important part of the everyday private and public life, as they were found in all living places, in temples, in baths, etc. Oil lamps represented the objects most frequently found in graves, certainly in the entire 1st and 2nd centuries. In the 3rd century their number respectively diminished and the presence of imports remarkably diminished.

It can be presumed that the general crisis of the Marcomanic wars influenced the economic situation, although Italic imports were quite noticeable also in the 2nd half of the 2nd century. And as is well known, the crisis of the Italic economy was essentially already under way in the 1st century.

Also the political and economic crisis, followed by temporary moving of populations onto hilltop shelters in the 3rd quarter of the 3rd century, was not synchronous with the disappearance of the oil lamp and could not be the main reason for general import diminution. That can be recognized already in the beginning of the 3rd century.

If agreeing with the statement that economical events alone did not provoke that situation, we should think of synchronous political events. Taking into account that oil lamps were a Roman state product for supplying Italic political, military and administrative elite, which in the Early Imperial period lived in provinces, in order to reinforce its *romanitas*, then we should have in view the Antonine Edict from the year 212. The political event legally equaled all freeborn inhabitants in the Roman Empire, therefore the needs for identity instrumentalization disappeared. Oil lamps remained in everyday life during the next centuries. The immediate change happened in the changes in massive production and exportation from Italic centers, which completely broke down. Characteristic types with relief decoration and inscriptions vanished. From that time on output from the provincial local workshops prevailed and wheel-made lamps reappeared together with glass lamps. All over the expanded use of olive oil for lighting had been limited. Undoubtedly also other oils, i.e., linseed, rapeseed or hemp oil were employed. Animal fats were also brought into use, as well as tallow and wax, which has been proven by individual finds of candlesticks.

Late Roman oil lamp mass production was mostly limited to African and rare Eastern products. Compared to Early Roman types, the lamps of African production were not found so frequently. Their distribution was scattered along the Mediterranean coastal area and was limited to a few individual imports in the hinterland.

Bibliography

Agustoni, C. 2005. Ex pavone lux, in *Lychnological Acts 1, Actes du 1er Congrès International d'études sur le luminaire antique*, Nyon-Gèneve, 29. IX-4.X.2003 Monographies instrumentum 31, 15-16, pls 2-3. Montagnac: Mergoil.

Atlante I, 1981. *Atlante delle forme ceramiche I. Ceramica fine romana nel bacino Mediterraneo (medio e tardo impero). Enciclopedia dell'arte antica, classica e orientale.* Rome: Istituto della Enciclopedia Italiana.

Bailey, D. M. 1996. *A Catalogue of the Lamps in the British Museum, IV. Lamps of metal and stone, and lampstands.* London: The Trustees of the British museum.

Bailey, D. M. 1998. *Excavation at El Ashmunein. vol. 5: Pottery, lamps and Glass of the Late Roman and Early Arab periods.* London: British Museum Publications.

Bitenc, P., Knific, T., Lauko, T., Čeferin, G. and Tepina, B. 2001. *Od Rimljanov do Slovanov.* Ljubljana: Narodni muzej Slovenije.

Bonifay, M. 2004. *Etudes sur la céramique romaine tardive d'Afrique.* Oxford: Archaeopress, International series 1301.

Bratož, R. 1999. Zadnji vzpon rimskega cesarstva, in B. Aubelj (ed.) *Zakladi tisočletij*, 295. Ljubljana.

Bratož, R. 2000. *Slovenija in sosednje dežele med antiko in karolinško dobo: Začetki slovenske etnogeneze = Slowenien und die Nachbarländer zwischen Antike und karolingischer Epoche: Anfänge der slowenischen Ethnogenese.* Ljubljana: Narodni Muzej Slovenije.

Carandini, A. 1999. La villa romana e la piantagione schiavistica, in A. Giardina and A. Schiamone (eds) *Storia di Roma*, 775-804. Torino: Einaudi.

Christie, N. 1991. The Alps as a Frontier (A.D. 168-774). *Journal Roman Archaeology* 4, 410-430.

Ciglenečki, S. 1999. Results and Problems in the Archaeology of the Late Antiquity in Slovenia, *Arheološki vestnik 50*, 287-309.

Cunja, R. 1994. Poznorimski in zgodnjesrednjeveški Koper. *Arheološko izkopavanje na bivšem Kapucinskem vrtu v letih 1986 1987 v luči drobnih najdb* 5. do 9. stol. Ljubljana.

Frelih, M. 1988, Komunikacijski sistem v prazgodovini in antiki na primeru Logaške kotline, *Zbornik poročil raziskovalnih skupin* '88, 10–51.

Guštin, M. and Knific, T., 1973. Halštatske in antične gomile z Javorja. *Arheološki vestnik* 24, 731- 847.

Hayes, J. W. 1972. *Late Roman Pottery.* London: British School at Rome.

Horvat, J. 1990. Nauportus (Vrhnika). *Dela SAZU 33.* Ljubljana.

Horvat, J. 2003. Poetovio. Develepment and Topography, in M. Šašel Kos and P. Scherrer. The Autonomous Towns of Noricum and Pannonia, 153-189. Ljubljana: Narodni muzej Slovenije.

Istenič, J. 1999. *Poetovio, the western cemeteries I: grave-groups in the Landesmuseum Joanneum Graz.* Ljubljana: Narodni Muzej.

Istenič, J. 2000. *Poetovio, the western cemeteries II: grave-groups in the Landesmuseum Joanneum Graz.* Ljubljana: Narodni Muzej,

Istenič, J. 2004. Poetoviona–središče izdelave in distribucije keramičnih izdelkov, in *Catalogue of the Exhibition: Rimljani. Steklo, glina kamen/ Die Römer : Glas, Ton, Stein*, 108-111. Celje: Pokrajinski muzej; Maribor: Pokrajinski muzej; Ptuj: Pokrajinski muzej.

Istenič, J. and Tomanič-Jevremov, M. 2004. Poetovian wasters from Spodnja Hajdina near Ptuj, with a contribution by M. Daszkiewicz and E. Bobryk. *Arheološki vestnik* 55, 313-341.

Klemenc, J. 1953. *Zgodnjekrščanska svetilka s Ptuja*. Ptuj: Ptujski zbornik.

Knific, T. 1979. Vranje pri Sevnici. Drobne najdbe z Ajdovskega gradca (leto 1974). *Arheološki vestnik* 30, 732-763.

Knific, T. 1991. *Pismo brez pisave, arheologija o prvih stoletjih krščanstva na slovenskem*. Ljubljana: Narodni Muzej.

Knific, T. and Sagadin, M. 1991. *Pismo brez pisave. Arheologija o prvih stoletjih krščanstva na Slovenskem*. Ljubljana: Narodni Muzej.

Korošec, J. 1956. Arheološke ostaline v Predjami. *Razprave IV*.

Korošec, P. 1980. Starokrščanska svečnika iz Rogoznice v Ptuju. *Arheološki vestnik* 31, 55-61.

Korošec, P. 1982. Predjama vom Ende des 4, bis zur Mitte des 5. Jhrts. *Arheološki vestnik* 33, 84-115.

Lazar, I. 2001. *Celeia. An Archaeological Image of the Town*. Celje: Pokrajinski Muzej.

Lazar, I. 2002. Celeia, in M. Šašel Kos and P. Scherrer (eds) The Autonomous Towns of Noricum and Pannonia. *Situla* 40, 71-101.

Mackensen, S. 1993. *Die Spätantiken Sigillata-und-Lampentöpfereien von El Mahrine (Nordtunesien). Studien zur nordafrikanischen Feinkeramik 4. bis 7. Jahrhunderts*. München: Beck'sche Verlagsbuchhandlung.

Mason, P. 1998. Late Roman Črnomelj and Bela krajina. *Arheološki vestnik* 49, 285-313.

Menzel, H. 1969. *Antike Lampen in römisch-Germanischen Zentralmuseum zu Mainz*. Mainz: Philipp Von Zabern.

Mikl-Curk, I. 1976. *Poetovio I, Catalogi et Monographie 13*. Ljubljana: Narodni muzej v Ljubljani.

Osmuk, N. 1978. Nove najdbe iz časa preseljevanja narodov v spodnji Vipavski dolini. *Arheološki vestnik* 29, 464-467.

Osmuk, N. 1997. Ajdovščina – Castra. Forschungstand (1994). *Arheološki vestnik* 48, 119-130.

Petru, S. 1972. *Emonske nekropole: (odkrite med leti 1635-1960)*. Ljubljana: Narodni muzej.

Petru, P. and Ulbert, T. 1975. *Vranje pri Sevnici. Starokrščanske cerkve na Ajdovskem gradcu, Catalogi et Monographie 12*. Ljubljana: Narodni muzej.

Petru, P. 1976. *Zaton antike v Sloveniji*. Ljubljana: Narodni muzej.

Pflaum, V. 2004. Poznorimski vojaški in obrambni sledovi 5. stoletja na ozemlju sedanje Slovenije. Unpublished PhD dissertation, University of Ljubljana.

Pirkmajer, D. 1994. *Rifnik: Arheološko najdišče vodnik = archäologische Fundort Führer*. Celje: Pokrajinski muzej.

Plesničar-Gec, L. 1972. *Severno emonsko grobišče, Catalogi et Monographie 8*. Ljubljana: Narodni muzej; Mestni muzej.

Plesničar-Gec, L. 1983. *Starokrščanski center v Emoni, Catalogi et Monographie 21*. Ljubljana, Narodni muzej; Mestni muzej.

Plesničar-Gec, L. 1999. *Urbanizem Emone: The Urbanism of Emona*. Ljubljana: Mestni muzej.

Pröttel, P. M. 1996. *Mediterrane Feinkeramikimporte des 2. bis 7. Jahrhunderts n. Chr. im oberen Adriaraum und in Slowenien: Philipp M. Pröttel*. Espelkamp: M. Leidorf.

Sagadin, M. 2000. Poznoantična steklarska delavnica v Kranju. *Kranjski zbornik*, 13-22.

Sotinel, C. 2001. L'utilisation des ports dans l'arc Adriatique à l'époque tardive (IVe-VIe siècle), in C. Zaccaria, *Strutture portuali e rotte marittime nell'Adriatico di età romana: [atti della XXIX settimana di studi aquileiesi, 20-23 maggio 1998]*. Trieste: Centro di Antichità Altoadriatiche.

Stare, V. 1980. *Kranj, nekropola iz časa preseljevanja ljudstev. Catalogi et Monographie 18*. Ljubljana: Narodni muzej.

Svoljšak, D. 1985. Solkan. *Varstvo spomenikov* 27, 278.

Šašel, J. 1989. Prispevki k zgodovini Kopra, in M. Guštin, *Koper med Rimom in Benetkami. Capodistria tra Roma e Venezia, etc*. Piran: Medobcinski Zavod za Varstvo Naravne in Kulturne Dediscine.

Šašel, J. and Petru, P. 1971. *Claustra Alpium Iuliarum I, Fontes*. Ljubljana: Narodni muzej.

Šašel, J. 1974. Koper. *Arheološki vestnik* 25, 446-461.

Šubic, Z. 1975. Rimske oljenke v Sloveniji. *Arheološki vestnik* 26, 82-99.

Tomanič-Jevremov, M. 2004. Obrtniške delavnice Poetoviona, in *Catalogue of the Exhibition: Rimljani. Steklo, glina kamen/ Die Römer: Glas, Ton, Stein*, 94-99. Celje: Pokrajinski muzej; Maribor: Pokrajinski muzej; Ptuj: Pokrajinski muzej.

Ulbert, T. 1981. *Ad Pirum (Hrusica): Spätrömische Passbefestigung in den julischen Alpen; der deutsche Beitrag zu den slowenisch-deutschen Grabungen 1971-1973*. München: Beck.

Vidrih-Perko, V. 2005. Seaborne trade routes in the north-east Adriatic and their connections to the hinterland in the late antiquity, in *L'adriatico dalla tarda antichità all'età carolingia : atti del convegno di studio Brescia 11-13 ottobre 2001*. Firenze: All'Insegna del Giglio, 49-77.

Vidrih-Perko, V. and Žbona-Trkman, B. 2005. Trgovina in gospodarstvo v Vipavski dolini in Goriških Brdih v rimski dobi. *Goriški letnik* 30/31, 17-70.

Vomer-Gojkovič, M. 1996. Grobišče pri Dijaškem domu v Rabeljčji vasi na Ptuju. *Ptujski zbornik 4/1, zgodovinsko društvo Ptuj*, 229-312, T. 20, 17.

Vomer-Gojkovič, M. 1996a. Rimski jantarni predmeti s Ptuja. *Arheološki vestnik* 47, 307-322.

Vomer-Gojkovič, M. 2008. Oil lamps from the territory of Eastern Slovenia: Poetovio, in C.-A.Roman and N. Gudea (eds), *Lychnological Acts 2. Trade and Local Production of Lamps from the Prehistory until the Middle Age. Acts of 2nd International Congress on Ancient and Middle Age Lighting Devices, Zalău/Cluj-Napoca, 13th-18th of May 2006*, 227-231. Cluj-Napoca: Editura Mega.

Wunderlich, Cr. H. 2006. Éclairage antique high-tech, in Crhrzanovski, L., *Lumière! L'éclairage dans l'Antiquité*, 40-45. Milano: Edizioni Et.

Žerjal, T. 2008. Rimska vila rustika v luči drobnih najdb: primer najdišča Školarice pri Spodnjih Škofijah. Unpublished PhD dissertation, University of Ljubljana.

A Light is On in the Hut.
Light and Lighting Equipment in Medieval Everyday Life

Thomas Bitterli

Swiss Castle Association and University of Basel, History Department

Abstract: Prior to the introduction of electricity, open fires using a variety of fuels including solids, liquids and gasses were the only sources of heat and light in people's everyday lives. In medieval castles daylight could only enter through narrow window slits and the castle inhabitants were dependent on artificial light from various devices. Besides archaeological evidence, literary and iconographic sources also provide vital clues on lighting equipment used in castles. Lighting devices could be as simple and small as a splint, or as large as a torch. Archaeological excavations at castles often uncover small bowl-shaped lamps, which would have been filled with tallow or oil and utilized a wick. Candles made of wax were only lit on special occasions. Tallow candles were more commonly used. Candlesticks and chandeliers were also used. Suspended oil lamps made of glass are not known from secular castle contexts, but iconographic sources often show them in sacred spaces. Courtly literature frequently describes special lighting in a hall or at table and use of a candle as a nightlight in the courtly bedchamber.

Keywords: Pine splint, torch, tallow candle, candle light, wick trimmer, oil lamp, fire striker, nightlight.

Introduction

Fire was and still is an important element in our lives. Prior to the general introduction of electricity, only open fires from different sources (such as solids, liquids or gas) brought warmth and light into people's everyday existence. As the darker times of autumn and winter are now upon us, the aspect of taking light from an open fire as part of medieval everyday life is once again being brought to the foreground.[1]

Although 'hut' in this context means any type of dwelling, we will see that the finds under consideration came primarily from the upscale everyday lives of the members of an aristocratic society that lived in castles. For details regarding the actual hut of a farmer or serf we are largely reduced to drawing conclusions from this upper-class lifestyle.

When we look at the imposing façade of a high-medieval castle or tower, we naturally wonder how such mighty buildings would have influenced people's lives. Large window openings are rarely seen, and if they are visible on a ruin or castle today, they usually date from late medieval or more recent times. Natural light enters the rooms only through small window openings. Narrow slits in the wall bringing scant light into the chambers are visible from the outside. These thin, usually rectangular openings high up in a masonry wall, are so-called slit windows (German: Lichtscharte; French: archère) or arrowslits.[2] The embrasure walls (German: Schartengewände; French: ébrasement) splayed inwards. In addition, the window ledge (German: Sohlbank; French: appui) was often bevelled, which allowed the odd ray of sunlight to penetrate through the thick walls into the interior of the castle. Occasionally the embrasures were whitewashed to better distribute the incoming light through diffuse reflection. In the winter, when these openings were boarded up against the wind and cold, it would have been really dark (and cold) in the castle.

It is known from archaeological and iconographic sources that window openings were occasionally fitted with so-called bullseye or crown glass.[3] Fragments of lead came and the associated glass are sometimes found at castle sites. However, before glass panes became more affordable, castle builders made do with cloth, oiled canvas, tanned skins, animal bladders, or wooden boards.[4]

Window recesses, often fitted with window seats (German: Sitzbank; French: coussièges) were built into the thick castle walls. Castle residents were happy to sit in these recesses when doing their chores.

If there was insufficient light for the purposes of work and everyday life despite the slit windows and window recesses, or if additional light was required in the early morning or late afternoon, residents would use lighting equipment. Artificial lighting –as opposed to natural daylight– provided only a slight improvement to the quality of life in the dark rooms and chambers of a castle, since using fire as a source of light was usually quite laborious.

Information about the lighting equipment used in castles can be obtained in various ways. A review of numerous publications on archaeologically examined castle sites in Switzerland (see the Appendix for a list) provides an overview of a diverse range of lighting devices. Since devices made of organic materials such as wood, textiles (for wicks) or wax and fat (for candles) are hardly ever discovered, we must instead refer to both literary and iconographic sources. Thanks to these additional sources, we can identify quite a range of

[1] This lecture, with a slightly different title, was held at the conference organised by the *Freundeskreis Bleidenberg* in Oberfell (Mosel) on 8th November 2014.

[2] For German and French technical terms for the elements of a window cf. Koepf 1985, 145-150 and Lavenu 1999.

[3] German Butzenscheiben; for medieval window forms cf. Schneider 2002, 36-54; Böhme *et al.* 2004, 122 f.

[4] Goll 1992, 279.

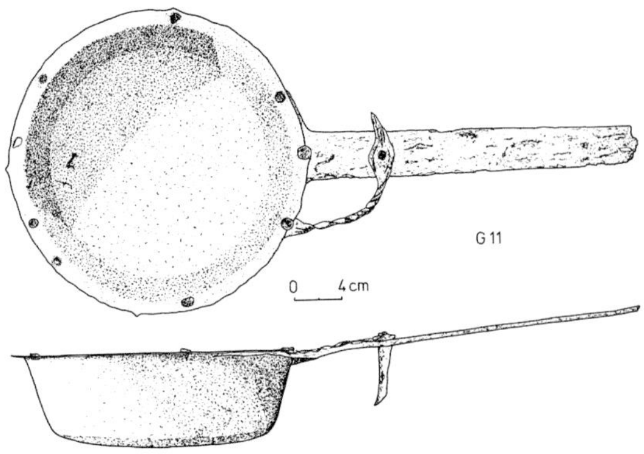

Figure 1. Example of an iron pan, which could be used both for cooking and for transporting embers. (Bischofstein, cat. no. G 11).

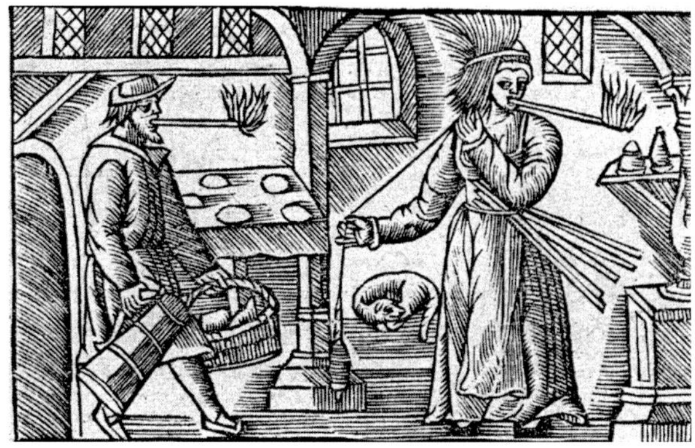

Figure 2. The busy couple. Both are holding a lit splint in their mouths. (From Olaus Magnus, De gentibus septentrionalibus, Rome 1555, lib., Sec., 77: de luminibus et tedis piceis).

lighting devices, the remains of which have either been found or which are highly likely to have been used in castles.

The light sources

Open fire

The most common source of light in a dwelling was an open fire in the kitchen or a log fire in one of the living rooms. This light source had the advantage that it also served as a source of heat and for cooking.

Fire was also portable, since embers could be carried around. Fire pans or ember baskets, equipped with lateral struts and a supporting leg fitted to the handle (Figure 1) were some of the implements used for this purpose.[5] This example shows that the objects used for lighting often had a range of other functions: the pan could also be used for cooking or for storing hot food.

Pine splints

A simple lighting device is a pinewood splint (German: Kienspan; French: tison). Splints are usually square or flat pieces of resinous wood of varying length mainly of pine, but also of fir, spruce, larch, stone pine, birch or cherry tree.[6] Splints were made of the inner core of the dead pinewood, where the resin had gathered during the process of the tree dying; these burnt more easily and for longer periods than splints made from beech and lime tree, though castle inhabitants sometimes had to make do with the latter.

A splint has certain advantages over a hearth fire, which is bright but can flicker wildly and be difficult to control. A splint burns more quietly and the burn rate can be regulated, depending on whether it is held in an upright or in an inclined position. However, the device still has obvious disadvantages: the splint has to be changed often, and a piece of about 20 cm in length burns only for a few minutes.

The burning properties of a splint can be improved significantly by adding a little fat or oil to it.

There are frequent mentions in the literature of lit splints being held between the teeth as a type of 'flashlight' (Figure 2). If one's mouth was needed for other things, for example eating food, the splint was placed into a clay holder (sometimes shaped like a human head with an open mouth)[7] or firmly clamped in a splint holder (German: Kienspanhalter; French: porte-tison). Other possible methods of storing a lit splint were stones or bowls full of sand. Burning splints could be dangerous because sometimes pools of resin in the wood could 'explode' (spit) and sparks would fly over distances of several meters. Since individual splint holders could have very different shapes, it is difficult to identify fragments of iron, which are usually heavily corroded, as such.[8]

Torches

Occasionally torches were also used. To this end, wooden sticks were wrapped in cloth and soaked in pitch or resin, lit and placed in torch holders.[9] This made it possible to light up even the larger rooms in a castle or the courtyard at night.[10] This form of lighting was quite cheap but had the disadvantage that it created biting smoke in the room and produced quite unpleasant smells.

Tallow candles

The rubble of castle ruins always contains inconspicuous, mostly grey potsherds, which can sometimes be restored to form small bowls.[11] They have a diameter of 6-12 cm and a

[5] See, for instance, Bischofstein. (Refer to the Appendix for references to these and other castle finds).
[6] Matz and Mehl 2000; Schneppendahl 1987, 200-233.

[7] The common German saying 'Maulaffen feilhalten' has its origins in these objects: 'Maulaffen' (mouth open).
[8] See, for instance, Bischofstein, Hallwyl, Heitnau, Hünenberg.
[9] For instance at Alt-Wädenswil.
[10] See Schulz 1880, 77f.
[11] See, for instance, Alt-Eschenbach, Alt-Regensberg, Alt-Tierstein, Alt-Wartburg, Alt-Wädenswil, Bischofstein, Dübelstein, Freudenau,

height of 2-4 cm. The rims, which can bear internal grooves, have a pinched spout: these are tallow candle bowls (German: Talglichtschale; French: lamp à suif, lamp coupelle), a form which was widely used from the 11th-14th centuries.[12] In addition to such ceramic bowls, there were also bowls made of soapstone or steatite, which differed only slightly in shape. In contrast to ancient oil or tallow lamps, the shapes of which were closed, the fuel bowls in medieval lamps were open. Lamp bowls were often made in large numbers and their shape changed over time, which allows us to deduce their relative chronology.[13]

Since the remains of these lamps have come to light at countless archaeological sites, not only at castles, we can assume that bowl-shaped tallow lamps were probably the most common form of artificial interior lighting. Remarkably, however, they are usually absent from archaeological finds at contemporary rural settlements and even the iconographic sources yield very little evidence.

The bowl was filled with tallow or oil and a wick (German: Docht; French: mèche) made of cotton, hemp cord or linen was placed in it. When the wick was lit, the fat which had soaked into the wick melted and burned. Because the fat or oil never burnt off completely, such oil and fat lamps gave off varying amounts of sooty smoke. Tallow (German: Unschlitt; French: suif) is obtained by melting the solid body fat of cattle or sheep. The fat produced by rendering pork and goose meat has a lower melting point and is called lard, which in medieval cuisine was mainly reserved for use in cooking and baking because of its quality. In addition to tallow, all types of vegetable oil (e.g., linseed or rapeseed oil) were also used as fuels depending on their availability.

One of the advantages of tallow or oil lamps, in addition to their low-cost fuel, was that they could be carried from room to room whilst they were lit. That is why they were often fitted with a handle.[14] Castle walls sometimes include niches for such lamps (Figure 3).[15]

In addition to ceramic tallow lamps, iron forms also existed. Most consisted of a plate filled with tallow and had a suspension device with a hook and spike,[16] which allowed them to either be hung from a beam or be inserted into a timber wall. In contrast to ceramic tallow bowls, iconographic sources often depict iron tallow lamps hanging from the ceiling (Figure 4).

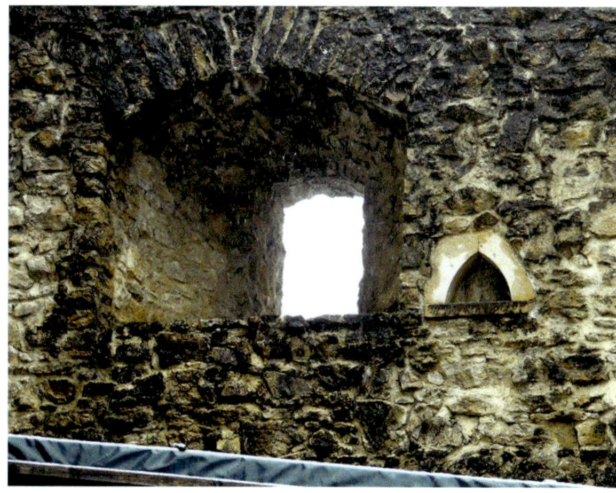

Figure 3. Gilgenberg castle ruins (Canton Solothurn). A niche next to the window for lighting devices such as candles or tallow and oil lamps (photo by the author 2014).

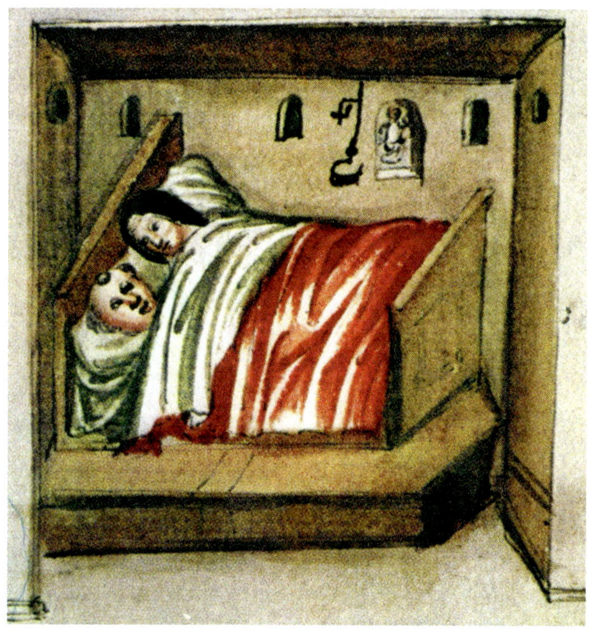

Figure 4. Lamp hanging from the ceiling. *Decameron*, Ninth Day, Second Story (Branca 1973, 804, note 34).

Candlelight

Wax candles (German: Wachskerze; French: bougie, cierge) were only lit by the castle inhabitants on special occasions. Wax was precious and was reserved for tithe payments to the church. Wax candles were used in church services.[17]

Tallow candles (German: Talgkerze; French: chandelle) served as substitutes for wax candles for everyday use at castles. However, the cheap raw material also had its disadvantages: tallow has a low melting point, which causes candles to drip; also, tallow candles are too brittle to be placed on a pricket candlestick.

The shape of a candlestick therefore allows us to deduce the type of candle used. A candlestick with a pricket served as a wax candleholder, a specimen with a depression or spout was meant for tallow candles.[18] An intact candlestick under

Frohburg, Gesslerburg, Habsburg, Hallwyl, Hasenburg, Hünenberg, Königstein, Madeln, Mülenen, Obergösgen, Scheidegg, Schiedberg, Zwing Uri.

[12] See also Höllhuber 1995.

[13] For a good typological and chronological overview, see Frey 2009.

[14] See, for instance, Dübelstein, Lägern.

[15] Gilgenberg castle ruins (Canton Solothurn).

[16] See, for instance, Alt-Regensberg, Alt-Wartburg, Freudenau, Hallwyl, Madeln, Schnabelburg, Urstein.

[17] See Elmshäuser 2000, vol. V, col. 1116.

[18] For candleholders in general, see Thomas Dexel 1980, 110-121; for

archaeological examination poses no problems with regard to its interpretation.[19] However, if it is only partially preserved, the question as to whether it was part of a candlestick, and if so, which part, is much more difficult to answer.[20]

Candlesticks can be made of different materials. In addition to iron candleholders, lead or brass examples are also found. The identification of a brass fragment from the ruins of Alt-Wädenswil as part of a candlestick was only possible by comparison with a preserved candlestick of Dutch provenance.[21]

An example from Alt-Eschenbach shows that a very simple object can serve as a stand for a candle.[22] A wooden object from a well at Friedberg/Meilen, previously interpreted as part of a piece of furniture (bed, wardrobe), could also have been part of a wooden candlestick.[23] The piece is, however, only roughly worked and does not seem to have been finished.

In medieval romances such as *Parzival* (English Percival) by Wolfram von Eschenbach candlesticks are usually called *kerzstal*.[24]

A candlestick holds a single candle, a candelabrum (German: Armleuchter; French: chandelier) can hold several. Chandeliers (French: lustre) were suspended from the ceiling and because of their ring shape are often referred to as *krônen* (crowns) in medieval German usage.[25] Of course, candlesticks, candelabra and chandeliers could also be equipped with tallow bowls instead of candles. In this case, the devices would have borne ring-shaped elements for the tallow bowls to rest in.[26] Chandeliers were usually made of metal or wood, sometimes also of horn (antler). There were also ring-shaped devices that could be fitted around pillars or columns. In all these cases, it is very difficult to identify them as chandeliers if the components are preserved in very small fragments.

Wick trimmers

The wicks of tallow candles, which were the most commonly used types of candle prior to the beginning of the 19th century, had to be trimmed constantly to prevent dripping and soot from developing. For the purpose of 'cleaning the lamps', special wick trimmers were used to remove the so-called 'snuff'.[27]

candleholders in liturgical contexts, see Springer 2000, vol. V, col. 1916-1918.

[19] See, for instance, Alt-Regensberg, Alt-Wartburg, Bischofstein, Lägern.

[20] See, for instance, Alt-Wädenwil, Clanx, Frohburg.

[21] Dexel 1980, Plate XV: '*Scheibenleuchter*' brass candlestick, Netherlands, 15th century.

[22] Rickenbacher 1995, no. C230/231

[23] For similar wooden candlesticks, see Dexel 1980, 110; Friedberg, Müller 1981, no. 118.

[24] For example Parzival 34, verse 26: '*Guldin wârn ir kerzstal*'; Schulz 1880, 76-78.

[25] Parzival 636, verse 9: '*manec tiuriu krône Was gehangen schöne Alumbe ûf dem palas. Diu schiere wol bekerzet was.*' Parzival 229, verse 23: '*Hundert krône dâ gehangen war, Vil kerzen drûf gestôzen* (...)'.

[26] Walter 1993, 113.

[27] See, for instance, Hallwyl, Zug.

Ceramic lanterns

A special form of candleholders were ceramic lanterns, whose shapes corresponded to common pots but whose walls were perforated. If found at an excavation, this lighting device can only be identified if a piece of considerable size has been preserved. Otherwise they are very difficult to distinguish from fragments of cooking pots.[28]

Lanterns

This also applies to lanterns where the light (the flame) was protected from the wind by a wooden or metal case. The light penetrated through transparent horn plates or crystal panes, which were fitted into the walls of the casing. These are mentioned in the courtly literature, but I am not aware of any archaeological lantern finds dating from the Middle Ages.[29]

Oil lamps

Suspended glass oil lamps are not found in the secular areas of castles, but iconographic sources often depict them in sacred spaces (Figure 5).[30]

The oil lamp in medieval symbolism is an attribute of the 'foolish virgins', as can be very clearly seen in the figures depicted on Strasbourg Cathedral, where they are shown as bowl-like vessels with cylindrically or conically elongated bases which served as handles; pictorial sources show that these glass bowls were placed in special devices. When fragments of oil lamps come to light at castle excavations, they can be assumed to have originated from the chapel.

Fire lighting

All medieval lighting fixtures had one thing in common: an open flame as a source of light (thermal light source). To this end, a fire had to be lit, which was a rather cumbersome task, especially in the dark.[31] In order to light a fire, a striker or fire steel (German: Feuerstahl; French: briquet à battre) (Figure 6) had to be struck against a piece of flint (silex, chert, rock crystal or pyrite).[32] The resulting sparks ignited the tinder (tinder fungus, German: Zunderpilz; French: amadouvier) which then generated an open flame. Since lighting a fire was a very basic everyday activity, it is not surprising that many assemblages from excavations carried out at medieval castles include fire strikers and flints.[33]

In general, lighting fixtures in everyday scenes are rarely found in iconographic sources. Even in countless illustrations for Giovanni Boccaccio's *Decameron*, which include quite a few bed and table scenes, suspended tallow lamps or candlesticks are not very numerous.[34]

[28] See Gross 1999, 299-301 and fig. 169.

[29] See also Schulz 1880, 79.

[30] Baumgartner 1988, 436f.

[31] Actual wooden matches first appeared in 1827. What we know as a lighter today has only existed since the beginning of the 20th century.

[32] See, for instance, Alt-Regensberg, Alt-Wartburg, Attinghausen, Bischofstein, Hallwyl, Mülenen, Rickenbach.

[33] Such fire strikers have been found at Frohburg, Zwing Uri; for finds from Alpine valley settlements see Bitterli-Waldvogel, 1998, 174-201, especially 193, pl. 218,21.

[34] Boccaccio 1973.

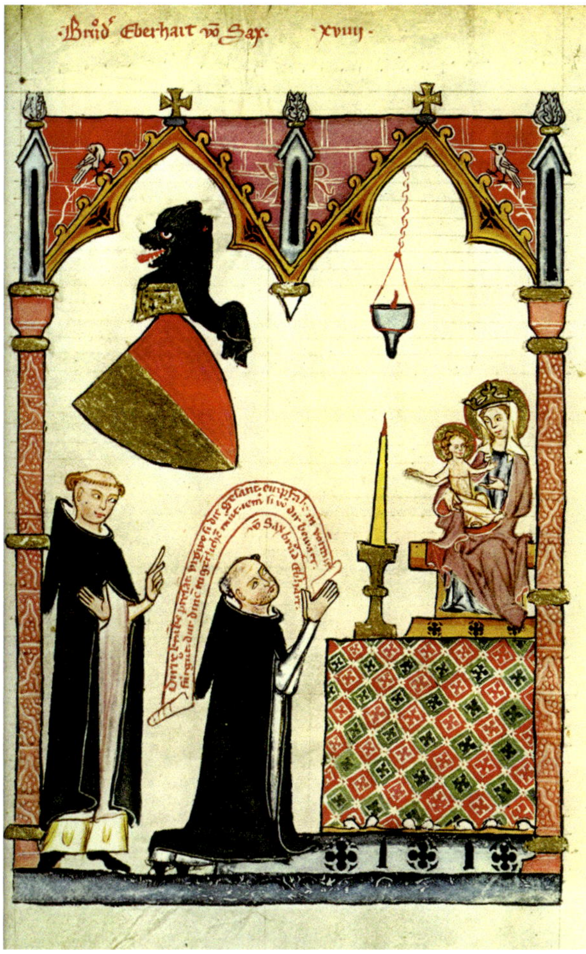

Figure 5. Brother Eberhard von Sax (Great Heidelberg Songwriter [Codex Manesse], University Library Heidelberg, Cod. Pal. Germ. 848, fol. 48v).

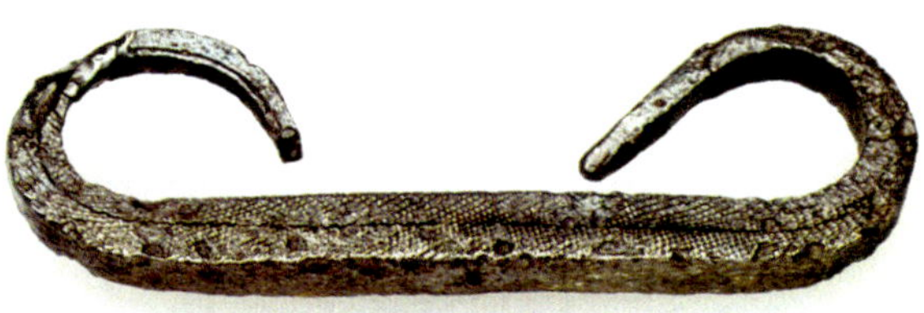

Figure 6. Fire striker. (Alt-Wartburg, cat. no. C 140/141).

There are many reasons why depictions of lighting equipment are lacking and I will refrain from entering the realm of speculation here. The limited options of illuminating a room by today's standards meant that the residents of a castle spent a lot of their time outdoors and also carried out all the tasks that required sufficient lighting there. Daily routine therefore centred around how much daylight was available. Nocturnal activities in artificial light hardly existed –though there was one exception: eating, drinking and being entertained was definitely possible in half-light and was obviously a popular pastime– which is now considered very romantic. Special lighting in a hall or at table is mentioned time and again in literary sources, as seen in *Parzival* during the first visit to Gralsburg Castle: '*Hundert krônen dâ gehangen war, Vil kerzen drûf gestôzen ob den hûsgenôzen. Kleine kerzen umbe an der want*' [A hundred crowns (chandeliers) were suspended, many candles were lit. Small candles hung on the wall].[35]

In contrast and surprisingly, table lighting is not usually seen in iconographic sources with table scenes.

Nightlights

A special usage of artificial light, which cannot be detected archaeologically, but is known from literary sources, was the nightlight. Whether a guest's night-time way to the bedchamber was lit up by torches or candles was –to put it in modern terms– a question of etiquette or protocol. Several passages in Wolfram von Eschenbach's *Parzival* mention a candle burning as a nightlight in the bedchamber. Gahmuret, for instance, mentions that large candles burned brightly in his bedchamber.[36] Also, young Percival, when going to his bedchamber in the besieged city of Pelrapeire, is eager to see if straw torches will be lit instead of candles: 'Oh no, there were indeed decent candles.'[37] And, a little later that night, when Queen Konwiramur sneaks up to the youthful Percival, she states with conviction, 'Candlelight bright as daylight was set up in front of his sleeping-place.'[38] Elsewhere, Gawain notes: 'Then daylight began, so the glow of the large candle in his chamber made little impact.'[39] Apparently it was the custom in aristocratic everyday life to have a candle burning as a nightlight in the bedchamber. This was apparently due to an earlier tradition, because according to Benedictine Rule, permanent illumination in the dormitory was mandatory.[40]

A light for the dead and the bird of death

In medieval everyday life, it was customary to leave a light burning near a deceased person over night. If there was no moonlight, night time in the Middle Ages was really dark – there was no scattered ambient light like there is today. As a result, the owl as a nocturnal animal saw such lights from afar and would approach them. The call of the female 'coo-witt, coo-witt' outside a window was said to mean 'come with, come with', and was ascribed to the bird inviting the dying person to follow it. In this way, the owl became the bird of death.

Conclusion

A final note: in courtly literature, a bright, calmly burning candlelight occasionally served as a metaphor for the beauty of the adored one. This is seen in two further passages from *Parzival*: 'A glow came from the face of Frau Heyezeloide [Percival's mother] –if the candles had suddenly gone out, her countenance would have been bright enough.'[41] In

[35] Parzival 229 verses 24-27, Parzival 638 verses 13-14. Stapfel 1937.
[36] Parzival 35, verse 15, Stapfel 1937, 23.
[37] Parzival 191, verses 17-19; Stapfel 1937, 99.
[38] Parzival 192, verses 25-30; , Stapfel 1937,100
[39] Parzival 588, verses 7-9; Stapfel 1937, 301.
[40] Jaritz 1838: Benedictine Rules, Chapter 22: How they are to Sleep, Sentence 4L A lamp shall be kept burning in the dormitory until morning.
[41] Parzival 84, verses 10-15; Stapfel 1937, 46.

another passage, Gawain describes the face of Queen Arvine '... the Duchess herself was so aglow that even if she had not lit any candles, no night would have been around her...'.[42] Bright candlelight was very familiar to the noble listeners of romances and heroic sagas.

In summary, we can state that there were many ways to illuminate the dark and almost windowless rooms and chambers in a medieval castle using lighting equipment; we know these from different sources. In general, however, they were only makeshift devices when no better options were available. As a rule, people in the Middle Ages used natural light for as many of their activities and chores as possible. Exceptions were hall and table lighting at festivities and parties in the evening, and nightlights, i.e., burning candles, which were often left in the bedchambers.

Appendix

Site	Report	Catalogue, page or image nos.
Alt-Eschenbach	Rickenbach 1995	C 227/228. 230, 231
Alt-Regensberg	Schneider 1979	C 108, 111, 13C, 114, E1
Alt-Tierstein	Erb 1935	Fig. 20,10
Alt-Wädenswil	Bitterli and Grütter 2001	202-210, 311, 448, 449
Alt-Wartburg	Meyer 1974	B206-B258, C141, 142
Attinghausen	Meyer 1984a	A 30
Basel	Rippmann 1987	
Bischofstein	Müller 1980	A102-148, F2, 42, 43, 78, G11
Clanx	Obrecht 2005	112
Dübelstein	Dubler 2006	
Freudenau	Baumann and Frey 1983	A140-143, E51, 52
Friedberg	Müller1981	118
Frohburg	Meyer 1989	A269-282, D29, H10
Gesslerburg	Schneider 1984	B11
Glätrich VS	Bitterli 1989	22
Habsburg	Frey 1986	B66-68
Hallwyl	Frey and Kühn 2007	S191-194, 224, 675-717, 732-747
Hallwyl	Lithberg 1932	Pl 99A,B,F,G;100D,E,H-M;102M, 104, 15, 107A
Hasenburg		
Heitnau	Knoll-Heitz 1957	28/13
Hünenberg	Schneider 1948	Pl III, 2.11, 2.42
Königstein	Matter 1997	A147-179
Lägern	Schneider 1946	Fig. 8.29
Madeln	Marti 1988	73-88, 218
Mülenen	Meyer 1970	E29, A168-169, A176
Nivagl	Meyer 1980	
Obergösgen	Bitterli-Waldvogel 1991	12-14
Rickenbach	Meyer-Hoffman 1972	B15
Scheidegg	Ewald 1975	A131-162
Schiedberg	Meyer 1977	B6
Schnabelburg	Schneider 1957	pg. 16
Urstein	Knoll-Heitz 1985	Fe18
Wieladingen	Bitterli-Waldvogel 1988	27
Willisau	Egggenberger 2005	26, Cat 6/12
Zug	Grünenfelder 2003	471
Zwing Uri	Meyer 1984b	19

[42] Parzival 638, verses 15-20; Stapfel 1937, 326

Bibliography

Note: SBKAM = *Schweizer Beiträge zur Kulturgeschichte und Archäologie des Mittelalters*, published by the Swiss Castle Association, Olten/Basel from 1974.

Baumann, M. and Frey, P. 1983. *Freudenau im untern Aaretal. Burganlage und Flussübergang im Mittelalter.* Stilli: M. Baumann; Untersiggenthal: Gemeindeverwaltung.

Baumgartner, E. and Krueger, I. 1988. *Phönix aus Sand und Asche: Glas des Mittelalters:* [Rheinisches Landesmuseum Bonn, 3. Mai - 24. Juli 1988, Historisches Museum Basel, 26. August - 28. November 1988]. Munich: Klinkhardt and Biermann.

Bitterli, T. and Grütter, D. 2001. *Alt-Wädenswil - vom Freiherrenturm zur Ordensburg (SBKAM 27)*, Basel.

Bitterli, T. 1989. Glätrich, Wiler (Lötschen) VS 1989.1990, in Meyer, W. *et al.*, *Heidenhüttli - 25 Jahre archäologische Wüstungsforschung im schweizerischen Alpenraum (SBKAM 23/24)*, Basel.

Bitterli-Waldvogel, T. 1988. Grabungen und Lesefunde auf der Ruine Wieladingen. *Badische Heimat 1988, Heft 1*, 109-119.

Bitterli-Waldvogel, T. 1991. Burg Obergösgen, in *Archäologie des Kantons Solothurn 7*, 77-103.

Boccaccio, G. 1973. *Dekameron, 1. bis 10. Tag. Mit einer Einführung von Vittore Branca*, Lucerne / Freudenstadt / Zurich.

Böhme, H. W., Friedrich, R. and Schock-Werner, B. 2004. *Wörterbuch der Burgen, Schlösser und Festungen.* Stuttgart: Philipp Reclam.

Dexel, T. 1980. *Das Gebrauchsgerät Mitteleuropas von der römischen Zeit bis ins 19. Jahrhundert Bd. 1.* Braunschweig: Klinkhardt and Biermann.

Dubler, R. *et al.* 2006. *Vom Dübelstein zur Waldmannsburg. Adelssitz, Gedächtnisort und Forschungsprojekt* (*SBKAM* 33), Basel.

Eggenberger, P. 2005. *Funde Aus Den Archäologischen Forschungen.* Lucerne: Kantonaler Lehrmittelverlag.

Elmshäuser, K. 2000. Kerze, in *Lexikon des Mittelalters.* vol. V., Munich /Zurich

Erb, H. 1935. Ausgrabung Tierstein 1934. *Argovia* 47,13-94.

Ewald, J. and Tauber, J. 1975. Die Burgruine Scheidegg bei Gelterkinden. *Berichte über die Forschungen 1970-74 (SBKAM 2)*, Olten.

Frey, J. 2009. Lampen und Leuchter im Kanton Bern. *Archäologie Bern*, 217-232.

Frey, P. 1986. Die Habsburg im Aargau. Bericht über die Ausgrabungen von 1978-83. *Argovia* 98, 23-116.

Frey, P. and Kühn, M. 2007. *Das Stammhaus der Herren von Hallwyl: Die archäologischen Untersuchungen auf dem Wasserschloss Hallwyl 1995-2003*. Baden: Hier + jetzt, Verlag für Kultur und Geschichte.

Goll, J. 1992. Baumaterial, in M. Flüeler (ed.) *Stadtluft, Hirsebrei und Bettelmönch: Die Stadt um 1300*, 267-280. Stuttgart: Theiss.

Gross, U. 1999. Beleuchtung, in *Burgen in Mitteleuropa. Ein Handbuch, Bd. 1*. Stuttgart: Konrad Theiss.

Grünenfelder, J. *et al.* 2003. Die Burg Zug. *Archäologie - Baugeschichte - Restaurierung (SBKAM* 28), Basel.

Heid, K. 1948, Die Keramik der Burg Hünenberg. *Zuger Neujahrsblatt* 1948, 60-64.

Höllhuber, A. 1995. Mittelalterliche Öllampen. Aus dem Fundgut von Burgruinen, Burgställen und Plätzen ehemaliger Holzburgen im unteren Mühlviertel. *Öberösterreichische Heimatblätter* 49, 208-224.

Koepf, H. 1985. *Bildwörterbuch der Architektur*. Stuttgart: Kröner.

Jaritz, G. 2000. Beleuchtung, in *Lexikon des Mittelalters, Bd. I*, Munich /Zurich.

Knoll-Heitz, F. 1957. Burg Heitnau 1950-56. *Thurgauische Beiträge zur Vaterländischen Geschichte* 93, 7-80.

Knoll-Heitz, F. 1985. Urstein, die grösste Burg in Herisau. *Appenzeller Jahrbücher* 113, 4-143.

Lavenu, M. and Mataouchek, V. 1999. *Dictionnaire d'architecture*. Paris: Jean-Paul Gisserot.

Lithberg, N. 1932. *Schloss Hallwil, Fundkatalog 3*. Stockholm.

Magnus, O. 1555. *De gentibus septentrionalibus*, Rome.

Marti, R. and Windler, R. 1988. *Die Burg Madeln bei Pratteln - BL: Eine Neubearbeitung der Grabungen 1939-40*. Liestal: Amt für Museen und Archäologie des Kantons Baselland.

Matter, G. and Reding, C. 1997. Funde aus der Schutthalde der Ruine Königstein, Gemeinde Küttigen AG. *Argovia 109*, 1-122.

Matz, J. and Mehl, H. 2000. *Vom Kienspan zum Laserstrahl: Zur Geschichte der Beleuchtung von der Antike bis heute*. Husum: Husum Druck- und Verlagsgesellschaft.

Meyer, M. 1970. Fundkataloge, in: *Die Wasserburg Mülenen (Mitteilungen des Historischen Vereins des Kantons Schwyz* 63, 105-264.

Meyer, W. 1974. Die Burgruine Alt-Wartburg im Kanton Aargau. *Bericht über die Forschungen 1966/67 (SBKAM* 1), Olten.

Meyer, W. 1977. Die Ausgrabungen der Burg Schiedberg, in M.-L. Boscardin *et al.*, *Burgenforschung in Graubünden. (SBKAM 4)*, Olten, 51-175.

Meyer, W. 1980. Nivagl GR - provisorischer Bericht über die Sondierung 1980. *Nachrichten des Schweizerischen Burgenvereins* 53, Heft 5, 93-105.

Meyer, W. 1984a. Attinghausen, in idem. *et al.*, Die bösen Türnli. *Archäologische Beiträge zur Burgenforschung in der Urschweiz (SBKAM* 11), Olten, 7-36.

Meyer, W. 1984b. Zwing Uri, *Die bösen Türnli. Archäologische Beiträge zur Burgenforschung in der Urschweiz (SBKAM 11)*, Olten, 63-88.

Meyer, W. 1989. *Die Frohburg, Ausgrabungen 1973-1977 (SBKAM 16)*, Olten.

Meyer-Hofmann, W. 1972. Die Burgstelle Rickenbach. *Jahrbuch für Solothurnische Geschichte* 45, 316-409.

Müller, F. 1980. *Der Bischofstein bei Sissach, Kanton Baselland: Die hochmittelalterl. Funde. Mit e. histor. Beitr. von Werner Meyer*. Derendingen-Solothurn: Habegger.

Müller, F. 1981. Die Burgstelle Friedberg bei Meilen am Zürichsee. *Zeitschrift für Archäologie des Mittelalters* 9, 7-90.

Obrecht, J. *et al.* 2005. Burgen in Appenzell. Ein historischer Überblick und Berichte zu den archäolog. Ausgrabungen auf Schönenbüel und Clanx (*SBKAM* 32). Basel.

Rickenbach, J. 1995. *Alt-Eschenbach: Eine spätmittelalterliche Stadtwüstung*. Lucerne: Kantonsarchäologie Luzern.

Rippmann, D. *et al.* 1987. *Basel, Barfüsserkirche (SBKAM* 13). Olten.

Schneider, H. 1950. Die Eisenfunde aus der Burg Hünenberg. *Zuger Neujahrsblatt* (1950), 55-66.

Schneider, H. 1984. Gesslerburg, in Werner Meyer *et al.*, *Die bösen Türnli. Archäologische Beiträge zur Burgenforschung in der Urschweiz* (*SBKAM* 11), 89-128. Olten.

Schneider, H. 1979. Die Buirgruine Alt-Regensberg im Kanton Zürich. *Bericht über die Forschungen 1955-57 (SBKAM* 6). Olten.

Schneider, H. and Heid. K. 1946. Das Fundmaterial aus der Burgruine Lägern. *Zeitschrift Für Schweizerische Archäologie Und Kunstgeschichte / Schweizerisches Landesmuseum*, 29-46. 29-46.

Schneider, H. 1957. Die Schnabelburg. *Blätter der Vereinigung Pro Sihltal* 7, 2-16.

Schneppendahl, R. 1987. Von Kienspänen, Fackeln, Öllampen und Kerzen, Lichter und Leuchter, in *Entwicklungsgeschichte und Technik eines alten Kulturgutes*, 200-233. Arnsberg: Neheim-Hüsten.

Schneider, J. 2002. Fenster und Fassaden im Alten Zürich. *Mittelalter: Zeitschrift Des Schweizerischen Burgenvereins*, 36-54.

Schulz, A. 1880. *Das höfische Leben zur Zeit der Minnesänger*, Leipzig, ND (repr. Kettwig 1991).

Springer, P. 2000, Leuchter, in *Lexikon des Mittelalters*, vol. V, Munich /Zurich.

Stapfel, W. 1937. *Wolfram vom Eschenbach, Parzival, Prosa* übertragen, ND (repr. Munich 2001).

Walter, P. 1993. *Le vieux château de Rougemont, site medieval*. Rougemont: Edition Deval.

Medieval Lighting Devices from the Collection of the State Historical Museum of Russia

Svetlana Avdusina
State Historical Museum, Moscow

Abstract: The article describes the medieval lighting devices from the collections of the State Historical Museum of Russia in Moscow. The collection of lighting devices includes several categories: lamps, candleholders and candles, as well as splinter holders. The devices described in this article are dated from the 10th-17th centuries AD and were found during excavations of ancient cities such as Kiev, Chernigov, Novgorod, Smolensk, Staraya Ryazan, Moscow and the cities of Volga Bulgaria: Bulgaria and Suvar.

Keywords: Lighting devices, Ancient Russia, museum collection, archaeology, Middle Ages.

Lighting devices were important elements of the Ancient dwelling. Natural light penetrated through a little window and the interior of a house remained dark most part of the day. This article focuses on lighting devices from the 10th century to the 18th century AD from excavations of the towns of Ancient Rus' and the Volga Bulgaria.

Lighting devices include such items as lamps which burned liquid oil, candlesticks which were used as a stand for wax candles and splinter holders.

Lamps were used in Rus' from early times. They were usually made of ordinary potter's clay. They were of different forms and sizes, on stacks and without them. In the Historical Museum collection there are lamps of the so-called Kiev-type (the first finds of such lamps were made in the town of Kiev). Made of light clay on a potter's wheel, the lamp itself consists of a stack opened at the bottom to which two saucers were fixed –one at the top, the other at the upper third of the stack. When there was fire in the lamp, the oil flew down from the upper saucer to the lower one. Stacks were decorated with linear-wave ornaments. In the collection of the State Historical Museum such lamps are presented by finds from Kiev (Figure 1), Smolensk, Chernigov and date from the 11th to the 13th centuries AD.[1] In the museum collection there are no completely intact examples of the type. These lamps were uncovered mostly during urban excavations. The shape dates back to the ancient tradition. Similar lamps have been found in the bottom horizons of a cultural layer of Moscow.[2]

In other towns of the Ancient Rus' oil lamps of a more primitive construction were used since the 11th century AD –a deep saucer was pasted onto a stack. The collection of the Historical Museum includes lamps found on the Old Ryazan site (Figure 2) not far from modern Ryazan. Probably such lamps could be put on clay saucers. Such saucers with a small depression in the middle are well represented in the museum collection (Figure 3). They could be used either as stands for lamps or as the holders for candles of small diameter.[3] Most

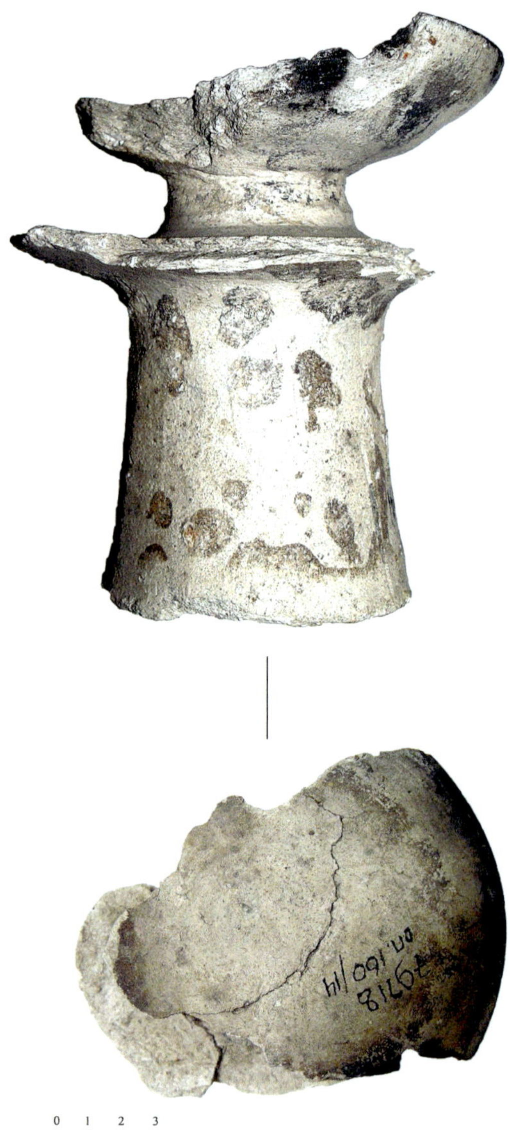

Figure 1. Lamp from Kiev.

[1] Karger 1958, 450-451.
[2] Rozenfeldt 1968, 9.
[3] Rozenfeldt 1997, 11.

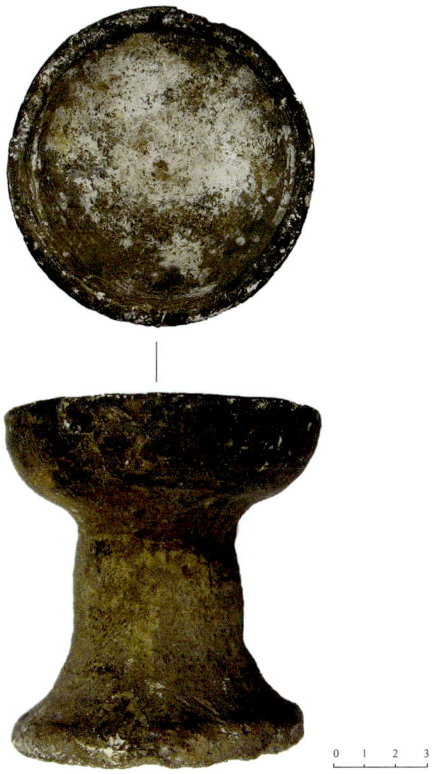

Figure 2. Lamp from Old Ryazan.

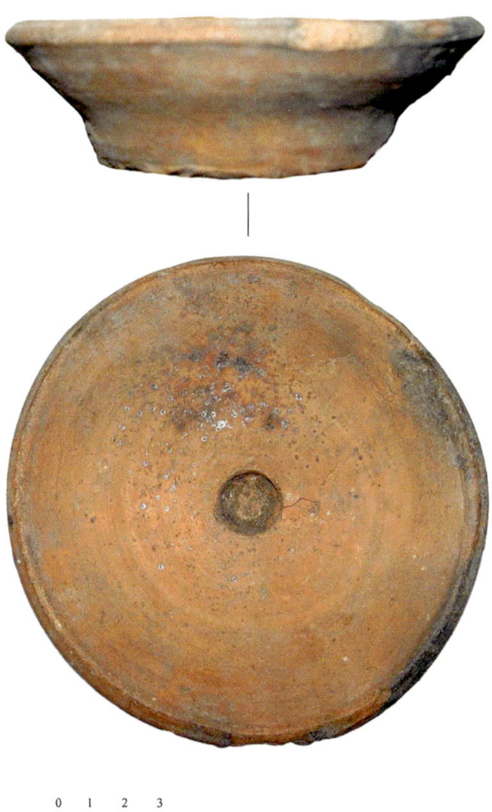

Figure 3. Saucer from Novgorod.

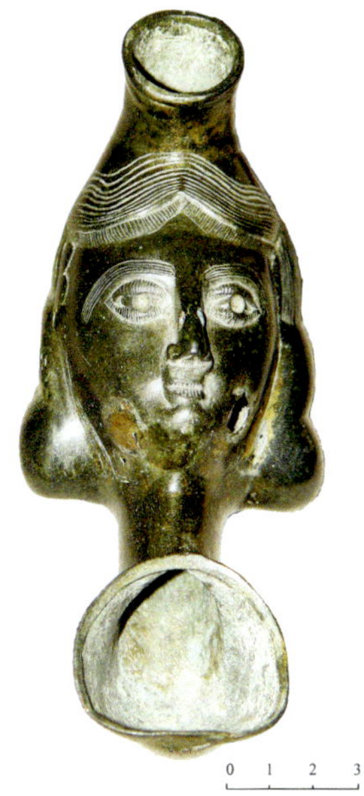

Figure 4. Lamp from Gnezdovo.

of the saucers in the collection were found in the Novgorod excavations.

The most interesting find in the collection is an item from the Gnezdovo site near the modern city of Smolensk which was situated on the road from Scandinavia to Byzantium. This bronze lamp was found in 1874 in a burial mound (Figure 4). It is dated to the 10th-11th centuries AD. It has the form of woman's head and has engraved hair, eyebrows and eyelashes. The lamp has dents and open-end holes. A turquoise bead was found in the nose. The pupils look like round depressions which were probably encrusted. Russian researcher Darkevich thought that this lamp had an Iranian origin.[4]

The Historical Museum collection includes a great number of lamps found in excavations in the Volga Bulgaria towns of Bulgar and Suvar. Lamps from the two towns have an identical form and look like low flat-bottomed saucers with a nozzle on one side and a handle on the opposite side (Figure 5). Some lamps have a ledge in the middle to affix a candle (Figure 6). Very often these lamps are covered with a glaze. They have no ornamentation and date from the 11th to the 15th centuries AD.

In the 16th century similar lamps but of a smaller size were used in Moscow. These small vessels have a sharp or rounded bottom and a small handle. Most of them have a linear pattern in the upper part. The lamps were covered with glaze. The diameter of such lamps varies from 3 to 5 cm (Figure 7). According to R. Rozenfeldt these small vessels were used as lighting devices and were attached to a wooden base by nail passing through the hole of the handle. According to him, due to the small size they could be used as icon-lamps.[5]

[4] Darkevich 1976, 52.

[5] Rozenfeldt 1968, 37.

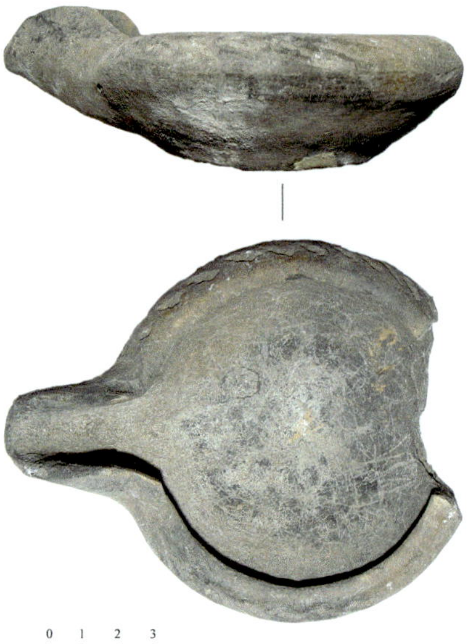

Figure 5. Lamp from Suvar.

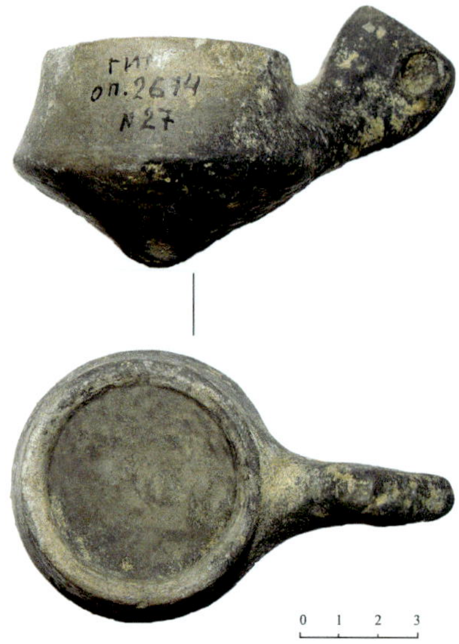

Figure 7. Lamp from Moscow.

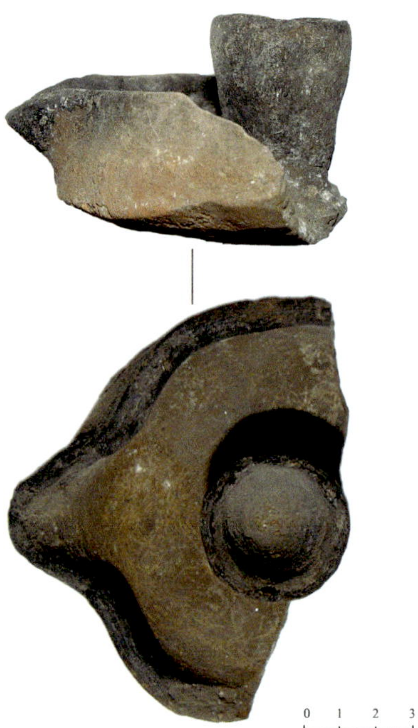

Figure 6. Lamp from Bulgar.

Medieval candles were found in excavations of many Ancient Rus' towns. In the collection of the Historical Museum there are candles from the Gnezdovo burial mound near Smolensk (Figure 8) and wax fragments from the Novgorod excavations. In addition, in the museum collection there is a cone-shaped candle, which had been used for ritual purposes, from the rich burial mound in Timerevo near Yaroslavl (Figure 9). This type of candle was found in the royal burial mounds by Elling and Mammen.[6] The Timerevo burial mound group dates back to the 10th century AD. It is assumed that the mounds with candles are associated with the early Christians.[7]

Candleholders made of clay or metal are also represented in the collections of the Historical Museum. Some could be put on a table; some hung on a wall. Candleholders are divided in two types: sleeve-type (where the candle is inserted into the sleeve) and spike-type (where the candle is put on a spike).

Bronze candleholders were used for lighting in rich houses and cult constructions. The museum collection includes several bronze candleholders. Two of them were found on the territory of the ancient town of Vschizh near Bryansk. They were found during the excavation of an construction in the church altar, built at the end of the 12th to the first half of 13th century AD. The first candleholder is made of bronze and has a circular stem decorated with two balls in the middle part. On both sides of the stem a grid divided by three ledges in the form of leaves is engraved. On the disk, which catches the melted wax, is the spike for installing candles. The bottom of the disk is supported by three figures of snakes. The foot of the candleholder has a pyramid shape and three legs in the form of dragons (Figure 10). The candlestick is 27.5 cm high. Darkevich relates this candleholder to objects of Westphalian

[6] Brøndsted, 1936, 106-107, 119-120.
[7] Pushkina 1997, 127.

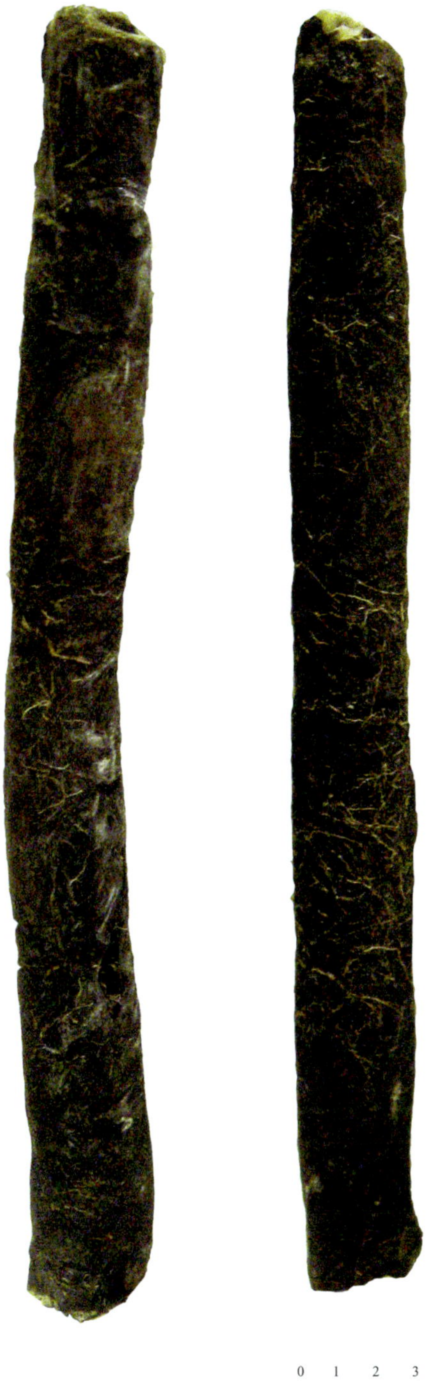

Figure 8. Candles from Gnezdovo.

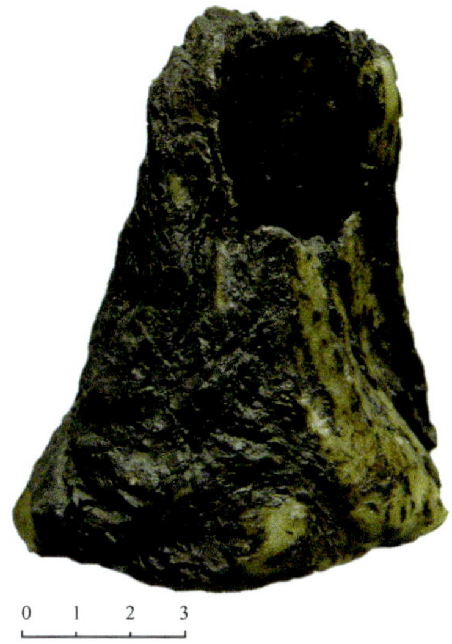

Figure 9. Candle from Timerevo.

workshops of the second half of the 12th century AD and thinks that there are analogous altar candleholders from the Eparchial museum and in Munster Cathedral.[8] The second bronze candleholder is made in the form of a round tower with a ball in the middle on the basis of a trihedral (Figure 11). At the top of the candlestick there is a disk to catch the melted wax and a spike for installing the candle. The footed base is formed as stylized animal paws ornamented with cast lines, zigzags and palmettes. Each of the three edges of the base is decorated with an enamel pattern of blue, red and green colors. The ball and the cup bottom are also decorated with enamel. The rest of the stem is covered with molded flake ornaments. The candlestick is 23 cm high. The researcher Darkevich thinks that this candleholder relates to objects of Limoges workshops and was made before 1238, when the church was destroyed during the Baty invasion.[9]

The Historical Museum collection also includes some simpler candleholders. For example, there is a three-legged base of a candleholder from the ancient settlement of Old Ryazan and two candleholders from the excavation of Old Ladoga. All the above-mentioned candlesticks are of the spike-type. Candlesticks from Old Ryazan have a trihedral base and three legs in the form of animal legs (Figure 12). Candlesticks from Old Ladoga consist of a stem on three legs and a disk to catch the melted wax in the middle (Figure 13). Some of these disks are stored in the museum's collection.

In ordinary dwellings, candleholders made of clay also could be used. All of them are sleeve-type. These candlesticks were found during excavations in Novgorod and entered the collection of the Historical Museum. They have a clay base and a sleeve for a candle (Figure 14).

In the materials from Bulgar there are also several types of clay candleholders. These candleholders have a cup in which there is a sleeve for a candle (Figure 15). Moreover, the museum's collection has a cone-shaped candleholder with a sleeve for a candle in the center (Figure 16).

Besides the above-mentioned material, a large group of lighting devices is presented by candleholders, which were used for candles of a smaller diameter. These devices consist of a metal rod of rectangular cross-section and sharpened

[8] Darkevich 1966, 21.

[9] Darkevich 1966, 21.

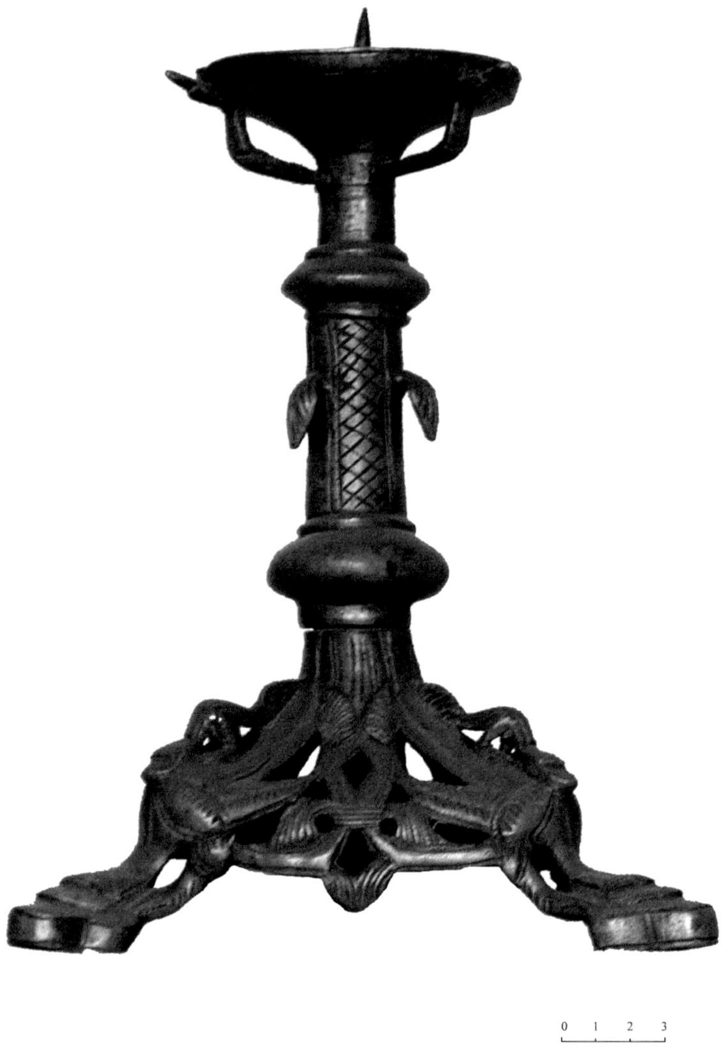

Figure 10. Candleholder from Vschizh.

on one side (for hammering in the wall). On the other side, there is a sleeve in which the candle can be installed. Such candlesticks were found in Novgorod in layers from the 13th century AD,[10] as well as in Bulgar[11] and in Vladimir burial mounds (Figure 17) dating from the 10th-13th centuries AD. The Historical Museum Collection also includes metal candleholders of this type from one of the Vladimir burial mounds. Its sleeve is fixed on the disk to catch the melted wax (Figure 18). Moreover, the museum collection has an interesting item found in Bulgar, which, to my mind, is also a candleholder. In the middle of the rod it has an oval rounding (Figure 19). In general, these candleholders existed for a long time on the territory of Ancient Russia. In Moscow, they have been found in layers of the 15th-17th centuries AD.

Since the 13th century AD houses in Ancient Russia were lit by splinters with the help of devices which held them. Structurally, splinter holders were of the same functional design – a splinter is clamped between two iron pins. But they were different in shape. There are three forms of splinter holders: for one splinter (Figure 20), for three splinters, and cranked form (Figure 21). Traditionally, it is believed that the one-splintered holder design is the oldest and dates from the 13th century AD.

From the 15th century AD, there appear two other types of splinter holders. According to B. Kolchin, splinter holders were put on wooden bases vertically in such a way that a burning splinter was always in a horizontal position and could be placed under a washing tub with water.[12] The collection of the Historical Museum includes many splinter holders. Most of them were found in Novgorod, and some were found in Bulgar in the cultural layers dating from the 13th to 15th centuries AD.[13] Splinter holders were also found in other

[10] Kolchin 1959, 97.
[11] Savtchenkova 1996, 61.
[12] Kolchin 1959, 98.
[13] Savchenkova 1996, 63.

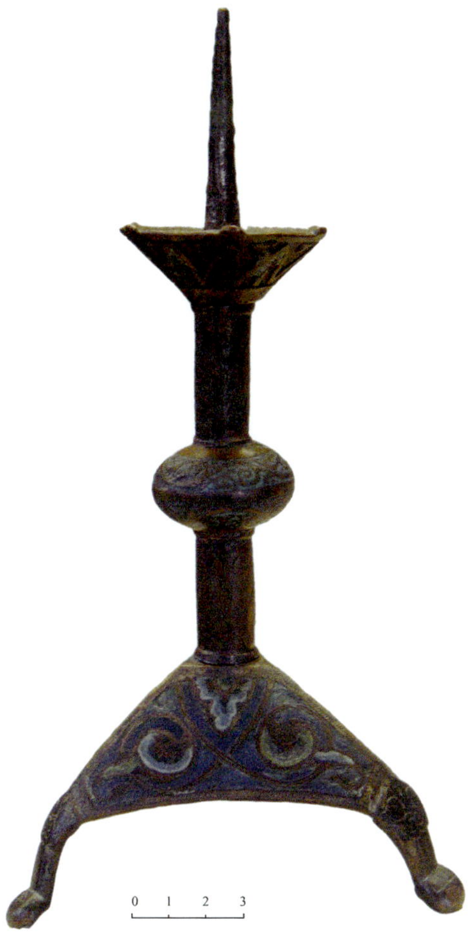

Figure 11. Candleholder from Vschizh.

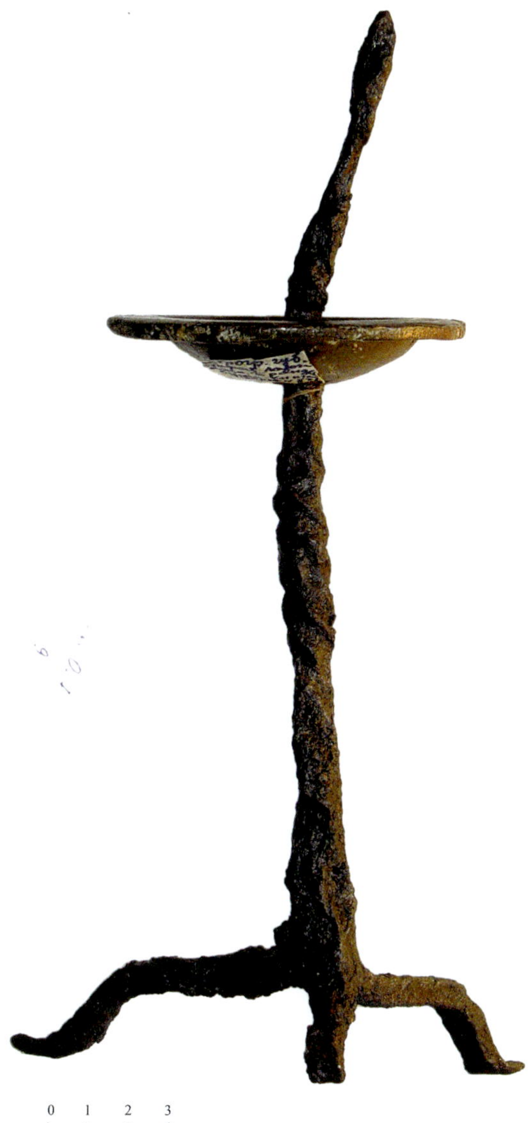

Figure 13 . Candleholder from Old Ladoga.

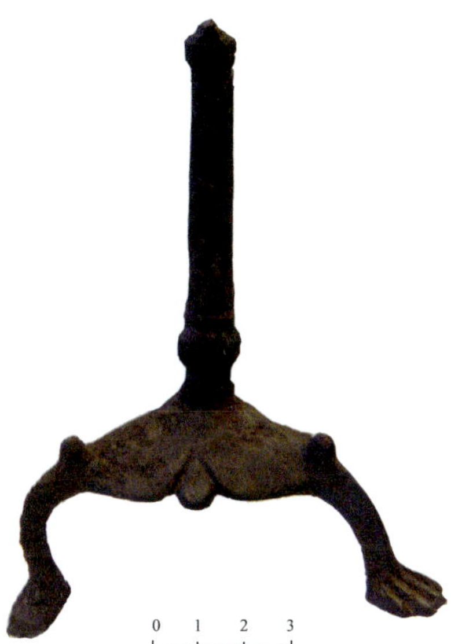

Figure 12. Candleholder from Old Ryazan.

cities of Ancient Russia. Many finds of splinter holders in such towns as Izborsk,[14] Beloozero,[15] Pskov[16] are well-known.

Since the 12th century branched candleholders of iron appear. Branching typically consists of two or three branches extending from the riser and the holder is equipped with a sleeve for candles typical of those distributed in Novgorod. These candleholders were found in Moscow in layers date from the 15th-17th centuries AD. Interesting items of this type of candleholder in the museum represent a compound splinter holder and candleholders which have four rails for clamping the splinter and one branch for setting a candle (Figure 22).

[14] Sedov 2007, 280-282.
[15] Zakharov 2004, 209, fig. 198.
[16] Lobanov 1986, 80-82, fig. 1.

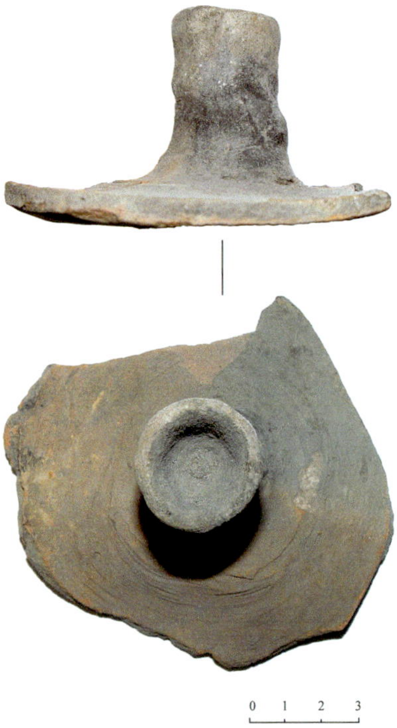

Figure 14 . Candleholder from Novgorod.

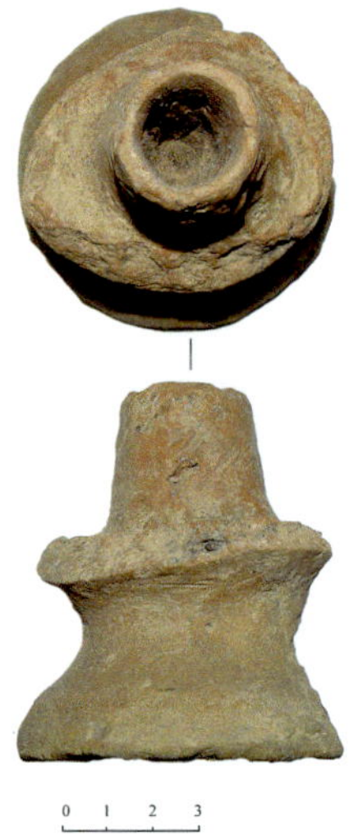

Figure 16 . Candleholder from Bulgar.

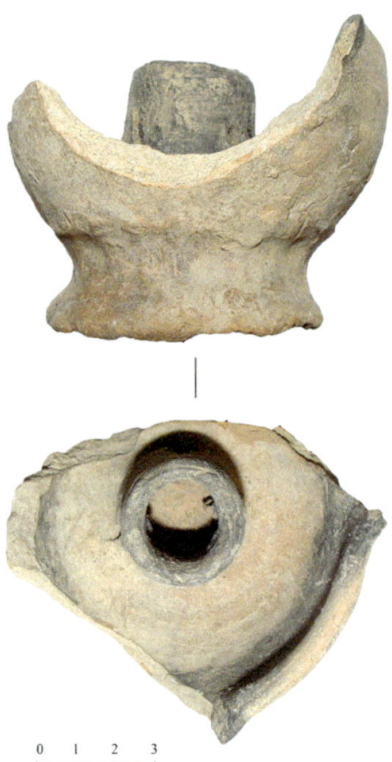

Figure 15 . Candleholder from Bulgar.

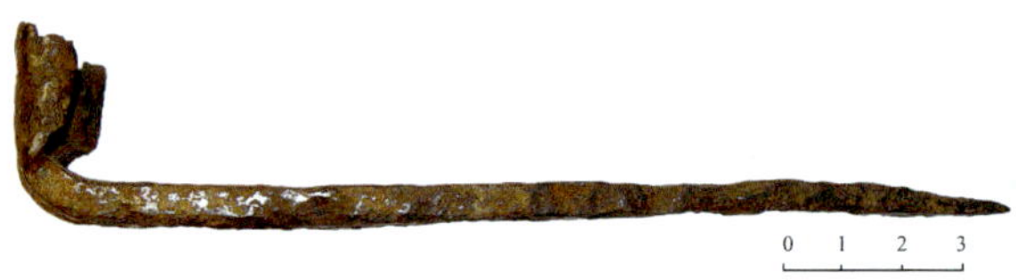

Figure 17. Candleholder from Vladimir burial mounds.

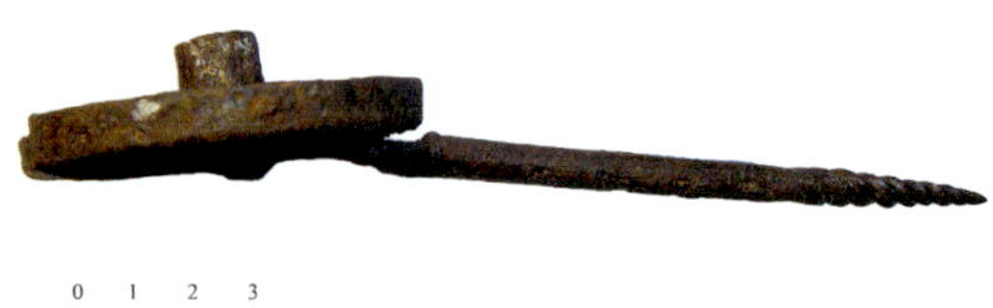

Figure 18. Candleholder from Vladimir burial mounds.

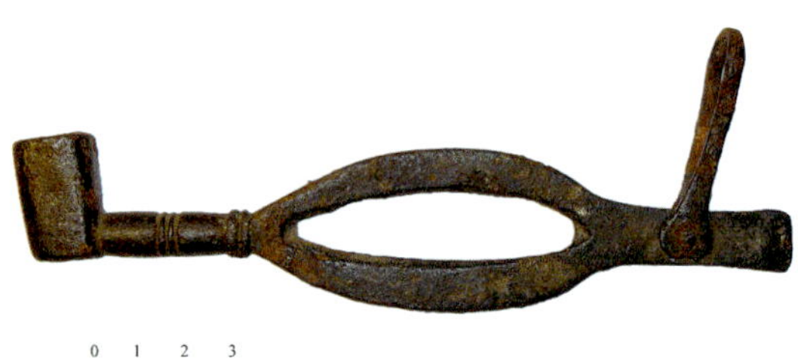

Figure 19. Candleholder from Bulgar.

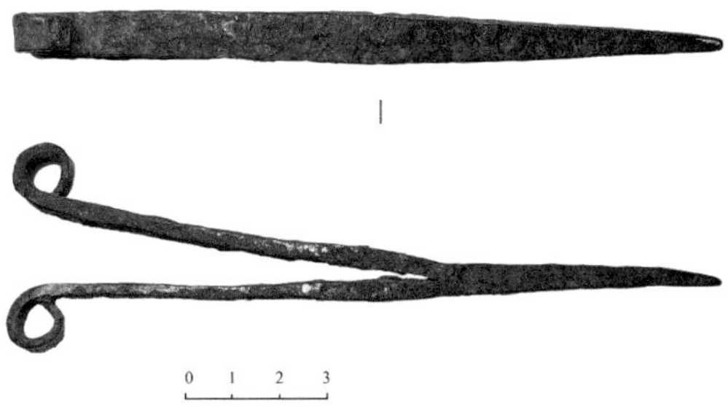

Figure 20. Splinter holder from Vladimir burial mounds.

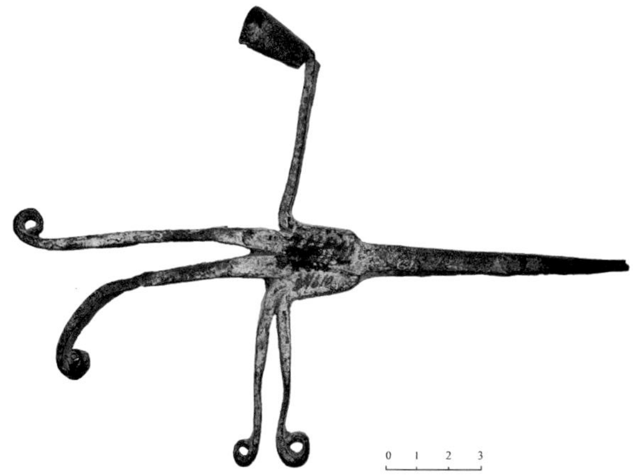

Figure 22. Splinter and candleholder from Novgorod.

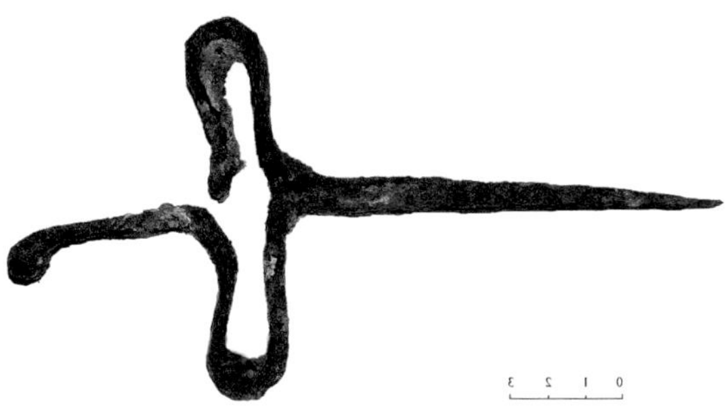

Figure 21. Splinter holder from Novgorod.

In conclusion it can be said that the majority of lighting devices in the Middle Ages have been found in town excavations. Judging by the small number of lamps found at the small settlements of the Ancient Rus', most houses at these sites were lit using splinters without any special devices.

Bibliography

Brøndsted, J. 1936. Danish Inhumation Graves of the Viking Age. *Acta Archaeologica* 7, 81-228.

Darkevitch, V. P. 1966. *Proizvedeniya zapadnogo khudozhestvennogo remesla v Vostochnoy Evrope (X-XIV vekov).* Arkheologia SSSR. Svod arkheologicheskikh istochnikov, vol. E-57. Moskva: Nauka.

Darkevitch, V. P. 1976. *Hudozhestvenny metal Vostoka VII-XIII vekov*. Moskva: Nauka.

Karger, M. K. 1958. *Drevniy Kiev*, Moscow-Leningrad, Izdatelstvo Akademii nauk SSSR.

Koltchin, B. A. 1959. Zhelezoobrabatyvaustchee remeslo Novgoroda Velikogo, in *Materialy I issledovaniya po arkheologii SSSR*, vol. 65, 7-120. Moskva: Izdatelstvo Akademii nauk SSSR.

Koltchin, B. A. and Rybina, E. A. 1982. Raskop na ulice Kirova, in *Novgorodsky sbornik. 50 let raskopok Novgoroda*, 178-238. Moskva: Nauka.

Lobanov, A. N. 1986. Nekotorye bytovye nahodki iz Pskova, in *Kratkie soobstchenya instituta arkheologii*, vol. 195, 80-85. Moskva, Nauka.

Pushkina, T. A. 1997. Vosk I svechi v Drevnerusskikh pogrebeniakh, in *Trudy Gosudarstvennogo istoritcheskogo muzeya*, vol. 93, 122-133. Moskva: Gosudarstvenniy istoricheskiy muzey.

Rozenfeldt, R. L. 1968. Moskovskoe keramitcheskoe proizvodstvo XII-XVII vekov. *Arkheologia SSSR*. Svod arkheologitcheskikh istochnikov, vol. E1-39. Moskva: Nauka.

Rozenfeldt, R. L. 1997. Osvetitelnye pribory, in Drevnaya Rus. Byt I kultura, Arheologiia SSSR, 10-12. Moskva: Nauka.

Savchenkova, L. L. 1996. Tcherny metal Bolgara. Tipologya, in *Gorod Bolgar. Remeslo metallurgov, kuznetzov, liteystchikov*, 6-87. Kazan: Akademiya nauk Tatarstana.

Sedov, V. V. 2007. Sedov. *Izborsk v rannem srednevekovye.*Moskva: Nauka.

Voronin, N. N. 1956. Arkheologicheskie zametki, in *Kratkie soobstcheniya instituta istorii I materialnoy kultury*, vol. 62, 17-32. Moskva: Izdatelstvo Akademii nauk SSSR.

Zakharov, S. D. 2004. *Drevnerussky gorod Beloozero*. Moskva: Indrik.

Lumière et luminaires dans la vie religieuse en Occident au Moyen Âge

Catherine Vincent
Professeur d'Histoire du Moyen Âge, Université Paris Nanterre. Membre senior de l'Institut Universitaire de France

Abstract: While there is an abundance of research on the use and meaning of the lighting devices in the Eastern Church based on texts and material culture, very few attempts have been made to define the same phenomenon in Continental Europe, mainly due to the very poorly disseminated studies in the field and, of course, of the differences in all aspects of life of the believers living in the different parts of Continental Europe. We try here, from ancient Rome to the 13th century, to provide a brief panoptic on the elaboration of the lighting, the faith, the symbolism and the use of lighting devices in the French area.

Keywords: Lumière, Éclairage, Religion Chrétienne, Croyance, Moyen-âge, Europe Continentale.

La lumière a entretenu — et entretient encore — avec les différentes familles religieuses des affinités qu'il est aisé de constater et qui se situent aussi bien sur le plan des pratiques que sur celui, plus théorique, de la valeur morale et spirituelle accordée à « l'illumination intérieure ».[1] Sans nul doute, le sens de ce symbole était perceptible avec plus d'acuité en des temps où la production de la lumière artificielle reposait sur des techniques qui rendaient coûteux, donc beaucoup moins fréquent et abondant qu'il en va de nos jours, l'éclairage entretenu hors des périodes diurnes. Le christianisme ne fait pas exception : il sut accueillir de nombreux rites de lumière dans ses cérémonies, lesquels inspirèrent en retour les gestes de la dévotion privée. Longtemps considérés comme des éléments mineurs, quand ils n'étaient pas relégués d'un trait de plume au rang des composantes de la « religion populaire », ces rites n'ont été jusqu'alors l'objet d'aucune étude systématique.[2] Or le monde médiéval a souvent fait appel aux signes lumineux pour exprimer le mystère de la divinité, dans des œuvres écrites, figurées ou architecturales. La vie liturgique et les modalités d'expression de la piété — processions, vénération des reliques ou des « images » des saints (représentations en deux ou trois dimensions), cérémonies des funérailles — ont usé régulièrement, voire fréquemment, de lampes, cierges et autres torches.[3]

De ces différentes observations naquit le projet de tenter une première approche de l'histoire des pratiques associées aux luminaires, en limitant délibérément la perspective à la sphère religieuse, dans le monde occidental des derniers siècles du Moyen Age.[4] L'enquête s'appuie sur la confrontation de sources de nature variée parmi lesquelles figurent, d'une part, des sources réglementaires générales de la vie de l'Église : actes de conciles, statuts synodaux, livres liturgiques et commentaires de la liturgie — dont l'œuvre de Guillaume Durand dit l' « Ancien », évêque de Mende et compilateur de la fin du XIIIe siècle — , d'autre part, des sources de la pratique : comptes rendus de visites pastorales, archives de divers établissements religieux réguliers et séculiers, de collectivités, telles les confréries, les métiers et les villes, ou de particuliers, dont les princes. À cet ensemble, qui constitue le fondement de la documentation, s'ajoutent de brèves incursions dans le monde des sources narratives : ouvrages spirituels ou chroniques, entre autres, ainsi que dans quelques dossiers rassemblés à partir des comptabilités d'églises cathédrale, collégiale ou paroissiale.

1- Le service cultuel des lumières

Avant même d'aborder les données propres à l'époque médiévale, on rappellera en un bref propos préliminaire à la suite de quelle démarche les luminaires, en dépit de leur présence parmi les usages païens qui jetait sur eux un véritable discrédit, gagnèrent une pleine reconnaissance dans le culte chrétien qui suivit en cela la tradition du Temple. À un dieu qui s'était manifesté à Moïse sous l'apparence du buisson ardent, qui se présente lui-même par la bouche du Christ, dans l'Évangile selon saint Jean, comme la « Lumière du monde »[5] et dont les fidèles reçoivent une initiation qualifiée par les premiers Pères de l'Église d'« illumination baptismale », ne purent être longtemps refusés les honneurs des torches et des lampes qui entouraient la personne impériale dans la tradition romaine. La conjugaison de ces multiples références, sans compter l'intérêt qu'il y avait à recourir à un signe d'une valeur anthropologique universelle, permit la rapide intégration des luminaires dans les cérémonies chrétiennes, au point qu'ils en devinrent un élément jugé indispensable.

Vers une période mal déterminée, alors que s'élaborent progressivement les éléments du culte, le droit canon érigea en obligation la présence d'au moins deux lampes ou cierges allumés lors de chaque célébration eucharistique.[6] Il définit également les fonctions du clerc qui en reçut la charge,

[1] *Le Thème de la lumière* 1976.
[2] La bibliographie sur le sujet demeure encore bien lacunaire (Postles 1999, est l'une des plus récentes études sur le sujet) ; on ne peut que se féliciter de l'organisation de ce colloque et de cette exposition qui comblent un authentique vide et devraient stimuler les études qui restent à mener sur ce thème.
[3] Dendy 1959.
[4] Vincent 2004.
[5] Jn 1, 4-9 (Prologue).
[6] « IV Pars. Nemo ecclesiam edificet ante quam civitatis episcopus veniat et ibidem crucem figat, publice atrium designet et ante prefiniat, qui edificare vult, que ad luminaria, et ad custodiam, et stipendia custodum sufficiat, et ostensa donatione sic domum edificet, et post quam consecrata fuerit, atrium eiusdem ecclesie sancta aqua aspergat » ; GRAT., *De cons.*, 1, 9 (Fr. 1296).

l'acolyte.[7] Enfin, il limita la nature des sources de lumière autorisées à l'huile d'olive ou à la cire d'abeille, en raison de leurs qualités intrinsèques (elles produisent une flamme moins charbonneuse et moins malodorante que les graisses animales) et au nom d'une symbolique qui exalte pureté et chasteté, à partir des conceptions zoologiques héritées de l'Antiquité, comme en témoigne ce passage des *Géorgiques* de Virgile :

« Parmi les habitudes chères aux abeilles, celle-ci est vraiment merveilleuse : elles ne s'abandonnent pas à l'accouplement, elles ne s'énervent point, indolentes, au service de Vénus, et ne mettent pas leurs petits au monde dans les douleurs. »[8]

Au cours de leurs inspections des divers établissements religieux, les visiteurs pastoraux veillèrent, plus qu'il n'aurait semblé de prime abord, au respect de ces préceptes.

Mais, dans la mesure où les luminaires demeurent situés à une place secondaire à l'intérieur du cérémonial chrétien, ils relèvent pour une large part de l'ornement et non des éléments considérés comme indispensables : ils furent en conséquence compris dans les débats qui, au Moyen Age, opposèrent divers courants de pensée quant à la légitimité de la splendeur matérielle apportée au culte. De fait, la limitation draconienne de l'emploi des luminaires ne concerna qu'un petit nombre de communautés régulières (chartreux et cisterciens, par exemple). Au contraire, la majorité des églises tenta de l'amplifier au fil des siècles considérés, quitte à consacrer à cette dépense une part non négligeable de leurs revenus. Le fait s'observe aussi bien de la part des monastères d'obédience clunisienne, entre autres, que de celle des chapitres cathédraux, dont les trésoriers cherchent à limiter leurs obligations en la matière, ou de celles des fabriques paroissiales qui se montrent fort attachées à la possession d'un « beau » cierge pascal.[9]

Le financement du luminaire cultuel reposa sur le concours de toutes les parties intéressées par le déroulement des cérémonies, clercs et laïcs. Dans les cathédrales, il revint en majorité aux chanoines, bien que les évêques aient dû y contribuer partiellement, dans des proportions propres à chaque lieu. Dans les paroisses, il se répartit entre fabriques et curés. La collecte des fonds usa de procédés variés : affectation de biens spécifiques à cette dépense souvent offerts par de puissants bienfaiteurs, redevances en cire attachées à certains dépendants ou dues par les paroisses du diocèse à l'église-mère, la cathédrale, quêtes et dons ; enfin, on n'aurait garde d'oublier que chaque église se livre à la récupération de la cire usagée, qui est refondue et traitée pour être de nouveau consumée.

En raison de son coût et du sens qui s'y attache, la lumière était distribuée à l'intérieur des églises selon des règles rigoureuses. Concentrés principalement dans le chœur, lieu le plus sacré de l'espace et le plus fréquenté par les « praticiens » de la liturgie, lampes et cierges étaient placés sur divers supports fixes ou mobiles : couronnes de lumière suspendues, chandeliers de tailles très différentes déposés à terre ou sur les autels, herses et appliques en bois ou, plus fréquemment, en métal, fixes ou mobiles, mises en place selon les besoins.[10]

L'agencement et l'entretien de ces sources de lumière ont compté au nombre des tâches qui étaient confiées à tout un petit monde de valets et marguilliers dont certains, auprès des établissements les plus gros consommateurs, abbayes ou cathédrales, sont allés jusqu'à assurer sur place la fabrication des cierges. L'utilisation de la lumière dans les églises, associée à celle des divers ornements liturgiques et autres pièces mobiles du décor, soulignait donc le rythme de l'année liturgique selon des modalités destinées à faire percevoir le sens véhiculé par ce signe aux fidèles qui, de près ou de loin, fréquentaient ces lieux.

2- Un riche symbolisme

Précisément, dans le sens qu'elle donna au « signe de la flamme ardente », l'époque médiévale n'a pas innové. Elle puisa aux sources des Écritures, de la tradition patristique et de la philosophie antique.[11] En revanche, elle manifesta sa singularité par un recours au signe lumineux plus systématique, semble-t-il, dans l'administration des sacrements et au cours de certaines cérémonies.[12]

Loin d'oublier le sens honorifique attaché à la lumière par la tradition impériale romaine, les derniers siècles du Moyen Age le développèrent selon une conception très hiérarchisée du temps et de l'espace liturgiques. Une escorte lumineuse accompagnait alors tous les déplacements, si courts fussent-ils, des vecteurs de la présence divine : les ministres du culte, le livre de la Parole, les reliques des saints. C'est dans une telle perspective qu'il convient de situer la multiplication des honneurs lumineux envers le Saint-Sacrement au cours des trois derniers siècles du Moyen Age, alors que le *Corpus Christi* devenait l'objet d'une dévotion croissante. Toute hostie consacrée ne pouvait être acheminée aux malades sans la présence d'une lampe que venait compléter la clochette signalant son passage. De même ne devait-elle pas être conservée sans une veilleuse placée à ses côtés, ni portée en procession (telle celle de la Fête-Dieu dont on connaît l'importance aux deux derniers siècles du Moyen Age dans la vie sociale urbaine notamment) sans lumière, ni présentée aux fidèles au moment de l'élévation sans que soit allumée la « torche a lever Dieu » qui a fait son apparition dans le cérémonial à partir du XIIIe siècle.[13] On aimerait savoir dans quelle mesure ces honneurs lumineux ont été le privilège de la seule « présence divine ». Celle-ci se serait ainsi distinguée de la personne des puissants, y compris de celle des souverains, qui ne se seraient emparés que très tardivement du signe pour manifester leur gloire, guère avant la fin de l'époque médiévale, en ce qui concernent les ducs de Bourgogne et pas avant l'époque moderne, pour les rois de France...[14]

Le point mérite d'autant plus d'attention que la lumière a incontestablement été couramment utilisée comme métaphore pour traduire le mystère divin, ce dont les

[7] « Ad acolitum pertinet preparatio luminariorum in sacrario ; ipse cereum portat, ipse suggesta pro eucharistia calicis preparat » ; Grat., 25, 1 (Fr. 90). Définition reprise d'Isidore de Séville.

[8] Virgile et Saint-Denis 1968, Livre IV, v. 197-199.

[9] Vincent 2004, chapitre III.

[10] Allemagne 1891.

[11] *Le Thème de la lumière* 1976.

[12] Vincent 2004, chapitre VI.

[13] Vincent 2009.

[14] Lecuppre-Desjardins 2008.

mystiques sont loin d'avoir eu l'exclusivité. La réalité de la flamme qui se subdivise sans s'amoindrir a servi à faire comprendre divers mystères chrétiens, entre autres celui de la présence divine infiniment démultipliée dans le sacrement de l'eucharistie et celui de la double nature du Christ, humaine et divine. C'est ce dernier qu'explique pour ses lecteurs Jacques de Voragine dans le chapitre qu'il a consacré à la fête de la Chandeleur dans la *Légende dorée* :

« Il y a trois choses dans le cierge, savoir, la cire, la mèche et le feu, qui sont la figure des trois substances qui existèrent en Jésus Christ : la cire est la figure de sa chair qui est née de la Vierge Marie sans la corruption de la chair, comme les abeilles composent la cire sans mélange ; la mèche cachée dans le cierge est la figure de son âme très candide cachée dans sa chair ; et le feu ou la lumière est la figure de la divinité, parce que notre Dieu est un feu qui consume.[15] »

La diffusion de la lumière du cierge pascal à tous les autres luminaires de l'église, lors de la vigile pascale, une cérémonie majeure du culte chrétien, a permis d'inculquer fort concrètement l'affirmation selon laquelle le Christ est la « lumière du monde », source de toute vie. De leur côté, récits et images accréditèrent le fait que les manifestations de Dieu aux hommes pouvaient être annoncées par une lumière à l'éclat insoutenable, beaucoup plus brillant que toute manifestation du soleil, préfiguration de la « lux perpetua » promise aux élus, ou accompagnées de phénomènes lumineux merveilleux : cierges miraculeusement allumés ou flamme produite sans consommation de matière première. Inversement, le monde infernal a été communément dépeint comme le « royaume des ténèbres », peuplé de sombres créatures en train d'attiser un feu qui brûle sans éclairer...

Signe de Dieu, la lumière a donc été en conséquence signe de l'union à Dieu : l'affirmation vaut pour toute créature humaine dotée par essence d'une parcelle de l'éclat divin, suivant une conception chrétienne qui est aussi nourrie de la pensée néo-platonicienne. Les apôtres ou les saints en constituent le meilleur exemple puisque, telles les sages jeunes filles de la parabole, ils surent garder leurs lampes allumées, une référence à laquelle l'Église médiévale fit largement appel.[16] Le magistère développa en effet un abondant discours moral relatif à la « lumière » que chaque fidèle doit être pour ses frères, par sa foi et ses œuvres, à commencer par les membres du clergé qui ont reçu mission de conduire leurs ouailles sur la voie du salut. La présence, accentuée semble-t-il au cours des trois derniers siècles du Moyen Age, d'un signe de lumière dans l'administration des sacrements ou lors de cérémonies qui marquent la communion à l'Église, dans son instauration (baptême), dans son renouvellement (entrée dans les ordres, mariage), enfin dans sa perte et sa restauration (excommunication et pénitence publique), aurait permis de rendre le propos encore plus intelligible.[17]

Pourtant, on ne saurait oublier la portée ambivalente du signe lumineux : le feu de la joie et de l'honneur peut aussi se muer en un feu de souffrance, de dévastation, de damnation...

[15] Jacobus de Voragine, Maggioni 1998, *De Purificatione beate Virginis*, 248.

[16] Mt 25, 1-13 (parabole des vierges sages et des vierges folles).

[17] Vincent 2000.

3- Des pratiques de plus en plus développées

Il est incontestable que la réception du message attaché par l'Église médiévale au signe lumineux s'est située à divers niveaux de compréhension. Cependant, le très vaste registre sur lequel le discours a pu se développer, où les plus hautes spéculations intellectuelles et philosophiques étaient susceptibles de se fonder sur les réalités les plus concrètes de l'expérience quotidienne, se trouve vraisemblablement à l'origine de la volonté largement partagée par la société occidentale, aux trois derniers siècles du Moyen Age, d'« avoir part à l'éclat ». On en veut pour preuve la participation massive des fidèles au financement du luminaire des églises et, mieux encore, à sa constante amplification dans les lieux de culte. De multiples donations, fondations et legs accomplis à titre privé ou public, individuel ou collectif, attestent à quel degré le luminaire s'est trouvé inclus parmi les éléments qui ont favorisé cette « augmentation du culte divin ». Une très large fraction de la population semble avoir contribué à cette œuvre d'illumination, à suivre les archives qui portent trace de nombreuses initiatives en ce sens de la part des souverains, des collèges de magistrats urbains, sans parler des confréries. Que ce soit à la suite d'une modeste offrande, du dépôt annuel d'un gros cierge auprès d'une statue ou de la réserve eucharistique, de l'accumulation d'ex-voto, les foyers lumineux ne durent guère manquer dans les églises médiévales, du moins lors des périodes fastes ; guerres et conflits virent au contraire les flammes se raréfier au grand dam des clercs.

Par-delà le souci de participer au soutien du culte, une obligation qui revenait en premier lieu aux autorités publiques et aux puissants, en raison de leur « état », mais que les clercs rappelèrent aussi à tous les fidèles, l'intérêt porté aux luminaires se fit l'expression d'authentiques manifestations de piété, comme en témoigne ce passage du célèbre maître parisien du début du XIIIe siècle, Pierre le Chantre :

« De même [...], lorsque de vieilles femmes ont déposé sur l'autel en offrande des chandelles, nous commettons un péché si nous les brûlons sur nos tables. On voit en effet que celles-ci ont offert des chandelles pour qu'elles brûlent dans l'église et que telle est bien leur volonté. Ne manquons-nous donc pas ainsi au respect de leur volonté ? On pourrait cependant dire que leur volonté peut être respectée, en ce que tant que ces vieilles femmes sont présentes dans l'église, les clercs gardiens de l'autel laissent brûler leurs chandelles sur l'autel, puis les éteignent lorsqu'elles sont parties.[18] »

Ce geste élémentaire de dévotion, que l'on soupçonne avoir été aussi familier aux fidèles qu'il l'est resté de nos jours tout spécialement dans les églises d'Orient, alimenta le commerce des « chandelières » qui vendaient leurs lumignons aux portes des églises. À cet égard, les lieux de pèlerinage ont enregistré

[18] « A simili (...) cum vetule offerunt candelas super altare, peccamus cum comburimus eas super mensas nostras. Videtur quod ideo offerunt candelas ut ardeant in ecclesia et hec est earum voluntas. Nonne ergo defraudamus voluntatem earum ? Quod autem hec sit voluntas earum perpendi potest ex eo quod dum presentes sunt vetule, dimittunt custodes altaris candelas earum, accensas super altare ; cum recesserint, extinguunt eas. » Pierre le Chantre et Dugauquier 1963, cap. XXXIX « De ecclesiis et cymiteriis », no 294 « Utrum oblata altari et ornamenta possint redigi ad communses usus », 330, l. 35-42.

une pratique soutenue, indicateur fiable, lorsqu'il peut être apprécié, des courants de ferveur. Venant à l'appui des prières votives et des processions exceptionnelles qui étaient organisées pour conjurer quelque fléau, le luminaire était manifestement doté de vertus propitiatoires, un fait entériné par le magistère ecclésiastique du moment qu'il s'agissait de cierges dûment bénits, tels les cierges de la Chandeleur. Cependant, comme il advint pour tout support concret de piété, la réglementation ne put empêcher l'existence de pratiques magiques auxquelles la cire prêtait à merveille sa couleur et sa malléabilité et dont les origines sont bien antérieures à l'époque médiévale.

Les fidèles, clercs et laïcs confondus, se sont enfin approprié le sens honorifique attachée de longue date à la lumière pour en faire un élément de distinction sociale, notamment lors des cérémonies funéraires et des diverses célébrations anniversaires du décès.[19] Le luminaire devint de la sorte un procédé d'affirmation identitaire — parfois jusque dans le refus ou la limitation de sa présence — et un moyen de transmission de la mémoire individuelle.[20] La pratique des luminaires aurait donc remporté un succès croissant à la fin du Moyen Age, conformément au goût prononcé de l'époque pour les cérémonies cultuelles. Aussi constate-t-on sans étonnement qu'elle fut bannie par la Réforme, notamment dans sa composante calviniste, avec l'ensemble des usages « papistes ».

Ce bref survol permet donc d'attirer l'attention sur un élément cultuel qui, en dépit de sa valeur de second ordre comparée à celle qui s'attache aux reliques ou aux espèces eucharistiques, n'en constitua pas moins l'une des composantes encore méconnue de la vie religieuse des derniers siècles du Moyen Age. À travers cet exemple, se laissent également saisir les lentes mutations qui ont pu affecter le cérémonial religieux. Or, pour ce qui est des éléments lumineux, ces transformations semblent avoir résulté d'un ajustement progressif, autant qu'on parvienne à le saisir, entre la perspective pastorale qui animait les autorités ecclésiastiques et les aspirations, voire les initiatives, des fidèles. Laïcs et clercs, à quelques exceptions près, auraient partagé une commune attirance pour cet humble support de piété, en un temps où l'objet voyait son rôle pleinement reconnu dans les cérémonies liturgiques et les gestes de la dévotion, sans que soient pour autant ignorés les risques de dérive que son usage pouvait engendrer. Ne constituait-il pas, et tout spécialement les vecteurs de lumière en raison de leur sens à la fois aisément perceptible et profondément ancré dans un riche substrat théologique et philosophique, un référent culturel commun ?

Bibliographie

Allemagne, H. R. d', 1891. *Histoire du luminaire depuis l'époque romaine jusqu'au XIXe siècle.* Paris : Picard.

Dendy, D. R. 1959. *The Use of Lights in Christian Worship.* London : SPCK.

Gaude-Ferragu, M. 1999. Le corps du prince. Le testament de Louis d'Orléans (1403), miroir de sa spiritualité. *Micrologus,* VII, 319-344.

Gaude-Ferragu, M. 2005. *D'or et de cendres : la mort et les funérailles des princes dans le royaume de France au bas-Moyen Age.* Lille : Presses universitaires du Septentrion.

Le Thème de la lumière dans le Christianisme, le Judaïsme et l'Islam, 1976. s. l. : Berg International éditeurs.

Lecuppre-Desjardins, É. 1999. Les lumières de la ville : recherche sur l'utilisation de la lumière dans les cérémonies bourguignonnes (XIVe-XVe siècles). *Revue Historique,* CCCI/1, 23-43.

Pierre le Chantre et J.-A. Dugauquier 1963. *Summa de sacramentis et animae consiliis.* Louvain : Nauwelaerts.

Postles, D., 1999. Lamps, lights and layfolk: "popular devotion" before Black Death. *Journal of Medieval History,* 55, no. 2, 97-114.

Vincent, C. 2004. *Fiat Lux : lumière et luminaires dans la vie religieuse du XIIIe au XVIe siècle.* Paris : Le Cerf.

Vincent, C. 2000. Rites et pratiques de la pénitence publique à la fin du Moyen Age : Essai sur la place de la lumière dans la résolution de certains conflits, in *Le règlement des conflits, Actes du 31e Congrès de la SHMESP (Angers, 2000),* 351-367. Paris : Publications de la Sorbonne.

Vincent, C. 2009. « Ob reverentiam dicti sanctissimi Corporis Christi... » : luminaires et dévotion eucharistique (XIIIe - XVe siècle), in Bériou, N. et B. Caseau, D. Rigaux (éds.), *Pratiques de l'eucharistie au Moyen Age,* 481-496. Paris : Études augustiniennes.

Virgile et E. de Saint-Denis 1956. *Géorgiques.* Paris : Les Belles Lettres.

Voragine, J. de, Maggioni, G. P. 1998. *Legenda Aurea.* Tavarnuzze : SISMEL - Edizioni del Galluzzo.

[19] Gaude-Ferragu 2005.

[20] Gaude-Ferragu 1999.

Olaus Magnus the Goth on Fire, Light and Lighting Devices of the Northern People

Arja Karivieri
Stockholm University / Institutum Romanum Finlandiae

Abstract: Olaus Magnus, Archbishop of Uppsala, published *Historia de Gentibus Septentrionalibus* in Rome in 1555. This encyclopedic, illustrated work on the customs and history of the Nordic people became popular and it was translated into several European languages during the following decades. In this article, a special focus is given to the various instances where Olaus Magnus comments on fire, light and lighting devices in Scandinavia, such as the symbolic meaning of fire and light at wedding ceremonies, lighting the guild feasts, light in Christian processions and in the church, lighting in wintertime, light and fire in summertime, fire and lights during warfare, birds and lights, and fishing by means of fire. Finally, the evidence provided by Olaus Magnus is compared with other sources, and what archival sources reveal about the use of light and lighting devices at Häme and Turku Castles in Finland during the same period.

Keywords: Olaus Magnus, Historia, Northern customs, fire, lighting devices.

Olaus Magnus' work on the customs and history of the Nordic People, *Historia de Gentibus Septentrionalibus* published in Rome in 1555, was based upon his journey to Norrland 1518-9. Olaus Magnus, Archbishop of Uppsala, the last Catholic bishop of Sweden, wrote his book during his voluntary exile in Rome, after the reformation in Sweden. While in Rome, he was the caretaker of the house of Saint Birgitta. He founded a printing house where he printed *The History*; several editions where made of the work and it was translated to Italian, French, Dutch, German and English during the following hundred years. During the first ten years after publication, the book was protected by license of Pope Julius III from unauthorised printing. The beginning of the long title of the book describes well the richness of Olaus Magnus' work: 'A Description of the Northern Peoples, their different positions, customs, habits, ways of life, superstitions, methods of instruction, activities, government, food, wars, buildings, implements, metal mines, and marvels,...'.[1]

Olaus Magnus' intention was to provide the educated upper class with an encyclopedic work to cover all relevant aspects, from mankind to insects, to satisfy a general curiosity about the North among foreigners.[2] In the preface, Olaus Magnus recalls ancient philosophers who travelled widely and recorded in their works their observations of peoples and places, and European humanists who preceded him or were contemporary with him. He describes the life and customs of ordinary people, work at home and in the fields, hunting and fishing, mining activities, and wars at sea and on land.[3] One of his special interests is the various kinds of animals, especially fish, birds and bears. In addition, he tells stories about infamous monsters and fantasy figures that seemed to have a special fascination for him. Many of his stories are more like folktales, aimed to catch the reader's interest, often accompanied with an illustration. He describes the old Nordic gods and their worship, as well as the magic arts that still existed in his own day, particularly in remote areas like Finland. Central for the discussions is Christian religion, especially the Church's role in society and for schooling the children. He describes many details of ceremonies, such as weddings, funeral processions, burials and the ritual use of candles.[4]

A central part of the work is the illustrations, including 461 different images. It seems probable that Olaus Magnus first produced sketches himself, but let his Italian engraver complete the pictures. Several illustrations are taken from other works, most, 124 pictures, from his own *Carta marina*, and some others are reused older prints with additions and adjustments.[5]

In the preface of his work, Olaus Magnus discusses the use of images in the book: 'An illustration not only gives satisfaction and a singular delight, but also preserves a record of the past and constantly brings glimpses of history before your eyes. Indeed, when we look at paintings where spectacular feats are represented we feel an urge to strive for fame, and embark on great tasks... An important advantage of such

[1] I wish to take this opportunity to express my congratulations for the organizers of Olten Round Table for giving us the possibility to experience the wonderful hospitality of the Historical Museum and the City of Olten. Furthermore, my sincere thanks go to the Berit Wallenberg Foundation for financing my participation in the Olten Round Table as well as the digital illustrations for the publication of my paper.
The editions used for this study are the Swedish translation of Olaus Magnus, *Historia de Gentibus Septentrionalibus*, originally published in five volumes between 1909-51 by Michaelisgillet, and the English edition of Peter Fisher and Humphrey Higgens, in three volumes, published between 1996-8 by the Hakluyt Society, which is also the main source for English translations of Olaus Magnus' original work in Latin cited in this paper. For the original Latin edition of the work, a facsimile edition by Rosenkilde and Bagger from 1972, with a commentary by John Granlund, was used, and additional illustrations for publication were provided by a Belgian edition of Olaus Magnus text, printed in 1562, *Historia de gentibus septentrionalibus, authore Olao Magno, Gotho, archiepiscopo Vpsaliensi, Suetiæ and Gothiæ primate. A Cornelio Scribonio Grapheo, præaclaræ vrbi Antuerpianæ à secretis, sic in epitomen redacta, vt non minus clarè quàm breuiter quicquid apud Septentrionales scitu dignum est, complectatur.*

[2] Fisher and Higgens 1996-1998, introduction, xxxix.
[3] Fisher and Higgens 1996-1998, introduction, xxxvii.
[4] Fisher and Higgens 1996-1998, introduction, xl-xli, xlix.
[5] Granlund, in Fisher and Higgens 1996-1998, introduction, 31-33; Fisher and Higgens 1996-1998, xlii-xliii; Gillgren 1999, 148.

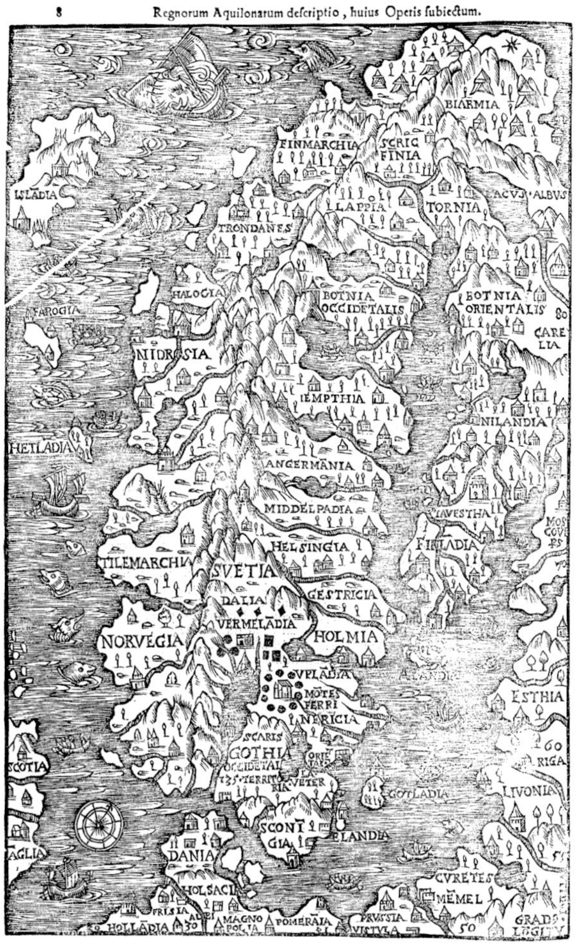

Figure 1. Map of Scandinavia in the beginning of the book *Historia de Gentibus Septentrionalibus,* page 8. (Courtesy of National Library of Sweden, Photo by Esbjörn Eriksson).

representation especially concerns princes and eminent men, since they are unable to depart from their own countries to gain experience of unknown lands and gaze on everything in person, with the help of such images they can at least grasp in their imagination what they have longed to see... there is in the works of painters,..., a wonderful power of instruction, produced by their charm and variety of their impressions, so that the viewer is in no way allowed to feel indifferent'.[6] Olaus Magnus recalls in the preface the importance of illustrating the text by comparisons with other regions, historical examples and Nature's secrets.

The richly illustrated work describes also the use of fire, light and the lighting devices used during the dark winter months in Scandinavia to light up the houses, during work, as well as during ritual practices in churches. In this article, a special focus is given to the various instances where Olaus Magnus comments on fire, light and lighting devices in Scandinavia (Figure 1).

[6] Transl. Fisher and Higgens 1996-1998, 5-6.

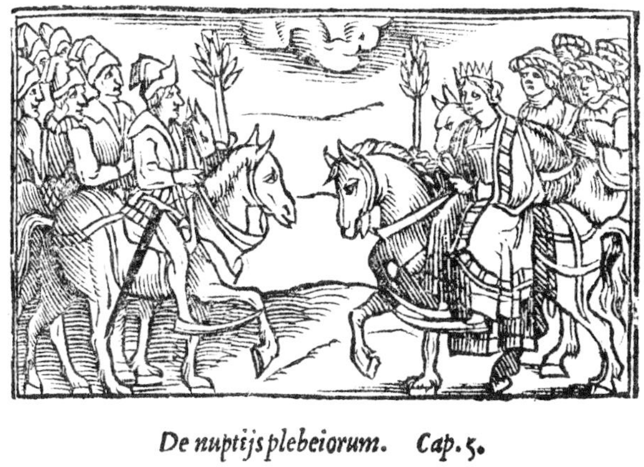

Figure 2. Woodcut illustrating the marriage procession of the common people, from the Belgian edition of *Historia de Gentibus Septentrionalibus* from 1562, page 125 (13.5),[7] showing a remade mirror-image of the first edition. (Courtesy of National Library of Sweden, Photo by Esbjörn Eriksson).

Light at wedding ceremonies

Fire and light had an important role during the wedding ceremony of the Lapps, when parents ratified the union of their children by means of fire by striking with iron out of flintstone (4.7: '*Parentibus amicis et sanguine iunctis, parentes per ignem foederant coniugia natorum...*'), because marriages were joined with fire and flint, which represented a wedded partnership, to make them stronger and more fortunate.[8] Olaus Magnus compared this custom with a similar custom of the Romans. Furthermore, when the groom and bride went to church for the priestly blessing, tall wax candles of various hues with coloured silken drapes hanging from them were carried before them. When the solemnization had been performed, the candles were left there together with their offerings, and the candle-bearers divided the pieces of silk among themselves as a prize.[9]

Olaus Magnus tells also about special customs of ordinary folk among the Götar and the Swedes ('*De nuptiis plebeiorum*'). Relatives by marriage and by blood came to the parish church on a fixed day in a mounted procession in their finest clothes, men led by an experienced man and women by an honorable woman. Here the bride, wearing a crown, was blessed by the priest and brought forward, with marvellously wrought torches burning,[10] to her bridegroom before the high altar

[7] The recut 'second generation' illustrations of the Belgian edition were printed in Antwerp in 1562. They have worn images and details, lack decorative borders, and have a different and italic typeface. Likely produced by another hand, but copied from an original copy of the book, the illustrations appear as mirror-images once the woodcuts were printed on the paper. Additionally the text and the illustrations are made more carelessly and the illustrations in the Antwerp edition appear in different locations relative to the text when compared to the original.

[8] Transl. Fisher and Higgens 1996-1998, 204.

[9] Transl. Fisher and Higgens 1996-1998, 204-205.

[10] '*ubi ardentibus faculis mirifice factis et incensis*'.

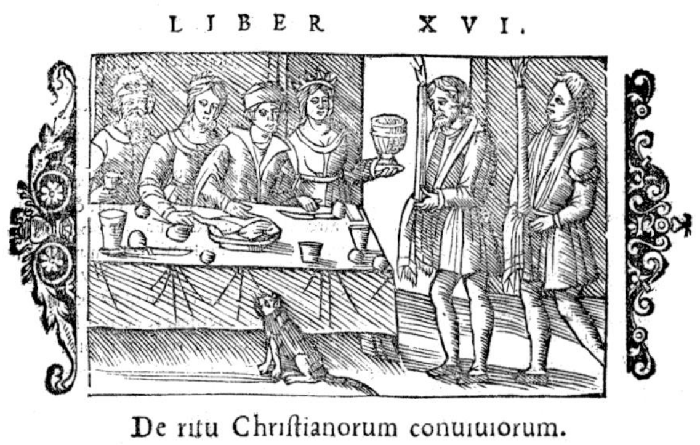

Figure 3. Woodcut showing two men carrying large candles, *Historia de Gentibus Septentrionalibus,* page 541, illustrating ch. 16.16. (Courtesy of National Library of Sweden, Photo by Esbjörn Eriksson).

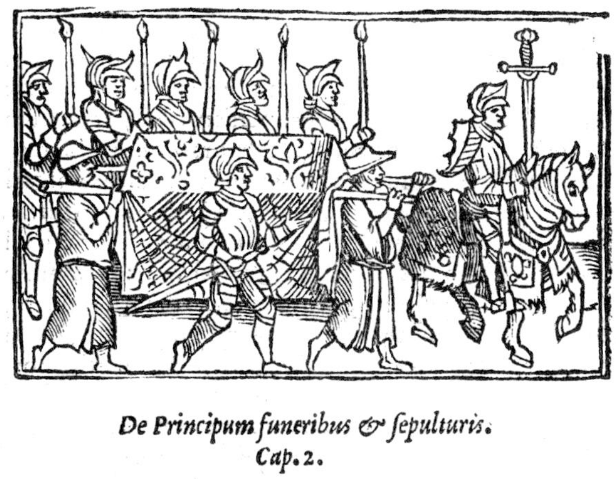

Figure 4. Woodcut showing the funerary procession of a nobleman, from the Belgian edition of *Historia de Gentibus Septentrionalibus* from 1562, page 130 (16.2), showing a remade mirror-image of the first edition. (Courtesy of National Library of Sweden, Photo by Esbjörn Eriksson).

(Figure 2). The ceremony was confirmed by putting on a ring and by a nuptial benediction (14.9).[11]

Lighting the guild feasts

One of Olaus Magnus' illustrations shows two men carrying large candles into a large hall (Figure 3), where the guild feast continued for three days during special occasions of the year. A large house was built for that purpose alongside the church (16.16-17). Those who did not follow the guild rules had to donate wax for the Holy Altar.[12]

Light in Christian processions and in the church

On the day of the Purification of the Blessed Virgin, Candlemas (2 February), Olaus Magnus tells us, the people retained the ceremony of bearing lighted wax candles, which had been a heathen custom, now blessed by the prayers of the priests; they made a reverential procession through churches and churchyards.[13] Olaus Magnus, who was, however, hostile against the Lutherans, continues by saying that 'the reckless misconception of certain innovators... thinks the ruddy splendour of this light offensive, and that the shining of faith and tokens of good works should not remain in the burning lamps and almsgiving hands of Christians'.[14] The sentence reveals the importance of light for Catholic ceremonies, and how Olaus Magnus criticizes Lutherans for being against the ceremonial use of light. Olaus Magnus adds: 'But not everywhere have they succeeded with their unholy intentions, for Christ, who is a light to lighten the Gentiles, preserves good ordinances and ensures that this ceremony remains unshaken in the Catholic Church'.[15] He points out that this consecration is performed, too, by all women after the childbirth when, after a period of forty days, carrying a burning candle, they must go to the church doorway to give the candle to the priest and to be purified by him (16.6).[16] In another chapter, Olaus Magnus writes that women who had born a child went after the birth with burning candles to church to give thanks to God, as all Christians were carried to the grave with lights flaming (Figure 4), and at baptism they entered the world with a lamp burning (4.7).[17]

When the procession during the day of the Blessed Virgin had passed through the church or churchyard, the blessed candle was laid at the altar, to be divided into small pieces and given to every family when the service ended. After taking these pieces, each person returned home and on the following day all the children assembled, fasting. They kneeled and, with the candles lit, their father bestowed a blessing on them from the Book of Numbers, Ch. 6. He singed the tips of the hairs of children's heads in the shape of a cross and then extinguished the candle after the blessing. If the smoke ascended, it was a token of a long, good life. But if the smoke twisted towards the floor, the youngster was shocked by fear into improving his ways (16.7).[18]

Olaus Magnus writes that the skins of wolves, lynx or fox or an animal of another species are donated to the church, so that candles could be bought with the proceeds and burned in veneration of the saints in the church (16.20; 18.24).[19] He writes that large numbers of candles were used and lit in Northern countries, mainly during worship in churches, during pilgrimages to the saints, and at the taking of the sacraments. However, candles blessed by the hand and mouth of the bishop were held in especially deep reverence. The strong virtue of the bishop's words uttered over the candles was believed to have the power to ward off thunderbolts and flashes of lightning if they were lit with that purpose in mind. When the people perceived the air tainted, they bore

[11] Fisher and Higgens 1996-1998, 693.
[12] Fisher and Higgens 1996-1998, 787-788; commentary, 833.
[13] Fisher and Higgens 1996-1998, 777.
[14] Transl. Fisher and Higgens 1996-1998, 777-778; comments, 831.
[15] Transl. Fisher and Higgens 1996-1998, 778.
[16] Fisher and Higgens 1996-1998, 778; Swedish translation, 718.
[17] *'Praeterea mulieres enixa prole post partum cereis accensis Deo gratias acturae accedunt: immo et cuncti Christi fideles luminibus accensis in tumulum reportantut, uti olim baptismo, et lampade accensa mundum sunt ingressi.'*; Fisher and Higgens 1996-1998, 205.
[18] Fisher and Higgens 1996-1998, 778-779; comments, 831.
[19] Fisher and Higgens 1996-1998, 791, 906.

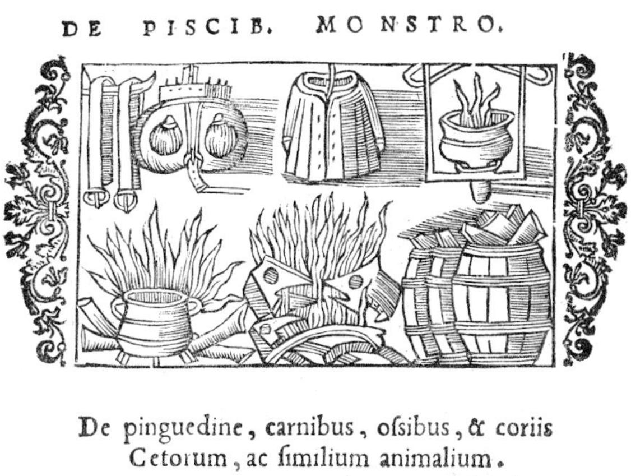

Figure 5. Illustration showing the various uses of blubber or oil from the whale, fish bones and heads; in the upper right corner a glass lamp is hanging, *Historia de Gentibus Septentrionalibus,* page 750; cf. Chrzanovski and Kaiser 2007, figs 33, 34, 105 and 106 for comparisons for the glass lamp. (Courtesy of National Library of Sweden, Photo by Esbjörn Eriksson).

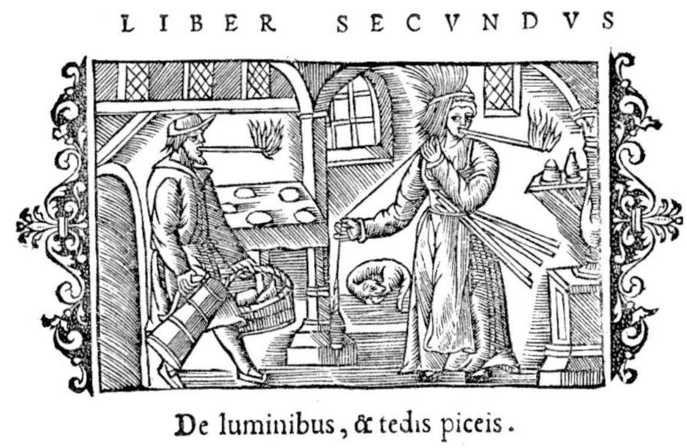

Figure 6. A woman carrying splinters under the belt and attached to the head-gear. Both the woman and a man coming to the room from the left hold a burning splinter in the mouth, *Historia de Gentibus Septentrionalibus,* page 77. (Courtesy of National Library of Sweden, Photo by Esbjörn Eriksson).

these candles in their hands, praying or in processions, so that the elements may give way to the efficacy of the blessing shed upon the candles. The Northern people averted with blessed candles the disasters that were harassing them and eventually put those same candles to use when they were about to breathe their last (16.8).[20]

Lighting in wintertime

Olaus Magnus was especially fascinated with the dark wintertime and what the Nordic people did after the winter solstice, during the period of six months without sunlight. He tells us that in place of the daylight the people set fire to the fat of sea beasts, and they attached a wick to it, and used it in their houses or in the fields and woods. Those who worked out in the country used this animal fat or fish oil, which was practical for working outdoors since the winds cannot put out the light (4.9).[21]

The blubber or oil from whale was stored in barrels and it was used for lamps, supplying unfailing light for altars and other holy places, night and day, as well as for private and domestic purposes (Figure 5). The smaller bones of whale and other fishes' heads served as fuel when the stack of logs for the family hearth ran out (21.20).[22] The blubber or oil from the whale was called *tran* or *lyse*, because it gave bright light in lamps, especially in larger halls and churches, where the flame always burned to honor the corpse of our Lord. To prevent the nocturnal birds, owls or bats from eating the *tran*, the mouth of the lamps had an iron grating (2.17).[23] Jan-Öjvind Swahn has pointed out that the tallow from the slaughter in the autumn was spared for special occasions such as the visit of a shoemaker when even the dark hours of the day were needed. According to Swahn, the lamps Olaus Magnus referred to were of a traditional type, low, flat stone lamps.[24]

Various fish, or rather monsters, by the Norwegian shores were caught for their fat, which was rendered down by roasting it over the fire and then sold to keep the light going in lamps during the season of uninterrupted darkness. The most prevalent of these monsters was one circular in shape, known in the Norwegian tongue as *Swamfisck*. This creature was so fat that when danger was imminent, then, like the hedgehog, it folded flesh, fat, and skin together over its head, which it concealed by rolling itself up tight (21.38).[25]

People in the North also used torches which included natural resin. When they had to use both hands, they carried splinters under the belt, put them one after another into the mouth or attached to the headgear with the other end burning (Figures 6, 7). Thus they could move around, without being disturbed by the wind and do what they wanted. In many places candles of an arm's length were used for lighting (2.17).[26] Jan-Öjvind Swahn tells us that the splinters could be attached to special supports, called *lysekärringar* or *stickedallor,* which were placed near the hearth that occasionally had a small niche for the splinters.[27]

Those who needed light during their travels through the woods (Figure 8) used pieces of decayed oak bark or mushroom Agaricus '*fungus ipse Agaricus appellatus*' that grows in '*summitate arboris glandiferae*', which give light during the night (2.16), and also could be used inside when the people needed to visit places that were full of material that could easily burn.[28]

[20] Fisher and Higgens 1996-1998, 779.
[21] Fisher and Higgens 1996-1998, 207.
[22] Fisher and Higgens 1996-1998, 1103.
[23] Swedish transl., 103.
[24] Swahn 2005, 141 and 143.
[25] Fisher and Higgens 1996-1998, 1123; commentary, 1151, no. 1, for interpretation of the word *Swamfisck*. The creature described by Olaus Magnus is illustrated in the upper right corner of the vignette.
[26] Swedish transl., 103-104.
[27] Swahn 2005, 141-143.
[28] '*ideoque frequentius quercum putidam agaricumque (se ipsos propria luce prodentes) colligunt*'; Swedish transl., 102.

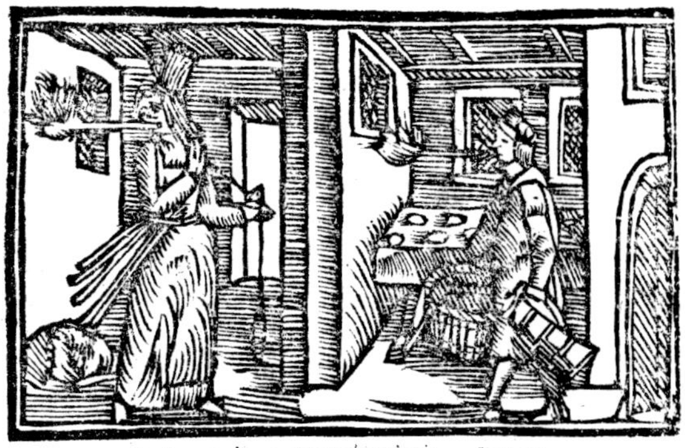

De luminibus, & tedis piceis. Cap.15.

Figure 7. A woman carrying splinters under the belt and attached to the head-gear. Both the woman and a man coming to the room from the right hold a burning splinter in the mouth, from the Belgian edition of *Historia de Gentibus Septentrionalibus* from 1562, page 19 (2.15), showing a remade mirror-image of the first edition. (Courtesy of National Library of Sweden, Photo by Esbjörn Eriksson).

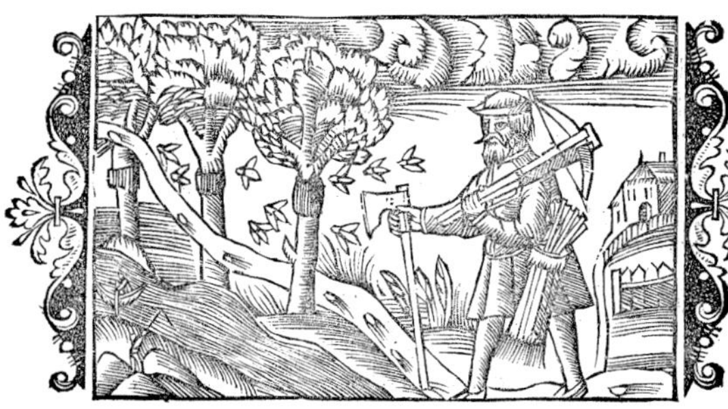

De tranſitu tenebroſo.

Figure 8. A man holding an axe and a crossbow walking through the woods during the night, his path being illuminated by decayed oak bark or mushroom Agaricus and fireflies, *Historia de Gentibus Septentrionalibus,* page 76. (Courtesy of National Library of Sweden, Photo by Esbjörn Eriksson).

LIBER XIII.

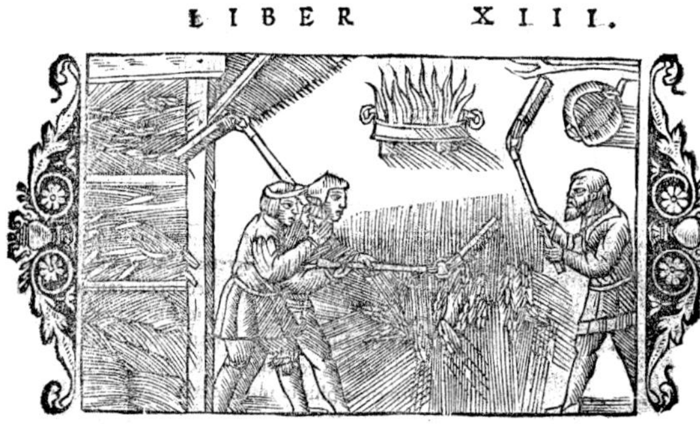

De trituratione per hyemem.

Figure 9. Threshing made in the light of a brazier, *Historia de Gentibus Septentrionalibus*, page 435. (Courtesy of National Library of Sweden, Photo by Esbjörn Eriksson).

LIBER XV.

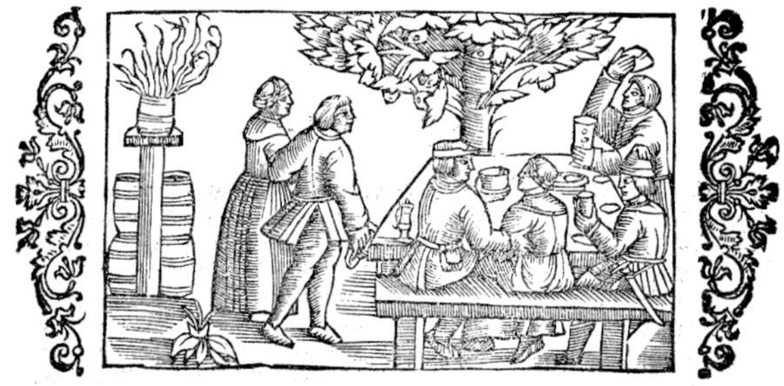

De Maialibus feſtis.

Figure 10. The associated illustration shows fire burning in a large barrel; the vignette was an adaptation of a print by Hans Sebald Beham (Fisher and Higgens 1996-1998, 764), *Historia de Gentibus Septentrionalibus*, page 505. (Courtesy of National Library of Sweden, Photo by Esbjörn Eriksson).

When corn had been gathered and was stored in large barns built for the purpose, all the threshing was done in wintertime, during the dark hours (Figure 9). The sheaves to be threshed were laid upon a wide floor of level beams and the men struck and cleaned the grains with flails when the flame from pitch pine gave light in the middle of the barn (13.7).[29]

Light and fire in summertime

When the sun was moving through Cancer, at the Midsummer day of St John the Baptist, all the people gathered in crowds in public open spaces or on a flat stretch of ground and everywhere lit great fires for dancing, sang and simulated in dancing the glorious feats of ancient heroes and the deeds of famous women (15.10).[30] The associated illustration in Olaus Magnus' text that shows burning fire in a barrel was an adaptation of a print by Hans Sebald Beham[31] (Figure 10).

Fire and lights during warfare

At a time of hostilities bright fires were seen on the mountaintops, which were used to signal for repelling attacks from the fleets of approaching enemies (7.10).[32] Before the impending siege, the people laid down and keep safe a sufficient store of firewood, and of coal especially. Besides this, 'burning lamps hang from roofs, towers, or windows towards the enemy, since these are either signals which the foe are pleased to see, or they implant in the loyal inhabitants some unpleasant suspicion that perhaps the castle's inmates are in collusion with the other side' (9.19).[33]

Birds and lights

According to Olaus Magnus, nocturnal birds could be caught 'by means of lamps lit in countrymen's barns, for they congregate in such places to hunt and feed on mice that hide

[29] Fisher and Higgens 1996-1998, 624; commentary in Swahn 2005, 150-151.
[30] Fisher and Higgens 1996-1998, 735; cf. commentary in Swahn 2005, 178-181.
[31] Fisher and Higgens 1996-1998, commentary, 764.
[32] Transl. Fisher and Higgens 1996-1998, 326-327.
[33] Transl. Fisher and Higgens 1996-1998, 429-430.

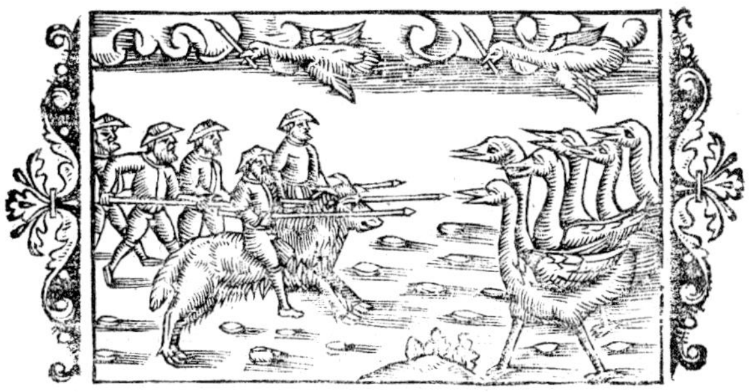

De Gruibus, & vigilia militari.

Figure 11. Cranes carrying candles in the beak, *Historia de Gentibus Septentrionalibus*, page 672. (Courtesy of National Library of Sweden, Photo by Esbjörn Eriksson).

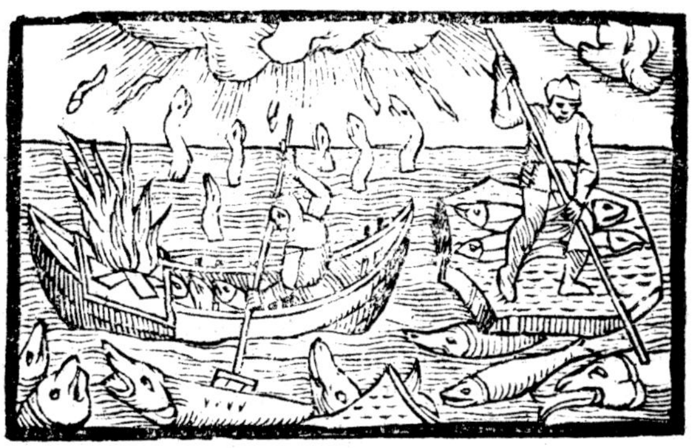

De piscatione per flammas. Cap.9.

Figure 12. Light and fire used during fishing, from the Belgian edition of *Historia de Gentibus Septentrionalibus* from 1562, page 166 (20.9), showing a remade mirror-image of the first edition. (Courtesy of National Library of Sweden, Photo by Esbjörn Eriksson).

in the crannies' (19.48: '*aves nocturnae lucernis capiuntur*').[34] One of Olaus Magnus' illustrations shows two flying cranes carrying candles in the beak, flying above dwarf-Greenlanders who fight against cranes (19.28) (Figure 11). There is however an explanation for the symbolism of this image: cranes were associated with light. The text does not mention cranes 'grues' carrying candles, but there is an old Swedish saying that the crane carries the light into the bed. The arrival of cranes to Southern Sweden was connected with the Day of the Annunciation of the Virgin Mary, 25 March, which people used to celebrate with a feast in the evening, when the crane carried the light to the bed. The same night well-behaved children could get candies, which they found under the pillow in the morning.[35]

Fishing by means of fire

Light and fire were also used during fishing. Fish-spearing by torchlight was referred to in medieval Swedish laws and

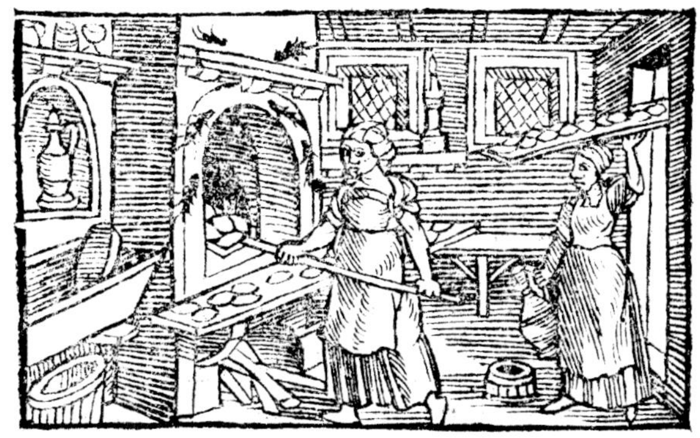

De pistrinis, & coctione panum. Cap.2.

Figure 13. Crickets cling to the chimney in the bakery; they were believed to give forecasts when they knock against lamps set on the tables; from the Belgian edition of *Historia de Gentibus Septentrionalibus* from 1562, page 120 (13.2), showing a remade mirror-image of the first edition. On the shelf between the windows, a candleholder can be seen; for comparisons, cf. Chrzanovski and Kaiser 2007, fig. 141. (Courtesy of National Library of Sweden, Photo by Esbjörn Eriksson).

is well-known elsewhere.[36] The fishermen caught pike and eel by a particular method (Figure 12): on the prows of their boats, which were constructed from a single piece of timber, they lit a brisk fire from torches coated with pitch, after it grew dark, so that they could see the vast numbers of fish as they approached in amazement, and then, having speared them with the iron prongs, drew them in to deal with as they needed. In the illustration a pike is seen in the foreground, eels responding to a thunderstorm in the background. The plate with the blazing fire on it projected forward from the bow of the boat. At times the fishermen piled up thickish logs on drifting ice floes and built a bonfire to obtain fish that became so hypnotized that they had no sense of danger (20.10).[37]

At night the eyes of herring shine like lamps in the sea, and when these fish move rapidly and the huge shoal turns back on itself, they resemble flashes of lightning in the churning water. This sight was therefore commonly known as herring lightning. Wherever they see a light shining above the seawaters, the entire shoal swims towards it, and by this piece of trickery they can be enticed into nets at certain times. The merchants had big fires blazing in front of their tents, as if it were a military camp, and herring approached the shore to gaze at the fires (20.29).[38]

Insects and lamps

Olaus Magnus writes in the chapter 13.15 that crickets could be seen hopping in the adjacent illustration. They were engraved to this picture because in the winter they would cling to the chimneys and set up chirping, which grated on the

[34] Transl. Fisher and Higgens 1996-1998, 1003.
[35] Swahn 2005, 44-45.
[36] Fisher and Higgens 1996-1998, 1071.
[37] Fisher and Higgens 1996-1998, 1041; commentary, 1071.
[38] Fisher and Higgens 1996-1998, introduction, lxv and 1062.

nerves, especially of strangers, for the natives were used to them and remained untroubled.[39] These insects were thought to give forecasts (Figure 13): according to Olaus Magnus, the more often these leapt up and knocked against lamps set on the tables '*saltu luminaria mensis imposita impetunt*', the greater the storms and snowfalls that will follow (13.15).[40] The image illustrating the text shows a candleholder on the shelf or table between the two windows at the entrance to the bakery, and large crickets hopping around the oven.

Comparing the evidence provided by Olaus Magnus with other sources – Light and lighting devices at Häme and Turku Castles in Finland

To complete this survey of fire, light and lighting devices in Scandinavia in the Medieval period, I have chosen to present as a comparison information concerning the lighting devices used at two Finnish castles, Häme Castle and Turku Castle, in the Late Medieval period, contemporary with the creation of Olaus Magnus' work. According to the study of Anna-Maria Vilkuna on the financial management of the Häme Castle in the mid-16th century, the archives tell us about the production of candles in the castle: for example, during slaughter, a part of the tallow was reserved for dipping candles, and hemp was used for the wick.[41] In January 1552, King Gustavus Vasa in his letter to the bailiff of Häme Castle, Isak Nilsson, criticized Isak for the unnecessary consumption of tax candles at the castle, for the castle was to get by with the candles that were produced in the castle.[42] Gustavus Vasa's criticism can be understood, as during his Russian War, even candles and tallow, which were delivered in the main to Viipuri and Helsinki, were collected in the province as a subsidy tax.[43]

Vilkuna points out that the year at Häme Castle continued in accordance with the festivals of the saints, despite the Reformation. The castle had its own chapel, from which, however, the valuable items had been delivered to King Gustavus' silver room. When Duke John, Gustavus' son, visited Häme in 1556, he was given a silver sanctuary lamp from the church at Pälkäne and the church at Renko.[44] The inventory of the Häme Castle's chapel reveals that in 1541 there were one silver sanctuary lamp, two small brass candlesticks and one small seven-branched iron chandelier.[45] The candles and the sanctuary lamp with the so-called eternal flame hanging in front of the altar illuminated the chapel, as well as the pans with glowing coals that were brought inside during the colder periods to warm up the room.[46]

In the mid-16th century, an ordinary Finnish everyday life was lived at Häme Castle, but a real Renaissance court was created by Duke John and his spouse Katarina Jagellonica at Turku Castle (1557-60), which is the only example of a luxurious continental court culture ever experienced in Finland.[47] Katarina imported the latest fashions in dress and table manners from Krakow to Turku and the inventories reveal that she also took with her a large number of lighting devices to embellish Turku Castle. Catarina had in her dowry from Poland candleholders to illuminate the dining halls, chambers and the castle chapel. She had also four chandeliers, two for 15 candles and two for 12 candles. The tables were embellished with 30 candleholders, and eight of them were made in silver. The largest candleholder in silver was seven-branched.[48]

The information provided by the inventories of Häme and Turku Castles complements well the description provided by the text and illustrations of Olaus Magnus. All in all, the illustrations and the text in Olaus Magnus' work support each other, sometimes better, sometimes not so well. However, *Historia de gentibus septentrionalibus* gives us valuable information concerning the use of fire, light and lighting devices in Scandinavia in the Medieval period. Some of the customs and lighting devices are also known from Central and South Europe, some are fantastic creations of Olaus Magnus, but there are also some local customs and lighting devices that would have stayed unknown for us without the encyclopaedia of Olaus Magnus.

[39] Transl. Fisher and Higgens 1996-1998, 630-631; illustration above ch. 13.13.
[40] See also Swahn 2005, 155.
[41] Vilkuna 2003, 57, 68.
[42] *Gustaf I:s Registratur XXIII 1552* 1905, 17-19; Vilkuna 2003, 75.
[43] Cf. Vilkuna 2003, 82.
[44] Vilkuna 2003, 78-79.
[45] 1541 VA 3683: 128v-129; Vilkuna 2003, 79.
[46] Vilkuna 2003, 79.
[47] Vilkuna 2003, 92.
[48] Gardberg 1986, 68-69.

Archival Sources

National Archive of Finland, Kansallisarkisto (VA), *Vanhempi tilikirjasarja, Hämeen lääni, vuodet 1539-1571,* 3668-4070.

Printed Sources

Gustaf I:s Registratur XXIII 1552. Handlingar rörande Sveriges historia med understöd af statsmedel. I tryck utgifna af Riksarchivet. Stockholm 1905.

Olaus Magnus, Historia de Gentibus Septentrionalibus. Romae 1555.

Bibliography

Chrzanovski, L. and Kaiser, P. (eds) 2007. *Dark Ages? Licht im Mittelalter/L'éclairage au moyen* âge. Milano: Edizioni Et.

Fisher, P. and Higgens, H. 1996-1998. Olaus Magnus, Historia de Gentibus Septentrionalibus, Romae 1555, Olaus Magnus, *Description of the Northern Peoples,* Rome 1555, transl. by P. Fisher & H. Higgens, ed. by P. Foote, with Annotation derived from the Commentary by J. Granlund abridged and augmented. London: Hakluyt Society.

Gardberg, C. J. 1986. *Tre Katarinor på Åbo slott*. Helsingfors: Schildts.

Gillgren, P. 1999. The artist Olaus Magnus, vision and illustration, in C. Santini (ed.) *I fratelli Giovanni e Olao Magno. Opera e cultura tra due mondi, Atti del Convegno Internazionale, Roma-Farfa (I convegni di classiconorroena),* 147–155. Roma: il Calamo.

Olaus Magnus, *Historia om de nordiska folken*. Stockholm: Gidlund (reprint 1982).

Swahn, J.-Ö. 2005. *Olaus Magnus bilder, utsnitt ur Olaus Magnus historia om de nordiska folken muntert kommenterade av Jan-Öjvind Swahn*. Lund: Historiska media.

Vilkuna, A.-M. 2003. Financial management at Häme castle in the mid-sixteenth century (1539-about 1570), in T. Mikkola and A.-M. Vilkuna (eds) *At Home within Stone Walls: Life in the Late Medieval Häme Castle (Archaeologia Medii Aevi Finlandiae VIII),* 15-132. Turku: Suomen keskiajan arkeologian seura.

Mines Illuminated – Reflections upon Lighting in Medieval Mining

Lena Berg Nilsson
Department of Archaeology and Classical Studies, Stockholm University, Sweden.

Abstract: Did the medieval Swedish miners light their workspace or was the daylight sufficient? According to common opinion no artificial light was needed and it has been assumed that in the small medieval open-cast mines lighting was of no need due to the available daylight. However, since that would imply very short daily hours of work, at least during the dark period of the year, from October to March, it doesn't seem to be an entirely satisfying argument. Artificial lighting was probably necessary also at the medieval Swedish mining operations, but what kind of lighting devices were used? Looking at the physical remains of medieval mining in Sweden there is no obvious answer to this question. However, there are a few written sources and examples from contemporary art –both domestic and European– that can give an entrance to reflections upon some aspects of lighting in the specific environment of the early mining industry in Sweden. Apart from torches and tar sticks there is a high probability that tallow lamps were also used by the medieval Swedish miners.

Keywords: Mining, Sweden, Medieval, miners, lychnology, lighting, torches, tar sticks, tallow lamps.

Introduction

This article is mainly based on my presentation at the International Lychnological Association Round Table in Olten 2007, although some minor changes and additions have been made. I strive to keep the essence of what I presented then, and not to try to repeat my words. By way of introduction, it is essential to emphasize that this presentation does not originate from a scientific study, it represents only initial thoughts on the subject by a complete novice in the lychnological field.

The thoughts came up during my ongoing PhD project in archaeology, 'From the bottom of the earth's interior', in which I am studying medieval ore mining in Sweden (*c.* 400-1500 AD), partly in relation to the contemporary European mining. One aspect of my work is to see in which contexts ore mining occurs in Sweden and to be open for the possibility of other chronological, spatial, and social contexts than previously have been accepted. Another aspect is to observe the techniques used in and around the medieval Swedish mines. The latter leads me on to thoughts about lighting, an issue in principal not much studied in relationship to medieval mining in Sweden.

To put my thoughts into context I start my contribution with a few words on medieval Swedish ore mining and from there I discuss my reflections of lighting within the medieval industry. Hopefully these thoughts can have bearing also on conditions outside of Sweden's boundaries.

Swedish medieval ore mining

My ongoing study is based mainly upon archaeological material, such as surveys, mappings, stray finds, and excavations, of both the actual mining sites and other remains from the medieval metal and mining industry. However, the approach is to a high degree interdisciplinary, and historical material such as written sources, historical maps and contemporary art are above all vital for understanding. In addition to these sources I also, to a lesser extent, use not only geological research in the form of ore analyses and analyses of sedimentary records, but also epic literature and folklore.

The starting point of the more extensive and organized mining activity in medieval Sweden occurs in the 12th century, but increasingly more indications point towards an earlier rise of ore mining in Sweden. Only recently the evidence for early ore mining in Sweden has been pushed back from the Middle Ages to the pre-Roman Iron Age.[1] In all there is evidence of more than 40 mines in operation during the Middle Ages in Sweden. If we also including the probable ones –the number is almost double.

During the period *c.* 400-1500 AD, as far as is known, only iron, copper and silver ore were mined and processed in Sweden. The ore was mined mainly in open cast pits, usually several small ones within a limited area. The technique of draining the mines set limits for their depth, which led to shallow open cast pits or sometimes slightly deeper pits with drainage ditches. Over time the open pits were getting deeper with galleries and shafts. The ore was mined through fire-setting, where wood was stacked against the rock-face and lit. The heat caused the rock to crack and the ore could be pried loose. Whether water was used to put the fire out or if the fire burned itself out is debated, but water was probably not used in most cases, except at the more work intensive, larger mines with several owners and work teams where time was an important factor.

One of the questions at hand is the lighting technology used in the mines. Did the medieval Swedish miners light their workspace or was the daylight sufficient? According to common opinion no artificial light was needed. In the

[1] Bindler, Karlsson, Rydberg *et al.* 2017.

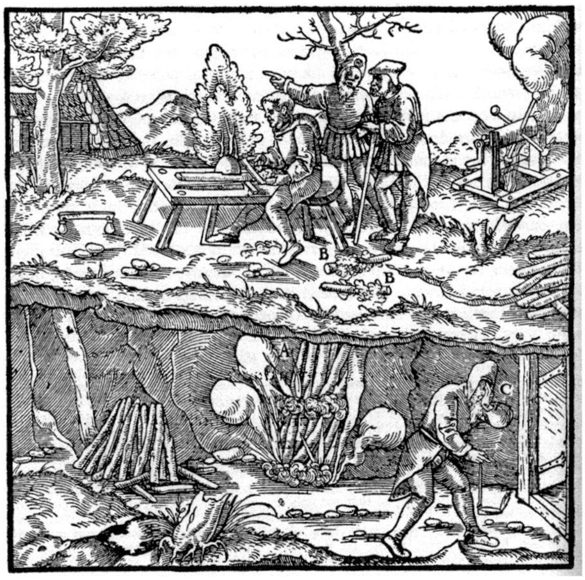

Figure 1. Wood-cut from Agricola´s 'De re metallica'[2] showing fire-setting. Note the man with a mine lamp with a handle at the bottom right corner of the picture. Is he covering his face to protect himself from the severe smoke or to keep his eyes adapted to the darkness, when passing a brightly lit area?

smaller open-cast mines or mines with sloping shafts it has often been assumed that artificial lighting was not needed because the daylight was sufficient. However, since that would imply very short daily hours of work -at least during the dark period of the year, from October to March, when mines in at least some parts of Sweden were worked– it does not seem to be an entirely satisfying argument. Especially at the larger, work intensive mines longer working hours ought to have been necessary. Also during winter, natural light was likely not enough. And what lighting was used when the open cast pits were getting deeper with galleries and shafts, which made artificial light necessary wherever the natural light could not reach?

Another, although rarer, opinion is that the fire-setting gave enough light in the mines. This is clearly a misconception since fire-setting can cause severe smoke and work would be impossible in the vicinity. This could be illustrated by a wood-cut in '*De re Metallica*' which shows fire-setting and a miner moving away from it covering his face from the smoke (Figure 1). Incidentally it could be mentioned that it might not only be the annoyance of the smoke that made the miner cover his face. Another theory, pointed out by Donald A. Trotter[3] is that the miner on the wood-cut is holding one hand over his eye when passing the brightly lit area, to keep his eye adapted to the darkness.

Consequently, artificial lighting was probably necessary at the medieval Swedish mining operation, but what kind of lighting devices were used?

Reflections from the Swedish material

If we look at the physical remains of the medieval mining sites there are no obvious answers to this question. There have been only few archaeological excavations of parts of mines and mine fields in Sweden. Most of the knowledge of Swedish medieval mining has instead so far emanated from historical sources, mainly concerning the dominate Sala Silver Mine and the Copper Mountain in Falun. Also, the contemporary Swedish material on medieval miner's lamps or lighting conditions in general at the medieval mines, is extremely meager and it is not an issue much studied. However, there are a few sources and examples that can give an insight into some aspects of lighting in the specific environment of early mining. I therefore want to share my initial thoughts on the subject, with the purpose of trying to find approaches and ideas on how to use and interpret the material at hand, to better understand how the medieval miners solved their lighting problems.

Torches and tar sticks

Torches of tar sticks or resinous pinewood seem to be closest at hand, and according to traditional opinion it also seems to have been the only possibility in Swedish mining before the 17th century. Was this really the only possibility of medieval mine lighting or is this only an assumption?

Torches of tar sticks or resinious pinewood were no doubt used in the mines from medieval times and onwards. One example from later art is an oil painting from 1784 showing bundles being made down in the Falu Copper Mine from the fire-setting wood. The bundles could also have handles to enable the miners to carry them in the mouth while climbing on ladders to keep their hands free (Figure 2).

It is also interesting to note that at, for example, the silver mine in Kongsberg, Norway (in operation from the early 17th century) the oldest lighting devices used were tallow lamps. Open bowls filled with tallow as fuel had a wick of cotton. The lamps were made to be carried in the hand but had a hook

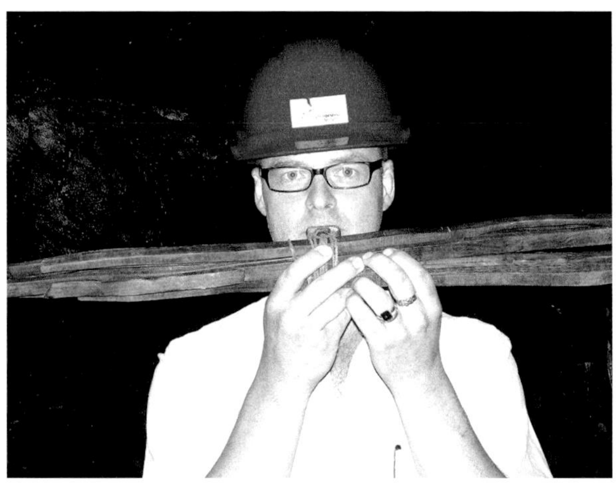

Figure 2. Demonstration of a mouth held bundle of tar sticks, suitable for climbing ladders. Copy at the Hornkullen Silver Mine, Sweden. Photo by Lena Berg Nilsson, 2007.

[2] Agricola 1912.
[3] Trotter 1983, 23.

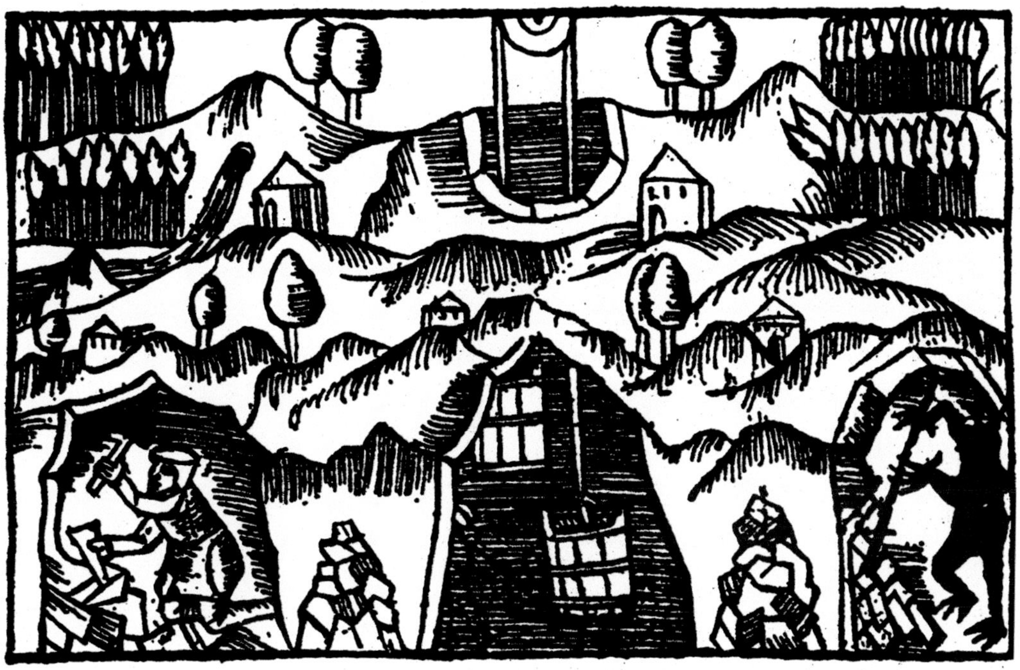

Figure 3. Illustration from Olaus Magnus[4] showing a troll at the lower right hand corner of the picture.

to enable them to also hang from the rock-face. Although torches were an earlier technique to achieve light it was originally forbidden to use in the Kongsberg mine, because of the fire hazard and since the smoke from the torches discoloured the rock face and made it more difficult to distinguish the ore veins. The lighting of mines using wooden sticks was also seen as a misuse of valuable wood, much needed for other purposes in the mines. In the 18th century however, the number of workers at Kongsberg increased and the difficulty of getting enough tallow led to an approval of the use of torches in that mine.[5] The downside of the use of torches ought to have been valid also in the Middle Ages in Sweden, but what evidence or indications do we have of other lighting devices at that time and place?

One way to find the answers could be to broaden the view and look also in the vicinity of the mines –at the settlements. A medieval mining village in Sweden at the Sala Silver Mine was archaeologically excavated in 1984-85. At this excavation, only one light holder and one candlestick were found and no mining lamps whatsoever.[6] Maybe the explanation to this meager lychnological material was that the mining equipment was kept close by the mine and not in the village.

Another source for medieval conditions in Sweden is Olaus Magnus '*Historia de Gentibus Septentrionalibus*' (History of the Northern People) from 1555. One part of his work concerns mining, but it contains no illustrations or descriptions of mining lighting devices. Although the work includes more than 400 illustrations, only three are of mining and none of them showing the lighting aspect in the mines. However, at two places in the text he does mention lighting. Regarding explorations of abandoned mines, he writes 'One goes down by torchlight'[7] and regarding the perils of mining and the courage of the miners, '...when they have taken up long galleries in the mountains and in the light of lamps...'.[8] The latter might indicate semi-permanent lamps at the current working areas in the galleries, being lit when arriving.

Tallow lamps

So far there is no evidence of sources of light other than torches of tar sticks or resinous pinewood in the medieval Swedish mines. But through the work of Olaus Magnus, as shown above, there is at least one indication of lamps being in use. What kind of lamps could he refer to? As seen from the Norwegian example above, tallow lamps was in use there at least in the early 17th century and in Sweden they are known to have been in use in mining from the same time. Could they in fact have been in use already in medieval times?

Another, although very far-fetched, indication of tallow lamps being a possible source of light in the medieval mines in Sweden is the folklore about entities in the mines. These are familiar to Olaus Magnus who states; 'It is an established fact that the Nordic inhabitants enjoys great services and assistance from the trolls' (Figure 3).[9]

As for European conditions Agricola in his '*De re Metallica*' puts emphasis on prayer and fasting to put the demons of the

[4] Magnus [1555] 1982.
[5] Berg 2002, 46.
[6] Bergold and Öhnegård 1987.
[7] Magnus [1555] 1982, 277, author's translation.
[8] Magnus [1555] 1982, 268, author's translation.
[9] Magnus [1555] 1982, 278, author's translation.

mines to flight,[10] which shows similar beliefs in underground entities among the European mining industry population. Another example are the Cornish 'knockers' who are said to have lived in the mines and who were thought to be spirits that was left out from both heaven and hell, spirits of old miners or of those who had crucified Christ. The knockers needed to be treated with respect and offerings of food or tallow should be left for them in the mine or it could become dangerous.[11] The offering of tallow indicates the use of it in the mines, probably as fuel in lamps. As for Sweden, I have so far not found any information whether offerings to the trolls were made or not, but it gives cause for more thought.

Reflections from contemporary European art

Partly because of the lack of research concerning Swedish conditions, the material at hand is extremely meager. In the process of starting to study it I could not help noticing and become fascinated by the different mine lamps that can be observed in European medieval art. Taking a closer look at the pictures the lamps in some cases even seem to have been specifically adapted to mining conditions.

By far the most useful source of pictures of medieval mining is the before mentioned '*De re Metallica*' by Georgius Agricola from 1556, a book of the art of how to find and mine minerals and ore and the smelting of metals. The book includes more than 300 wood engravings and the most depicted mine lamp in medieval art seem to be the handheld one. It can be seen in many of the wood-cuts in '*De re Metallica*' (Figure 4).

Two other examples among many, of art depicting the handheld mine lamp are an altarpiece in Annaberg by Hans Hesse from 1521 and a drawing by Heinrich Gross from La Rouge Mine the Saint Nicholaus in la Croix-aux-Mines, from *c.* 1530. On the latter, it is also visible that the bowls are open and filled with tallow as fuel.

There are also examples of more improved mine lamps in medieval art. In many cases the lamp is no longer made to be held directly in the hand. Instead it has an attached handle. Sometimes the handle is in one piece, and sometimes it is two pieces joined together (Figures 1, 4 and 6).

According to the images some mine lamps were even specifically adapted to mining operations. One of the finest of the medieval drawings of mining is the title page from the '*Kuttenberger Kanzionale*', a manuscript book on vellum with 252 pages from Kutná Hora in the Czech Republic, including both biblical and mining illustrations. It contains several mining songs in Latin, many with a religious tone. It was compiled, probably based on older manuscripts, by an unknown author in 1471.[12] A not too bold interpretation is that the song book was compiled in connection with the crowning of Vladislav II in Kutná Hora the same year, especially since it is his monogram on the title page.[13] The picture of the title page shows scenes from and around a mine, in three sections; the mining underground, the processing, and the trade. It is attributed, somewhat uncertainly, to Mattheus Illuminator who is said to have died before finishing it, leaving to other artists to complete it, which was done *c.* 1490.[15]

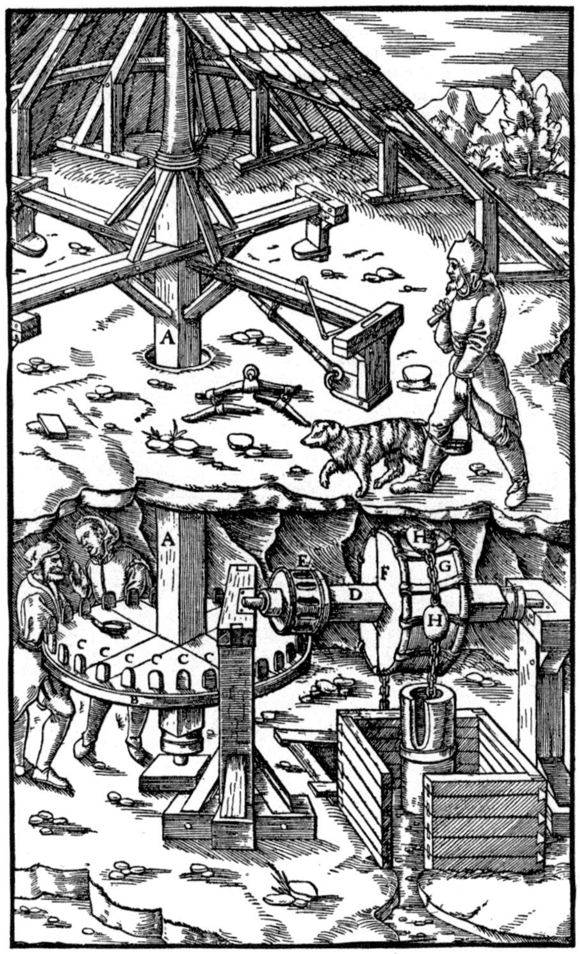

Figure 4. Wood-cut from '*De re Metallica*'[14] showing a man, at the middle right, holding a mine lamp with a joint handle.

The reason for mentioning the '*Kuttenberger Kanzionale*' is the section of it showing underground mining, in which several miners' lamps can be seen. Trevor D. Ford observes that some of the miners seem to have lamps of roman type[16] but as shown in the example above also tallow lamps were used in medieval mines. Of special interest are the miners who wear lamps on top of their heads (Figure 5). The hoods used by the miners was therefore maybe not only protection from dripping water, which is the common interpretation, but also protection from dripping tallow.[17]

There are also other examples in medieval art depicting headlamps on miners, as on the aforementioned alterpiece from Annaberg, where one can see a flame on top of the head on one of the miners, and on a wood-cut in Agricola's '*De re Metallica*' (Figure 6). On this specific wood-cut it seems like the handle of the headlamp is fastened in the miner's clothing,

[10] Topping and Lynott 2005, 183.
[11] Topping and Lynott 2005, 183-184.
[12] Ford 1994, 81.
[13] Berg Nilsson 2008, 36.
[14] Agricola [1556] 1912.
[15] Ford 1994, 81.
[16] Ford 1994, 81.
[17] Berg Nilsson 2008, 38.

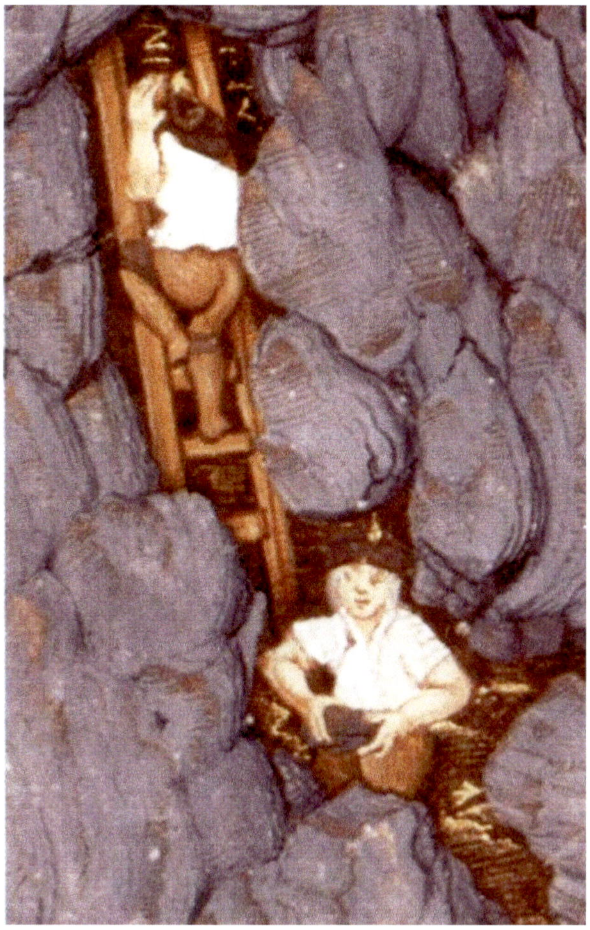

Figure 5. Detail from the front-page of '*Kuttenberger Kanzionale*' showing miners with headlamps. Picture from Berg Nilsson 2008. The original is kept at the National Library of Austria in Vienna.

at the back of his neck. Apart from the method of fastening it, this could be said to be an early variety of the modern headlamp!

Conclusion

Artificial lighting was probably necessary at the medieval Swedish mining operations, even though the pits mostly were shallow. Torches of tar sticks or resinous pinewood seem to be the closest solution at hand. However, it is important to question whether they were the only lighting devices used, or if it is merely an assumption and other methods of lighting could have been in use.

So far there is no hard evidence of the use of tallow lamps in medieval Sweden -handheld or headlamps. However it would be a most imaginable scenario, considering contemporary European technology and its influence on Swedish mining in other aspects. Since only a few archaeological excavations of mines in Sweden have been made, it would be a mistake to rule it out completely, only because of the lack of physical remains. One strong indication of lamps being used in medieval Swedish mines are the words of Olaus Magnus -'in the light of lamps'- and so far, nothing speaks against it.

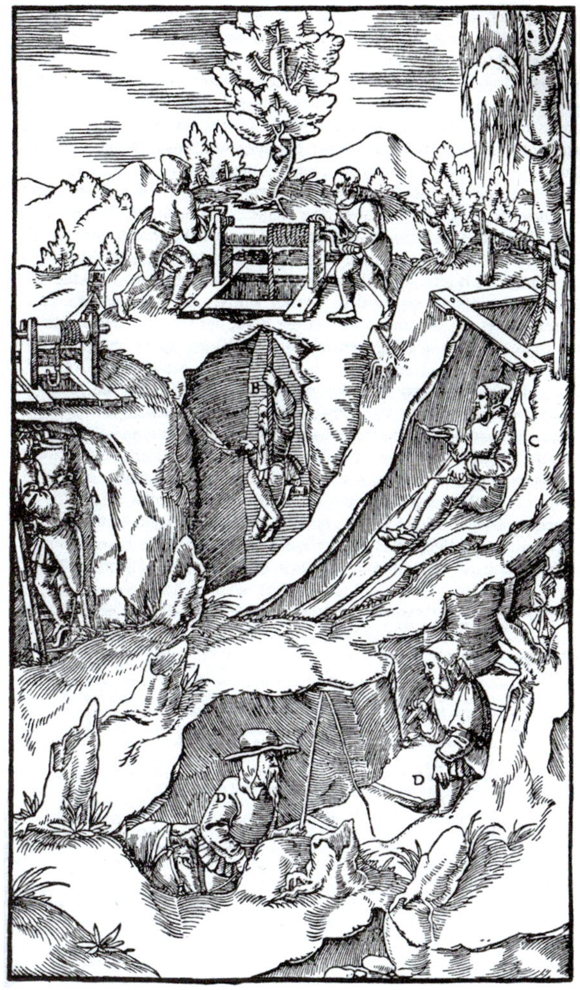

Figure 6. Wood-cut from '*De re Metallica*'.[18] Note the two miners in the middle, going down the mine with handheld mine lamps and the man in the bottom right of the picture holding a mine lamp with joint handle, as well as the man in the middle left of the picture with a headlamp.

Final thoughts

Reaching this conclusion, and bearing the medieval headlamp in mind, suddenly a Swedish coat of arms from 1621 becomes more understandable. The coat of arms in question (Figure 7) is one of the oldest known Swedish pictures of a miner. Having another look at it, it is a probable interpretation that a flame is visible on the head of the miner on the coat of arms. Is it in fact a head lamp like the ones in medieval European art? I argue it is!

With these final thoughts as a starting-point, it is time for me to move away from my initial reflections towards a scientific approach, towards deeper lychnological knowledge and conducting deeper studies of the Swedish material. I hope to return when this is accomplished.

[18] Agricola [1556] 1912.

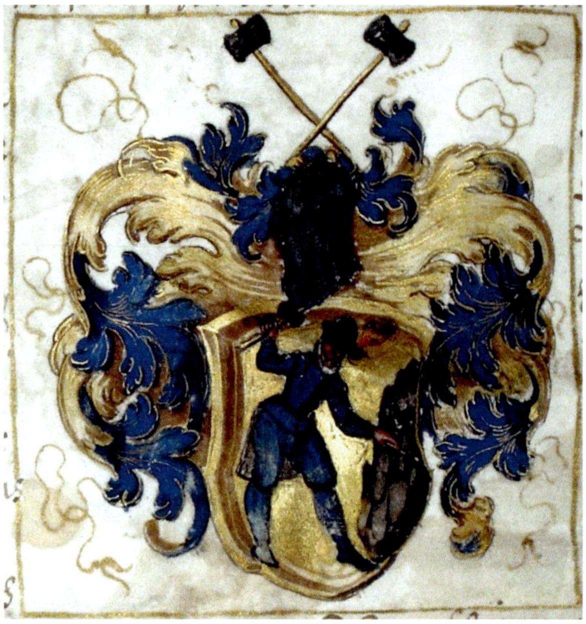

Figure 7. The earliest known Swedish picture of a mining lamp. Coat of arms of Christoffer Olsson (Gyllenknoster) from 1621. Photo: Göran Mörner, the House of Nobility, Stockholm, 2008.

Bibliography

Agricola, G. [1556] 1912. *De re Metallica.* The Hoover edition. London.

Berg, B. I. 2002: *Gruveteknikk ved Kongsberg sølvverk 1623–1958.* Kongsberg: Norsk Bergverksmuseum.

Berg Nilsson, L. 2008. Myller vid ett malmberg. En betraktelse kring titelsidan av en medeltida gruvkantional från Kutná Hora. *Med hammare och fackla XL*, 31-44.

Bergold H. and Öhnegård, W. 1987. *Sala gruvby. Ett industrisamhälle från 1500-talet.* Sala: Kommunen.

Bindler, R., Karlsson, J., Rydberg, J., Karlsson, B., Berg Nilsson, L., Biester, H. and Segerström, U. 2017. Copper-ore mining in Sweden since the pre-Roman Iron Age: lake-sediment evidence of human activities at the Garpenberg ore field since 375 BCE. *Journal of Archaeological Science: Reports* 12, 99-108.

Ford, T. D. 1994. 15th century mining as shown in the Kuttenberger Kanzionale, in T. D. Ford and L. Willies (eds) *Mining Before Powder*, 81-82. Matlock Bath, Derbyshire: Peak Districk Mining Museum.

Magnus, O. [1555] 1982. *Historia om de nordiska folken.* Malmö.

Topping, P. and Lynott, M. 2005. *Miners and Mines.* In: P. Topping and M. Lynott (eds) *The Cultural Landscape of Prehistoric Mines,* 181-191. Oxford: Oxbow.

Trotter, D. A. 1983. *The lighting of underground mines.* Clausthal-Zellerfeld: Trans Tech Publications; Houston: Gulf Pub. Co., Book Division.

Lighthouses in the Tabula Peutingeriana and their Importance in Late Antiquity

Yvonne Seidel
Independent archaeologist, Vienna, Austria

Abstract: The Tabula Peutingeriana is a reproduction of an ancient world map. It shows the main roads with the distances between cities and road stations. The original map can be dated in the middle of the 4th or the second half of the 5th century. Most of the buildings are represented by symbols, including the famous lighthouse of Alexandria and a lighthouse at the passage from the Bosporus to the Black Sea. The important lighthouse of Ostia is included in the city vignette of Rome. The tower next to Constantinople could also mean a lighthouse.

Keywords: lighthouse, Tabula Peutingeriana, Late Antiquity, Constantinople, Alexandria, Ostia.

A special parchment is kept in the Vienna National Library: The Codex Vindobonensis 324, which is commonly known as Tabula Peutingeriana, named according to its first explorer.[1] The parchment shows a street map of the ancient world. The preserved edition is a copy of the 12th or early 13th century, but it is based on an older Late Antique original, although attempts at dating it are controversial.

The map does not present a uniform picture of geographic conditions at a given time, as it was certainly composed from different sources. A terminus *post quem* has to be seen in the foundation of Constantinople in 330 AD, since the city is listed in the map under this name. Further concrete indications for dating are missing so far. The places mentioned suggest an origin between the middle of the 4th and the second half of the 5th century.[2]

The map shows mainly the most important traffic routes with details of distances and associated road stations. Major cities such as Rome, Constantinople and Antiochia are represented with pictograms of their city personifications and are thereby highlighted. This map was not suitable for navigation, but important ports and lighthouses are recorded, in order to point out their position along the traffic routes. Four buildings can be identified as lighthouses. Two of them are represented by standard vignettes of stepped towers with a beacon at the top. The remaining two lighthouses show individual architectures.

The tower of Chrysopolis (Figure 1) is located on the Bosporus opposite Constantinople, today's Istanbul.[3] In the area of the ancient Chrysopolis there was a tower with a light signal in the 18th century. Its name –Leandertower– refers to a

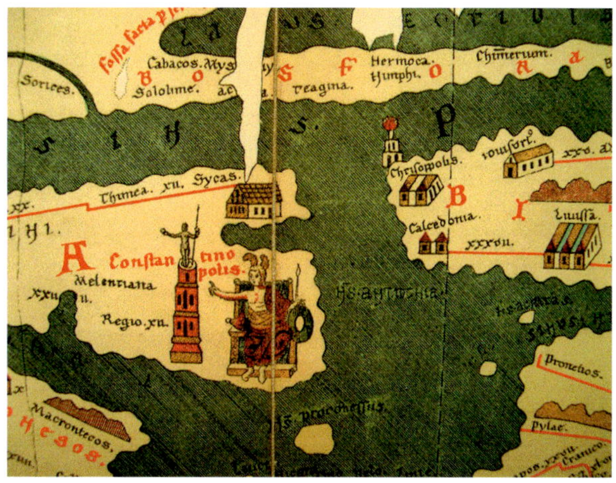

Figure 1. Detail of Tabula Peutingeriana, The tower of Chrysopolis (Miller 1888, Segment IX, 1).

Greek legend, which is best known in the version of Ovid. The present appearance of the tower dates back to the 18th century, but it would be an exciting question of research, if it still contains remains of Roman masonry.

The next tower to discuss is unmarked on the map. Probably it is the famous Pharos of Alexandria, which is one of the ancient Seven Wonders of the World (Figure 2), located on the eastern tip of the island of Pharos in front of the city of Alexandria. The Donjon of the Kaitbey fortress contains most likely the architectural remains of the lighthouse. Some nonspecific blocks documented during archaeological dives along the coast are also suspected of belonging to the tower.[4] Numerous historians and geographers provided evidence concerning the size and technical equipment of its monumental structure. Combined with some pictorial representations they form the basis of our knowledge.[5] Thiersch reconstructed a triple-

[1] Weber 1976; Miller 1916.

[2] The dates of decline of the cities marked on the map as significant: Rome 537 AD, Constantinople 626 AD, Antioch 526 AD, Lucera 663 AD. The map appears to be a compilation of several older predecessors, for example, the cities buried by the eruption of Vesuvius are reproduced in great detail, and the port of Ostia is apparently represented in the state of the construction phase initiated by Emperor Claudius.

[3] Identified by Miller 1916, 635. Due to a misunderstanding of the often hard-to-read Tabula Peutingeriana the place was often wrongly recorded at the exit of the Bosporus to the Black Sea on the right bank. Nowadays it is a district of Istanbul called Üsküdar.

[4] Empereur 1998; Empereur 1995, 743-760.

[5] Thiersch 1909; further: Pfrommer 1999, 10-16; Empereur 1998; Grimm 1998, 43-45; Bedon 1988, 54-66; Haas 1997, 25-26; Brodersen 1993, 207-211; Clayton 1990, 182-207; Ekschmitt 1984, 183-197; Forbes 1958, 180.

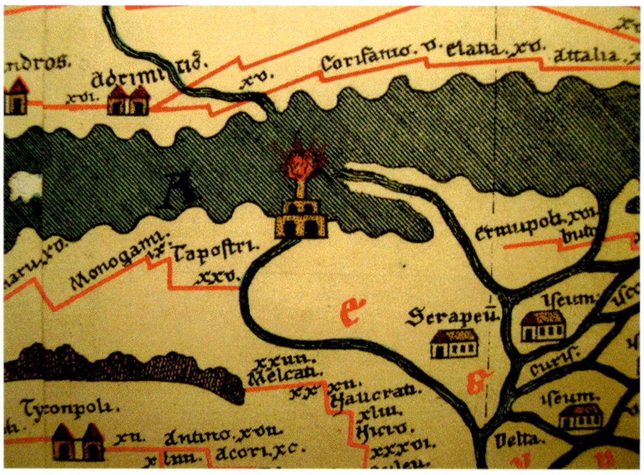

Figure 2. Detail of Tabula Peutingeriana, Probably the famous Pharos of Alexandria (Miller 1888, Segment IX, 3).

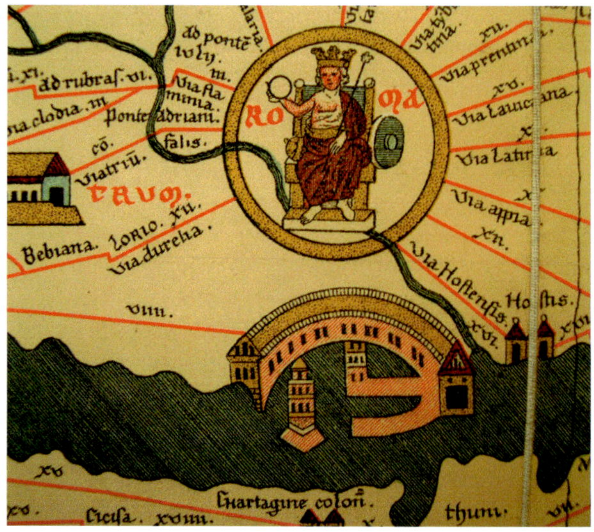

Figure 3. Detail of Tabula Peutingeriana, The port of Rome - Portus Ostiensis Augusti (Miller 1888, Segment V, 5).

stepped structure with a statue on the top on the basis of ancient and Arabic sources as well as depictions on coins. The tower is surrounded by a wall, which corresponds to the course of the modern enclosure. Thiersch reconstructed the internal structure as a large transport shaft surrounded by numerous small chambers. It has to be considered, if indeed, as much fuel was required, as he assumed. A technical solution could have reduced the amount of fuel by far. The high scientific standards of Alexandria make it plausible that there was a kind of optical system working with mirrors. This is mentioned in several late sources, unfortunately without more detailed explanations.[6] According to Flavius Josephus the light was visible to sailors up to 300 stadia (about 54 km) from the harbour.

The port of Rome –Portus Ostiensis Augusti– is depicted on the Tabula Peutingeriana below the city vignette of Rome (Figure 3). Here the lighthouse is a terraced tower with some storeys, placed in the middle of the harbour basin. Sueton and Pliny report that Emperor Claudius gave an order to sink a large ship, which brought an obelisk from Egypt to Rome, in the harbour basin.[7] Therefore it can be dated between 42 and 52/54 AD. The sunken ship was filled with a mixture of puzzolan, lime, tuff and rubble stone for building the foundation of the tower. Remains of the wreck were found in the nowadays completely dry harbour basin.[8] The lighthouse was situated in the area of the Ship's bow. A round structure with a diameter of 4.40 m has to be considered as the core of the Pharos. Insights into the external shape of this tower can only be hypothesized from coin pictures, reliefs and mosaics.[9] The tower consists of four to six storeys. The lower very broad steps seem to be arranged on a rectangular ground plan. The upper part is apparently smaller and round. An open flame was burning on the top. The entrance is marked in the middle of the lowest segment by an arched door. Some representations also show windows in the upper parts.

All other depictions deviate from the illustration in the Tabula Peutingeriana. Although the open fire on the top was a characteristic of this lighthouse, here the tower is shown with a closed roof. Probably one of the copyists misunderstood the original drawing. The harbour appears in the state of its construction between 42 and 64 AD. The representation of the buildings must not be over-interpreted, since all ports are depicted quite similarly. Standardized vignettes are also used for street stations and castra in the map. Still a certain unification took place in the reproduction of the port of Ostia.[10]

Another high tower is depicted on the Tabula Peutingeriana near the city vignette of Constantinople (Figure 1).[11] usually interpreted as the Column of Constantine. However, it has to be pointed out that this illustration shows a triple terraced architecture, which could also be understood as a lighthouse.[12] The ports south of Constantinople, as well as the entrance to the Bosporus, make lighthouses essential, as they have difficult winds and strong currents. The lighthouses of Ottoman times are well documented. Apart from the tower at Chrysopolis, older lighthouses are suspected in the so-called Ovid Tower and a tower at the wall of the sea front.[13] This tower is located south of the palace district, in the east of Bukoleon harbour (Figure 4). The massive tower is rising on a rectangular ground plan. It stands in front of the sea wall, but has no direct connection to the wall. Inside the tower is divided into three floors. A *terminus post quem* has to be seen

[6] Thiersch 1909, 35-75. 89-96. E. g. Ibn Chordabeh, Kitab al-masalik (de Goeje 1879 VI 115) and Ibn al-Faqih (de Goeje 1879 V 50).

[7] Suetonius, *The Life of Claudius*,20, 3; Pliny the Elder, *Natural History* 16, chapter 76.

[8] Giuliani 1992, 30 ff.; Fellmeth 1991, 7 ff.; Scrinari 1979; Testaguzza 1970 -with further records.

[9] Fondation pour le Lexicon 1992, 294, no. 19, s. v. Lupa Romana (Weigel); Boardman 1981,797, fig. 12, 644, s. v. Annona (D'Escurac); Dulière 1979, fig. 302; Becatti 1961, pl. 161, 45. 164, 320. 174, 104. 175, 85. 176, 123. 178, 92. 179, 105. 120. 183; Stuhlfauth 1938, fig. 3.

[10] Descoeudres 2005, 77-84.

[11] Bauer 1996, 173-177; Müller-Wiener *et al.* 1977, 256-257 (with further records); for the interpretation als Colum of Constantine e.g., Descoeudres 2005, 79; Ryll 1983, 166-180.

[12] Hauschild 1976, 247.

[13] Müller-Wiener 1994; Müller-Wiener *et al.* 1977, 60; Karnapp 1936, 8-12; Barth 1911, 137.

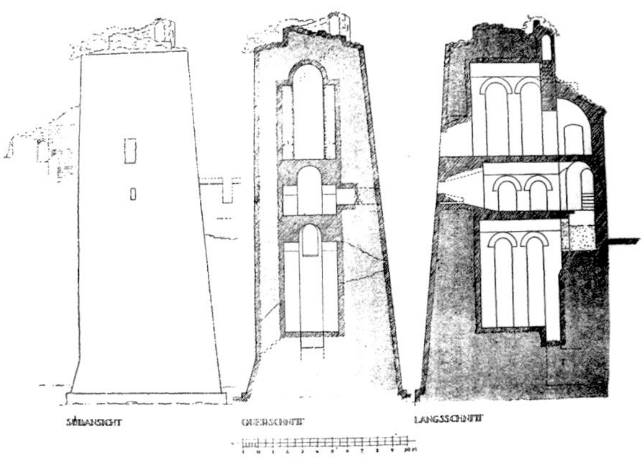

Figure 4. Ovid Tower and a tower located south of the palace district, in the east of Bukoleon harbour in Constantinople (Karnapp 1936, fig. 3. 4).

in the erection of the sea wall under Theodosius II between 439 and 447 AD. This dating does not match the remains of the tower, as the masonry dates to the late Byzantine period. Therefore it remains unclear if there was a predecessor in the same place or the potential lighthouse of the Tabula Peutingeriana stood elsewhere. Considering the character of the ancient world map, I would suggest an interpretation as a lighthouse has to be preferred. On the other hand the representation of the Column of Constantine makes little sense in a map not showing any single monument without use for travellers. The doors and windows in the edition of Miller would speak for a lighthouse, but they can be traced back to an older drawing and cannot be seen in the parchment.[14]

The Tabula Peutingeriana is full of gaps regarding lighthouses. Other towers are known in La Coruña,[15] Dover, Boulogne-sur-Mer, Classis, Salona, Messina and Laodicea. These lighthouses probably still existed in Late Antiquity. Older lighthouses must be disregarded here, since they are still little explored.

The term 'Pharos' was originally used exclusively for the Tower of Alexandria and is derived from the name of its location. From the 3rd century AD onwards this name became synonym for the entire building type. Isidore of Seville describes his task as follows: A lighthouse gives a sign for the night-time navigation of ships. It marks shallows and harbour entrances, so that ships, which were surprised by the night, do not run on cliffs.[16]

The towers of Alexandria and Ostia are the endpoints of one of the most important shipping lines, as Rome was supplied with grain from Alexandria. The other two lighthouses are important markings of the difficult sea route through the Bosporus. Why only these lighthouses are contained in the world map is uncertain. The lighthouses of Messina and La Coruna also mark difficult ship passages. Dover, Ravenna and Salona are important as naval ports. Although the lighting system of the lighthouses is nowhere preserved, two different systems can be distinguished from the representations: towers with an open fire at the top (e.g., Ostia) and those with a closed, partly windowed upper segment (e.g., Messina or Alexandria). The first group had an open fire on the top of the tower. Even if the type of light in the closed towers is uncertain, the construction might have imitated the form of a lantern, which would be very useful especially in view of the wind conditions at the coast.

[14] Weber 1984, 6, fig. 9; Weber 1976, Segment VIII 1 - against an interpretation as a lighthouse.

[15] The first segment of the map is missing, on which would be the lighthouse of La Coruna.

[16] Barney 2011, 402, XX.x.10.

Bibliography

Barney, S. A. 2011. *The etymologies of Isidore of Seville*. Cambridge: Cambridge University Press.

Barth, H. 1911. *Konstantinopel*. Leipzig: Seemann.

Bauer, F. A. 1996. *Stadt, Platz und Denkmal in der Spätantike*. Mainz am Rhein: Philipp von Zabern.

Becatti, G. 1961. *Scavi di Ostia IV. Mosaici e pavimenti marmorei*. Roma: La Libreria dello Stato.

Bedon, R. 1988. Les phares antiques. *Archeologia Paris* 231, 54-66.

Boardman, J. 1981. *Lexicon iconographicum mythologiae classicae: (LIMC)*. Zürich: Artemis.

Brodersen, K. 1993. Ein Weltwunder auf 'gläsernen Füßen', Der Pharos von Alexandria in neuem Licht. *Antike Welt* 24, 207-211.

Clayton, P. A. 1990. Der Pharos von Alexandria, in P. A. Clayton and M. J. Price, *Die sieben Weltwunder*, 182-207. Stuttgart: Reclam.

Descoeudres, J. P. 2005. La vignette d'Ostie dans la Tabula Peutingeriana. *Genava* 53, 77-84.

Dulière, C. 1979. *Lupa Romana II*. Bruxelles; Rome: Institut historique Belge de Rome.

Ekschmitt, W. 1984. *Die Sieben Weltwunder*. Mainz am Rhein: Verlag Philipp von Zabern.

Empereur, J. Y. 1995. Alexandrie (Égypte). *Bulletin de Correspondance Hellénique* 119, 743-760.

Empereur, J. Y. 1998. *Alexandria Rediscovered*. London: British Museum Press.

Fellmeth, U. 1991. Die Häfen von Ostia und ihre wirtschaftliche Bedeutung für die Stadt Rom. *Münstersche Beiträge zur antiken Handelsgeschichte* 10, no. 1, 1-31.

Forbes, R. J. 1958. *Studies in ancient Technology VI*. Leiden: Brill.

De Goeje, M. J. 1879. *Bibliotheca Geographorum Arabicorum*. Lugduni-Batavorum (Leiden): E. J. Brill.

Grimm, G. 1998. *Alexandria: Die erste Königsstadt der hellenistischen Welt; Bilder aus der Nilmetropole von Alexander dem Grossen bis Kleopatra VII*. Mainz am Rhein: Verlag Philipp von Zabern.

Giuliani, C. F. 1992. Note sulla topographia di Portus, in Mannucci, V. (ed.) *Il parco archeologico naturalistico del Porto di Traiano: Metodo e progetto*, 28-44. Roma: Gangemi Ed.

Haas, C. 1997. *Alexandria in late antiquity: Topography and social conflict*. Baltimore: Johns Hopkins University Press.

Hauschild, T. 1976. Der römische Leuchtturm von La Coruna. *Madrider Mitteilungen* 17, 238-257.

Karnapp, W. 1936. Der Leuchtturm der Seemauer, in A. M. Schneider (ed.) *Byzanz. Vorarbeiten zur Topographie und Archäologie der Stadt*, 8-12. Berlin: Istanbul des archäologischen Instituts des Deutschen Reiches.

Fondation pour le Lexicon iconographicum mythologiae classicae 1992. *Lexicon iconographicum mythologiae classicae (LIMC): [vol. VI, pts. 1-2]*. Zürich: Artemis.

Miller, K. 1888. *Weltkarte des Castorius, genannt die Peutinger'sche Tafel.* Ravensburg: O. Maier.

Miller, K. 1916. *Iteneraria Romana. Römische Reisewege an der Hand der Tabula Peutingeriana.* Stuttgart: Strecker & Schroder.

Müller-Wiener, W. 1994. *Die Häfen von Byzantion. Konstantinupolis. Istanbul.* Tübingen: Ernst Wasmuth Verlag.

Müller-Wiener, W., Schiele, R. and Schiele, W. 1977. *Bildlexikon zur Topographie Istanbuls: Byzantion, Konstantinupolis, Istanbul bis zum Beginn d. 17. Jh.* Tübingen: Wasmuth.

Pfrommer, M. 1999. *Alexandria im Schatten der Pyramiden.* Mainz am Rhein: Verlag Philipp von Zabern.

Ryll, A. 1983. Über Probleme der kunsthistorischen und schriftlichen Quellen zur Konstantinssäule in Konstantinopel, in: Historisch-archäologische Quellen und Geschichte bis zur Herausbildung des Feudalismus. Beiträge des 1. und 2. Kolloquiums Junger Wissenschaftler Archäologischer und Althistorischer Disziplinen der DDR, Berlin, 166-180.

Scrinari, V. S. M. 1979. *Le navi del Porto di Claudio.* Roma: Tipografía centenari.

Stuhlfauth, G. 1938. Der Leuchtturm von Ostia. *Römische Abteilung* 53, 139-163.

Testaguzza, O. 1970. *Portus.* Roma: Julia ed.

Thiersch, H. 1909. *Pharos. Antike Islam und Occident. Ein Beitrag zur Architekturgeschichte.* Leipzig und Berlin.

Weber, E. 1976. *Tabula Peutingeriana: Codex Vindobonensis 324.* Graz: Akademische Druck- und Verlagsanstalt.

Weber, E. 1984. Die Tabula Peutingeriana. *Antike Welt* 15, no.1, 3-8.

Lighting Devices in Byzantium: Comparisons in Time and Space

Ioannis Motsianos
Museum of Byzantine Cutlure, Thessaloniki

Abstract: This paper discusses common practices and differences in lighting devices used in Byzantium, as compared to those used during the Roman period, and to identify the interactivity between them and corresponding lighting devices used in Central and Northern Europe. Mold-made clay lamps of the Early Christian period preserve the same forms but they are decorated with Christian symbols. At the end of Late Antiquity the end 'parenthesis' of the era of terracotta mold-made lamps production is closed as a result of both financial and technological reasons. Use of wheel-made terracotta lamps, either glazed or not, superceded the earlier lamp styles. However, from the 4th century AD onward the use of glass *kandelai* for lighting purposes brought forth a great 'revolution' in artificial lighting. The use of glass kandelai had a lot of advantages in comparison with clay and metal lamps. In addition to the various type of glass *kandelai*, many other lighting devices were designed for supporting and hanging glass lamps, e.g., metal *polykandela*, *choroi*, etc.

Keywords: Roman period, Byzantium, mold-made lamps, wheel-made lamps, incense-burners, hanging lamps, *kandelai*, *polykandela*, *choroi*.

Introduction

This paper attempts to address common practices and differences in lighting devices used in Byzantium, as compared to those used during the Roman period, and to identify the interactivity between them and corresponding lighting devices in Central and Northern Europe.

The variety, magnificence and cost of using the various means of illumination were all so great in Roman times that the concept of light (*lux*) actually became the etymological root of the Latin word *luxus*, meaning luxury.[1] There were a large number of lighting devices, some for use by ordinary people with limited financial means, and others for the more affluent who had the means to pay not only for the most decorated and luxurious lighting devices themselves, but also for the considerable amount of fuel required. The Roman Empire became a huge marketplace. Trade flourished with standard measures and weights, a single currency as well as safe roads and waterways extending from Syria to Spain, and from England to Libya[2] (especially during the *Pax Romana* from 27 BC to AD 180). Thus, it was possible to purchase not only the lamps themselves but also the olive oil, even in areas where no olive trees were grown. These items could be acquired from traders or through the supply system which provided necessities for the Roman legions.[3] However, there remained some remote parts of the Empire where it was not easy to purchase olive oil, a liquid fuel, and where people would opt instead for open lamps which could burn solid fuel, like animal fat.

A number of the papers presented at the Round Table in Millau, in March 2007[4] demonstrated that, particularly in

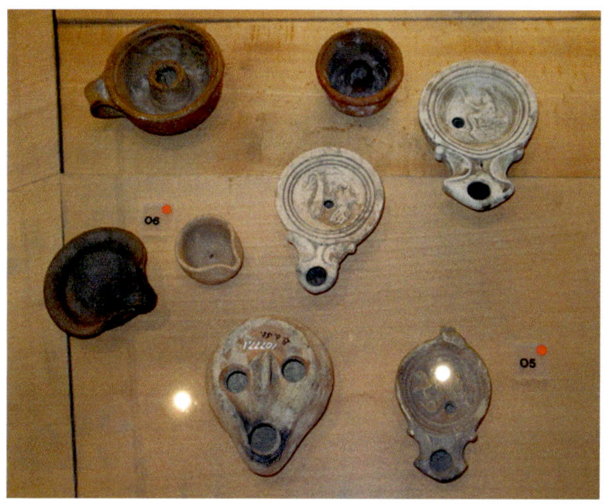

Figure 1. Roman clay lamps, Swiss National Museum of Zurich. (Photo by the author).

Roman times, in the northern areas of Gaul there were not only numerous mass-produced mold-made lamps, but also wheel-made and hand-made lamps produced locally. These lamps were usually open and intended to be used with solid fuel, such as animal or other fat. The same applies to the Roman lamps of Helvetia (Modern Switzerland) (See Figure 1).

Thus, already in the Roman period, the existence or absence of olive oil was a factor which differentiated the method of lighting in the Eastern, Western and, mainly, Northern provinces of the Empire. In fact, the distinction should perhaps be drawn between regions which had access to olive oil, either because they produced or because they imported

[1] Wunderlich 2003, 251.
[2] Flutsch and Rey 2006, 59.
[3] Verena Perko confirmed in her presentation the second method of supply. See Perko and Tratnik 2007.
[4] Archéologie et histoire de l'éclairage en Gaule, de la Préhistoire au début du Moyen-âge, 2nd International Round Table of ILA, Musée de Millau (Aveyron), 22-23-24 May 2007.

it, and regions which had no such access and were obliged to resort to other solutions –mainly the use of solid fuels.[5]

According to the definition of the *Oxford Dictionary of Byzantium*,[6] 'Byzantine Empire' or 'Byzantium' is the conventional name of a medieval state that existed for more than one thousand years. The date of its beginning remains a subject of discussion; most scholars prefer the date 324 (or 330), when Constantinople was founded by Constantine the Great, or 395, when the Roman Empire was divided between the sons of Theodosius I. It ceased to exist in 1453 when Constantinople was captured by the Ottomans. Byzantium can be viewed as a continuation of the Roman Empire, inasmuch as its legal and administrative systems retained numerous Roman features. Of course, the same can be said with respect to the history of lighting devices. The Byzantines themselves called their state 'Roman Empire' (*basileia ton Rhomaion*) rather than Byzantium, using the name *Byzantion* only for their capital, renamed Constantinople. At the same time, the Empire underwent significant transformations, evolving into a Christian and primarily Greek-speaking state centered in the Balkans and Eastern Mediterranean. Both these elements influenced the form and use of lighting devices. At the dawn of the new era, the Christian state was a decisive factor in choosing forms of lighting devices, types of fuel and the decorative symbols associated with them. At the same time, a large number of objects were destined exclusively for use in Christian churches and in Christian burial customs. The majority of surviving depictions of light fittings are associated with Christian rites and ceremonies.[7] On the other hand, the increasing prevalence of Greek in the Byzantine world meant that it was the main language used for the inscriptions associated with means of illumination. Thus, regarding the lighting devices used in the western part of the Roman Empire (Figure 2), an obvious and significant difference is the Greek inscriptions which appear on the lighting devices of Byzantium (Figure 3).

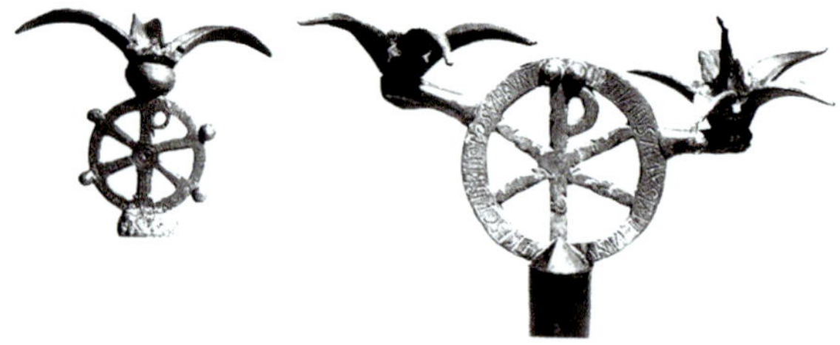

Figure 2. Lighting devices with Latin inscriptions and Christian symbols. Candelabra from Poetovio (Slovenia), discovered in 1858 at Rogoznica near Ptuj. (Photos thanks to Verena Perko).

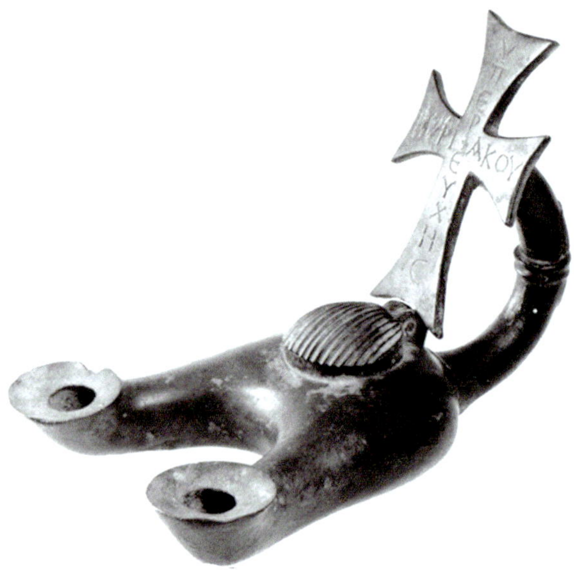

Figure 3. Metal lamp with Greek inscription (Campbell 1985, 50, no. 44).

Clay oil lamps[8]

Although they preserve the same form, the main evolutionary difference between clay oil lamps of the Byzantine period and those of the Roman era lies in the decoration, and specifically in the use of Christian symbols (Figure 4).[9] The same symbols are also found, of course, on lamps in the western part of the Empire.

Another important change is the shift towards production of closed wheel-made lamps, probably by local manufacturers. Or rather, the gradual decline in the production of mold-made lamps. In fact, the discontinuity is linked to the gradual decline and disappearance of mold-made lamps in the end of Late Antiquity (3rd-7th centuries). Indeed, the era which had begun in the 1st century BC, with the systematic adoption of

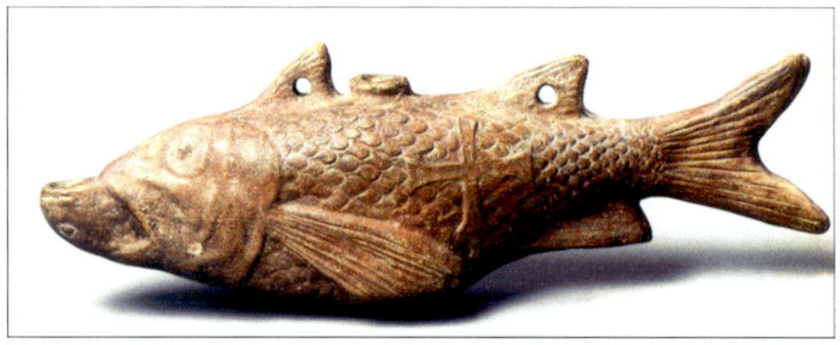

Figure 4. Clay lamp decorated with a cross (Papanikola-Bakirtzi 2002, 296, no. 313).

[5] Dorin Alicu mentions in his abstract that the oil needed for the lamps arrived in Dacia from other Roman provinces, mainly from Asia Minor. See Alicu 2007.
[6] Kazhdan 1991, vol. 1, s.v. Byzantium, 344-345.
[7] The role of lighting in the Christian Church in the West is discussed in Vincent 2007, and in more detail in Vincent 2004.
[8] For a panorama of the mold-made lamps in the Mediterranean region during the Late Roman Period, see Chrzanovski 2006.
[9] See, also Papanikola-Bakirtzi 2002, 299, no. 318.

the mold in the manufacture of lamps, seems to reach its end in the Late Antiquity period.[10]

The ability to produce large numbers of clay lamps by using molds was in fact a key reference point for the Roman period. As far as the Greek area is concerned, during the Roman period, there were major workshops manufacturing clay lamps, like the workshops of Corinth,[11] Patras[12] and Athens.[13] Some of them continued their creations into the Early Byzantine period -when production was dominated by the workshops of Asia Minor and North Africa.[14] However, it is worth noting that in recent years a significant number of wheel-made lamps dating from the 1st to the 6th-7th centuries AD have been uncovered in excavations in the Greek and Mediterranean regions as well as in northern countries directly or indirectly associated with the Roman Empire. Wheel-made lamps dating from the 1st to the 3rd century AD have been also found in Britain[15] and Romania.[16]

Wheel-made lamps from the entire period of Late Antiquity, i.e., from the 3rd to the 7th century AD, have been found in the regions of modern Romania (Figure 5),[17] Southern France, Bulgaria,[18] Egypt,[19] Algeria,[20] Palestine[21] and Cyprus,[22] as well as in a range of archaeological sites in the Greek world, such as Crete,[23] Thessaloniki (see Figure 6) and elsewhere.[24]

It should be noted that, in many cases, one should be cautious about the dating of wheel-made lamps manufactured in imitation of similar models. This is because, while the process of reproducing lamps using a mold leads to very similar if not identical replicas, the creators of wheel-made lamps were not able to reproduce the original lamps precisely, but only attempted to render their general form. Moreover, it is almost inevitable that the wheel-made lamps would be similar in form, since the craftsmen producing them had only limited choices as they worked the wheel, and since all lamps were intended to meet more or less the same specific needs. Thus, only in cases where the wheel-made lamps come from systematically dated excavations can we be absolutely sure of their date. Indeed, it is the methodical excavations of recent years in Greece and in areas dominated by Byzantium which have yielded evidence confirming that wheel-made lamps were produced continuously throughout the Roman and Late Roman periods.[25]

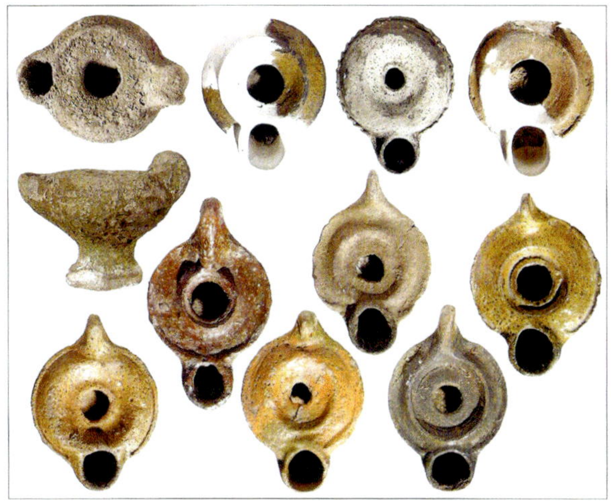

Figure 5. Wheel-made lamps from Romanian territory, 4th century AD (Isac, Roman 2006, 79-80, 132-133, nos. 240-251).

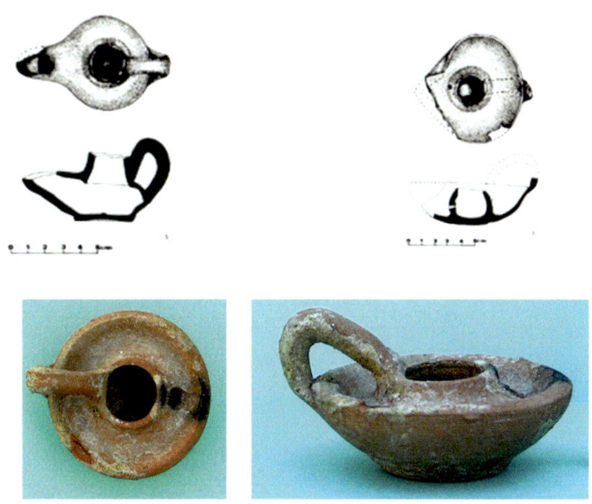

Figure 6. Wheel-made lamps from Thessaloniki (photo thanks to Damianos Kommatas).

We can assume that some of the following reasons help explain the decline and disappearance (at the end of Late Antiquity) of the use of the mold.

1. Organized and specialized production -which had favored the development of the manufacture of lamps by mold- declined.
2. Fewer new types of lamps arrived in certain areas -which led to a decline in interest in their replication.
3. There was a fall in demand for the more luxurious, mold-made lamps, not only because of the cost of the lamps themselves, but also because of the cost of fuel,

[10] For more details, see Motsianos 2010.
[11] Broneer 1930; Bruneau 1971; Bruneau 1977; Broneer 1977; Williams 1981.
[12] Petropoulos 1999.
[13] Karivieri 1996; Perlzweig 1961; with respect to exchanges and imitations of lamps, see Pétridis 2000.
[14] For a recent bibliography, see Chrzanovski *et al.* 2000.
[15] Bailey 1988, 170, Q 1633 PRB-Q 1636 PRB; Yangaki 2005, 235.
[16] Isac and Roman 2006, nos. 41-49, 48-49, 102.
[17] Scorpan 1973, σ. 209; Isac and Roman 2006, nos. 240-251, 79-80, 132-133.
[18] Kuzmanov 2002, 257-260, nos. 53-87.
[19] Bailey 1988, 274, Q2272MLA, Q2273MLA, Q2275MLA, pl. 56, Q2272MLA, Q2273MLA, Q2275MLA.
[20] Bussiere 2000, pl. 135.
[21] Macalister 1912, pl. CI (a), no. II; Rosenthal and Sivan 1978, 122-123, variation B, no. 508.
[22] Oziol 1977, 45-46, nos. 96-98, pl. 7, 96-98, pls 57, 98.
[23] Poyloy-Papadimitriou 2000, 224-225, fig. 20.
[24] Kallipolitis and Petrakos 1963, 51-52, pl. 58b; Lazaridis 1965, 19, pl. 18a; Abadie-Reynal, Sodini 1992, fig. 35, L88, pl. XIII, n (L88), o (L89), p (L90); Triantafyllos 1972, 542, pl. 480, d; Roymeliotis 2001, 262, 266, no. 18, pls 6, 18; Bailey 1988, 418, Q3339, Q3340, Q3341, pl. 125, Q3339, Q3340, Q3341.

[25] See footnotes 11-19.

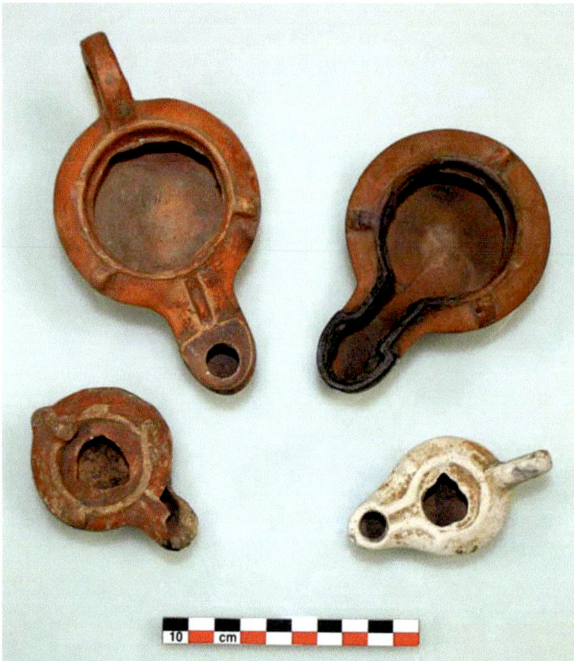

Figure 7. Lamps having their disk intentionally cut, Musée de Bavay, North France (photo thanks to Alice Hanotte).

i.e., olive oil, which was beginning to disappear from certain areas as trade exchanges decreased.[26]

4. The preference for other types of fuel instead of olive oil, for example solid fuel, made it difficult to use mold-made lamps, and this probably led people either to use wheel-made lamps, which usually had a broader aperture through which solid fuel was inserted, or to other solutions (Figure 7) as the one supposedly represented by the *Firmalampen* in the Museum of Bavay, Northern France: their disks seem to have been removed on purpose, in order for the lamps to be used with solid fuel.[27]
5. The isolation of certain regions from the major centers of production of mold-made lamps, or the ability of regions to meet their own lighting needs through local production.
6. Greater difficulty in movement of craftsmen owing to the break-up of the Roman Empire and the upheavals occurring in both eastern and western parts. In the third century AD, political instability and warlike confusion led to an economic crisis. In the Early Middle Ages trading decreased geographically.[28]
7. A shift to the use of glass *kandelai* and metal *polykandela* (chandeliers) (Figure 8). Their use seems to become frequent from the 4th-5th centuries, especially in Christian churches (Figure 9).[29]

In each separate case, there may well have been more than one factor involved in the declining use of the mold and in the increased production and use of wheel-made lamps. And,

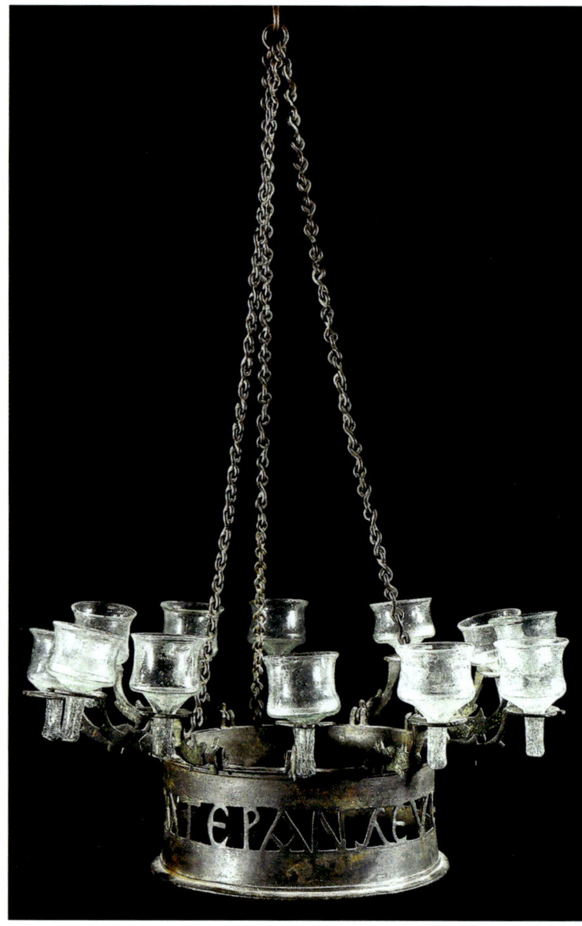

Figure 8. Glass *kandela* and *polykandelon* (Papanikola-Bakirtzi 2002, 283, no. 296).

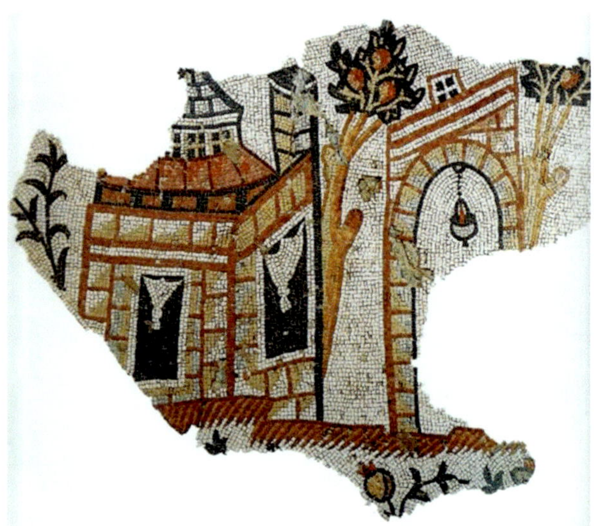

Figure 9. Representation of suspended glass oil lamps, i.e., glass *kandelai*. Mosaic from St. John the Baptist, Jerash Piccirillo *et al.* 1993, 274.

[26] Gerolymatoy 2001, 348, 355.
[27] Hanotte 2008. See, also Hensen 2009, fig. 6.
[28] Flutsch and Rey 2006, 59.
[29] See also, Piccirillo 1993, 274, fig. 505.

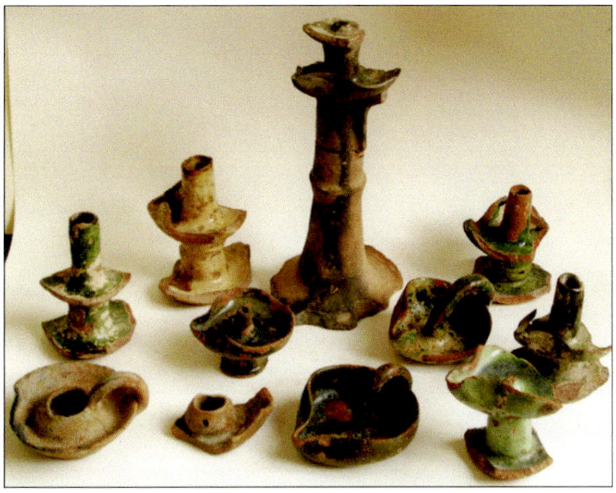

Figure 10. Byzantine glazed Wheel-made Lamps, Thessaloniki, Museum of Byzantine Culture.

of course, the combination of such factors may well have been different in different regions.

Finally, another significant turning point in the history of clay lamps was the evolution from unglazed wheel-made lamps[30] to glazed wheel-made lamps (Figure 10) –a development which occurred during the final years of the ancient world (7th century).[31] This practice –common in Western Europe[32] as well– was to continue during later periods, lasting in fact until the 19th century.

Glazed, wheel-made clay lamps can be classified into the following types, on the basis of the information available to date:[33]

1. Lamps with a flat, abrupt, untrimmed base and a hemispherical top.[34]
2. The next type is the same as that of the previous one, but the hemispherical top is smaller and the walls of this smaller oil bowl are almost vertical, with a hole at the base through which the wick went through to the nozzle, where it burned.[35] The handle would probably have been opposite the hole and the nozzle.
3. Lamps with a single oil bowl and a flat, abrupt, untrimmed base.[36]

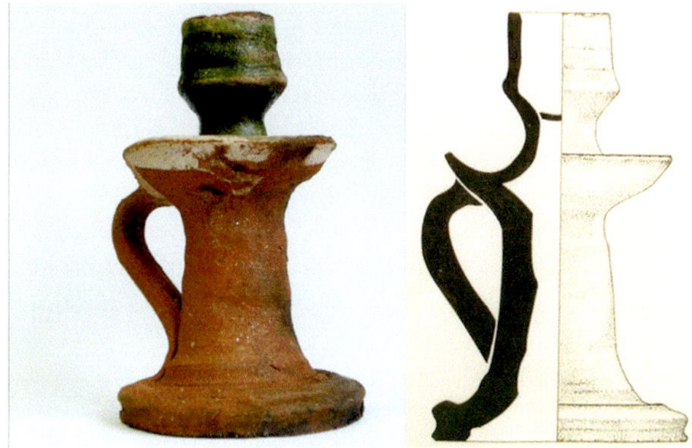

Figure 11. Wheel-made clay candlestand, Post Byzantine Period, Thessaloniki, Museum of Byzantine Culture.

4. Lamps with a single oil bowl and a stem.[37]
5. Lamps with two receptacles, one being an oil bowl and the other, at the base, serving to catch spilt oil.[38]
6. Lamps with two oil bowls, the lower one being at the base of the lamp.
7. Lamps with two oil bowls and a stem,[39] either without a handle or with a handle attached to the rim of the upper oil bowl or, at the other end, to the rim of the lower oil bowl or the stem.
8. Lamps with three receptacles, two being oil bowls and the lowest one, at the base, serving to catch spilt oil.[40]

Candlestands

Also, there is variation among the clay glazed wheel-made lamps which take the form of glazed, wheel-made candlesticks.[41]

9. Some candlestands with a base and a stem[42] are similar in shape to the lamps (Figure 11). However, in this case the stem terminates in a candle-socket and the dish below was intended to collect the dripping wax.[43]

[30] Yangaki 2005, fig. 67, pl. XVII.

[31] According to new data, even some of the types identified by Broneer (Broneer 1930, 125, type XXXV, pl. XXIV, nos. 1518, 1519), which were considered a reference until recently for Middle Byzantine period lamps, have been rightly revised and dated to the 7th -8th century.

[32] For glazed wheel-made lamps of Continental Europe, see Manser 2007; Frey 2007a; Frey 2007b; Frey 2007c.

[33] For more details about Byzantine glazed wheel-made lamps, see Motsianos 2003 and Motsianos 2005. For types 1-9, see Motsianos 2003, 249, pls 115-116, figs 2-12.

[34] Broneer 1930, 125, type XXXV, pl. XXIV, no. 1519; Orssaud and Sodini 1995, fig. 1; Djuric 1995, 104-6, cat. nos. C 340-4, C 346-8; François *et al.* 2003, 333, fig. 9.

[35] Broneer 1930, 125, pl. XXIV, no. 1530; Orssaud and Sodini 1995, fig. 5; Djuric 1995, 105-7, cat. nos. C 345, C 349-53; Filadelfeus 1924, 46, nos. 41-3.

[36] Papanikola-Bakirtzi 1996, 200, cat. nos. 127–30, pl. LXXI, fig. XVII:10.

[37] Broneer 1930, type XXXVI, 125, pl. XXIV, nos. 1548, 1555; Gourgiotis 1992, 42; Vavylopoyloy-Haritonidoy 1988, 196, nos. 68, 114, fig. 10; Keramopoyloy 1929, 127–8, figs 4b, 4e; Papanikola-Bakirtzi 1996, 137, no. 162, pl. XXIX, fig. IX:13.

[38] Bakirtzis 1980, 148, fig. 16, pl. 9, 12; Keramopoyloy 1929, 127-8, fig. 4a; Gourgiotis 1992, 42; Papanikola-Bakirtzi 2002, 302, cat. no. 327; Papanikola-Bakirtzi 1996, 137, no. 163, Pl. XXIX, fig. IX :14.

[39] Keramopoyloy 1929, 127-8, fig. 4g; Gourgiotis 1992, 42; Papanikola-Bakirtzi 2002, 302-3, cat. nos. 328-30; Frantz 1938, 449, 453, 463, 465, figs. 10, 14, 27, 29.

[40] Vavylopoyloy-Haritonidoy 1988, 196, no. 70, fig. 10; Papanikola-Bakirtzi 1996, 137, no. 164, pl. XXIX, fig. IX: 15.

[41] It should be noted that the remark 'Even though literary texts continue to mention lamps through the entire Byzantine period, archeological evidence shows that clay lamps practically disappeared after the 7th century they were replaced by candles. Certainly, palaces and rich houses were brightly illuminated, especially during banquets' (Kazhdan 1991, vol. 2, s.v. Lighting, Ecclesiastical, 1227-1128) is only valid with respect to the fact that sources continue to mention lamps throughout the Byzantine period. However, it is not true that clay lamps were replaced by candles.

[42] Gourgiotis 1992, 42.

[43] Wunderlich 2003, 253; Gourgiotis 1992, 42.

In many cases, the dating of glazed, wheel-made clay lamps in the Late Byzantine and even Post Byzantine periods has conveyed the impression that there was a large interval between the period in which the mold-made clay lamps were disappearing and the first appearance of glazed, wheel-made lamps. However, I consider that more methodical study of this material in the years to come may lead us to the conclusion that wheel-made lamps were produced without interruption from the earliest years of Byzantium until the Post Byzantine period.[44]

Metal lighting devices

Metal lighting devices could be categorized into four groups: metal lamps with a nozzle[45] (Figure 3), open metal hanging-lamps (*kandelai*)[46] (Figure 12), candelabra[47] (Figure 13) and *polykandela*[48] (Figure 8), which are present throughout the Early Christian and Byzantine period. This is particularly important with respect to metal lamps with a nozzle, about which we have assumed that they were not produced after the 7th-8th century. Indeed, lamps with a nozzle are attested in the Middle Ages, both in metal and clay form. Moreover, one can see the direct and interactive relationship between metal and clay lamps. This relationship is obvious in the form, decoration and technique.[49]

Metal lamps with a nozzle

The luxurious closed metal lamps of the Roman era were still used during the Byzantine period. They were widely used up to the 7th-8th century and were, like their clay counterparts, adorned with symbols of the new religion. But from this point on, they might have been less frequent because they evolved, taking other forms -as far as we can tell from the evidence available to date. These changes are probably related to the preference for lighting using glass lamps over polykandela, as well as to a probable shift to the use of candles. If we attempt to establish a general typology, we can distinguish between metal lamps intended to be suspended (Figure 12)[50] and lamps manufactured to be placed on lamp stands (Figure 13).

More modest, utilitarian-styled lamps made of cheaper metals, usually iron, (Figure 14) were usually found in places where lighting was necessary even during daylight hours, such as mines, and where less extravagant and cost-effective solutions were preferred. These lamps were usually open in form. The fact that the material used for metal lamps, in contrast to clay lamps, could be recycled allows us to consider that this is one of the reasons why they survived in great numbers into the Middle and Late Byzantine periods. However, this view is less convincing if we reflect that, following the same reasoning,

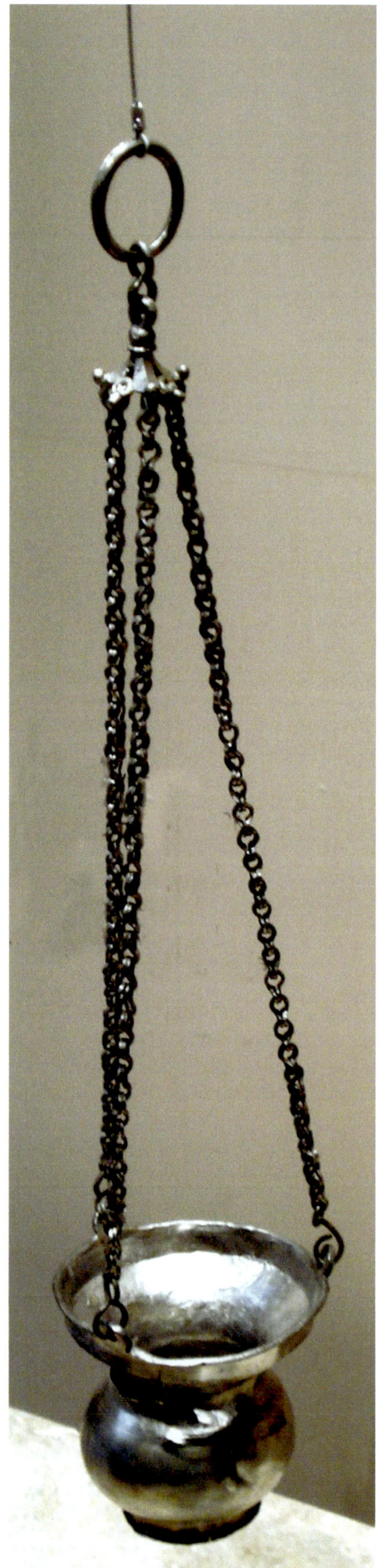

Figure 12. Silver hanging lamp (*kandela*), 7th century AD, Athens, Byzantine and Christian Museum, The Kratigos, Mytilene Treasure (Photo by the author).

[44] Laurent Chrzanovski during the works of the *3rd International of International Lychnological Association Round-Table*, Dark Ages? History and archaeology of lighting devices in Continental Europe, from late Antiquity to late Medieval Ages, *Historisches Museum Olten*, Olten (Switzerland), 24-29 September 2007 mentioned the wheel-made lamps found in Saint Denis. Their presence is constant from AD 600-1400.

[45] Xanthopoulou 1997, vol. 1, 6-55.

[46] Xanthopoulou 1997, vol. 1, 94-108.

[47] Xanthopoulou 1997, vol. 1, 66-72.

[48] Xanthopoulou 1997, vol. 1, 109-121.

[49] Xanthopoulou 1997, vol. 1, 231.

[50] For a photo see Wamser 2004, 228, no. 339.

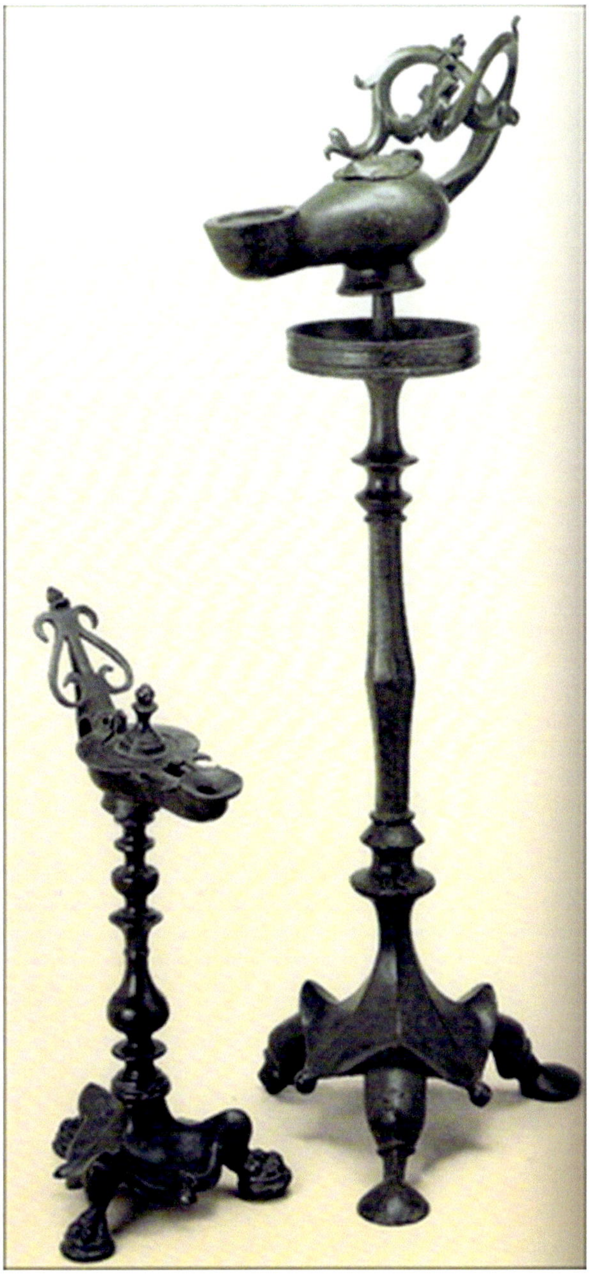

Figure 13. Metal lampstands (Papanikola-Bakirtzi 2002, 288, nos. 303, 304).

the metal lamps of the Early Christian period might have been recycled, as well. It therefore seems very likely that, as suggested above, metal lamps with a nozzle were rarer and took different forms in the years following the 8th century AD.

Open metal hanging-lamps (*kandelai*)[51]

These[52] represent a separate category, since –especially in the early centuries of the Byzantine era (4th-7th)– their shape was similar to that of the metal incense-burners (Figure 15),

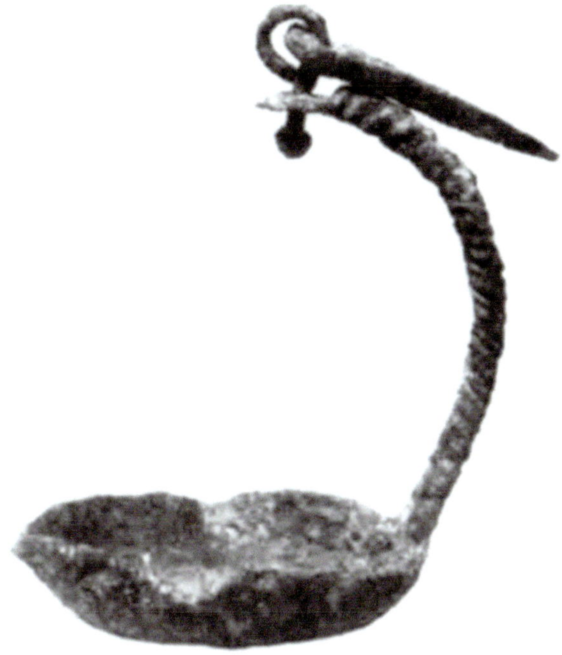

Figure 14. Mine lamp, Bulgaria.

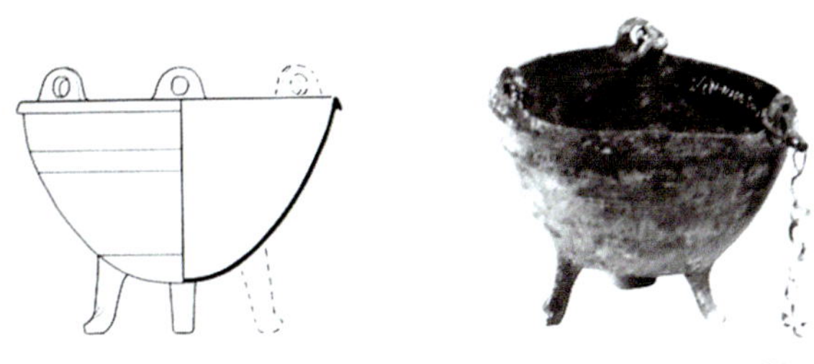

Figure 15. Incense-burner. Eleftherna in Crete (Themelis 2004b, 102, fig. 79a, b).

and further research is definitely necessary into their use and dating.

There are a number of cases in which there is obvious doubt as to whether the item presented in a publication or displayed in a museum is a metal incense-burner or a metal oil lamp, as, for example, in the case of an item from Eleftherna in Crete (Figure 17), or in various cases of items featured in the Malcove Collection catalogue.[53]

There are a number of observations and representations[54] (Figures 16)[55] from which we can set definite criteria, or a combination thereof, allowing us to determine with accuracy in each case whether these items were used as incense-

[51] For more details, see Motsianos 2007.
[52] For a photo Chrzanovski and Kaiser 2007, pl. 9a.

[53] Campbell, Malcove, 88-93, nos. 114-119.
[54] See, Piccirillo 1992, fig. 9; Weitzmann 1979, 578, no. 517;
[55] This representation is an excellent example as depicts at the same time two candlesticks, a hanging oil lamp and a censer.

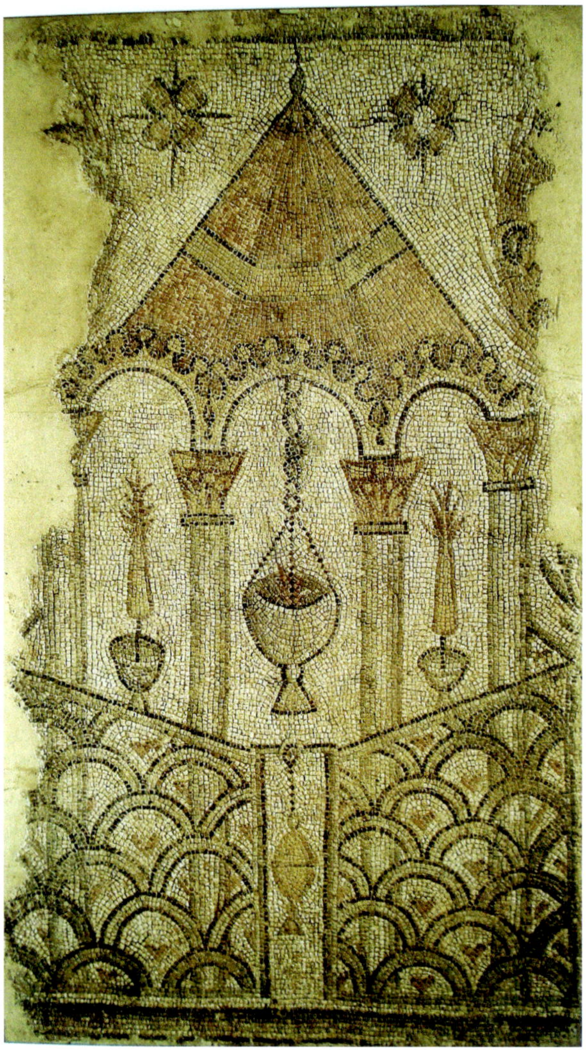

Figure 16. Mosaic from Syria with depiction of candlesticks, hanging oil lamp and a censer, Copenhagen, National Museum, 5th-6th century (Biddle 1999, 25, fig. 23).

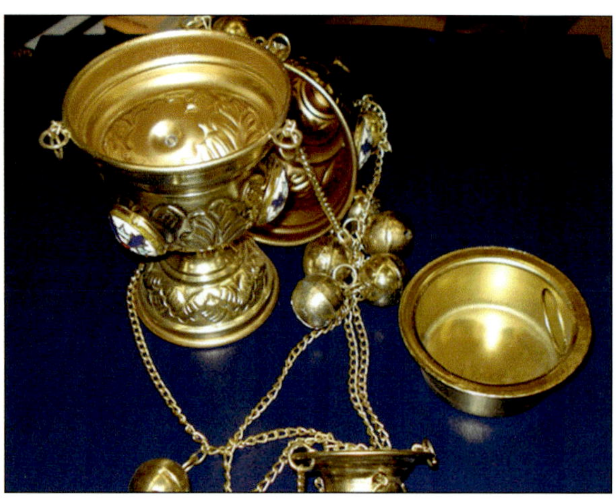

Figure 17. Contemporary censer with internal metal cup.

burners or hanging lamps. These observations are linked for the most part to details of the items' manufacture, such as:

1. The degree of thickness or thinness of the sides, in association with the material used in the manufacture of the items;
2. The presence of a glass or metal internal cup.[56] This difference can be observed even today (Figure 17);
3. The deep or shallow body;
4. The reinforced or otherwise hanging chains;
5. Whether the body of the item is perforated or not;
6. The shape of the base;
7. The presence of an inscription[57] (Figure 18) which gives an indication about the possible use of the vessel;
8. The absence of a base or feet which are indications that the item is a *kandela*.

These criteria can be combined with a study of the fuel deposits,[58] in some cases, to establish whether the items were used as incense-burners or oil lamps, thereby providing a yardstick for identifying other items.

Openwork silver and copper lamps[59] were employed in churches along with lamps of solid metal.[60] From the 12th-13th centuries onwards it becomes easier to distinguish between the two, since both oil lamps and incense-burners acquired a more distinctive shape.

Candelabra (Figures 13, 16)[61]

Regarding candelabra the tendencies of the Roman period continued. Their use is associated, to a great extent, directly with metal lamps, and indirectly with earthenware lamps. The candelabrum from the Kaiseraugst treasure[62] dates to the period AD 294-350 and could be described as the link between candlesticks of the Roman period and those of the Early Christian period.[63] It is especially reminiscent of the candlesticks in a sixth-century mosaic from Syria[64] (See Figure 16). The drip-pan is shaped like a cup.

Nevertheless, the term 'candelabra' covers both lampstands and candlesticks. In cases where the metal support is coupled with a matching metal lamp and the two have been found together, it is easy to determine that the object is a lampstand. This is confirmed by this miniature in a Syriac Gospel written in the Monastery of St John at Zagba in 586.[65]

This distinction can most definitely be associated with the use of lamps burning liquid fuel in every separate region. It

[56] See Evans 2001, 38.
[57] See also Stiegemann 2001, 146, no. 148.
[58] See Buckton 1994, 105, no. 113 (b). Analysis of the deposit within the bowl shows it to consist of ash, metal corrosion products and traces of gum resin. The resin may be derived from a typical incense material such as *myrrh* or *olibanum*.
[59] See, for example, Ševčenco and Pattterson-Ševčenco 1984, fig. 16; Papadopoyloy 2000, 94, no. 154.
[60] Kazhdan 1991, vol. 2, s.v. Lighting, Ecclesiastical, 1227-1128.
[61] For more details, see Motsianos 2008.
[62] For a photo see Chrzanovski 2003, 33, fig. 27-28.
[63] Mango1982, 255.
[64] Biddle 1999, 25, fig. 23.
[65] Dauterman-Maguire *et al.* 1989, 36, fig. II.

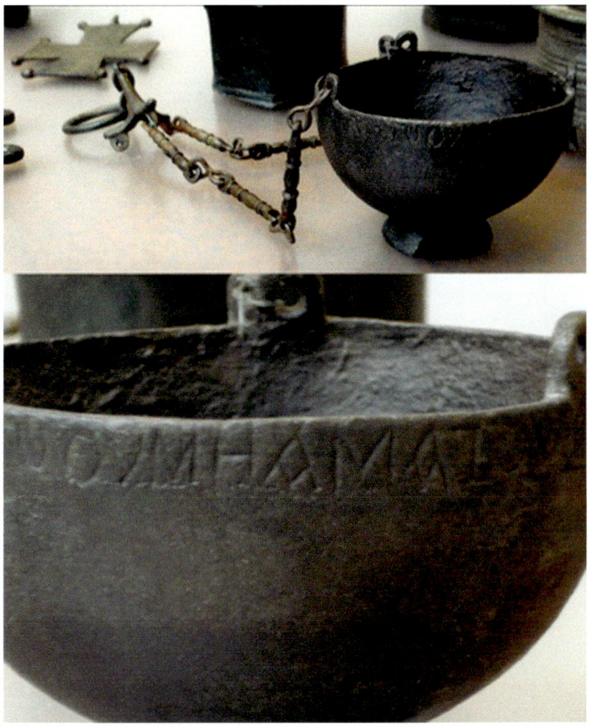

Figure 18. Censer with inscription declaring its use, 6th century AD. Athens, Byzantine and Christian Museum. Photo by the author.

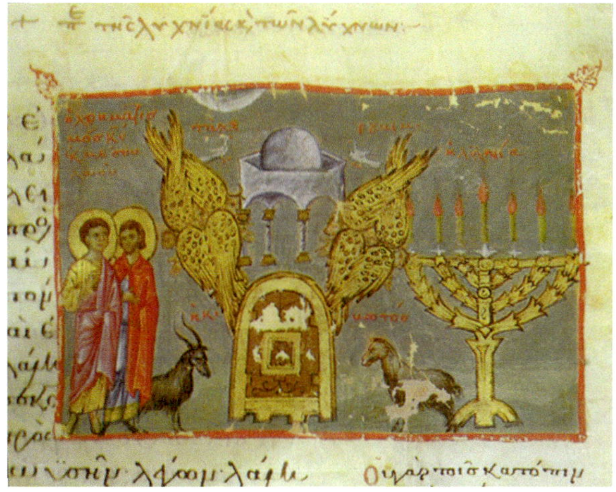

Figure 19. Miniature depicting a seven-branched candelabrum under the name *λυχνία* (*lychnia*). (Pelekanidis 1973, vol. 4, fig. 59).

could be said that, in those regions where there was no local production of olives, the preference was for candlesticks. However, it may be that this preference was associated with local liturgical needs and customs, since we find both depictions of candlesticks[66] and actual candlesticks[67] in regions with a tradition of olive production, such as Thessaloniki.

The candelabra were usually of silver or bronze and varied in height: the shorter ones[68] (Figure 13, left), between fifteen and fifty centimeters tall, were meant to stand on a table or other elevated surface, while the taller ones[69] (Figure 13, right), between eighty and a hundred thirty centimeters tall, stood on the floor.[70]

Apart from the candelabra with the three-legged base and long shaft, the seven-branched candelabrum, known as a *menorah*,[71] also continued to be used or at least to be depicted in the Early Christian period and later, both by Christians and Jews.[72] This menorah is depicted in a ninth-century miniature in a Christian context and is called *λυχνία* (lychnia) (Figure 19).

From the 9th century onward, ecclesiastical lighting increasingly relied on candles. Floor candelabra in pairs, *μανουάλια* (manoualia), were employed in front of votive icons, sometimes equipped with disks with extra candlesticks for the major feasts. Metal strips supplied with candlesticks were employed above the *Templon* epistyle and Icon Frames.[73] Candelabra were usually placed in front of some venerated icons and consisted of a candlestick with an adjustable disk with six, seven, eight or twelve candleholders. Most of the candelabra recorded in inventories were bronze with the exception of a few iron ones. However, candelabra of jasper and rock-crystal are listed in an inventory of Hagia Sophia of Constantinople, dated to 1396. Perhaps the most elaborate examples of this type are the bronze candelabra 'of the Amalfitans' in the Grand Lavra Monastery at Mt. Athos.[74]

I should also underline that candlesticks continued to be similarly depicted in manuscript miniatures, wall paintings, and wooden icons in the Middle and Late Byzantine periods and into the Palaeologan era.[75] However, in many cases we cannot be certain whether they are made of metal, wood, stone,[76] or clay.

Furthermore, although candlesticks are frequently depicted in paintings, there are still very few extant examples of either candlesticks or lampstands from the Late Byzantine period.[77]

In the Post Byzantine period, apart from metal candlesticks, wooden ones were also extremely common.[78] Of course, wooden candlesticks similar to those of the Post Byzantine

[66] Motsianos 2011, figs 31, 32i-ii.
[67] Motsianos and Bintsi 2011, 170, no. 61.
[68] Dauterman-Maguire *et al.* 1989, 70, no. 16.
[69] Dauterman-Maguire *et al.* 1989, 71, no. 17; Ross 1962, 37-38, no. 39.
[70] Dauterman-Maguire *et al.* 1989, 58; Ross 1962, 37-38, no. 39.
[71] For the menorah in general, see Israeli 1999.
[72] Pelekanidēs 1973, vol. 3, 149, fig. 229; Israeli 1999, 115, fig. 5-6.

[73] Kazhdan 1991, vol. 2, s.v. Lighting, Ecclesiastical, 1227-1128; Bouras 1981, 480; Xanthopoulou 1997, 169-170.
[74] Bouras 1981, 480.
[75] Christou *et al.* 1991, fig. 5; After Byzantium 1996, no. 48 (detail); Papanikola-Bakirtzi 2002, 226-227.
[76] Tsigaridas 2003, 10-11.
[77] The City of Mystras 2001, fig. 106.
[78] *With Faith and Fantasy* 2004, 64-65, 68-69, 76-77, 102-103, nos. 2B, 3, 4B, 8B; *Ιερά Μητρόπολις Μόρφου 2000*, 411, no. 83; *Treasures of Mount Athos* 1997, 356; Ιστορίες θρησκευτικής πίστης στη Μάνη 2005, 140, 141, nos. 61, 62; For an equivalent in the occident, see Chrzanovski and Kaiser 2007, pl. 17.

period may well have existed in earlier times and disappeared, because wooden objects burn easily, disintegrate easily, and survive as excavation finds only under ideal conditions and in rare cases. Rare are the examples of wooden candelabra survived in medieval Western Europe.[79]

The Post Byzantine period saw the appearance of wheel-made clay candleholders (Figure 11) similar in shape to the wheel-made lamps.[80]

One can say that supports for lighting devices were ever present, as they were always needed to raise the devices. They ensured better illumination from higher up, without inconvenient shadows, and also helped to prevent mice from drinking the oil or knocking the lamps down. Their shape further contributed to this purpose: the baluster-shaped stand must have made it difficult for mice to climb up to the disk on which the lamp stood. The objects themselves, their representations, and the information provided by the sources all indicate that candelabra of this kind served a dual purpose, supporting both lamps and candles, especially in the Early Christian period. The very thin, long, round spike shows that they were used as candlesticks (Figure 20), while the discovery of the metal supports together with the lamps shows that they were used as lampstands. From the seventh to eighth century onwards they were predominantly used as candlesticks, rather than as lampstands, at least as far as we can judge from representations. This observation is also related to the major issue of preference for the use of oil lamps or candles, and provides an excellent occasion for a discussion of the whole issue of the use of candles in Byzantium.[81]

It is a fact that the use of candles for liturgical and processional purposes can be attested well before the 7th century.[82] In Western Europe, when imports of oil from Africa and elsewhere were disrupted in the Early Middle Ages, there was a general shift to the candle, either of wax or tallow: hence the appearance of the *cerarii* (Latin words for candlemakers) who were unknown before the end of the Merovingian period. Could the same have happened in Byzantium? It can hardly be coincidental that the profession of *κηρουλάριος* (candlemaker) is first attested in the 7th century.[83]

In relation to the same issue there is a whole series of other questions which require answers; for example: When was the use of candles adopted in each region? At what point did the use of candles begin to supersede that of olive oil? At what period did the two practices co-exist? What was the cost of each method? In what cases and for what reasons was a candle preferred over an olive oil lamp?

Glass oil lamps/glass *kandelai*[84]

The first use of glass lamps can be dated to the end of the Roman and beginning of the Byzantine eras. From the western parts of the empire, we do know of some very rare glass examples that imitated the wheel-made clay lamps, which are generally dated to the second to third centuries AD.[85]

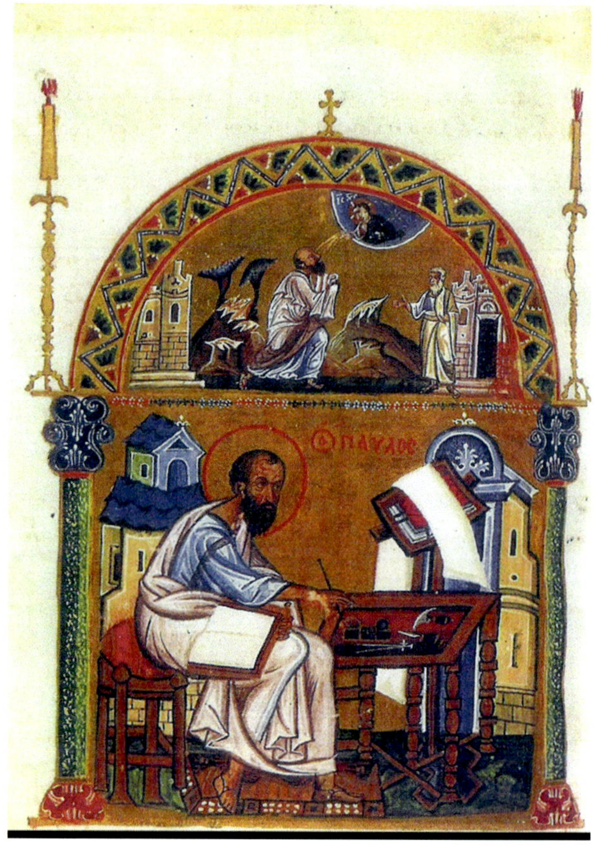

Figure 20. Depiction of candlesticks. The Conversion of St Paul, fol. 312v, MS Auct. T infra 1.10 (Misc. 136), 1125-1150 AD, Oxford, Bodleian Library (Galavaris 1995, 242, no. 139).

However, it appears that, as early as the 2nd-3rd centuries AD, open-form drinking vessels were used as oil lamps. There are certain factors which would have led to a preference for glass oil lamps, namely:

1. The fact that the glass surrounding would not limit the light shed, as was the case with earthenware and metal lamps, but would actually increase the degree of illumination.
2. Filling the glass *kandela* partially with water transformed the hollow surface into a magnifying glass.[86]
3. Provided illumination from above without creating shadows.
4. It is not necessary to frequently replenish the supply of fuel, since the oil container is usually capacious.
5. In some types of glass *kandelai*, the wick probably follows the descent of the level of fuel as it burns.

[79] Chrzanovski and Kaiser 2007, pl. 17.
[80] Motsianos 2005, 249.
[81] On this subject see Sorochan 2002, 113.
[82] Mango 1982, 256.
[83] Mango 1982, 255.
[84] Glass oil lamps are mentioned here because they led to the development of the *polykandela* and associated chains and hooks that were used to suspend them.
[85] Beretta, Pasquale 2006, 286-288, nos. 3.9, 3.10, 3.11.
[86] This can be seen very clearly in later examples where a glass ball is used to increase the artificial light (19th century AD). See Chrzanovski *et al.* 2012, 204, figs 246, 248, 249.

All these advantages easily outweigh the disadvantage of the high cost of glass *kandelai* and their fragility. Where these considerations were not an obstacle,[87] there was a clear preference for such lamps.

A study of glass artifacts in Thessaloniki and the surrounding region provides valuable information on glass lighting devices in Roman and early Christian times.[88] Similar studies of other areas will gradually help us to complete the mosaic of our knowledge of lighting devices across broader regions and periods. Glass *kandelai* -albeit in fragmented form- are very frequently found and can be grouped in the following categories:[89]

1. Conical goblets
2. Stemmed goblets: i. without handles, ii. with handles
3. Hemispherical bowls: i. without handles, ii. with handles
4. Calyx-shaped with handles
5. With a high cylindrical base
6. Cylindrical with a small knob
7. Cylindrical with a large knob

In general, it can be said that glass *kandelai*, which were often drinking vessels or vessels that evolved, slightly altered, from drinking vessels, can be found in quite a wide variety of shapes in the Late Roman and Early Byzantine periods, meeting different lighting needs and representing different forms of lighting devices, either free-standing or suspended, alone or in groups of several examples burning together. Whatever the case, they were always a solution that related to specific spaces, rituals, and social and economic status.[90]

Another category of lamp which until now has not been found amongst the excavation data of the Thessaloniki area includes lamps with a glass cylinder fused, in a separate process, to the interior floor of the vessel (Figure 29). This glass cylinder is believed to have been designed to hold the wick of the lamp. Examples of this type of glass kandelai have been found in Egypt, the East Mediterranean region and, recently, in Athens.[91]

Finally, a special category of glass vessel includes those called *θήλαστρα* (thilastra),[92] or suckling devices, by analogy with the earthenware vessels used, according to ancient sources, to feed babies. It is certain, however, that the glass vessels were not used for such a purpose. In fact they were most probably used for filling oil lamps -glass and clay and metal lamps- a process greatly facilitated by their low and conical spout. A conventional name as 'oil-jugs' has been proposed for these vessels.[93] I personally suggest that these specific items were used mainly for filling closed earthenware and metal lamps -since it is these which had a narrow aperture for receiving the fuel.

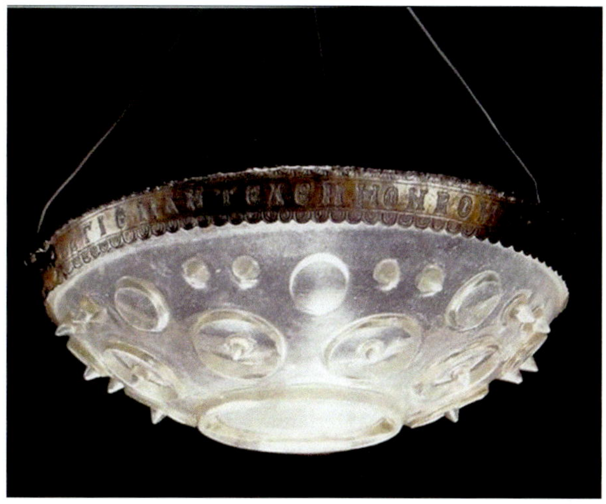

Figure 21. Glass hanging *kandela* with a silver mount, Venice, Treasury of San Marco, Provenance: Constantinople, 11th century AD (Cassanelli 2000, 224).

In the Middle Byzantine period (8th-12th century), not only the ordinary lamps but also the most luxurious ones were made of glass (Figure 21).[94] The use of glass *kandelai* has been continues even into the Late Byzantine Period (13th-15th century AD).[95]

From the Early Christian period the use of this new means of illumination led to the parallel production of *polykandela* for glass *kandelai* (Figure 8), as well as metal parts, chains and hooks used to suspend them.[96] Thus their appearance in this role must be dated later than the appearance of glass *kandelai*.

Polykandela (chandeliers)

The placing of glass *kandelai* in the *polykandela*[97] allowed for the production of abundant light. *Polykandela* with glass *kandelai* were the dominant lighting devices before the 8th century. Originally, *polykandela* included mainly oil lamps, but eventually a mixed type developed, combining oil lamps and candles.[98] The earliest types are crown-shaped with dolphin-brackets supporting glass *kandelai*. Three other types of *polykandela*, like the ones found in the Sion Treasure, were circular,[99] cross-shaped,[100] or in the form of a rectangular tray.[101] In Hagia Sophia, Constantinople, cross and disk-shaped *polykandela* have been recorded, along with boat-shaped lamps and tree-shaped chandeliers. A variation of these tree-

[87] Antonaras 2006, 224-225.
[88] Antonaras 2009.
[89] The first typology for glass oil lamps was proposed by Crowfoot and Harden 1931. Among more recent studies, Uboldi's work is one of the more detailed (Uboldi 1995). For Dalmatian coast findings, see Chevalier 1997-98. For a summary description of the Thessaloniki area findings, see Antonaras 2008, Antonaras 2009 and Antonaras 2010.
[90] Antonaras 2008.
[91] For this subject see in this volume Motsianos, Difficulties and pre-conditions for utilitizing glass in lighting in Byzantium.
[92] Isíngs 1957,118.
[93] Themelis 2004a, 720-721.
[94] Bouras 1981, 479.
[95] See, for example, Parani 2005, figs 5, 6.
[96] For the terminology of these elements of suspension see Bouras 1981, 480-481 and Xanthopoulou 1997, 174-175.
[97] For a photo see Bénazeth and Rutschowscaya 2000, 183, no. 193.
[98] Bouras 1981, 480.
[99] Durand 1999, 44.
[100] Ševčenco and Pattterson-Ševčenco 1984, fig. 15.
[101] Kazhdan 1991, vol. 2, s.v. Lighting, Ecclesiastical, 1227. For a photo see Mundell Mango 2003, 69, fig. 9.13.

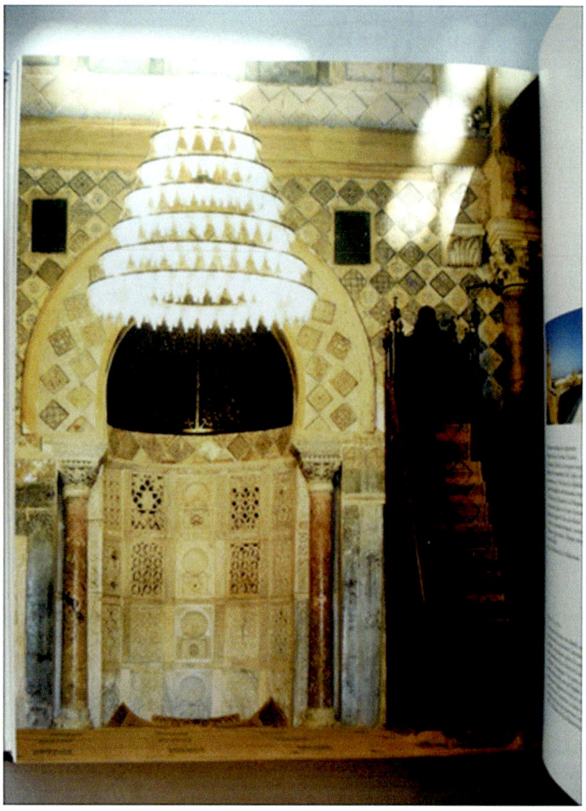

Figure 22. Interior of a contemporary mosque (Carbonelli Esteller *et al.* 2001, 236).

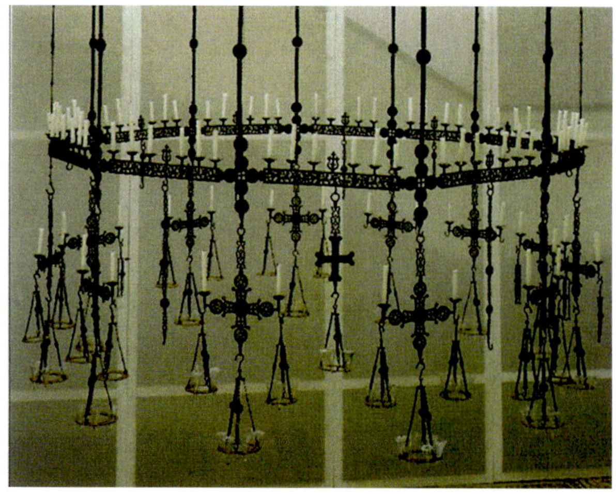

Figure 23. Byzantine *choros*, 13th-14th century AD (Stiegemann 2001, fig. 7).

shaped chandeliers, as described by Paul Silentiarios,[102] can still be seen today (Figure 22) –in another form and using a different source of energy– in Muslim mosques.

***Choroi* (Figure 23)**

These represent the most sophisticated lighting devices in the Middle Byzantine period. They consisted of a polygonal structure suspended from the dome of a church and carrying several lamps and small *polykandela* (chandeliers)[103] and candles, and were specially designed to fit domed churches. Though the term clearly recalls Silentiarios' description of the choir of lamps hanging from the dome cornice of Hagia Sophia in Constantinople, the *choros* is not attested before the twelfth century.[104] Similar devices are still in use in Mt. Athos.[105]

Conclusions

Studies of lighting devices from the Byzantine period to the present day have largely focused on the lamps of the Early Christian period, both the earthenware lamps manufactured using a mold and those made of metal. It is worth noting the research which has been conducted into the use of glass *kandelai* during the same periods –research which focuses mainly on grouping them into specific types and exploring their methods of manufacture. The approach taken by Laskarina Bouras, in 'Byzantine Lighting Devices',[106] has been of seminal influence to date in the study of lighting devices of the middle Byzantine period, and has provided the basis for a distinction between devices intended for secular use and those intended for use in church. The work of Maria Xanthopoulou contributed to the study of metal lamps of the Early Byzantine and Byzantine periods.[107] In relation to ecclesiastical use, a number of monastic Typika provide explicit instructions for the lavish illumination of churches on major feasts and the anniversaries of the deaths of the founders.[108]

Also worthy of further investigation is the question of how local economic conditions may have affected the form and use of lighting devices, and the question of what exactly was the status and development of other means of lighting such as lanterns, torches, beacons, flambeaux, etc. and objects such as the ones which were presented in the Exhibition at Olten[109] and were published in the Exhibition Catalogue.[110]

The Round Table at Olten offered us an excellent opportunity to explore whether, and to what extent, the inhabitants of the two parts of the Roman Empire, in the years following AD 395, pursued different paths in their choice of lighting devices, or whether the influential shared Roman tradition led them to make similar choices.

[102] Kazhdan 1991, vol. 2, s.v. Lighting, Ecclesiastical, 1227.
[103] Thomas *et al.* 2000, 1680.
[104] Bouras 1981, 480.
[105] Bouras 1981, 481.

[106] Bouras 1981, 479-491.
[107] Xanthopoulou 1997.
[108] For the monastic Typika, see Thomas *et al.* 2000; also Xanthopoulou 1997, 223-229, Annexe V. 4: Passages relative aux luminaires et a l'éclairage dans les typika médiobyzantins.
[109] Dark ages? Licht im Mittelalter/L'éclairage au moyen âge, Historiches Museum Olten 2007.
[110] Chrzanovski and Kaiser 2007.

Bibliography

Abadie-Reynal, C. and Sodini, J. P. 1992. *La céramique paléochrétienne de Thasos: Aliki, Delkos, fouilles anciennes*, Études Thassiennes XIII. Paris: Boccard.

Alicu, D. 2007. L'évolution de la production de lampes dans la Dacie intra-carpatique, depuis l'Antiquité classique jusqu' au début du moyen âge, Communication at the 3rd International of International Lychnological Association Round-Table, Dark Ages? History and archaeology of lighting devices in Continental Europe, from late Antiquity to late Medieval Ages, Historisches Museum Olten, Olten (Switzerland), 24-29 September 2007.

Antonaras, A. 2006. Γυάλινα ρωμαϊκά και παλαιοχριστιανικά αγγεία στη Θεσσαλονίκη και την περιοχή της. Unpublished PhD dissertation, University of Ioannina.

Antonaras, A. 2008. *Glass lamps from Late Roman Thessalonica*, in C.-A. Roman, and N. Gudea (eds) *Lychnological Acts 2. Trade and Local Production of Lamps from the Prehistory until the Middle Age. Acts of 2nd International Congress on Ancient and Middle Age Lighting Devices, Zalău/Cluj-Napoca, 13th-18th of May 2006*, 23-30. Cluj-Napoca: Editura Mega.

Antonaras, A. 2009. *Ρωμαϊκή και παλαιοχριστιανική Υαλουργία. Αγγεία από τη Θεσσαλονίκη και την περιοχή της* (Roman and Early Christian Glassworking. Vessels from Thessaloniki and its region). Athens: I. Sideris.

Antonaras, A. 2010. Glassware in Late Antique Thessalonikē, in L. Nasrallah, Ch. Bakirtzis and S. J. Friesen (eds) *From Roman to Early Christian Thessalonikē: Studies in Religion and Archaeology*, 301-334. Cambridge, Mass.: Harvard University Press.

Bailey, D. M. 1975. *A Catalogue of the Lamps in the British Museum, I, Greek, Hellenistic, and Early Roman Pottery Lamps*. London: Trustees of the British Museum Publications.

Bailey, D. M. 1988. *A Catalogue of the Lamps in the British Museum, III. Roman Provincial Lamps*. London: Trustees of the British Museum Publications.

Bakirtzis, Ch. 1980. Didymoteichon: un centre de céramique post-byzantine. *Balkan Studies* 21, 147-153.

Bénazeth, D. and M. H. Rutschowscaya (eds) 2000. *L'art copte en Égypte. 2000 ans de christianisme*, Exhibition Catalogue, Paris, Institut du monde arabe, May-September 2000. Paris: éditions Gallimard.

Beretta, M. and G. di Pasquale (eds) 2006. *Le verre dans l'empire Romain. Arts et sciences*, Exhibition Catalogue, Paris, Cite des sciences et de l'industrie, 31 January-27 August 2006. Florence: Giunti.

Biddle, M. 1999. *The Tomb of Christ*. Stroud: Sutton Publishing.

Bouras, L. 1981. Byzantine Lighting Devices, Akten II/3 der XVI. Internationaler Byzantinistenkongress, Vienna 4-9 October 1981, in *Jahrbuch der Österreichischen Byzantinistik* 32/3, 479-491.

Bouras, L. 1991. Three Byzantine Bronze Candelabra from the Grand Lavra Monastery and Saint Catherine's Monastery in Sinai. *Δελτίον Χριστιανικής Αρχαιολογικής Εταιρείας*, period Δ΄, vol. ΙΕ΄, 19-26.

Broneer, O. 1930. *Corinth vol. IV, II: Terracotta Lamps*. Cambridge, Mass. 1930.

Broneer, O. 1977. *Isthmia III: Terracotta Lamps*. Princeton, N.J.: American School of Classical Studies at Athens.

Bruneau, P. 1971. Lampes corinthiennes. *Bulletin de Correspondance Hellénique XCV: 2*, 437-501.

Bruneau, P. 1977. Lampes corinthiennes (II). *Bulletin de Correspondance Hellénique* 101:1, 249-295.

Buckton, D. (ed.) 1994. *Byzantium. Treasures of Byzantine Art and Culture*, Exhibition Catalogue, British Museum. London: British Museum Press.

Campbell, S. D. (ed.) 1985. *The Malcove Collection. A Catalogue of the Objects in the University of Toronto*. Toronto; Buffalo: University of Toronto Press.

Carbonell i Esteller, E., R. Cassanelli, D. Arasse and Ch. De Montclos (eds) 2001. *De Mahomet à Charlemagne: la Méditerranée et l'art*. Paris: Citadelles & Mazenod.

Cassanelli, R. (ed.) 2000. *La méditerranée des croisades*. Paris: Citadelles & Mazenod.

Chevalier, P. 1997-1998. Les luminaires paléochrétiens de Dalmatie romaine. *Vjesnik za arheologiju i historiju dalmatinsku* 90-91, 161-199.

Chrzanovski, L. 2003. *Lumière! L'éclairage dans l'antiquité*. exhibition catalogue, Musée romain de Nyon, 15 May 2003-30 April 2004. Milan: Edizioni Et.

Chrzanovski, L. 2006. L'illuminazione artificiale nel mondo tardoantico. *Dacia* 50, 361-388.

Chrzanovski, L. and P. Kaiser (eds) 2007. *Dark ages? Licht im Mittelalter/L'éclairage au moyen âge*, Exhibition Catalogue, Historiches Museum Olten. Milano: Edizioni Et.

Chrzanovski, L., M. Martiniani-Reber and J.-Y. Marin (eds) 2012. *À la tombée de la nuit: Art et histoire de l'éclairage;* [à l'occasion de l'exposition ... présentée au Musée d'Art et d'Histoire de Genève du 24 février au 19 août 2012]. Milan: 5 Continents.

Chrzanovski, L., Sandoz, C., Piccamiglio, A. and Rémy B. 2000. *Lampes et monnaies romaines de Carthage. Catalogue de la collection Dr. A. Devant. Musée de Viuz-Faverges*. Faverges: Amis de Viuz-Faverges.

Christou *et al.* 1991: *Οι θησαυροί του Αγίου 'Ορους*, σειρά Ά: Εικονογραφημένα Χειρόγραφα (παραστάσεις, επίτιτλα, αρχικά γράμματα), vol. Δ΄: Μ. Βατοπειδίου, Μ. Ζωγράφου, Μ. Σταυρονικήτα, Μ. Ξενοφώντος. Athens: Ekdotike Athenon.

Crowfoot, G. M. and Harden, D. B. 1931. Early Byzantine and Later Glass Lamps. *Journal of Egyptian Archaeology* XVII, 196-208.

Dauterman-Maguire, E., Maguire, H. and Duncan-Flowers M. J. 1989. *Art and Holy Powers in the Early Christian House*, Exhibition Catalogue, Kranert Art Museum of the University of Illinois at Urbana-Champaign, 25 August - 1 October 1989 and The Kelsey Museum of Archaeology of Michigan in Ann Arbor, 27 October 1989-29 April 1990. Urbana and Chicago: University of Illinois Press.

Djuric, S. 1995. *Ancient Lamps from the Mediterranean, The Anawati Collection*. Toronto: Eika.

Durand, J. 1999. *L' Art Byzantin*. Paris: Terrail.

Duval, N. and V. Popović (eds) 1984, *Caričin Grad I. Les basiliques B et J de Caričin Grad, Quatre objets remarquables de Caričin Grad, Le tresor de Hajdučka Vodenica*. Belgrade: Institut archéologique de Belgrade; Rome: École française de Rome.

Evans, H. C. 2001. *The Arts of Byzantium*. New York: Metropolitan Museum of Art.

Fehérvári, G. 2000. *Ceramics of the Islamic World in the Tareq Rajab Museum, chapter six: Glazed Lamps and Lanterns (Tenth to Thirteenth centuries)*. London; New York: I. B. Tauris.

Filadelfeus, A. 1924. Βυζαντινά αγγεία εκ των ανασκαφών Παλαιάς Κορίνθου. *Δελτίον Χριστιανικής Αρχαιολογικής Εταιρείας*, period Β΄- vol. Α΄, 22-47.

Flutsch, L. and Rey, T. 2006. *Unearthing the past: From the beginning to 800 AD: a guide to the exhibition*. Zurich: Swiss National Museum.

François, V., Nicolaïdes, A., Vallauri, L. and Waksman Y. 2003. Premiers éléments pour une caractérisation des productions de céramiques de Beyrouth entre domination franque et mamelouke, in Ch. Bakirtzis (ed.) *Actes de VIIe Congrès International sur la Céramique Médiévale en Méditerranée*, Thessaloniki, 11-16 October 1999, 325-340. Athens: Caisse des Recettes Archéologiques.

Frey, J. 2007a. *Lumières et éclairage sacré, dans les sources écrites et illustrées*, in L. Chrzanovski and P. Kaiser (eds) *Dark ages? Licht im Mittelalter/L'éclairage au moyen âge*, Exhibition Catalogue, Historiches Museum Olten, 81-85, figs in pp.76-80. Milano: Edizioni Et.

Frey, J. 2007b. *Lampes dans le canton de Berne*, in L. Chrzanovski and P. Kaiser (eds) *Dark ages? Licht im Mittelalter/L'éclairage au moyen âge*, Exhibition Catalogue, Historiches Museum Olten, 287-292, figs in pp. 284-286. Milano: Edizioni Et.

Frey, P. 2007c. *Luminaire médiéval dans le canton d' Argovie*, in L. Chrzanovski and P. Kaiser (eds) *Dark ages? Licht im Mittelalter/L'éclairage au moyen âge*, Exhibition Catalogue, Historiches Museum Olten, 315-316, figs in pp. 305-314. Milano: Edizioni Et.

Galavaris, G. 1995. *Ζωγραφική βυζαντινών χειρογράφων*. Athens, Ekdotike Athenon.

Gerolymatoy, M. 2001. *Εμπορική δραστηριότητα κατά τους Σκοτεινούς Αιώνες του Βυζαντίου* (7th-9th), in *Οι Σκοτεινοί Αιώνες Του Βυζαντίου: (7ος-9ος αι.) = the Dark Centuries of Byzantium: (7th-9th century)*, 347-364. Athens: National Hellenic Research Foundation.

Gourgiotis, G. 1992. Εξέλιξη του πήλινου λύχνου από την παλαιοχριστιανική εποχή ως το τέλος των βυζαντινών χρόνων. *Αρχαιολογία* 43/2, 40-42.

Hanotte, A. 2008. Les collections du musée/site archéologique de Bavay conservent deux lampes témoignant d'une ingénieuse transformation, in C.-A. Roman and N. Gudea (eds) *Lychnological Acts 2. Trade and Local Production of Lamps from the Prehistory until the Middle Age. Acts of 2nd International Congress on Ancient and Middle Age Lighting Devices, Zalău/Cluj-Napoca, 13th-18th of May 2006*, 135-138. Cluj-Napoca: Editura Mega.

Hayes, J. W. 1992. *Excavations at Saraçhane in Istanbul, vol. 2: The pottery*. Princeton, N.J.: Princeton University Press; Washington, D.C.: Dumbarton Oaks Research Library and Collection.

Hensen, A. 2009. Öllampen Der Römischen Nekropole Von Heidelberg - Indikatoren Einer Energiekrise in Der Provinz, in J. Biel, J. Heiligmann and D. Krause (eds) *Landesarchäologie, Festschift Für Dieter Planck Zum 65. Geburtstag*, 425-441. Stuttgart: Theiss Konrad.

Ιερά Μητρόπολις Μόρφου, 2000 χρόνια Τέχνης και Αγιότητος. Nicosia 2000.

Isac, A. and Cr.-A. Roman (eds) 2006. *Lychnus et Lampas*, Exhibition Catalogue. Cluj-Napoca: Editura Mega.

Isings, C. 1957. *Roman Glass from dated finds*. Groningen; Djakarta: J. B. Wolters.

Israeli, Y. 1999. *In the Light of the Menorah, Story of a Symbol, From the Collection of the Israel Museum, Jerusalem*, Exhibition Catalogue. Jerusalem, The Israel Museum. Philadelphia: Jewish Publication Society.

Ιστορίες θρησκευτικής πίστης στη Μάνη 2005. Athens: Kapon Editions.

Kallipolitis, B. G. and Petrakos, B. H. 1963. Αίγινα. *Αρχαιολογικόν Δελτίον* 18, 51-52.

Karivieri, A. 1996. *The Athenian lamp industry in late antiquity*. Helsinki: Finnish Institute at Athens.

Kazhdan, A. P. (ed.) 1991. *The Oxford Dictionary of Byzantium*, Prepared at Dumbarton Oaks, vols. 1-3. New York-Oxford: Oxford University Press.

Keramopoyloy, A. D. 1926. Παλαιαί χριστιανικαί και βυζαντιακαί ταφαί εν Θήβαις. *Αρχαιολογικόν Δελτίον* 10, 124-136.

Kuzmanov, G. 2002. Roman and late antique lamps from Ratiaria. *Annuary of the Institute of Archaeology with Museum*, vol. II. Sofia 2002.

Lazaridis, P. 1965. Ανασκαφή Νέας Αγχιάλου. *Πρακτικά Αρχαιολογικής Εταιρείας* (1965), 10-23.

Macalister, S. R. A. 1912. *The Excavation of Gezer, 1902-1905 and 1907-1909*. London: J. Murray.

Mango, C. 1982. Addendum to the Report on Everyday Life, Akten II/1 der XVI. Internationaler Byzantinistenkongress, Vienna 4-9 October 1981, in *Jahrbuch der Österreichischen Byzantinistik* 32/1, 252-257.

Manser, J. 2007. Lampes médiévales provenant d'une décharge de céramique du bas moyen age a la Stadthofstrasse 16 à Lucerne, in L. Chrzanovski and P. Kaiser (eds) *Dark ages? Licht im Mittelalter/L'éclairage au moyen âge*, Exhibition Catalogue, Historiches Museum Olten, 263-264, figs in pp. 261-262. Milano: Edizioni Et.

Motsianos, I. 2003. Wheel-made Oil Lamps in the Museum of Byzantine Culture. *Museum of Byzantine Culture* 10, 58-79 (also in Greek and French).

Motsianos, I. 2005. *Some comments on Byzantine wheel-made lamps*, in L. Chrzanovski (ed.) *Lychnological Acts 1. Actes du 1er Congrès international d'études sur le luminaire antique (Nyon-Genève, 29.IX - 4.X.2003)*, 247-251. Montagnac: M. Mergoil.

Motsianos, I. 2007. Communication under the title: *Θυμιατίζοντας με κανδήλια; Προβληματισμοί για τον διαχωρισμό των μεταλλικών θυμιατηρίων και κανδηλιών της πρωτοβυζαντινής εποχής (4ος-8ος αι. μ.Χ.). Paper presented at the* 27th Symposium Symposium of Byzantine and Post-Byzantine Archaeology and Art organized by the Christian Archaeological Society, Athens, 11-13 May 2007.

Motsianos, I. 2008. Candlesticks, candlestands, and lampstands. Observations and questions on their use from the Early Christian to the Postbyzantine period, in C.-A. Roman and N. Gudea (eds) *Lychnological Acts 2. Trade and Local Production of Lamps from the Prehistory until the Middle Age. Acts of 2nd International Congress on Ancient and Middle Age Lighting Devices, Zalău/Cluj-Napoca, 13th-18th of May 2006*, 177-182. Cluj-Napoca: Editura Mega.

Motsianos, I. 2010. Ύστερη Αρχαιότητα: Το λυκόφως των κατασκευασμένων με μήτρα λυχναριών (Late Antiquity: the twilight of mouldmade lamps), in D. Papanikola-Bakirtzis and D. Koussoulakou (eds) *Κεραμική της Ύστερης Αρχαιότητας από τον Ελλαδικό χώρο*, Επιστημονική Συνάντηση, Θεσσαλονίκη 12-16 Νοεμβρίου 2006, 270-281. Thessaloniki: Archaeological Institute of Macedonian and Thracian Studies (in Greek with English abstract).

Motsianos, I. 2011. Artificial Lighting during Byzantine and Post-Byzantine Period, in I. Motsianos and E. Bintsi (eds) *Light on light, an illuminating story*, Exhibition Catalogue, 90-125. Thessaloniki: Folklife & Ethnological Museum of Macedonia-Thrace.

Motsianos, I. and E. Bintsi (eds) 2011. *Light on light, an illuminating story*, Exhibition Catalogue. Thessaloniki: Folklife & Ethnological Museum of Macedonia-Thrace.

Mundell-Mango, M. 2003. Three illuminating objects in the Lampsacus treasure, in C. Entwistle (ed.) *Through a Glass Brightly. Studies in Byzantine and Medieval Art and Archaeology Presented to David Buckton*, 64-75. Oxford: Oxbow Books.

Orssaud, D. and Sodini, J. P. 1995. Les lampes tournées de Qal'at Sem'an et leurs parallèles dans le bassin méditerranéen, in *La céramique médiévale en méditerranée, Actes du VIe Congrès International sur la Céramique Médiévale en Méditerranée*, Aix-en-Provence (13-18 November 1995), 63-72. Aix-en-Provence: Narration Editions.

Oziol, Th. 1977. *Salamine de Chypre VII, Les lampes du musée de Chypre*. Paris: De Boccard.

Papadopoyloy, V. 2000. Βυζαντινά μεταλλικά αντικείμενα, in O. Μεντζαφού-Πολύζου (ed.) Συλλογές Ευάγγελου Αβέρωφ: ταξιδεύοντας στο χρόνο, Exhibition Catalogue, Πινακοθήκη Ευάγγελου Αβέρωφ, Metsovo, 9 September 2000-15 January 2001, 71-97. Αθήνα: Ίδρυμα Ευαγγέλου Αβέρωφ-Τοσίτσα.

Papaggelos, I. 2005. *Η Αθωνική Μονή Ζυγού*, Thessaloniki 2005.

Papanikola-Bakirtzi, D. 1996. *Μεσαιωνική Εφυαλωμένη Κεραμική της Κύπρου. Τα εργαστήρια της Πάφου και Λαπήθου.* Θεσσαλονίκη: Εκδόσεις Ιδρύματος Α. Γ. Λεβέντη.

Papanikola-Bakirtzi, D. (ed.) 2002. *Everyday life in Byzantium*, Exhibition Catalogue, Thessaloniki, White Tower, October 2001-January 2002. Athens: Kapon Editions.

Parani, M. G. 2005. Representations of Glass Objects as a Source on Byzantine Glass: How Useful Are They? *Dumbarton Oaks Papers* 59, 147-171.

Pelekanidis, S. (ed.) 1973. *Οι θησαυροί του Αγίου Όρους*: Εικονογραφημένα Χειρόγραφα (παραστάσεις, επίτιτλα, αρχικά γράμματα), 4 vols Athens: Ekdotike Athenon.

Perko, V. and Tratnik, V. 2007. *An overview of late-antique Roman lamps in Slovenia*, Communication at the *3rd International of International Lychnological Association Round-Table*, Dark Ages? History and archaeology of lighting devices in Continental Europe, from late Antiquity to late Medieval Ages, *Historisches Museum Olten*, Olten (Switzerland), 24-29 September 2007.

Perlzweig, J. 1961. *Lamps of the Roman Period. First to Seventh Century after Christ*. The Athenian Agora 7. Princeton, N.J.: American School of Classical Studies at Athens.

Pétridis, P. 2000. Échanges et imitations dans la production des lampes romaines et paléochrétiennes en Grèce centrale, in F. Blondé and A. Muller (eds) *L'artisanat en Grèce ancienne. Les productions, les diffusions. Actes du Colloque de Lyon, December 1998*, 193-240. Villeneuve d'Ascq: Université Charles-de-Gaulle - Lille 3.

Petropoulos, M. 1999. *Τα εργαστήρια των ρωμαϊκών λυχναριών της Πάτρας και το Λυχνομαντείο*. Athens: Archaeological Receipts Fund.

Piccirillo, M. and P. M. Bikai and T. A. Dailey (eds) 1993. *The Mosaics of Jordan*. Amman: American Center of Oriental Research.

Poyloy-Papadimitriou, N. 2000. Οι Λύχνοι, in P. P. Betancourt, N. Marinatos *et al.* (ed.) *Το Σπήλαιο της Αμνισού: Η έρευνα του 1992*, 214-227. Αρχαιολογική Εφημερίς 2000 (in Greek).

Rosenthal, R. and Sivan, R. 1978. *Ancient Lamps in the Schlöessinger.* Jerusalem: Institute of Archaeology; Hebrew University of Jerusalem.

Roymeliotis, N. H. 2001. Παλαιοχριστιανικά λυχνάρια από την Αλάσαρνα (σημ. Καρδάμαινα) της Κω, in G. Kokkorou-Alevras, A. Laimoy and E. Simantoni-Bournia (eds) *Ιστορία - Τέχνη - Αρχαιολογία της Κω, 1ο Διεθνές Επιστημονικό Συμπόσιο, Κως 2-4 Μαΐου 1997*, 255-276. Αθήνα: Πανεπιστήμιο Αθηνών, Φιλοσοφική Σχολή.

Scorpan, C. 1973. Tipi sconosciuti di lumini a olio romani e il problema di certe transmissioni nel primo feudalismo. *Pontica* 6, 209-228.

Ševčenco, I. and Pattterson-Ševčenco, N. 1984. *The Life of Saint Nicolas of Sion*. Brookline, Mass.: Hellenic College Press.

Sorochan, S. 2002: Sorochan, Light for life and death in early Byzantine Empire, in D. Zhuravlev (ed.) *Fire, Light and Light Equipment in the Graeco-Roman World*, 111-119. Oxford: British Archaeological Reports.

Stiegemann, Ch. (ed.) 2001. *Byzanz. Das Licht aus dem Ostem. Kult und Alltag im Byzantinischen Reich vom 4. bis 15. Jahrhundert*, Exhibition Catalogue, Erzbischöflichen Diözesanmuseum Paderborn. Mainz: Verlag Philipp von Zabern.

The City of Mystras. Byzantine Hours: Works and Days in Byzantium 2001. Exhibition catalogue, *Mystras*, August 2001 - January 2002. Athens: Kapon Editions.

Themelis, P. 2004a. Ονοματολόγιο σκευών και αγγείων, in ΣΤ' Επιστημονική Συνάντηση για την Ελληνιστική κεραμική. *Προβλήματα χρονολόγησης, κλειστά σύνολα-εργαστήρια.* Volos 17-23 April 2000. Athens: Archaeological Receipts Fund.

Themelis, P. G. (ed.) 2004b. *Πρωτοβυζαντινή Ελεύθερνα*, τομέας I, vol. I. Athens.

Thomas, J. P., A. Constantinides-Hero and G. Constable (eds) *Byzantine Monastic Foundation Documents: A Complete Translation of the Surviving Founders' Typika and Testaments.* Washington D.C.: Dumbarton Oaks Research Library and Collection.

Treasures of Mount Athos 1997. *Treasures of Mount Athos*, Exhibition Catalogue, Thessaloniki, Museum of Byzantine Culture. Thessaloniki: Organization of Cultural Capital of Europe 'Thessaloniki 1997'.

Triantafyllos, D. 1972. Αρίσβη. *Αρχαιολογικόν Δελτίον* 27, *Β'2*, 541-542.

Tsigaridas, E. 2003. Manuel Panselinos, From the Holy Church of Protaton. Thessaloniki: Agioritiki Hestia.

Uboldi, M. 1995. Diffusione delle lampade vitree in età tardoantica e altomedievale e spunti per una tipologia. *Archeologia Medievale XXII*, 93-145.

Vavylopoyloy-Haritonidoy, A. 1988. Βυζαντινά κεραμικά από τη συλλογή του Βυζαντινού Μουσείου Χίου. *Αρχαιολογικά Ανάλεκτα εξ Αθηνών* XXI, 185-197.

Vincent, C. 2004. *Fiat Lux. Lumière et luminaires dans la vie religieuse du XIII au XVI siècle*. Paris: Éditions du Cerf.

Vincent, C. 2007. Conference inaugurale, Communication at the *3rd International of International Lychnological Association Round-Table*, Dark Ages? History and archaeology of lighting devices in Continental Europe, from late Antiquity to late Medieval Ages, *Historisches Museum Olten*, Olten (Switzerland), 24-29 September 2007.

Wamser, L. (ed.) 2004. *Die Welt von Byzanz- Europas Östliches Erbe. Glanz, Krisen und Fortleben einer tausendjährigen Kultur*, Exhibition Catalogue. Stuttgart: Theiss.

Weitzmann, K. (ed.) 1979. *Age of Spirituality: Late Antique and Early Christian Art. Third to Seventh Century*, Exhibition Catalogue, New York, The Metropolitan Museum of Art, 19 November 1977 - 12 February 1978. New York: The Metropolitan Museum of Art.

Williams, H. 1981. *Kenchreai, Eastern Port of Corinth 5: The Lamps*, Leiden: E.J. Brill.

With Faith and Fantasy. Ecclesiastical Woodcarvings of Western Taygetos 2004. Exhibition catalogue, Kalamata, Naos Agioy Geōrgioy, July-September 2004. Athens.

Wunderlich, Ch.-H. 2003. Light and Economy: An Essay about the Economy of Prehistoric and Ancient Lamps, in L. Chrzanovski (ed.) *Nouveautés lychnologiques = Lychnological news*, 251-263. Hauterive NE: LychnoServices,

Xanthopoulou, M. 1997. Les luminaires en bronze et fer aux époques paléochrétienne et byzantine: typologie, technologie, utilisation. Unpublished PhD dissertation, University of Sorbonne.

Yangaki, A. G. 2005. *La céramique des IVe-VIIIe siècles ap. J.-C. d' Eleutherna*. Athens: Εκδόσεις Πανεπιστημίου Κρήτης.

Lamps from the Excavation at 45 Basileus Heracleiou Street, Thessaloniki

Sofia Akrivopoulou, Periklis Slampeas and Paraskevi Leventeli
Ephorate of Antiquities of Thessaloniki City

Abstract: More than 300 sherds of terracotta lamps have been collected during the excavation at 45 Basileus Heracleiou Street in Thessaloniki, Greece. The majority have been dated from the 4th to the beginning of the 7th century AD. The material can be classified in five groups. The first group consists of a few examples of imported Roman lamps of the 1st-2nd centuries AD. The second group includes examples of Attic types. Because of their poor quality together with the lack of workshop marks, we presume that they are all imitations. The third group includes only one piece of the so-called 'Aegean workshop' and is probably imported. The fourth group includes African type lamps the majority of which should also be imitations, although some pieces are imported, especially among sherds small and difficult to identify, and decorated with a cross on the disk. The last group includes the majority of the finds, which are the Asia Minor types. Asia Minor types were very common in Thessaloniki as well as in the rest of continental Greece and the Aegean area. This is enough reason to presume that all of them are imitations. The poor quality of clay and the vertical narrow face formed at the junction of the two parts of the lamps (upper part and bottom), the result of surmoulage, confirms this suggestion. Both of these characteristics, especially the second, are not usually found in Asia Minor lamps from Ephesus. For these reasons, we suggest that Thesssaloniki was a production center of these imitations, taking under consideration that clay molds were found in the Thessaloniki's Agora excavation. Nevertheless, no kiln or workshop waste have been found yet.

Keywords: Thessaloniki, mold-made lamps, wheel-made lamp, *kandela*, Attic lamps, Aegean workshop, African lamps, Asia Minor lamps, chi-ro.

A significant section of Late Roman Thessaloniki, with parts of two urban streets and two *insulae*, adjacent to the southern border of the present *Hagia Theodora* monastery, has been revealed in a rescue excavation conducted in 2009-2010 by the 9th Ephorate of Byzantine Antiquities (Figure 1).[1] The bigger street, aligned NE-SW (Figure 1:9) ran between two *insulae* of private houses dating from the Imperial period at the latest. Nevertheless, the main phase of the preserved construction for both of these *insulae* can be linked to the 5th-6th century AD.

In the eastern *insula* underneath a later coarse mosaic pavement (Figure 1:7), a mosaic pavement of black-and-white technique has been preserved (Figure 1:6) dating the foundation of the house before the 3rd century AD. In this *insula* a rather short phase of abandonment of private life can be traced, at the end of the Late Roman period, when for some decades the southern wing became a glass workshop (Figure 1:8).[2] A storage area, a so-called *pitheon* of re-used *Late Roman Amphora* 4 amphorae was found parallel to the external face of the southern wall of the workshop (Figure 1:4).[3] The *pitheon* was covered by discarded potsherds so that a street could be laid over this fill after the abandonment of the *pitheon*.

In the western *insula* a significant part of a luxurious private Late Roman *domus* has been preserved, including a square courtyard with lateral corridors (Figure 1:1) and part of a room with a mosaic pavement with a *venatio* scene adjacent to the west corridor of the yard (Figure 1:5). The house was found above the remains of an earlier building, probably contemporary with the one with the black-and-white mosaic pavement at the opposite side of the street. The Imperial building extended not only underneath the Late Roman house of the western *insula* but also to its north, in a space that more or less has been preserved as an empty lot for centuries (Figure 1:2).

The courtyard of the Late Roman house was in continuous use as an open-air space and into the Byzantine period. We presume that during the Byzantine period the already dilapidated Late Roman house became the property of the neighboring convent of *Hagia Theodora* (9th century onwards). Such a suggestion would explain why no later private buildings were found on its grounds.[4] In full contrast with the house at the opposite side of the street (i.e., eastern *insula*), which probably never ceased to be in use, the Late Roman *domus* at the western *insula* was abandoned, gradually ruined and finally transformed into a mine for building materials, stones, bricks and marble parts, cruelly robbed from its remains. Moreover, its yard was used as a garbage dump, mostly for potsherds (Figure 1:1).

The main street (Figure 1:9) was probably laid out during the Imperial Period, when the older buildings were erected, or even earlier. Nevertheless, during the Late Roman 'building boom', when the older dwellings were fully reconstructed, the street was also renovated and a vaulted evacuation channel was constructed beneath it (Figure 1:10). A ten-meter length of the street was unearthed. Its southeastern edge meets another street which was laid down at a much later date and

[1] Bakirtzis 2003, 54, fig. 11.
[2] Antonaras 2014.
[3] Akrivopoulou and Slambeas 2014.
[4] Akrivopoulou 2010.

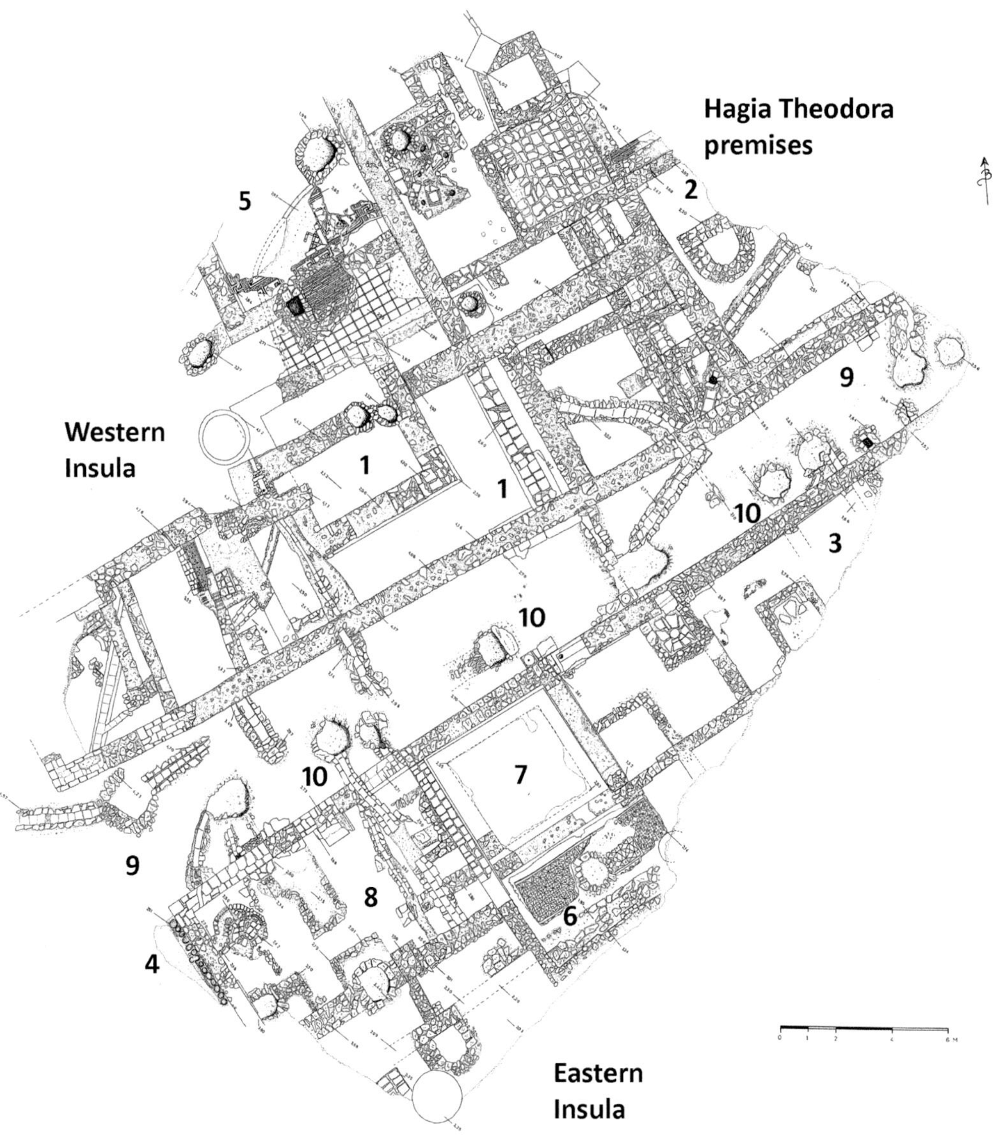

Figure 1. Ground plan of the excavation in 45 Basileus Heracleiou Str.: 1.1. Courtyard of the Late Roman building 1.2. Ruins of Early Roman building 1.3. Early remains in the Eastern Insula 1.4. Pitheon 1.5. Venatio mosaic 1.6. Black-and-white mosaic 1.7. Coarse mosaic 1.8. Glass Workshop 1.9. Main street (Kastilya Havrasi) 1.10 Sewer.

positioned parallel to the southern wall of the glass workshop, over the *pitheon*. This ten-meter long street was surfaced with a mixture of earth, pebbles, grit and pottery chips, a rather simple method dating back to Hellenistic times. More than ten layers of this mixture were found, as the street was regularly renewed and its pavement level was consequently raised significantly over the centuries. The street ceased to be used only after the great fire of 1917, when a great part of the city was planned afresh, more or less as it is today. The main street has been identified as the Ottoman *Kastilya Havrasi*, the road which separated a small Greek neighborhood called *Kizlar Manastir* (the successor of *Hagia Theodora* monastery) from two Jewish neighborhoods called *Aguda* and *Pulya*.

At least 300 fragments of pottery lamps have been collected, the majority dating from the 4th through the beginning of the 7th century AD. Most are Asia Minor lamps. Less are examples representing Attic workshop styles while quite a few are imitations of North African styles. The vast majority of the material consists of sherds, sometimes too small to be identified and lacking an outstanding variety of types. From the 300 pieces only 11 are intact or almost intact (3,3% of the total) and 17 more are half preserved (5,5%). Also, two exceptional wheel-made pieces have been found: a one-nozzle lamp with a handle, and an open lamp this is cup-like, probably a *kandela*.

1. Roman Lamps

A handle-shield in the form of crescent,[5] probably of the first half of the 1st century AD, is one of the earliest dated pieces

[5] Broneer type XXI, Broneer 1930, 75, no. 396. Bailey type D, Bailey 1980, no. 997 (group i), 1002 (group ii), 1007 (group iii), 1028, 1031

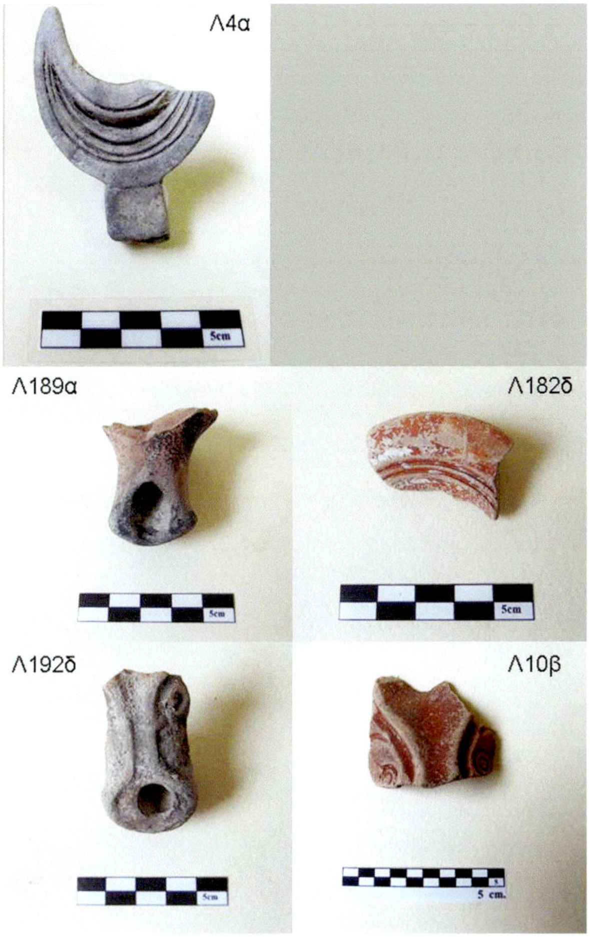

Figure 2. Imported Early to Middle Roman Lamps.

found (Figure 2. Λ4α). Mold-made of light colored, almost white clay, it is decorated with grooves that follow the outline of the crescent. Also dated to the 1st to 2nd centuries AD are: a nozzle with a pointed tip and remnants of black gloss paint (probably the oldest sherd of the group, Figure 2, Λ189α); a fragment of a plain rim with a fine red slip (2nd century AD, Figure 2, Λ182δ); and two fragments of volute nozzles, one covered with a red slip (1st century AD, Figure 2, Λ192δ, Λ10β).[6]

2. Attic Lamps

Approximately 23 sherds of Attic or Attic type lamps were found, dated from the 4th to the beginning of the 7th century, together with one intact lamp badly damaged by overheating. A sherd with one row of globules on the rim and a handle with three grooves above and below can be dated at the end of 4th century AD (fig 3, Λ191).[7] Some of the earliest pieces in this group can be dated mid-4th to the mid-5th century AD and bear a shiny slip (Figure 3, nos. Λ195, 181β, 185, 113, 143). There is also a fragment preserving a small part of a grooved handle and plain rim (Figure 3, Λ185) that bears two little clay pearls on each side of the handle, indicating the use of a plaster mold.[8] Only one sherd preserves part of its disk decoration (Figure 3, Λ143).

A sherd preserving a rectangular disk with framing grooves and herringbone pattern, scattered clay pearls resulting from use of a plaster mold, and wavy lines on the rim can be dated in the second half of the 4th century AD[9] or maybe a little later, possibly as late as the mid-5th century (Figure 3, Λ60).[10]

Two more sherds, covered with a red slip, can be dated to the middle of 5th century (Figure 3, Λ176β, Λ181α). A sherd with herringbone pattern on the rim, a rosette with concave petals on the disk and three grooves along the handle can also be dated to the mid-5th century (Figure 3, Λ90).[11] The same date can be given to a small sherd with a hatched band along the handle (Figure 3, Λ172β).[12]

Six fragments with herringbone patterns on the rim have been found, and can be dated between the mid-5th and mid-6th centuries (Figure 3, nos Λ36α, 196γ, 161β, 67, 114δ, 29.[13] The latest appear to be no Λ114δ,[14] with strokes on the disk, and no Λ29 with a rosette of concave petals on the disk and a grooved handle.

Of Attic or Attic type could also be a badly preserved fragment, with two grooves along the rim, a rosette of 12 concave petals on the disk, an air hole between the nozzle and disk and a solid handle with three grooves (Figure 3, Λ190β).[15] The rest of the Attic or Attic type lamp-sherds can be dated between the mid-5th and mid 6th centuries.

The single intact lamp in the group is an imitation of an Attic lamp that can be dated between the end of 5th and the beginning of 6th century (Figure 3, Λ155).[16] It is rather plain and badly preserved because of over-exposure to heat. It has a narrow shallow channel opens from the disk to the wick hole and its rim is decorated with a herringbone.

3. 'Aegean workshop' Lamp

An unexpected find was the top of a lamp of the so-called 'Aegean workshop', which can be dated in the 6th century (Figure 4, Λ176α).[17] It is an almond-shaped lamp, of reddish yellow clay with a buff surface. The rim is decorated with a set of three waved lines between two rows of dotted circles. The concave disk is framed by a raised grooved ring, which opens

(group iv), 217-218, nos. 1036-1055, pl. 34 (group vii and similar handle-ornaments). Bailey 1985, nos. 411-414. For the derivation of handle-shields from bronze lamps, see Perlzweig 1961, 4-5, pl. 2, no. 30.

6 Broneer type XXIV, Broneer 1930, 80-83, nos. 454, 459.

7 Bovon 1966, nos. 467, 468, 473 (4th century) with one row of globules on the rim. Compare with Karivieri 1996, nos. 114-116 (mid-4th to mid-5th century), with double row of globules on the rim.

8 Karivieri 1996, 23.

9 Bovon 1966, no. 569 (second half of 4th century).

10 Compare with Karivieri 1996, nos. 152-153 (late 4th to early 5th century).

11 Karivieri 1996, no. 212 (mid-5th century).

12 Karivieri 1996, no. 84 (second quarter to mid-5th century); Bovon 1966, no. 534, proposed a rather earlier date for a similar piece (4th century).

13 From mid-4th to mid-6th century, according to Perlzweig 1961, 154, pl. 12, no. 1905.

14 Karivieri 1996, no. 170 (second half of 6th century).

15 Karivieri 1996, no. 210.

16 Garnett 1975, 194, pl. 43, nos. 13-15.

17 Geroussi 2003.

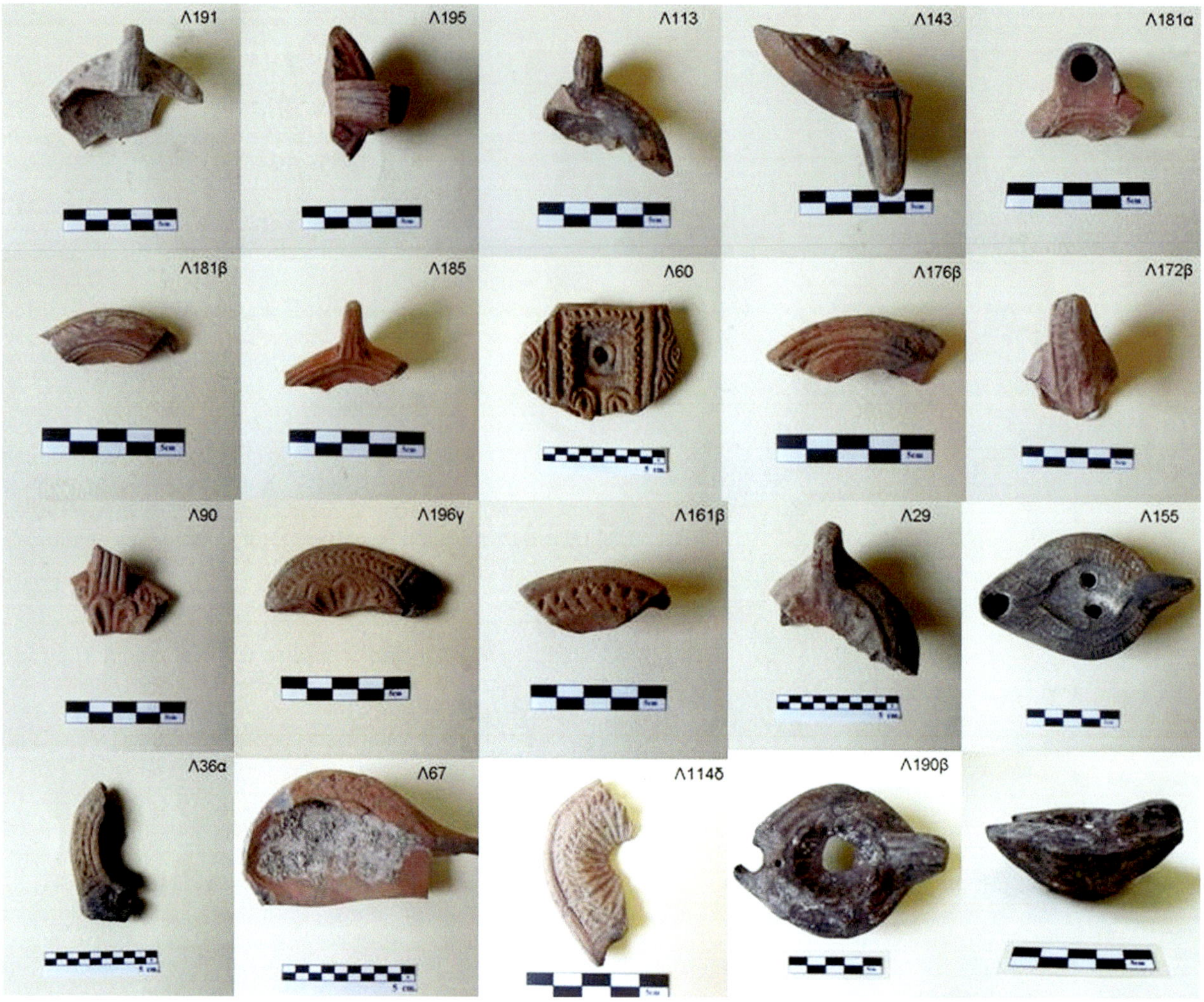

Figure 3. Attic types.

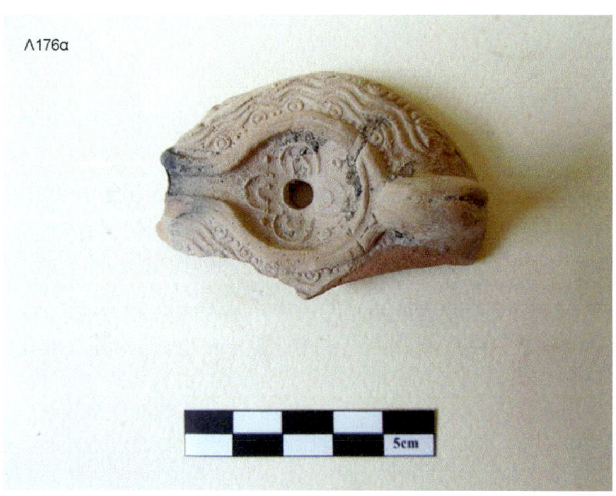

Figure 4. 'Aegean' workshop lamp.

to a narrow channel to the nozzle. It has a central fill hole, around which is arranged a quatrefoil with double petals and bullets among them. The knob-like handle bears a single groove. An almost identical lamp (top) has been found in the Athenian Agora where it was dated to 5th or 6th century.[18]

4. North African Lamps

Twenty fragments of North African type lamps, together with one badly preserved intact lamp, have been found (Figure 5). The earliest in date is a fragment of a nozzle with a fine red slip and a hatched-band of oblique grooves on the small preserved part of the rim (Figure 7, Λ196).[19] The rest can be dated from the 5th to the beginning of the 7th centuries. Eight of them have a disk decorated with a gemmed cross or a cross monogram (Figure 6).[20] On the rim they bear either three lines of bullets, a herringbone, grooves, or alternating triangles and squares. The handle is generally pointed. On two sherds the trace of a fill hole, punched marked but not pierced, has been preserved (Figure 7, Λ116, Λ114ε).

[18] Perlzweig 1961, no. 374 (5th-6th century, classified as imported).
[19] Ennabli 1976, pl. XLII, no. 779.
[20] Ennabli 1976, 177-221, pls XLVII-LIX, nos. 858-1132.

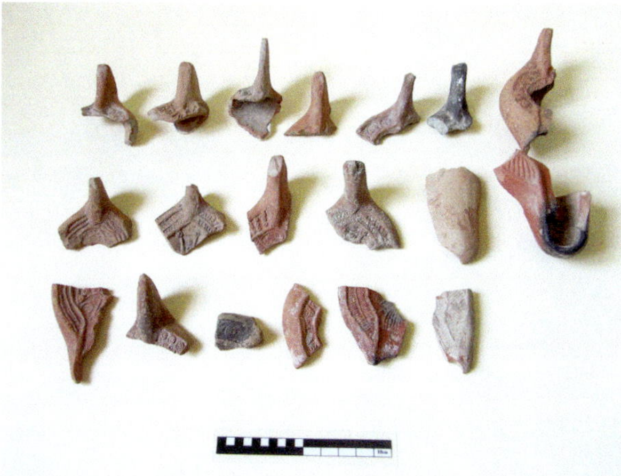

Figure 5. African types.

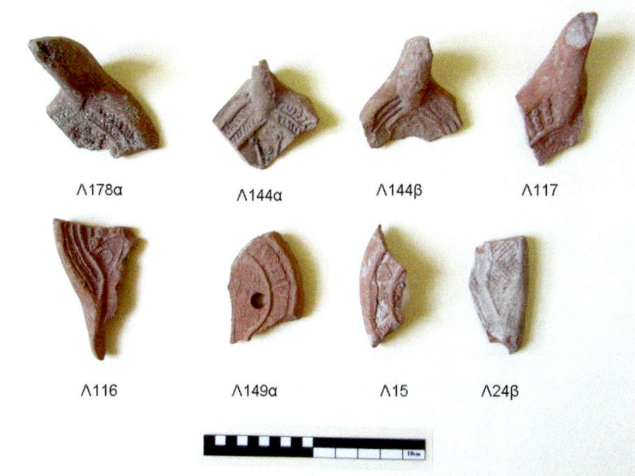

Figure 6. African types with gemmed cross or *chi-ro* on the disk.

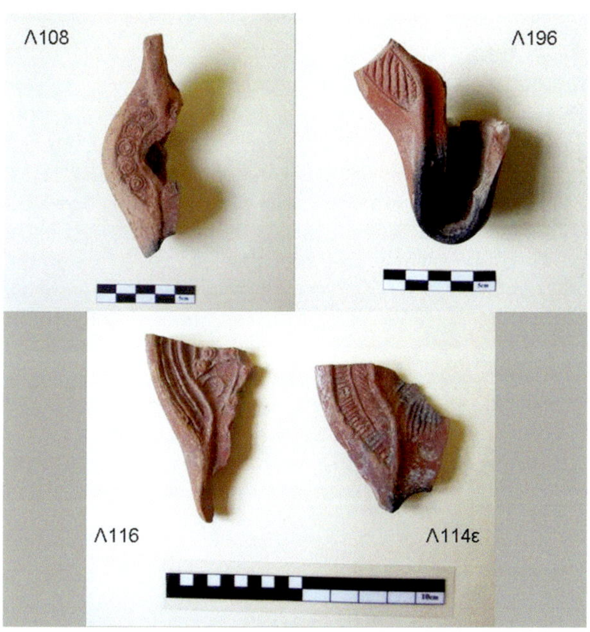

Figure 7. Various African types.

One lamp, almost half of which is preserved, has a row of concentric circles on the rim and likely is of the same late date (Figure 7, Λ108).[21]

5. Asia Minor Lamps

Asia Minor lamps constitute the vast majority of the material in question. With one exception, which can be dated to the 4th or 5th century (Figure 14, Λ52β), all the lamps are of 5th century. They have been classified in five groups, according to the arrangement of the top, i.e., the disk decoration (plain, various, cross, rosette) and presence of a channel.

Group A: Lamps with non-geometric decoration on disk (scenes, animals, etc.)

Group B: Lamps with plain disk and channel

Group C: Lamps with plain disk without channel

Group D: Lamps with cross motif on disk

Group E: Lamps with rosette motif on disk

The decoration on the bottom of these lamps may be elaborate but is more or less standardized, with a fishtail at the base of the handle, oblique grooves marking off the nozzle, and a raised ring, usually circular, on the base, sometimes with a small impressed mark at the center (*planta pedis* or other) (Figure 9, Λ156α bottom). Nevertheless, these symbols signify the workshop rather than the date of the lamp.

GroupA: Lamps with non-geometric decoration on disk (Figure 8)[22]

The lamps of this group have figural or narrative decoration on the disk and diverse themes on their rims (Figure 8). Of the four characteristic pieces of the group, three have vine-scroll decoration on rim (Figure 8, Λ55, Λ24γ, Λ5) and only the fourth has oblique strokes (Figure 8, Λ152α). The disk decoration, although generally not preserved, would have been rather sophisticated, as we can assume from the examples of Λ5 and Λ24γ.[23]

Group B: Lamps with plain disk and channel (Figure 9-10)[24]

The most popular type has rows of globules on the rim, usually four or five (Figure 9, Λ156β, Λ156α, 146; Figure 10, Λ118α, Λ4).[25] Significantly fewer lamps have oblique strokes on the rim (Figure 10, Λ10γ). In some of these lamps the disk is damaged in the center, where the fill hole should have been (Figure 10-Λ118α).[26] This damage is probably not accidental.

[21] Compare with Garnett 1975, 197-198, pl. 44, no. 20. The molding of the shoulder in both lamps looks alike. The rim decoration is slightly different, though. The Corinthian lamp was dated to the second half-end of the 6th century. See also Bonifay 2004, Type Atlante X-C3, 382-386, fig. 214.
[22] Poulou 1986, nos. 1-7.
[23] Compare with Bailey 1988, no. 3111, decoration on disk (but no channel).
[24] Poulou 1986, nos. 8-10.
[25] Bailey 1988, nos. 3185-3186, 3192-3194.
[26] Perlzweig 1961, no. 349; Karivieri 1996, no. 305.

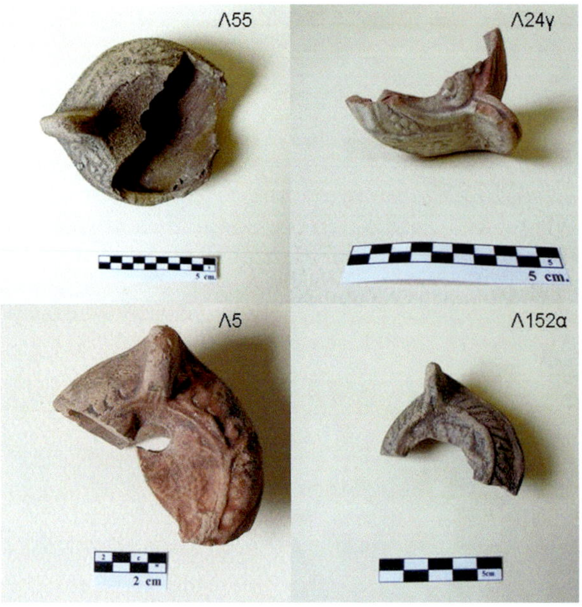

Figure 8. Asia Minor types, Group A, Lamps with non geometric decoration on the disk (scenes, animals, etc.).

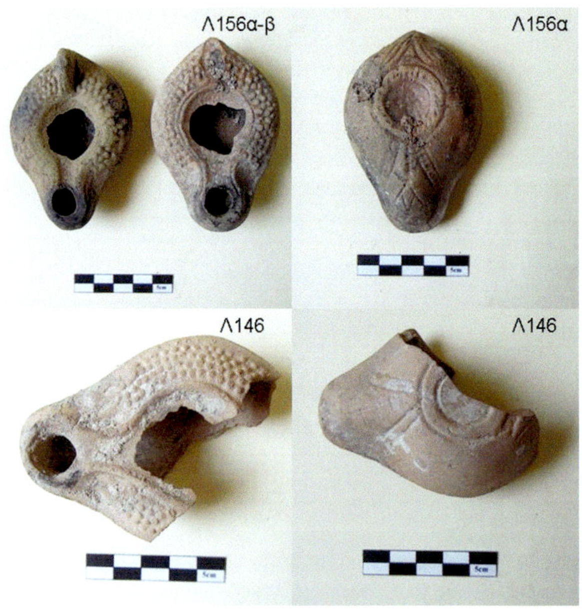

Figure 9. Asia Minor types, Group B: Lamps with plain disk and channel.

It could have been done on purpose, in order to enlarge the fill hole and possibly facilitate the fill of the lamp with oil.[27]

A lamp of the same type, probably an imitation, has a shallow channel with faint grooves (Figure 10, Λ4). A similar one from Yassi Ada confirms the late date. A second lamp can be classified as a Balkan imitation of the Asia Minor type,

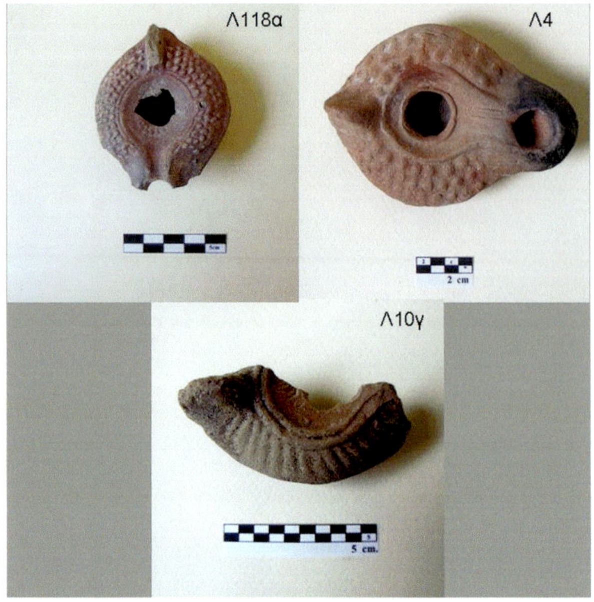

Figure 10. Asia Minor types, Group B: Lamps with plain disk and channel.

and the type has also been found in Tomis, Novae, and Constantinople.[28]

Group C: Lamps with plain disk and without channel (Figure 11-12)[29]

The lamps of this group are strongly reminiscent of the lamps of Group A. They have the typical rows of globules on the rim, usually four or five, but instead of a channel, fluted tongues are arranged around the nozzle.[30] On the base there may bear a simplified fishtail that usually continues the grooves of handle (Figure 11, Λ64, Figure 12, Λ96). Ladder patterns may mark off the nozzle (Figure 11, Λ64, Λ111, Figure 12, Λ129γ) or there may be even more elaborate motifs.

Group D: Lamps with cross motif on disk (Figure 13)[31]

There are only two lamp fragments with a cross on disk. The first fragment (Figure 13, Λ159) preserves part of the top, rim disk and nozzle. The rim is decorated with oblique strokes. On the disk there is a cross with flattened edges and a fill hole is preserved beneath the horizontal arm. Originally, a similar fill hole would have been pierced on the disk on the other side of the vertical arm. The second sherd (Figure 13, Λ192α) also preserves part of the top, rim disk and handle. This one is more elaborate, although similar. The rim is decorated with a row of dot-in-circle motifs. This is a highly symbolical yet simple ancient decorative motif that seemed to offer protection from the 'evil-eye'. A sun-like symbol is placed over the horizontal arm of the cross; there must have been another symbol (perhaps a crescent) at the other side of

[27] Chrzanovski and Zhuravlev 1998, nos. 55, 56; Motsianos 2010, 278, fig. 3.

[28] Bass and van Doornik 1982, 196, fig. 9-4, no. L12.

[29] Poulou 1986, no. 12; Bailey 1988, nos. 3152-3158.

[30] Perlzweig 1961, no. 349; Karivieri 1996, no. 305.

[31] Poulou 1986, nos. 15-17; Bailey 1988, nos. 3123, 3127, 3130, 3131, 3134, 3135, 3138.

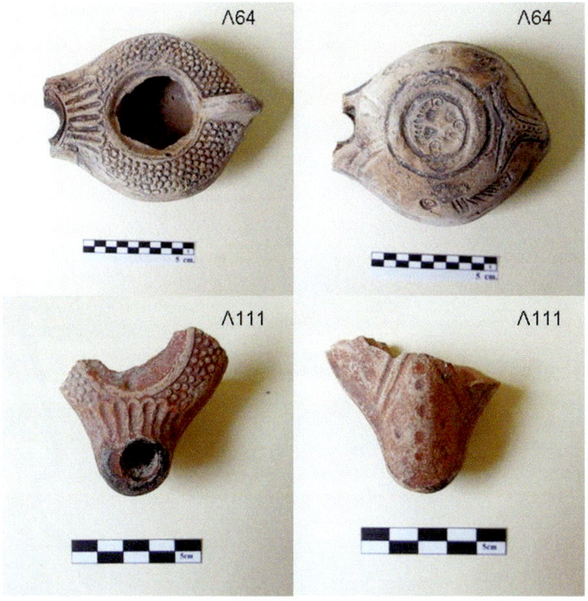

Figure 11. Asia Minor types, Group C: Lamps with plain disk without channel.

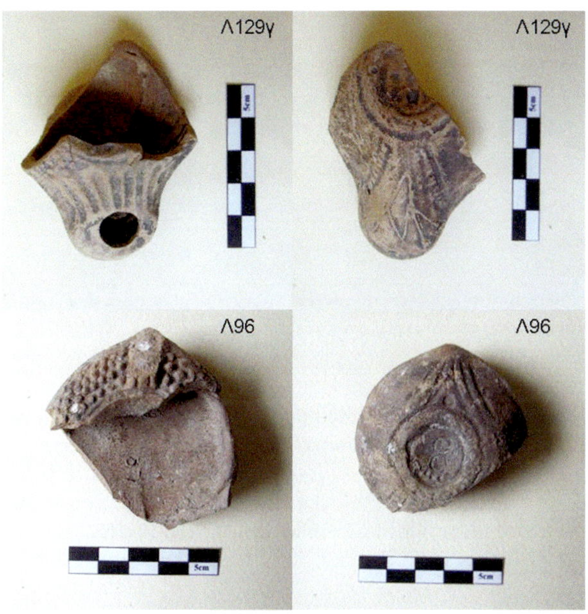

Figure 12. Asia Minor types, Group C: Lamps with plain disk without channel.

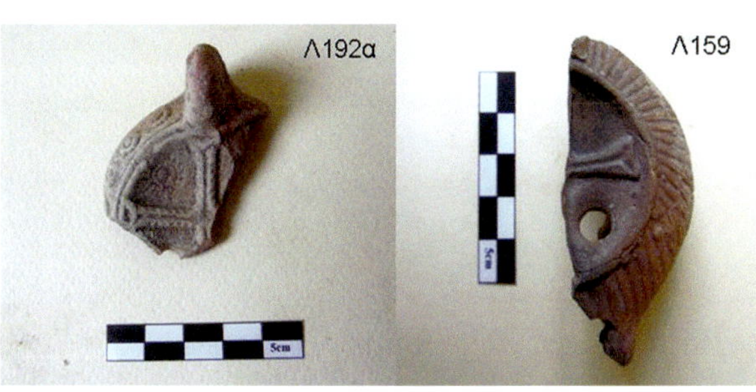

Figure 13. Asia Minor types, Group D: Lamps with cross motif on disk.

the vertical arm.[32] Below the horizontal arm, there must have been two fill holes, as on Λ159. In both lamps the crosses have flattened arm ends, indicating an imitation of bronze crosses, which similarly had the ends flattened by hammering.[33]

Group E: Lamps with rosette motifs on disk (Figure 14)[34]

This group consists of three lamp sherds, two of which are alike (Figure 14).

The first lamp fragment (Figure 14, Λ148α) preserves rim, disk and nozzle. The rim is decorated with a single row of globules. The disk is decorated with a rosette of concave petals, arranged around a central fill hole, and there are tongues arranged around the nozzle. The second fragment (Figure 14, Λ31) is very similar, with a rosette of concave petals arranged around a central fill hole. It has a vague chain-like scheme on the remarkably sloping rim, instead of globules. The third fragment (Figure 14, Λ52β) is completely different, though. It has a buff surface, raised dotted ovules on the rim, a rosette with arrows between the petals on the disk and a knob-like handle with two grooves.[35]

Wheel-made Lamp (Figure 15)[36]

An almost intact wheel-made lamp was found.[37] It has a round body, a chipped open conical rim standing on top a high cylindrical neck, and a handmade handle of triangular profile attached opposite the nozzle. It also has a rather big nozzle and a six-pointed motif incised on the right side of the body.

Wheel-made lamps of this type have striking similarities with bronze ones of the same period, i.e., the late 6th-7th centuries.[38] Nevertheless, the general shape of the type also reminds one of much earlier examples from the Hellenistic era,[39] as indicated by three lamps from Egypt in the British Museum.[40] Type I 'a fiacheta' from Gortyn is probably a local product very common in Crete, as was the case in Eleutherna.[41] Some rather vague information, dated in the troubled period of the Greek-Turkish war of 1920-1922, implies that a rather large quantity of such lamps (a variation rather) were found during the excavation of the Saint John Theologos church in Ephesos by the Greek Archaeological expedition, headed by G. Sotiriou.[42] The incised symbol on the body of many of these lamps, which may be either symmetrical, like a chi-ro, or somehow merged, resulting more like the letter A,[43] has been proposed to be a workshop mark rather than a Christian symbol.

These lamps, which can be dated between the 6th and 7th centuries or even a little latter, as is the case in our excavation,

[32] Compare with Bailey 1988, no. 3123.
[33] Abadie-Reynal and Sodini, pl. XVIIa, c, d, XVIIIa.
[34] Poulou 1986, no. 18 ; Bailey 1988, nos. 3140-3148.
[35] Abadie-Reynal and Sodini, no. L59a.
[36] Yangaki 2005, 232-241.
[37] Yangaki type 2.a/Eleutherna, Yangaki 2005, 233, pl. XVII.8. Baldini and Parello type I 'a fiachetta'/Gortyn. Baldini and Parello type I 'a fiachetta'/Gortyna. Baldini and Parello 2001, 167-71, pls. XXXI-XXXIII.
[38] Yangaki 2005, 235. For the bronze lamps see Xanthopoulou 2010.
[39] Yangaki 2005, 240 (fn. 1517).
[40] Bailey 1975, nos. 530-532.
[41] Yangaki 2005, 232-241. Vogt 2000, fig. 46.
[42] Sotiriou 1921-22, 189-90, fig. 63.
[43] Stories from the Hidden Harbor, no. 49 (110).

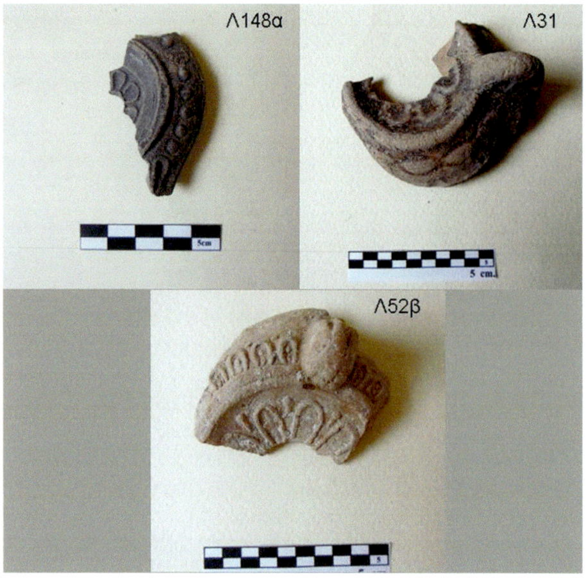

Figure 14. Asia Minor types, Group E: Lamps with rosette decoration on disk.

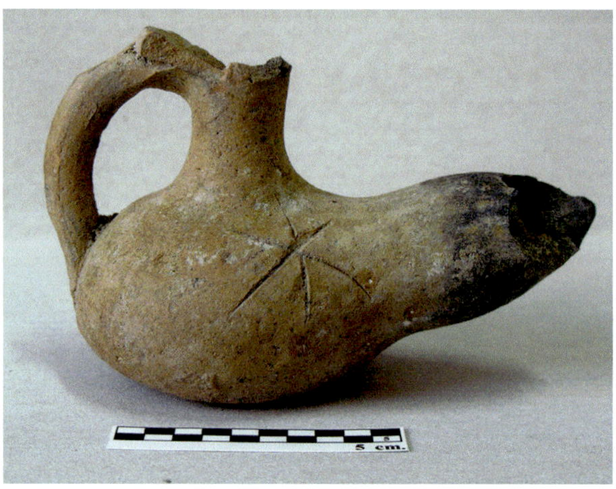

Figure 15. Wheel-made lamp.

are to be found in the Eastern Mediterranean but not with the same frequency as in Crete. To the examples mentioned by A. Yangaki,[44] we can add a recent find from Constantinople,[45] as well as a couple of finds from Thessaloniki, the one presented here and another one from the excavation of Agogos Hydreysis II.[46]

After a short and rather limited survey we have traced the appearance of this mark on other categories of pottery vessels only twice: in a jar from Albania (medium size, almost 40 cm high), probably a grave-good dated between the 7th and 9th centuries,[47] and on a Keay 55B amphora from Spain,

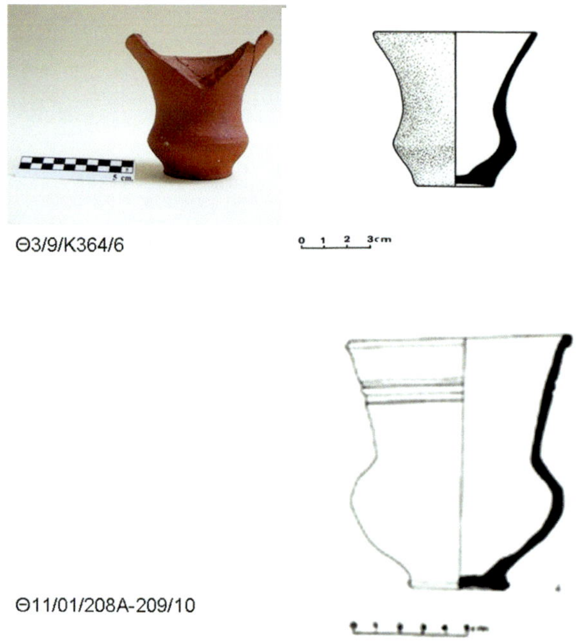

Figure 16. Clay *kandela.*

dated probably to the 6th century.[48] On the jar the mark was incised on the body, whereas on the amphora it was incised on the neck. We presume that the sign on the jar could also be a workshop mark, but for the amphora more than one hypotheses can be made, as such *graffitis* could have also been connected with the contents of the amphora or with the merchant or owner.

Nevertheless, the form of the mark is rather provoking. It could be interpreted as a naive reproduction of the *chi-ro* symbol. We have traced such improvisations made by hand among our own material: on two Asia Minor lamps, where a Latin cross and a cross monogram were faintly incised after casting and before the lamps were put in the oven (Figure 9, Λ146, bottom, enclosed in the ladder patterns). If such is the case, the symbol is in full accordance with the date and the place of production, as Crete by the 7th century had already been conquered by the iconoclast Arabs. An even later date, in the 8th century, is not at all unlikely, and would place the object into the official Byzantine iconoclastic period, when the cross was a main decorative pattern.

APPENDIX

Calyx-shaped cup-kandela (Figure 16)[49]

A rare find, difficult either to date or classify, is a calyx-shaped cup (Figure 16, K364/6) of micaceous orange clay with a flat base, bi-conical rather than spherical body, and a broad open rim. It is 6 cm high, with a base diameter of 3.5 cm and a rim diameter of 7 cm. An initial dating might assign it to the Ottoman period, alongside small items of glass tableware, not comparable but similar.

[44] Yangkaki 2005, 239, 240.

[45] *Stories from the Hidden Harbor*, no. 49 (110).

[46] Bakirtzis 2006, D, pl. 136.5; Vassileiadou and Tzevreni 2011, no 163.

[47] Hoti 1999-2000, 291, 295, fig. I.6.

[48] Revilla-Calvo 2011, 148, fig. 12.92.

[49] Akrivopoulou and Slambeas 2014.

The context in which it was found contained mainly fragments of vessels from the 5th through the 7th centuries. This cup could indeed be a later artifact that found its way into an earlier context. On the other hand, it might also be compared to a Roman calyx-shaped *kandela* from a third-century closed grave set from Thessaloniki, whose similarity to Ottoman ware has frequently been remarked.[50] The only close parallel found is a slightly bigger but similar clay cup, with two incised ring-lines under the external lip, found in Thessaloniki 10 years ago in a fifth-century context. It has not been classified.[51]

Although we can cite no bibliographical corroboration for our view that this vessel does in fact date to the Late Roman period, since we have been unable to find a published parallel, we have nonetheless chosen to present it with a proposed dating in the 5th to 6th century or, given the contextual finds, perhaps a little later. Besides, its shape makes its use as a lamp more likely than a table vessel.

Conclusion

Five groups of lamps and lamp sherds have been presented here. The first group includes a few examples of imported Roman lamps of the 1st to 2nd centuries AD. The second group includes examples of Attic types, which, because of their poor quality and the lack of workshop marks, we presume were all imitations. The third group consists of only one fragment of a lamp produced by the so called 'Aegean workshop', probably imported. The fourth group includes North African type lamps, many of which were also imitations. Nevertheless, it cannot be excluded that some sherds, bearing cross motifs, likely come from imported lamps. The last group includes the majority of the lamp finds, consisting of Asia Minor types.

Asia Minor types are very common in Thessaloniki as well as in the rest of continental Greece and the Aegean area. This is enough reason to presume that they are all imitations. The poor quality of clay and the vertical strip formed at the junction of the two parts of the lamps (upper part and bottom), a result of *surmoulage*, confirms this hypothesis. These characteristics, especially the second are not to be found in Asia Minor lamps from Ephesos. For all these reasons, and taking into consideration the clay molds found in the Archaia Agora excavation,[52] we suggest that Thesssaloniki was a production center for lamp imitations, Nevertheless, kilns or workshop waste have yet to be found.

Bibliography

Abadie-Reynal, C. and Sodini, J. P., 1992. *La céramique paléochrétienne de Thasos.* Paris : Boccard.

Akrivopoulou, S. 2010. Η ανασκαφή στην οδό Βασιλέως Ηρακλείου 45 στη Θεσσαλονίκη (The Excavation in 45, Vassileos Irakeliou Street in Thessaloniki). *Το Αρχαιολογικό Έργο στη Μακεδονία και τη Θράκη* 24, 255-262.

Akrivopoulou, S. and Slambeas, P. 2014. Late Roman Coarse Ware from a rescue excavation in Thessaloniki, in N. Poulou-Papadimitriou, E. Nodarou and V. Kilikoglou, *LRCW4: Late Roman Coarse Wares*, Cooking Wares and Amphorae in the Mediterranean, Archaeology and Archaeometry. Mediterranean: a Market without Frontiers, 287-295. Oxford: Archaeopress.

Antonaras, A. 2008. Glass Lamps of the Roman and Early Christian Periods. Evidence from the Thessaloniki Area, in C. A. Roman and N. Gudea (eds) *Lychnological Acts 2. Acts of the 2nd International Congress on Ancient and Middle Age Lighting Devices, Zalău-Cluj-Napoca, 13th-18th of May 2006. Trade and Local Production of Lamps from the Prehistory until the Middle Age,* 23-30. Cluj-Napoca: Editura Mega.

Antonaras, A. 2014. An early Christian glass workshop from Thessalonica, in D. Keller, J. Price and C. Jackson (eds) *Neighbours and Successors of Rome: traditions of glass production and use in Europe and the Middle East in the later first millennium AD*, 95-113. Oxford: Oxbow Books.

Bailey, D. M. 1975. *A Catalogue of the Lamps in the British Museum, 1. Greek, Hellenistic and early Roman pottery lamps.* London: British Museum Publications.

Bailey, D. M. 1980. *A Catalogue of the lamps in the British Museum, 2. Roman lamps made in Italy.* London: British Museum Publications.

Bailey, D. M. 1985. *An Excavation at Sidi Khrebish Benghazi (Berenice), 3.2: The Lamps.* Tripoli: Department of Antiquities.

Bailey, D. M. 1988. *A catalogue of the lamps in the British Museum, 3. Roman provincial lamps.* London: British Museum Publications.

Bakirtzis, Ch. 2003, The Urban Continuity and Size of Late Byzantine Thessalonike, *Dumbarton Oaks Papers* 57, 35-64.

Bakirtzis Ch. (ed.) 2006. *Σωστική ανασκαφή αγωγού ύδρευσης Θεσσαλονίκης II*, Thessaloniki: 9th Ephorate of Byzantine Antiquities.

Baldini, I. and Parello, G., 2001. Lucerne, in A. di Vita (ed.) *Gortina V.3, lo scavo del Pretorio (1989-1995), I, I Materiali*, 114-89. Padova: Bottega d'Erasmo, Aldo Ausilio ed.

Bass, G. and van Doornick, F. H. 1982. *Yassi-Ada I, A seventh-century Byzantine shipwreck.* College station: Texas A & M University Press.

Bonifay, M. 2004. *Etudes sur la céramique romaine tardive d'Afrique.* Oxford: British Archeological Reports International Series 1301.

Bovon, A. 1966. *Les lampes d'Argos.* Paris: Librairie Philosophique J. Vrin.

Broneer, O. 1930. *Corinth IV.2: Teracotta lamps.* Cambridge Massachusetts: Harvard University Press.

Chrzanovski, L. and D. Zhuravlev 1998. *Lamps from Chersonesos in the State Historical Museum Moscow.* Roma: L'Erma di Bretschneider.

Ennabli, A. 1976. *Lampes Chrétiennes de Tunisie. Musées du Bardo et de Carthage.* Paris : Ed. du Centre national de la recherche scientifique.

Garnett, K. S. 1975. Late Roman Corinthian Lamps from the Fountain of the Lamps, *Hesperia* 44, 173-206.

Geroussi, E. 2003. «Αιγαιακό» εργαστήριο παραγωγής λυχναριών του 6ου αιώνα μ.Χ. ['Aegean' workshop of lamps (6th century)], *7ο Διεθνές Συνέδριο Μεσαιωνικής Κεραμικής της Μεσογείου*, Θεσσαλονίκη 11-16 Οκτωβρίου 1999, 513-516. Αθήνα.

Hoti, A. 1999-2000. Some features of the Early Medieval Pottery in Albania (7th-11th centuries). *Ilyria* XXIX, 283-295 (in Albanian with English summary).

Karivieri, A. 1996. *The Athenian Lamp Industry in Late Antiquity.* Helsinki: Finnish Institute at Athens.

[50] Antonaras 2008, 25-26, pl. 3:4.

[51] Bakirtzis 2006, 1:257, 3:141.

[52] Velenis and Zachariadis 2011, nos. 79-81.

Motsianos, I. 2010. Ύστερη Αρχαιότητα: Το λυκόφως των κατασκευασμένων με μήτρα λυχναριών (Late Antiquity: the twilight of mouldmade lamps), in D. Papanikola-Bakirtzis and D. Koussoulakou (eds) *Κεραμική της Ύστερης Αρχαιότητας από τον Ελλαδικό χώρο*, Επιστημονική Συνάντηση, Θεσσαλονίκη 12-16 Νοεμβρίου 2006, 270-281. Thessaloniki: Archaeological Institute of Macedonian and Thracian Studies (in Greek with English abstract).

Motsianos, I. and E. Bintsi (eds) 2011. *Light on light, an illuminating story*, Exhibition Catalogue. Thessaloniki: Folklife & Ethnological Museum of Macedonia-Thrace.

Perlzweig, J. 1961. *The Athenian Agora VII: Lamps of the Roman Period*. Princeton, N.J.: American School of Classical Studies at Athens.

Poulou-Papadimitriou, N. 1986. Lampes paléochrétiennes de Samos. *Bulletin de Correspondance Hellénique* 110, 583-610.

Revilla-Calvo, V. 2011. Contextos cerámicos del siglo VI d.C. de Iluro (Hispania Tarraconensis), in M. Ángel Cau, P. Reynolds, M. Bonifay (eds) *Late Roman Fine Wares 1. Solving Problems of Typology and Chronology. A review of the evidence, debate and new contexts*, 129-154. Oxford: Archaeopress.

Σωτηρίου, Γ. Α. 1921-22. Ανασκαφαί του Βυζαντινού Ναού Ιωάννου του Θεολόγου εν Εφέσω. *Αρχαιολογικόν Δελτίον* 7, 89-226.

Stories from the Hidden Harbor. Shipwrecks of Yenikapi 2013. Istanbul: Istanbul Archaeological Museums Press.

Vassileiadou, S. and Tzevreni, S. 2011. Lamps from the rescue excavation of the new main water supply system in Thessaloniki (Agogos Idrefsis II), in I. Motsianos and E. Bintsi (eds) 2011. Light on light, an illuminating story, 228-236. Thessaloniki: Folklife & Ethnological Museum of Macedonia-Thrace.

Velenis, G. and Zachariadis, S. 2011. Lamps from Thessalonikis's Agora, in I. Motsianos and E. Bintsi (eds) 2011. Light on light, an illuminating story, 185-210. Thessaloniki: Folklife & Ethnological Museum of Macedonia-Thrace.

Vogt, Ch. 2000. The Early Byzantine Pottery, in Π. Θέμελης, *Πρωτοβυζαντινή Ελεύθερνα, τομέας I*, II, 39-199. Ρέθυμνο: Εκδόσεις Πανεπιστημίου Κρήτης.

Xanthopoulou, M. 2010. *Les lampes en bronze à l'époque paléochrétienne*. Turnhout: Brepols.

Yangaki, A. 2005. *La céramique des IVe-VIIIe siècles ap. J.-C. d'Eleutherna : sa place en Crète et dans le bassin égéen*. Athènes: Εκδόσεις Πανεπιστημίου Κρήτης.

Lighting Artifacts from the Episcopal Complex at Louloudies Pierias

Evangelia Angelkou
Museum of Byzantine Culture of Thessaloniki
Maria Cheimonopoulou
Ephorate of Antiquities of Imathia

Abstract: Excavation of the episcopal complex in the years 1993-1999 yielded a total of 230 intact lamps and lamp fragments, both wheel-made and mold-made, representing the major chronological phases of the complex. From the establishment phase and times of prosperity in the late 5th and 6th centuries come 24 lamps of African origin, 40 (intact and fragments) Asia Minor lamps and another 57 partial or whole lamps, which may be attributed to an unspecified workshop imitating African products. In addition to evidence on use of clay lamps for lighting, the excavation gave also evidence for the use of copper alloy lamps and more complex metal and glass lighting devices. From material study we conclude that, during the last phases of the episcopal complex in 7th-8th century, newly manufactured lamps, morphologically peculiar to the productions of the 5th-6th century but of an inferior quality, were used together with older African lamps of the 5th-6th century.

Keywords: Louloudies Kitrous, lamps, episcopal complex, Late Antiquity, multi–hooked suspension system, *polyangistron*.

Louloudies Kitrous, lies in Northern Pieria between Alykes Kitrous and Korinos, where, according to N. Hammond and L. Heuzey, the battle of Pydna in 168 BC occurred. Construction work for the new twin-track railway line between Athens and Thessaloniki in this area has revealed a square fortified enclosure with four projecting square corner towers (quadriburgium), consisting of a basilica, an episcopal residence and extensive wine and oil producing installations (Figure 1).[1]

The area had already been settled by the Mycenaean era, and continued to be inhabited through the Early Christian period. In the 2nd century AD the location was upgraded to an intermediary station of the Thessaloniki-Larissa route, which probably coincides with the *mansio* or *mutatio* 'Anamon' mentioned between Pydna and Atira in Antonine itineraries.[2] From the early fourth century there are architectural remains of a Greco-Roman house with a peristyle and mosaic floor decoration, as well as its warehouses and rural facilities. When also considering the wealth available from the neighboring salt-works, circumstances are created under which the area could have functioned as a rural stock-concentration center.

In AD 479 an episcopal complex was erected in the same location, the foundation of which was a consequence of the move of the bishop and the inhabitants of Pydna to Louloudies, due to the establishment of the Goths of Theodoric at Pydna and five other Macedonian cities in exchange for the termination of Thessaloniki's siege, according to the historian of the Goths Jordanes. Perhaps the bishop and the inhabitants of Pydna did not want to live with the Goths and therefore asked to move to a nearby area which preserved still enough of the building material of its previous buildings. So upon the ruins of the *mutatio* baths and the fourth-century villa, a four-towered episcopal complex was founded probably with imperial sponsorship.

This fifth-century episcopal complex in its original design comprised a moat and gate with a four-sided tower in the middle of its west side. Inside there were arcades carried on piers, accommodation facilities for the garrison, a basilica and an episcopal palace.

In the era of Justinian the episcopal complex expands and loses its defensive role, making the most of the agricultural wealth of the region, for the exploitation of which special large scale processing installations were constructed. After a devastating earthquake that could be dated in the third quarter of the 6th century the basilica, reduced to its central aisle, acquired a cemetery use, while a new seismic disaster, probably in the 7th century, contributed to the abandonment of the area, and its gradual transformation during the 7th and 8th centuries into an extensive workshop of recycled and recovered materials.

Excavation of the complex in the years 1993-1999 yielded a total of 230 intact lamps and lamp fragments, both wheel-made and mold-made, representing its major chronological phases.

The 52 Hellenistic and Early Roman lamps, are associated with the small rural hamlet which was founded on this site in the Mycenaean age and then in Roman times relocated 2 kilometers eastward. Five middle Roman lamps (2nd–3rd century) and one with a seven-branch menorah (3rd–4th century) (Figure 2)[3] are possibly related to the phase of the Anamon mansio.

[1] Main publications for the excavation are the following: Marki 1997b; Marki 1993; Marki 1994; Marki 1995; Marki 1996; Marki 1999b; Marki 1997a; Marki 1999a; Marki 2001; Marki 2002a; Marki 2002b; Marki 2004; Marki 2008; Marki 2013.

[2] Miller 1916, 574.

[3] Intact lamp BK 4513/91, published in Marki *et al.* 2010, 209. Similar

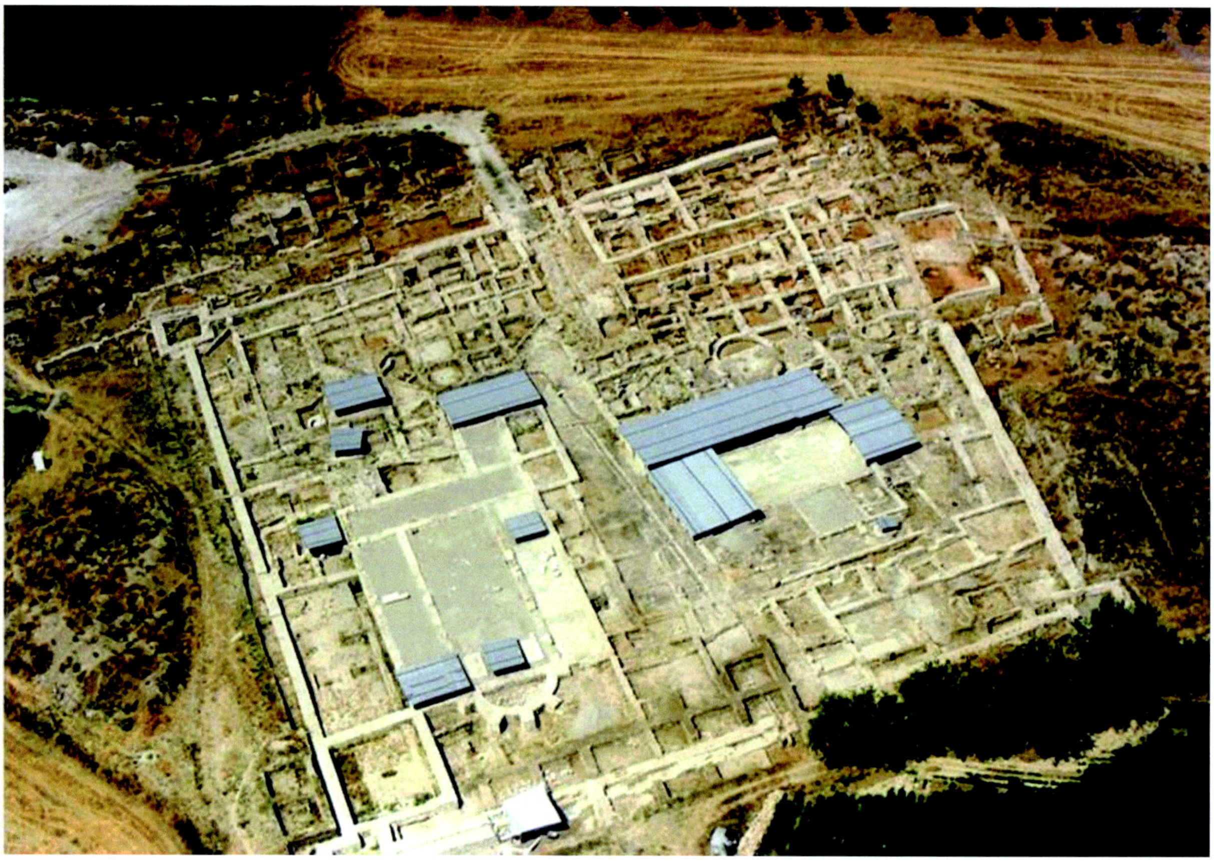

Figure 1. Aerial photograph of the archaeological site of the Episcopal Complex at Louloudies Pierias.

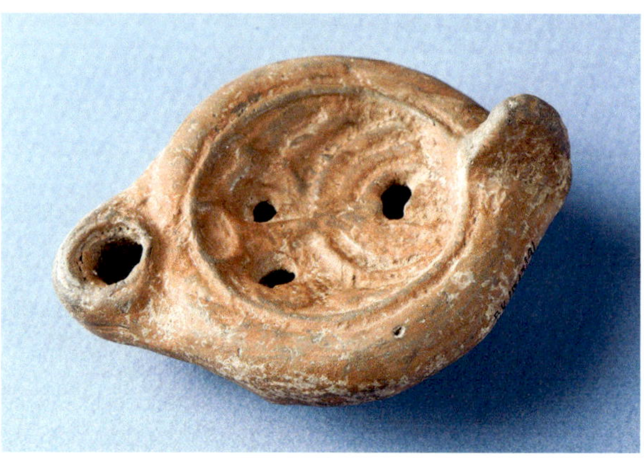

Figure 2. Mold-made clay lamp (inv. no. BK 4513/91).

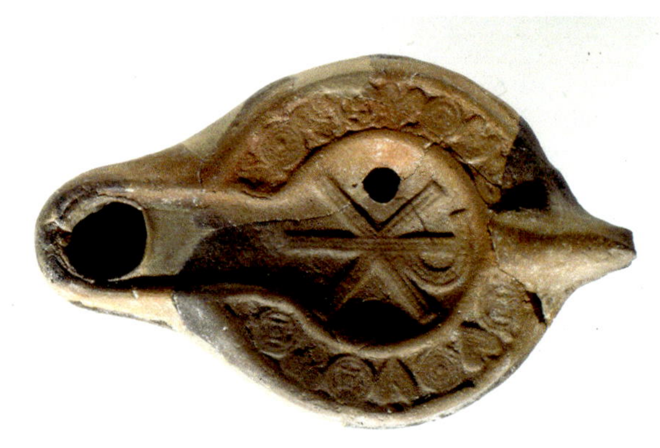

Figure 3. Mold-made clay lamp (inv. no. BK 4513/50). © Museum of Byzantine Culture. Photo by M. Skiadaresis.

From the phase of establishment and prosperity of the episcopal complex in the late 5th and 6th centuries come: 24 lamps of African origin, which by their decoration seem to have been part of the episcopal palace furnishings;[4] 40 (intact and fragmentary) Asia Minor lamps; and another 57 partial and whole lamps, which may be attributed to an unspecified workshop imitating African products.

Nine of the North African lamps belong to type Hayes II[5] (Atlante X, Bonifay C and D), have a round body with a strongly elongated and protruding nozzle connecting with a broad straight channel to the disk, a plain stump handle and a low base ring, connected by a ridge to the handle. They are mainly decorated with representations of saints (BK 4513/111,[6] BK 4513/128 and BK 4513/6)[7], the Chi-Rho Christogram (Figure

African type lamp, dated between the 3rd and 4th century in Hoff 1986, 90-91.

[4] Marki 1998.

[5] Hayes 1972, 313-314.

[6] Marki 1998, 25; Cheimonopoulou and Angelkou 2000, 130-131; Marki *et al.* 2010, 210.

[7] Marki 1998, 26; Cheimonopoulou and Angelkou 2000, 130; Marki *et*

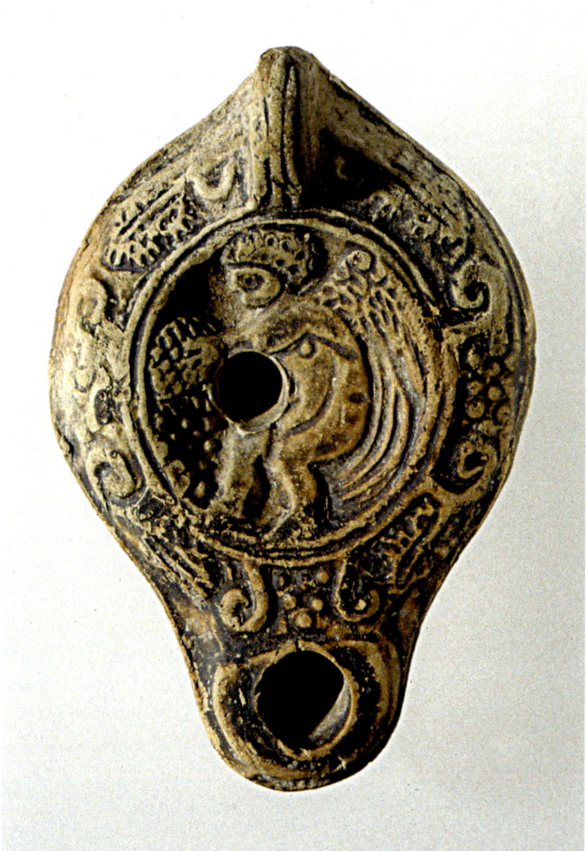

Figure 4. Mold-made clay lamp (inv. no. BK 4513/2). © Museum of Byzantine Culture. Photo by M. Skiadaresis.

3) or Cross-monogram (BK 4513/50, BK 4513/131, BK 4513/30, BK 4513/62, BK 4513/73).[8] The 40 lamps of Asia Minor origin, representing a significant percentage of the total, are characterized by a round or almond-shaped body, protruding round nozzle, solid handle ending in a fish tail and a low base ring bearing an impressed mark of a foot (planta pedis). The Asia Minor lamps are a very popular type of the 6th and 7th centuries (500-600), with precise chronological limits set by the excavation of the double basilica in Aliki Thassos (late 5th and 6th centuries) and the 7th century shipwreck of Yassi Ada (625/626).[9] Most have a circular body and the impressed bare foot (planta pedis) motif as a trademark on the base, are decorated with figurative scenes of an Eros turning over a basket of fruit (Figure 4)[10] or venison,[11] embossed rosettes[12] and crosses,[13] while the rim is covered with stylized grape clusters and tendril (Figure 5). [14]

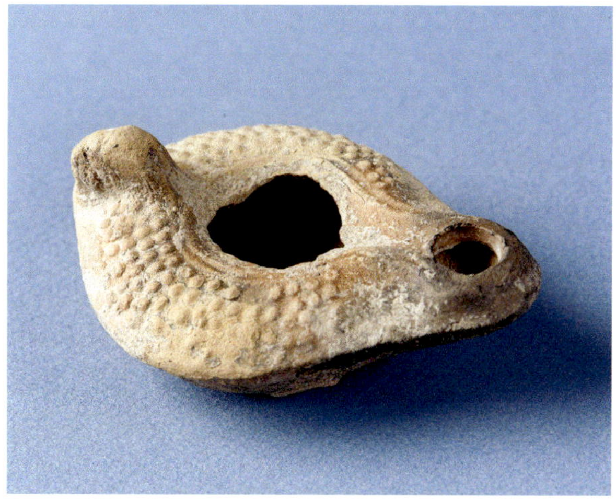

Figure 5. Mold-made clay lamp (inv. no. BK 4513/43).

The 57 intact and fragmentary lamps with eclectic decoration also form an interesting group that imitates elements of both the African and Asia Minor workshops with influences from famous production centers of the time. Major decorative motifs of the time are common: the cross (Figure 6) and Christogram for the disk (BK4513/112, BK4513/56) or the herringbone for the relief of the rim (BK 4513/6).[15]

In addition to the practice of lighting with clay lamps, the excavation also gave evidence for the use of copper alloy lamps and more complex metal and glass lighting devices.

The only close-shaped metal lamp from the excavation, which has a handle in the shape of a cross with forked arms, and a fill hole with a shell-shaped lid (BA 51/3) (Figure 7), comes probably from the equipment of the episcopal palace.[16] The lamp has a round body, a long spout and discoid base. Preserved at two points, on the center of the cross-shaped handle and behind the spout, are chain rings that were used for hanging the lamp. Lamps of this type were extremely popular during the 5th through the 7th centuries. Manufactured in many centers of the Eastern Mediterranean, they are represented by a very large number of surviving examples. Similar lamps are found in private collections in German and Italian museums, the Agora of Athens, Romania, Bulgaria and Ukraine, Pergamon, Smyrna and Constantinople, Syria and Palestine, while a large number of them are kept in the Cairo and British Museums.[17] The metal lamp from the excavation at Louloudies, a mass-produced imported artifact of the 5th-7th centuries, was probably part of a larger group of metal lamps and other lighting devices associated with the establishment and habitation of the episcopal complex at the end of the 5th and during the 6th century. The lamp could be mounted on a metal lampstand, hung from a wall, or set on a table.

Another group of findings associated with lighting are the bronze multi-hooked suspension systems (*polyangistra*), consisting of thin plates with respective hooks mounted at

al., 2010, 210.

[8] Marki 1998, 23-24; Cheimonopoulou and Angelkou 2000, 130; Marki *et al.* 2010, 210.

[9] Bass and van Doornick 1982, 190-196.

[10] Cheimonopoulou and Angelkou 2000, 131-132; Marki *et al.* 2010, 211, footnote 16. Similar in Bailey 1988, 382, no. Q3108, pl. 105.

[11] Cheimonopoulou and Angelkou 2000, 133; Marki *et al.* 2010, 211.

[12] Similar in Bailey 1988, 385, no. Q3141, pl. 108.

[13] Marki 1998, 22.

[14] Similar in Bailey 1988, 391, nos. Q3192-3194, pl. 112.

[15] Marki *et al.* 2010, 212.

[16] Marki 1998, 27.

[17] Xanthopoulou 2010, 6, 100-105.

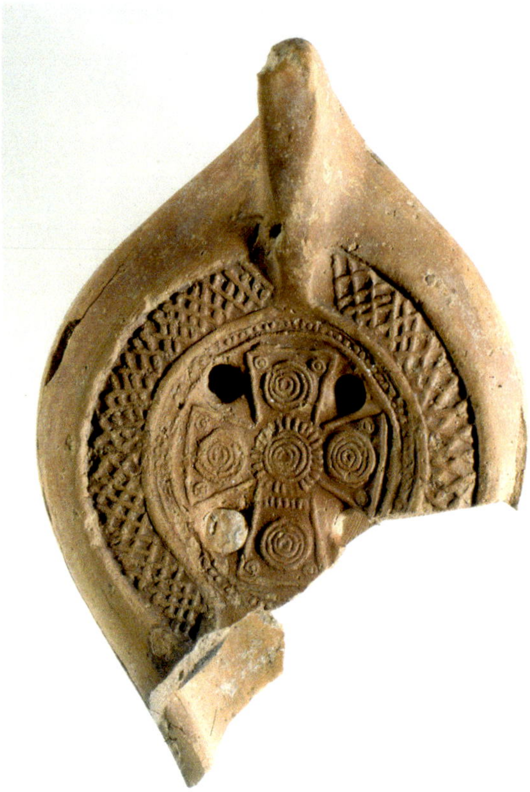

Figure 6. Mold-made clay lamp (inv. no. BK 4513/112). © Museum of Byzantine Culture. Photo by M. Skiadaresis.

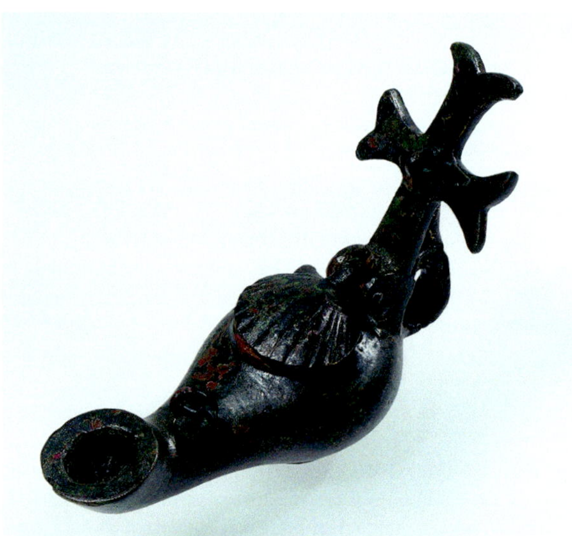

Figure 7. Copper alloy lamp (inv. no. BA 51/3). © Museum of Byzantine Culture. Photo by M. Skiadaresis.

their ends, for the hanging of glass lamps. One of the best surviving examples from the excavation shows that at least three such thin plates with hooks were used for hanging oil lamps by their respective handles. Those metal plates were attached to a ring passed to the lower edge of the vertical arm of a Latin cross formed of flattened metal strips, while a thin stem ending in a suspension hook is attached to their upper end (Figure 8). [18] The excavation revealed the surviving fragments of five metal plates with hooks and three intact suspension systems, which, despite the fragmentary state of their preservation, provide a more complete picture of such objects, the restored form of which is retained in a *polykandelon* (chandelier) of unknown origin exhibited at the Museum of Byzantine Culture[19] in Thessaloniki. Bronze multi-hooked suspension systems form valuable evidence for the illumination of the basilica with glass oil lamps, either suspended from bronze polykandela or installed in various parts of the basilica such as the arcades of the aisles or the ciborium of the altar.

For the use and arrangement of copper alloy chandeliers or *polykandela* and glass lamps in early Christian basilicas, the study of data coming from the church in Nahariya is useful.[20] According to the excavation testimony from a sealed destruction layer, it was discovered that 25 glass lamps were hung in the eastern half of the church, six were discovered on the floor of the north aisle arcade in front of each column, 14 were found in the central part of the nave, and four were in front of the chancel screen and inside the sanctuary, while there was no evidence for the illumination of the south aisle. In front of the entrance to the sanctuary at the basilica extra muros in Dağ Pazarı, a glass oil lamp and its metal suspension system were also found.[21]

Illustrations showing the setting and functioning of glass lamps are retained in a mosaic panel from Syria with a representation of the Holy Sepulcher, now in the National Museum in Copenhagen,[22] on an ivory plaque with St Menas flanked by camels from Castello Sforzesco,[23] and on the Stuma silver paten.[24]

A very large number of fragments from glass lamps coming mainly from the excavation of the basilica at Louloudies consist of the widespread types of the 5th-6th century, such as the hemispherical shoe shaped with three looped glass handles, stemmed beakers, hemispherical bowls with a long, tubular and narrow lower body, and cylindrical or slightly conical beakers with a knob on the base. These are commonplace in archaeological finds from Thessaloniki, Dion, Philippi, Corinth, Thasos, Samos, Egypt and Constantinople.[25] The glass lamps suspended from their looped handles attached to multi-hooked systems, were hung up with wall hooks up in various parts of the church or they were used in association with more complicated multi-lighting devices, such as openwork polykandela, where parts of their round plates have been unearthed through excavation.

Evidence for the illumination of the basilica with bronze polykandela includes the unearthed strong chain parts used for hanging them (Figure 9).[26] The closest parallels for

[18] Angelkou and Cheimonopoulou 2011, 246-252.
[19] Xanthopoulou 2010, 292.
[20] Xanthopoulou 2010, 69; Dauphin and Edelstein 1984, 103-104 and 113-114.
[21] Xanthopoulou 2010, 69; Gough 1975, 161, fig. 5.
[22] Bouras and Parani 2008, 28, fig. 27.
[23] Bouras and Parani 2008, 25, fig. 23.
[24] Bouras and Parani 2008, 27, fig. 26.
[25] Xanthopoulou 2010, 56-57.
[26] Angelkou and Cheimonopoulou 2011, 252-253.

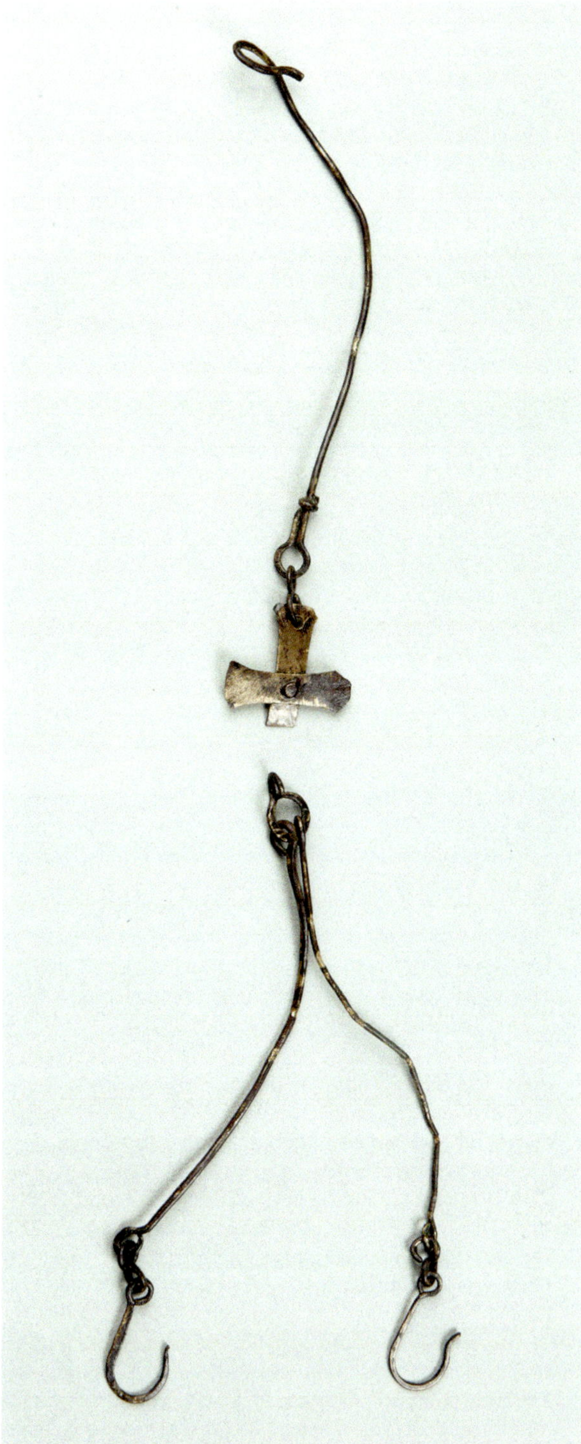

Figure 8. Copper polyangistron (inv. no. BX 253/616).

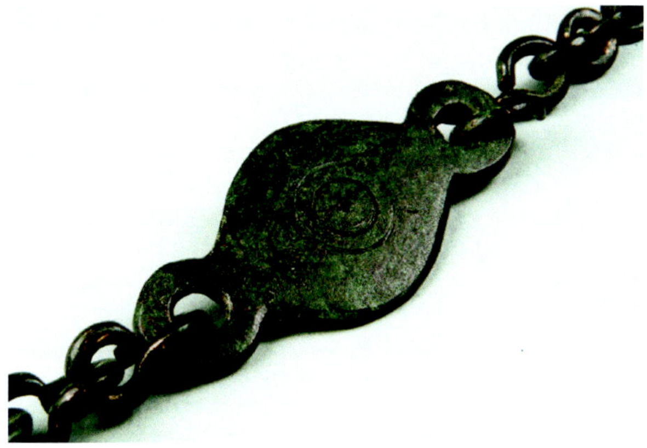

Figure 9. Part of the copper alloy chain of a polykandelon (inv. no. BX 253/716).

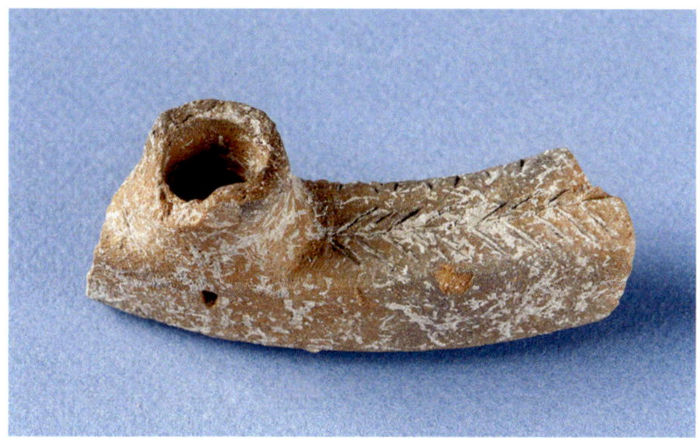

Figure 10. Fragment of a clay lamp (inv. no. BK 4513/135).

the surviving parts of the chains from Louloudies, formed of figure eight shaped links and interrupted by flattened elliptical plates for safer operation, are examples from Eleftherna, Gortyn and Porto Lagos, which also preserve data regarding their cross-shaped endings.[27]

Quite interesting is the finding of a multi-nozzled, ring-shaped clay lamp (Figure 10), [28] which suggests multiple influences from different materials and objects of similar use, and also works as a clay imitation of corresponding copper polykandelon. The multi-nozzled, ring-shaped lamp, was probably known as a *στέφανος*, as is indicated from a list of lighting devices on an Egyptian ceramic shred,[29] where among various denominations for lighting instruments the word *στέφανος* is mentioned referring perhaps to a circular ring clay lamp with multiple nozzles.

In an effort to identify the lighting methods used by the residents of the episcopal complex in the wider span of its chronological phases, the widespread use of earthen lamps for home lighting needs is easily discernible. The evidence, however, changed with the construction of the episcopal complex, as more complex methods were used for the illumination of the basilica and episcopal palace: with the suspension of metal lamps or their placement on lamp stands and, mainly for the basilica, with the use of copper alloy chandeliers in the form of polykandelon. The probable increase and variety in lighting methods suggests that while the episcopal complex flourished, there were increased energy requirements. This would account for the operation

[27] Xanthopoulou 2010, 54-55.

[28] Angelkou and Cheimonopoulou 2011, 246.

[29] Xanthopoulou 2010, 68.

of two oil mill installations, whose remains were found in the northwestern part of the complex. These mills might have been able to satisfy with relative ease the necessary quantity of oil.

From material study we conclude that, during the last phases of the episcopal complex in the 7th-8th century, newly manufactured lamps, morphologically peculiar to the productions of the 5th-6th century but of an inferior quality, such as the lamp with a flat disk decorated with a Latin cross (a deep oil container/lower part) and an appearance betraying incomplete firing conditions during manufacture, were used together with older lamps, such as the African lamps of the 5th-6th century, found in a destruction layer with eighth-century ceramic finds and clipped Late Roman (4th-5th century) numismatic finds.

Bibliography

Angelkou, E. and Cheimonopoulou M. 2011. Lighting devices from excavations in Pieria, in I. Motsianos and E. Bintsi (eds) *Light on light, an illuminating story, Exhibition Catalogue*, 237-253. Thessaloniki: Folklife & Ethnological museum of Macedonia-Thrace.

Bailey, D. M. 1988. *A catalogue of the lamps in the British Museum, 3. Roman provincial lamps*. London: British Museum Publications.

Bass, G. and van Doornick, F. H. 1982. *Yassi-Ada I, A seventh-century Byzantine shipwreck*. College station: Texas A & M University Press.

Bouras, L. and Parani, M. G. 2008. *Lighting in Early Byzantium*. Washington, D.C.: Harvard University Press.

Cheimonopoulou, M. and Angelkou, E. 2000. Λυχνάρια με παραστάσεις από το επισκοπικό συγκρότημα στις Λουλουδιές Πιερίας, *Θεσσαλονικέων Πόλις*, Second Period, v. 2, 127-134.

Gough, M. 1975. Dag Pazari: The Basilical Church 'extra muros', in Robertson G. and G. Henderson (eds) *Studies in Memory of David Talbot Rice*. Edinburgh: Edinburgh University Press.

Hayes, J. W. 1972. *Late Roman Pottery*. London: British School at Rome.

Hoff, V. 1986. Lampes romaines tardives et lampes chrétiennes en terre cuite, in Lyon-Caen, C. and V. Hoff (eds) *Musée du Louvre, Catalogue des lampes en terre cuite grecques et chrétiennes*. Paris: Éditions de la Réunion des musées nationaux.

Marki, E., Angelkou, E. and Cheimonopoulou, M. 2010. Τα λυχνάρια του επισκοπικού συγκροτήματος των Λουλουδιών, (The lamps from the bishop's complex at Louloudies), in D. Papanikola-Bakirtzis and D. Koussoulakou (eds) *Κεραμική της Ύστερης Αρχαιότητας από τον Ελλαδικό χώρο*, Επιστημονική Συνάντηση, Θεσσαλονίκη 12-16 Νοεμβρίου 2006, 208-216. Thessaloniki: Archaeological Institute of Macedonian and Thracian Studies (in Greek with English abstract).

Marki, E. 1993. Ανασκαφή στα Λουλούδια Κίτρους (Excavation at Louloudia, Kitros), *Το Αρχαιολογικό Έργο στη Μακεδονία και Θράκη* 7, 223-234.

Marki, E. 1994. Ανασκαφή στις Λουλουδιές Κίτρους (Excavation at Louloudia, Kitros). *Το Αρχαιολογικό Έργο στη Μακεδονία και Θράκη* 8, 151-158.

Marki, E. 1995. Ανασκαφή Λουλουδιών 1995 (Excavation at Louloudia, Kitros). *Το Αρχαιολογικό Έργο στη Μακεδονία και Θράκη* 9, 195-202.

Marki, E. 1996. Συμπεράσματα από τις ανασκαφές της 9ης Εφορείας Βυζαντινών Αρχαιοτήτων στη Βόρεια Πιερία. *Αρχαιολογικό Έργο στη Μακεδονία και Θράκη* 10A, 239-258.

Marki, E. 1997a. Λουλουδιές 1997. *Το Αρχαιολογικό Έργο στη Μακεδονία και Θράκη* 11, 289-296.

Marki, E. 1997b. Ανασκαφή στη θέση Λουλουδιές Κίτρους, in Ath. E. Karathanasis, (ed.) *Η Πιερία στα Βυζαντινά και Νεότερα χρόνια*, Acts of 1st Congress, Katerini 25-28/11/1993, 167-185. Thessaloniki.

Marki, E. 1998. Oil lamps from the Episcopal Complex at Louloudies. *Museum of Byzantine Culture* 5, Short studies, 23-29 (also in Greek and French).

Marki, E. 1999a. Λουλουδιές 1999. *Το Αρχαιολογικό Έργο στη Μακεδονία και Θράκη* 13, 425-433.

Marki, E. 1999b. Μία άγνωστη πόλη των παλαιοχριστιανικών χρόνων στην Πιερία, in *Πρακτικά 6ου Συμποσίου 'Αρχαία Μακεδονία', Thessaloniki 15-19 October 1996*, 723-733. Θεσσαλονίκη, Ινστιτούτο Βαλκανικών Σπουδών, v. 2.

Marki, E. 2001. *Κίτρος. Μια πόλη-κάστρο της βυζαντινής περιφέρειας, αρχαιολογική και ιστορική προσέγγιση*. Θεσσαλονίκη: Τα τραμάκια.

Marki, E. 2002a. Από το ρωμαϊκό στο βυζαντινό κάστρο. Δύο παραδείγματα από την Πιερία. Το τετραπύργιο στις Λουλουδιές και το κάστρο της Πύδνας (βυζαντινό Κίτρος), *Θεσσαλονικέων πόλις*, Second Period, vol. 7, 121-130.

Marki, E. 2002b. Λουλουδιές Κίτρους: Η βασιλική του επισκοπικού συγκροτήματος, in N. Γραίκος (ed.) *Η Πιερία στα Βυζαντινά και Νεότερα χρόνια, Πρακτικά του 2ου Συνεδρίου, Κατερίνη, 27-29/11/1998*, 692-706. Κατερίνη: Εστία Πιερίδων Μουσών Κατερίνης-Τέρτιος.

Marki, E. 2004. Χωροθέτηση παραγωγικών και εργαστηριακών δραστηριοτήτων στο επισκοπικό συγκρότημα των Λουλουδιών Πιερίας, in *Αρχαιολογικά τεκμήρια βιοτεχνικών εγκαταστάσεων κατά τη βυζαντινή εποχή 5ος-15ος αιώνας, Χριστιανική Αρχαιολογική Εταιρεία, Athens, 17-19 May 2002*, 27-45. Athens: Piraeus Bank Group Cultural Foundation.

Marki, E. 2008. Το Επισκοπικό Συγκρότημα στις Λουλουδιές *Κίτρους*. in *Αρχαιολογικές Τροχιοδρομήσεις, από τη Θεσσαλονίκη στον Πλαταμώνα*, 88-115, Αθήνα: IDEA A.E. (in Greek).

Marki, E. 2013. Το αρχαιολογικό εύρημα ως ιστορικό τεκμήριο. Η εγκατάσταση των Γότθων στην πεδιάδα της Κεντρικής. Μακεδονίας και η ίδρυση του επισκοπικού συγκροτήματος στις Λουλουδιές (The Archaelogical find as Historical Evidence. The Settlement of Goths in the Plain of Central Macedonia and the Establishment of the Episcopal Complex at Louloudies), *Δελτίον Χριστιανικής Αρχαιολογικής Εταιρείας* 34, 1-10. (in Greek with English abstract).

Miller K., *Itineraria Romana*. Stuttgart: Strecker & Schröder 1916.

Xanthopoulou, M. 2010. *Les lampes en bronze a l' époque paléochrétienne*. Turnhout: Brepols.

Bottoms Up!
Bases and Handles on Attic KY Shop Lamps from Corinth's Fountain of the Lamps

Karen S. Garnett
Morgan Hill, California

Abstract: Excavations in the Gymnasium area of Ancient Corinth from 1965-1972 uncovered a large deposit of over 4000 lamps in a bath complex now referred to as the Fountain of the Lamps. Digital cameras now make it economical to take and evaluate images of not only lamp disk and rim patterns, but also of base, wall and handle components. Examples from the Fountain of the Lamps help clearly identify patterns in the evolution of base and handle design for lamps common in the Peloponnese during the Late Antique / Early Byzantine era up to the time of the collapse of the Fountain roof late in the 6th century. Most lamps found in the Fountain of the Lamps are of Attic or local manufacture, and they fit comfortably into Broneer's typology. Post-Broneer research has identified additional subdivisions within the types. This article summarizes the main evolutionary characteristics of KY Shop lamp bases, with a focus on Fountain finds from this single Attic lampmaker. At least eight different base groove patterns illustrate the chronological evolution in design for the KY Shop, which produced a variety of Broneer 28 subtypes.

Keywords: Attic lamps, Late Roman lamps, KY Shop, lamp bases, Corinth, Gymnasium Excavations, Fountain of the Lamps, terracotta lamps.

Introduction

The Fountain of the Lamps is located in the Greek Peloponnese in the environs of Ancient Corinth, northeast of the Temple of Apollo at the edge of a natural terrace where a number of other sacred or healing springs and fountains exist.[1]

Excavations in the Gymnasium area of Ancient Corinth from 1965-1972[2] uncovered a large bath pool as well as a number of terracotta lamps deposited in a recessed grotto within an underground cliff hollow. The vaulted grotto contained six water catch basins and is now referred to as the Fountain of the Lamps (Figure 1). Most of the excavation was carried out under the water table, as the grotto was actually discovered from the back, i.e., by digging in a water channel (Figure 2). Much of the excavation was done by first pumping out water, then removing mud. That there is any stratification to the deposits is amazing.[3]

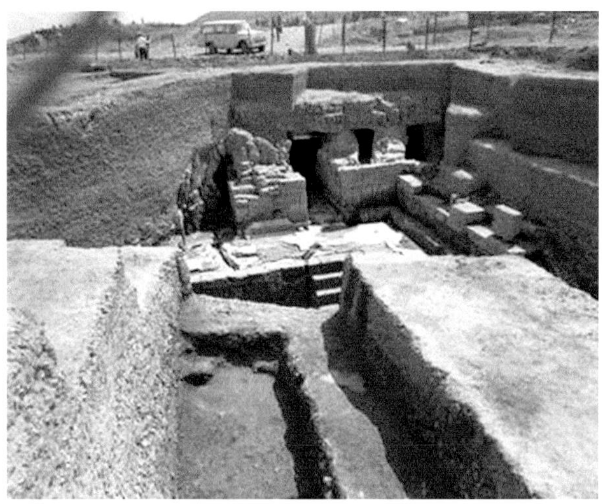

Figure 1. Fountain of the Lamps Entrance. Gymnasium area, view looking south, showing partially excavated pool area and entrance to subterranean bath complex, where the major lamp deposits were located. (Corinth photo 1970-152-06)

Deposit Chronology

The associated deposits contained over 4000 lamps, approximately 2000 intact. An overview of the locally produced Late Roman lamps from the deposit was published.[4] The lamps from the Fountain are well preserved Late Antique / Early Byzantine examples, which date up to the time of the collapse of the Fountain roof, estimated by context coins to have occurred late in the 6th century. There are few coins associated with the lamp deposits. Seven lots from the basin area and entrance corridor had large numbers of intact lamps.

So many lamps were found that not all were inventoried. Those which appeared to be duplicates of already inventoried examples were stored with their lot. The largest of these lamp storage lots was Lot 6103, which extended from bench level to the floor in the basin area (Figure 3). Over 750 intact lamps were found in this lot, but the lot produced only two coins, only one of which was partially datable (Coin identification provided by Dr James Dengate and Dr Michael Ierardi). The range of datable coins for the ten largest lamp deposits is AD 341-547 (Figure 4). The earliest datable coin is C1969 612

[1] Landon 2003, 47, fig. 3.1 and 49, fig. 3.2.
[2] Wiseman 1969, 1970, 1972.
[3] I would like to thank especially Director James R. Wiseman for his past and present support and helpful advice. Thanks are also due many other lychnological and archaeological colleagues who expand my horizons and suggest new avenues of investigation.
[4] Garnett 1975.

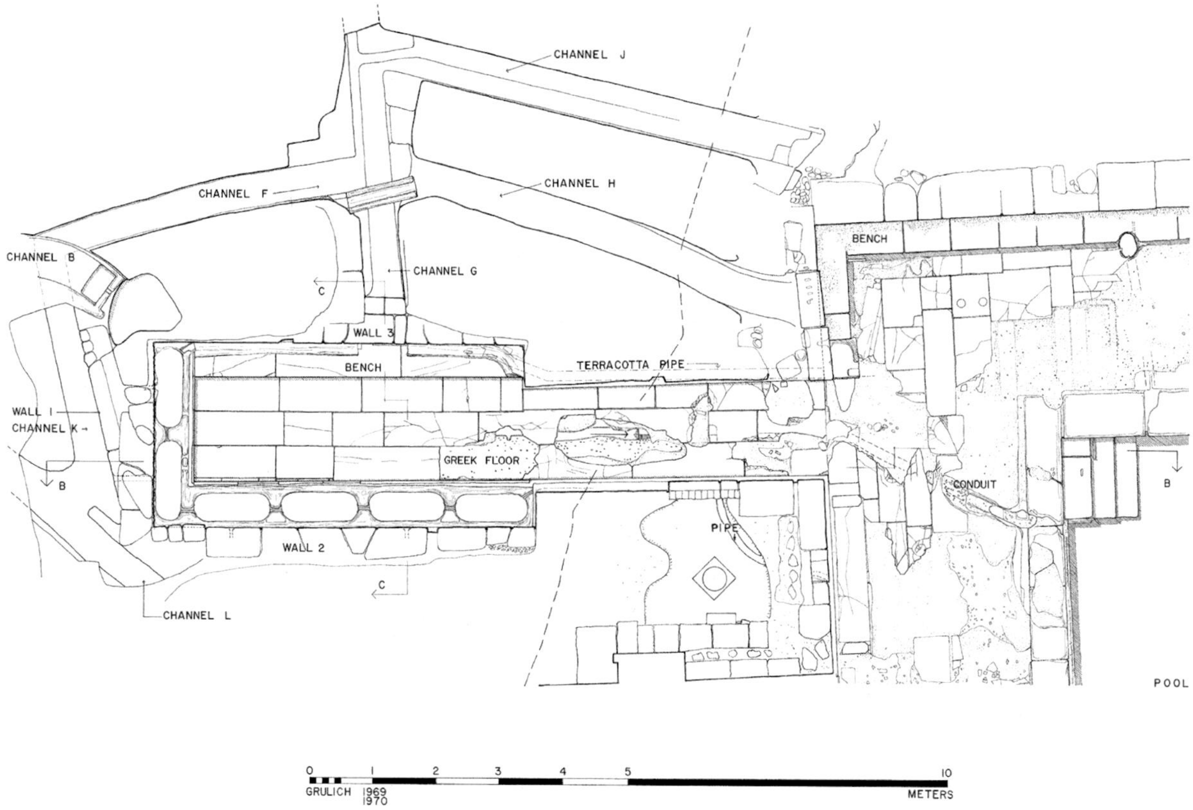

Figure 2. Fountain of the Lamps. Plan showing pool edge, water channels and bath catch basin area where lamps were found (Corinth Gymnasium Drawing 831).

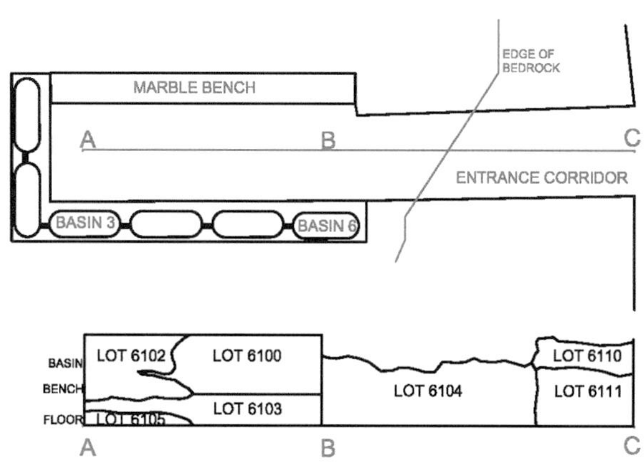

Figure 3. Fountain of the Lamps Section showing lot locations for the large lamp lots.

from Lot 6103, dated to the reign of Constantius II, 351-361 AD. Other coins from the lower level lots (Lots 5419 and 6105) can also be dated to the reign of Constantius II. There are no other legible coins in the large lamp lots that date to the reign of Constantius I or earlier. On this basis we might assume that the lamp deposits did not commence before the reign of Constantius II; from Lot 6105, see especially C1969 660, dated 341-345 AD Constantius, mint Alexandria or Nicomedia, and also C1969 530, 534, 535, and 634. The latest coin from the low level Lot 6105, the only coin in that lot dated to the 5th century, is assigned to the reign of Honorius and dated 408-423 AD. As with any of the coins in the fluid deposits of the Fountain, this coin could represent a migration from above. The latest coins are found in mud from a basket of Lot 6104 and include C1969 561, dated 546-547 AD in the era of Justinian I, and a coin of Anastasius I, C1969 611 dated to 491-498 AD. Numerous unclassified or 'vandal' coins suggest use well into the 6th century.

Lamp Types found in Fountain of the Lamps

Most lamps from the Fountain of the Lamps are of Attic or local manufacture, and they fit comfortably into Broneer Types 28, 29, 31 and 32 (Figure 5) of the typology established by Oscar Broneer.[5] These are broad categories, and post-Broneer research has identified additional subdivisions within the types.[6]

With the advantages provided by digital photography, it is now economical to take and evaluate images of not only lamp disk and rim patterns, but also of the less interesting components of a lamp's design, such as the base, wall and

[5] Broneer 1930.

[6] Broneer 1977; Slane 1990; Slane 2008; Slane and Sanders 2005.

Lot	Relation	Coin count Total (# legible)	Coin Date Range	KY count Total (glazed, unglazed, copy)	Whole Lamp Count
6100	Over 6103	1 (0)	unclassified	1 (1c)	58
6102	Over 6103	1 (0)	unclassified	1 (1u)	40
6098	FOL Basins Fallen Fill	3 (2)	388-392	2u	20+
6101	FOL Basin /Bench	3 (2)	378-408	2 (1g/1u)	19+
6110/11	FOL entrance	10 (8)	364-378	0	72
6104	FOL Corridor	9 (7)	364-547	1g	48
6103	Over 6105	2 (1)	341-361	2 (1u/1c)	750
6105	FOL floor	10 (8)	341-423	3 g	228
5419	underwater	45 (29)	341-408	18(9g/9u)	183

Figure 4. Table of Ten Largest Lamp Deposits, Coin date range and count of KY Shop lamps.

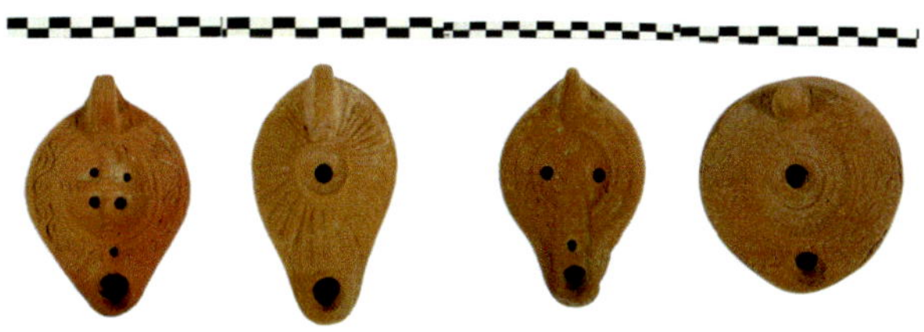

Figure 5. Typical Broneer Types 28, 29, 31 and 32 lamps from the Fountain of the Lamps.
Left: Attic Type 28 L1969 83
Middle and Right: Corinthian: Type 29 L1969 116, Type 31 L1969 162, and Type 32 L1969 135.

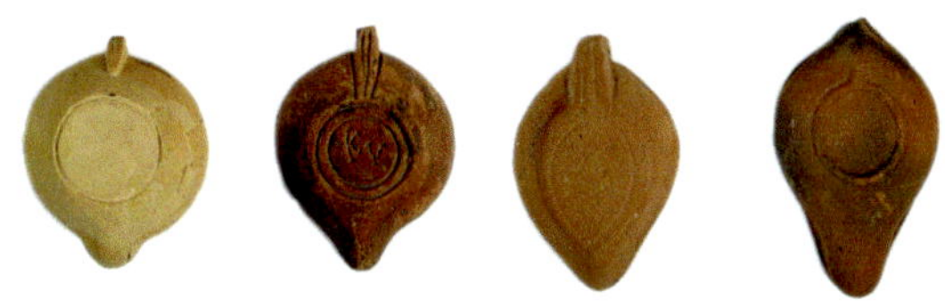

Figure 6. Major Base Styles from Corinth.
Left: Single circular groove. Type 27.
Middle left: Double circular grooves Type 28 L1969 215.
Middle right: Double almond grooves Type 28. L4592.
Right: Base ring Type 31 local L1969 177.

handle. Reference material incorporating images of these lamp components can facilitate accurate field identification of lamp fragments by subtype. Ultimately, this will help provide refined date ranges useful when evaluating context pottery. Looking at examples from the Fountain of the Lamps, we can recognize patterns in the evolution of base and handle design for lamps common in the Peloponnese during Late Antique / Early Byzantine times.

Base Typology Transitions from Broneer Type 27 to Type 31

The major steps in the evolution of the base and handle treatments for lamps of the late 4th and 5th centuries can be seen in Figure 6.

- On the left is a Broneer Type 27 lamp, a lamp type not found among the Fountain lamps, but is provided here for comparative reference. A Type 27 lamp base and handle can be characterized by its unglazed, round body, its large base with a single circular groove, and a grooved handle that is pierced and does not reach to the base. The fabric is usually thin walled, well levigated and evenly fired, and of a light pink buff hue. Many examples are signed.
- The lamp base on the right in Figure 6 represents a base of Broneer Type 31, or North African Hayes II type.[7] The example lamp is contemporary with some of the late Type 28 lamps. Broneer Type 31 lamps usually have a ring foot, no signature, and a ridge connecting the ring foot base to a solid fin-like handle at the rear of the lamp.
- The two lamps in the middle of the row in Figure 6 show some of the shape evolution in lamp bases from the late 4th into the 6th century. They are both Broneer Type 28 lamps. Type 28 is the most numerous type found in the Fountain (536 examples), and Type 28 subtypes include lamp bases which have been variously described as: single circular groove, double circular grooves, tear-drop shaped grooves, single almond groove, double almond shaped grooves, heart-shaped grooves, leaf shaped grooves, ring foot, oval ring foot, and more.

This paper attempts to refine the typological changes and chronological flow for Broneer Type 28 lamps of the late 4th through mid-6th centuries AD, by looking specifically at the bases and handles produced by a single Athenian lampmaker well represented by examples from the Fountain of the Lamps. This lampmaker is the KY Shop, one of the larger Athenian lampmakers of the Late Antique, and a shop which made neither Type 27 nor Type 31 lamps. Detailed discussions of the shop have been previously published.[8] Karivieri's more recent dating has the shop's activities range from the end of the 4th century till early in the 6th century, a time window

[7] Hayes 1972.

[8] Perzlweig 1961, 41; Karivieri 1996, 110.

that is slightly narrower than, but well within, the time span of lamp deposit activity in the Fountain. Lists of disk and rim motifs are provided by both Perzlweig and Karivieri, and will only be mentioned in this article in their general relationship to design changes in bases and handles.

KY Shop Base Typology

KY Shop lamps from the Fountain provide a sample of good working size with which to identify the shop's development flow. Some 37 intact or inventoried examples which might be products from the KY Shop were found in Gymnasium and Fountain of the Lamp deposits. Sixteen of these have the shiny slip glaze which groups them in the category referred to as Broneer Type 28 Attic glazed. There are five signature variants in the samples:

- **KY**: typical and clearly recognizable signatures: 18 (eight unglazed).
- **KV**: six signatures with KV (three unglazed).
- **KY + branch**: two examples (one unglazed)

The following lamps have a K on their base but do not appear to be of Attic manufacture, and are expected to be later fifth or sixth-century unglazed surmoulage copies of lamps by the KY Shop or produced by another as yet unknown K- lampmaker. These examples are excluded from the typology.

- **K**: One fountain example was signed with a K within a small single circular groove and appears to be an import from an as-yet unidentified locale.
- **K***: one in the local Corinthian fabric.
- **K + two circles**: two in the local Corinthian fabric.

The remaining seven lamps are preserved fragments which may be KY Shop, but do not provide enough base or profile to add value to this typological investigation.

The range of the KY Shop is representative of the range found for Broneer Type 28 lamps, not only those of Attic manufacture but also lamps of local manufacture which mimic trends visible in the KY Shop repertoire. The basic trends include:

a) migration from glazed (technically a slip, but long referred to in the literature as a glaze, in this paper a slip glaze) to unglazed,
b) migration from figurative disk motifs to plain or Christian motifs, and
c) shape migration from round bodied lamp to a longer more elliptical boat-like shape with more vertical walls.

These trends are well represented by the early slip glazed lamp in Figure 8 (L1969-215) and the later unglazed lamp with almond base grooves represented in Figure 14 (L 4592).

The uppermost and outermost layers of the Fountain deposits had no KY Shop lamps. Unglazed KY Shop lamps are found below the latest deposit layers of the Fountain, while most of the slip glazed examples derive from Lots 5419 and 6105, i.e., from the innermost area and floor in the Fountain. Only four of the large lamp lots had any of the early slip glazed KY Shop products (Lots 5419, 6105, 6101 and 6104).

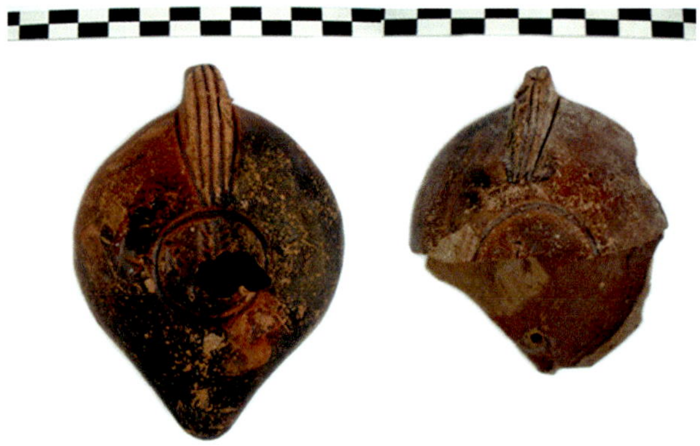

Figure 7. KY Base Group 1. (single circular groove, slip glaze) L4595, L4699.

Based on the Fountain of the Lamps examples, eight different Base Groups can be differentiated for the KY Shop. The first three groups are found on the slip glaze lamps. These are:

Group 1 – single circular groove
Group 2 – two circular grooves
Group 3 – two small circular grooves

Slip Glazed Lamps of the KY Shop

KY Base Group 1: single circular groove, slip glaze

KY Base Group 1 (Figure 7 and Figure 8, top) is characterized by a base which has a single circular groove. KY Shop lamps with this base have a solid handle, and the lower portion of the handle is flanked by grooves. There are two or three vertical grooves on the lower portion of the handle. Often the handle grooves terminate almost in a point at or very near the edge of the single circle. The KY letters are positioned horizontally across the base of the lamp. Unfortunately, the examples from the Fountain of the Lamps are not complete, though they are characteristic Attic in their clay and typical of KY workmanship, and have at least part of what looks like the letter K on their bases.

KY Base Group 2: two circular grooves, slip glaze

KY Base Group 2 (Figure 8, middle) is characterized by a base with a double circular groove. The handle is solid as in the previous base group. The handle may have three grooves, but more frequently, it has two grooves on the upper portion, and the termination of the handle where it meets the base grooves is trending wider.

KY Base Group 3: two small circular grooves, slip glaze

KY Base Group 3 (Figure 8 bottom and Figures 9, 10) lamps have a double circular groove on the base but the circles are smaller and even less concentric than the double grooves found in Group 2. The handle is solid, wider below, and the termination rarely meets the circular grooves. This group is the most common among the slip glazed KY Shop lamps found in the Fountain. There are subtle differences within the lamps

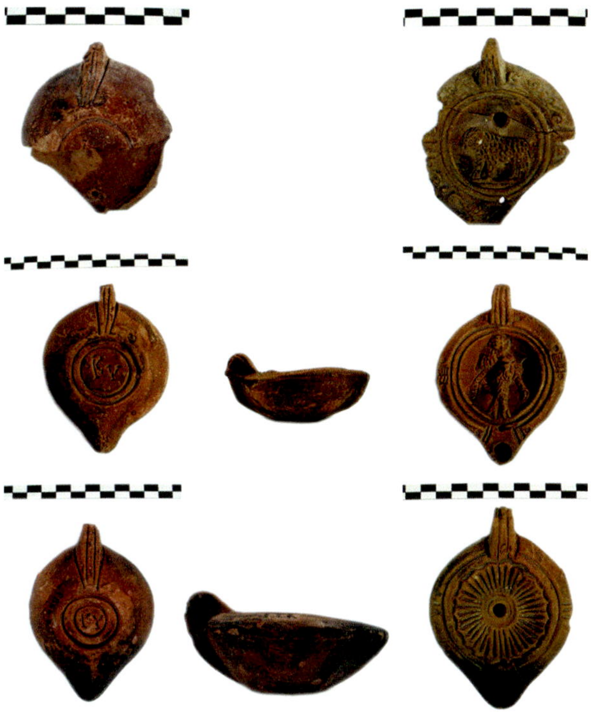

Figure 8. KY Base Groups 1-3.
All with slip glaze. Top to bottom:
Group 1: L4699 (single circular groove, slip glaze)
Group 2: L1969 215 (two circular grooves, slip glaze)
Group 3: L4583 (two small circular grooves, slip glaze)

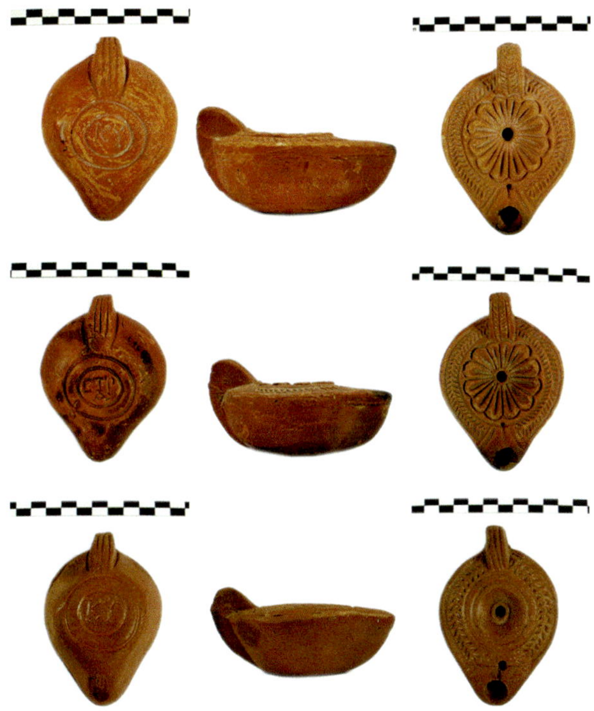

Figure 10. KY Base Group 3 (cont.)
Top: L4582
Middle: L4613 (Stratolaos lamp shop)
Bottom: L1969 24

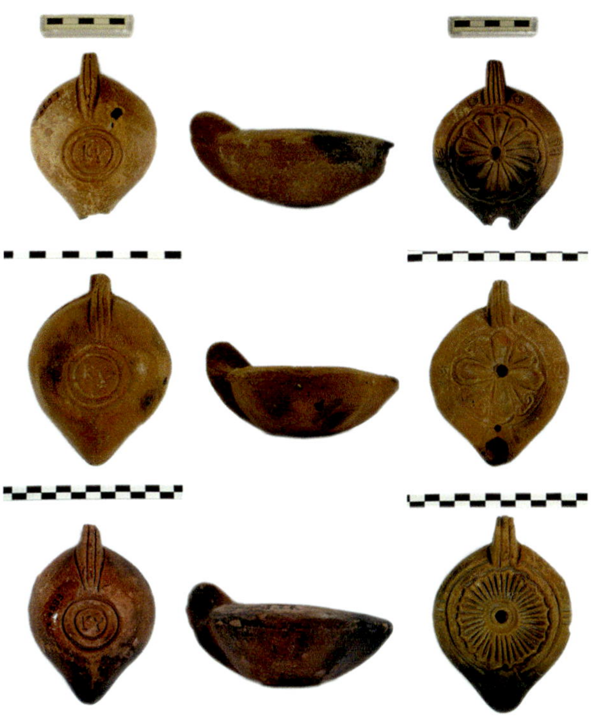

Figure 9. KY Base Group 3.
(two small circular grooves, slip glaze)
Top: L4392 (found in Gymnasium area, not in FOL)
Middle: L1969-391
Bottom: L4583

of this base group, and it is possible these are chronological differences, but they could also represent simple stylistic differences in expression among the works of two brothers, or a father and son.

When looking at the profiles of the three KY Base Groups for slip glazed lamps (Figure 8), one notices that all groups share a round body shape, but for Group 3, the nozzle is lower than the rim area, while Group 2 has a slightly elevated nozzle. The handles for all groups are solid and the angle at the rim increases so that it becomes closer to 45°. The later examples appear to have less mass and not as sharp a turn on the lower portion.

The glaze lamp products of the KY Shop have disks that are figured, plain or rosette. The motifs may be worn or over-incised, indicating the shop is not working from fresh molds. The rims may have panels, a remnant from Type 27 lamps, though herringbones and wavy lines start to appear. The nozzle is outlined by single or double kite-shaped grooves. The wick mover hole appears at the rim, but there is no groove from it to the nozzle as is common in later examples.

KY Base Group 3 (Figure 9) has more examples from the Fountain and they are more complete than examples from the earlier two groups, possibly indicating a more robust or at least consistent export program. Notice again how the nozzle is lower in profile, and the handle angle is consistent. There is one fill hole, and one air/wick adjusting hole. The disk motifs for Group 3 are usually plain or rosette, and rarely figured.

In KY Base Group 3 (Figure 10) we start to see crisp designs for non-figured lamps such as rosette disks and wavy line or herringbone rims. Some sort of craze seems to start, and it includes herringbone rims. Panels disappear, though some rims have circles positioned where the panels would be. The shape begins to elongate, but very subtly, and the walls are more vertical, especially in the lower example L1969 24. Notice how the base of the handle is much wider now and there is enough room for three grooves. That there is close partnership and interaction between the Athenian lampmakers is noticeable when we compare the KY Shop lamp L4582 (Figure 10 top) with an almost identical Stratolaos lamp from the Fountain L4613 (Figure 10 middle).

Unglazed lamps of the KY Shop

At some point the KY Shop, along with most other Attic lampmakers, ceased using the red brown slip glaze with which they were 'glazing' their lamps. Later Type 28 lamps are unglazed. This transition has been difficult to precisely date, and the literature provides dates ranging from *c.* 360 AD[9] to the middle of the 5th century.[10] Whenever the transition did occur, it probably occurred over a number of years beginning sometime late in the 4th century and completing sometime near the end of the first quarter of the 5th century, as inventories of and requests for 'old-style' slip glaze lamps disappeared.

Five Base Groups can be differentiated for the unglazed lamps of the KY Shop:

Group 4 – Two circular grooves
Group 5 – Two teardrop grooves
Group 6 – Experimental grooves
Group 7 – Two almond grooves
Group 8 – Two almond grooves set back

KY Base Group 4: two circular grooves, unglazed

KY Base Group 4 (Figure 11) represents the transition from slip glazed to unglazed lamps for this shop. The designs tend to be crisp, and the overall shape is similar to the later crisp shapes seen in Group 3 lamps. The circular grooves on the base may be large or small. The handle may be pierced or partially pierced. The disks have multiple holes, a central larger hole and accessory holes, and the KY Shop examples found to date never have a figured representation. The rims never have raised panels, though grooved lines or circles in the panel area do appear, albeit very infrequently.

There are some characteristics common to all the Base Groups for the KY unglazed lamps. They have no slip glaze, no figurative motifs, and multiple fill/air holes. The handle is usually pierced or semi-punched; in fact, a solid handle is a good indicator that the lamp may be an imitation, instead of a true product of the Athenian KY Shop. There is an air of experimentation that is evident in the earlier groups of unglazed lamps, but this is absent by the time the shop ceases production.

[9] Perlzweig 1961, 64.
[10] Slane 2008, 488-489.

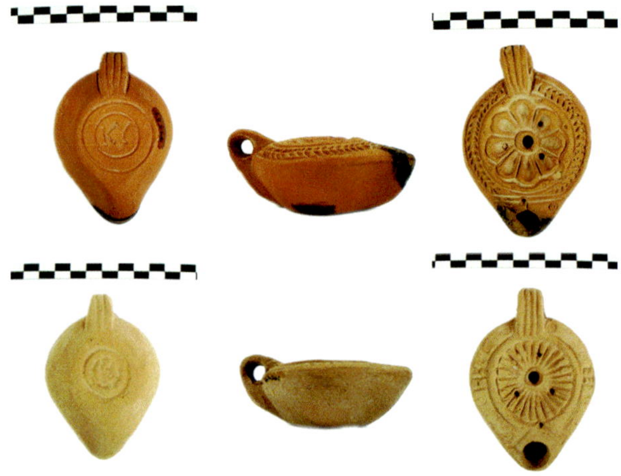

Figure 11. KY Base Group 4.
(two circular grooves, unglazed)
Top: L4566
Bottom: L1969 264

KY Base Group 5: two teardrop grooves, unglazed

KY Base Group 5 (Figure 12) is represented by only two examples from the Fountain of the Lamps. Two teardrop shaped grooves, with the points meeting at the nozzle end is a base motif found on a variety of Attic and local lamps, but it is never as popular a design as the double almond grooves found in the later types. The handle is partially pierced and seems to settle down to two grooves below. The angle between rim and handle is greater. The nozzle is now usually set off by a double groove, and often has a groove from rim to wick hole. The trend towards a simpler design using quality clay can be seen.

KY Base Group 6: experimental grooves, unglazed

KY Base Group 6 is not represented by Fountain examples, but excavations in Athens provide excellent examples for this group.[11] These base motifs seem experimental: 'let's try this, let's try that'. The signatures on these examples vary more than the KY Shop examples from the Fountain.[12] Perhaps these represent the product of a youthful lampmaker who is excited about taking over the business and full of ideas. Or the business is currently very profitable and attracting good talent. All the lamps mentioned above are unglazed except for Perlzweig 1431. Even handle piercing varies; of the examples, two are solid (1423 and 1431), one pierced (1141), one uncertain (1761) and one not preserved (2213). Why were there no examples of these bases in the Fountain of the Lamps? Were these truly experimental designs and not fit for export? Or was there a disruption in the export traffic at the time this experimentation was going on?

[11] See Perlzweig 1961, pl. 35: 1141 (leafy teardrop), 1431 (two heart), 1761 (one almond), and 2213 (one teardrop).
[12] See Perlzweig 1961, pl. 35: 1431 and 2213 (KY /incised branch), 1761 (K with no Y), and 2213 (letter bottom at handle end instead of nozzle).

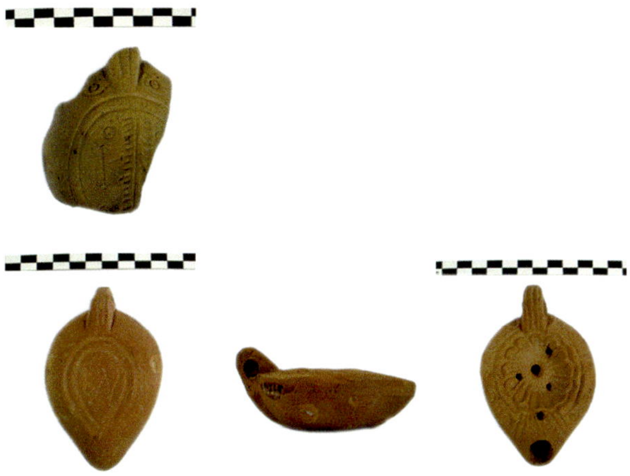

Figure 12. KY Base Group 5.
(two teardrop grooves, unglazed)
Top: L4535
Bottom: L1969 265

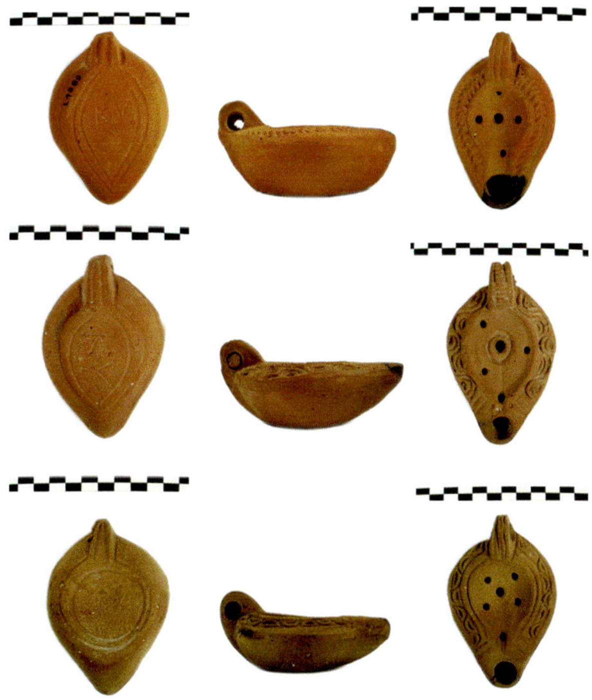

Figure 13. KY Base Group 7A.
(ovoid disks, varied bases, unglazed)
Top: L4589
Middle: L4590
Bottom: L1969 258

KY Base Group 7: two almond grooves, unglazed

KY Base Group 7A (Figure 13) is a transitional grouping for the typology flow and includes the ovoid and channeled disk lamps produced by the Attic KY Shop. The base of these lamps is often consistent with Base Group 7, but can sometimes be more experimental as in Group 6 or can be closer to the base motif for Group 8. The plain disk may be ovoid in shape (unlike the lamps of all the other Base Groups which have a circular disk). Rim patterns are limited to herringbone or wavy line.

KY Base Group 7 (Figure 14) includes those lamps with the typical and common double almond grooves on the base. The grooves are open at the handle, and usually meet at the nozzle end. The base on these lamps widens. The signature appears sideways on these lamps, when it is KY; however, KV and KΛ signatures are usually not sideways which provides an additional indication that these letters may represent lampmakers other than the KY Shop. The handle is two grooved below and partially pierced. The walls become more vertical. The nozzle is usually set off by double grooves.

Disk motifs found with this base group do not show creativity or experimentation. Interestingly, there are no rosette disks in the Fountain of the Lamps examples in this Base Group. Of the motifs that do appear there are three plain, three shell, one ray, and two cross. This is the first group with Christian disk motifs. Rims are herringbone, wavy line, or plain with circles.

KY Base Group 8: two almond grooves set back, unglazed

The last group is KY Base Group 8 (Figures 15 and 16). These lamps have a double almond groove base, with KY set sideways, similar to the previous group. However, on these later lamps the general position and shape of the lamp has evolved. Note how the base grooves are wide, and set back towards the handle. The walls of the lamp are closer to vertical than previous examples, and the general impression is more elliptical, similar to a boat-like shape. The lower part of the handle does not extend as much beyond the lamp body.

Christian motifs are common on lamps of this shape. Motifs appear limited to plain, shell and cross. There are no plain rims with this base group from the Fountain, only herringbone and wavy line rims are now in use.

The KY Shop miniature lamps (Figure 15, bottom) found in the Fountain can be grouped with Base Group 8. About two-thirds standard size, are these lamps being manufactured specifically as an inexpensive single-use votive? Or are they a child's toy? Or a mni 'travel' flashlight?

In Conclusion

These KY Shop examples from the Fountain of the Lamps illustrate that there is clear value to examining not only disk and rim motifs on lamps, but also to researching the evolution of base and handle characteristics. Trends are clearly identifiable. Hopefully, this research will help us refine our lamp timelines. At this writing, the detailed task of analyzing and coordinating the Fountain deposit chronology and coordinating it with modern research is still on-going. Even though there are a large number of KY Shop lamps in the Fountain deposits, given the liquid nature of those deposits, it is unlikely that the Fountain of the Lamps will provide the evidence which will allow us to isolate specific decades for the chronology of each base group. Any attempt to do so would

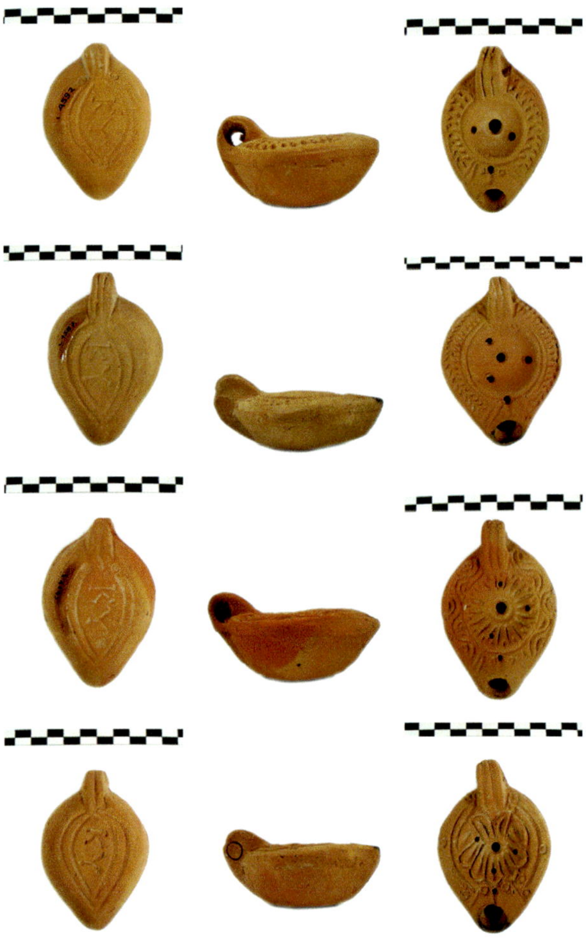

Figure 14. KY Base Group 7.
(two almond grooves, unglazed)
Top: L4592
Middle: L4587
Middle Lower: L4593
Bottom: L1969 243

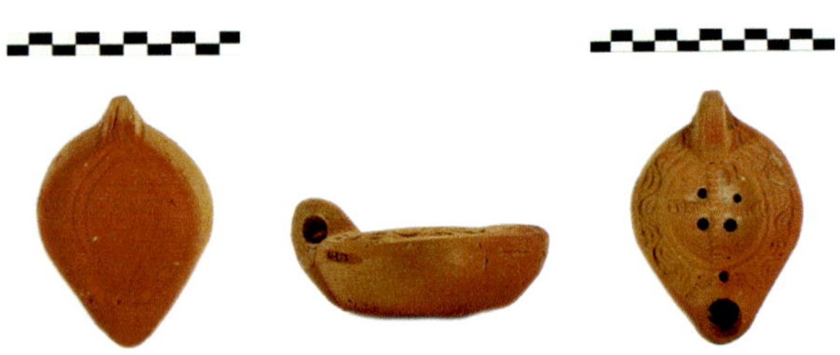

Figure 15. KY Base Group 8.
(two almond grooves set back, unglazed)
L1969 83

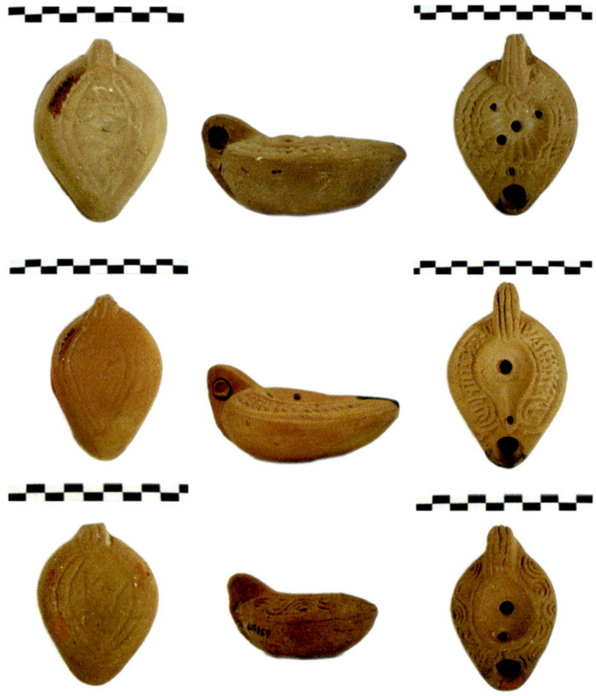

Figure 16. KY Base Group 8 (cont).
Top: L4586
Middle: L4588
Bottom: L1969 257 (miniature)

certainly be worthwhile, but at this point in time would also be premature.

Base and handle typologies might be especially useful in the field where only a lower fragment of a lamp base or handle is found. Useful when evaluating localized collections, such examinations provide an economical means to help further distinguish imports from local wares, and one lampmaker from another, even when lamps are not signed.

For quick reference Figure 17 provides base images of the lamps discussed in the paper, arranged in a array reflecting their likely evolutionary positions.

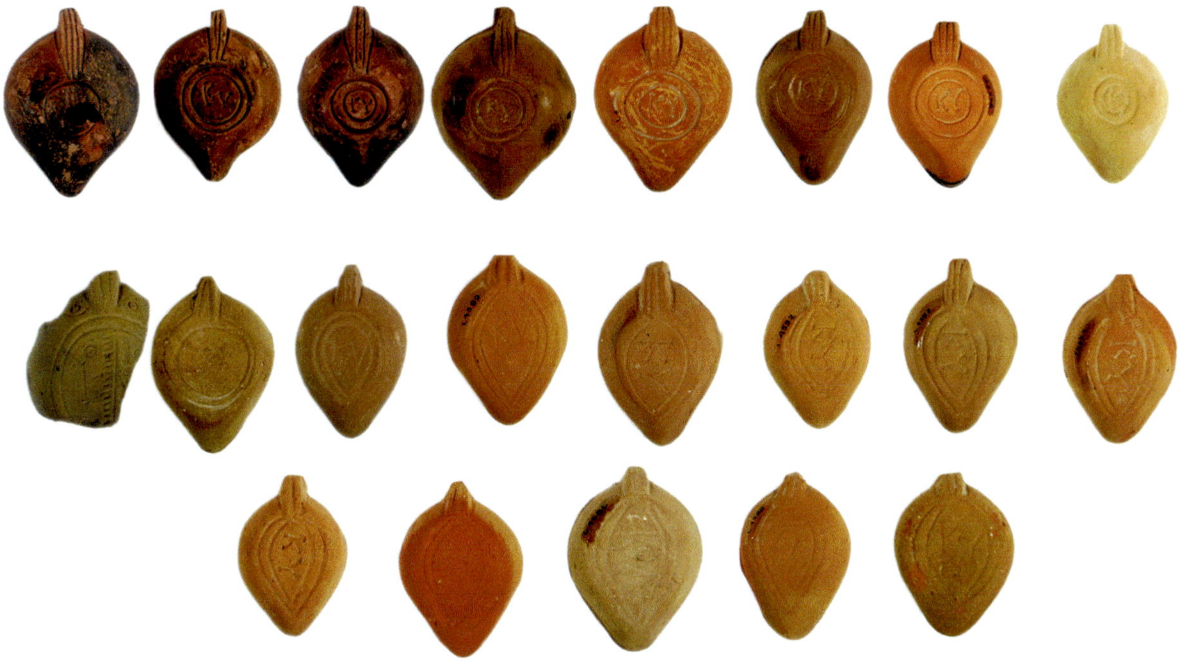

Figure 17. KY Base Evolution: Examples from the Fountain of the Lamps.

Bibliography

Broneer, O. 1930. *Corinth IV.2: Teracotta lamps.* Cambridge Massachusetts: Harvard University Press.

Broneer, O. 1977. *Isthmia III. Terracotta Lamps*. Princeton, New Jersey: American School of classical studies at Athens.

Hayes, J. 1972. *Late Roman Pottery*. London: British School at Rome.

Garnett, K. S. 1975. Late Roman Corinthian Lamps from the Fountain of the Lamps. *Hesperia* 44, 173-206. (q.v. for acknowledgements and additional references)

Karivieri, A. 1996. *The Athenian lamp industry in late antiquity*. Helsinki: Foundation of the Finnish Institute at Athens.

Landon, M. E. 2003. Beyond Priene: Toward a Broader View of Corinthian Water Supply, in C. Williams and N. Bookidis (eds) *Corinth XX: Corinth, The Centenary*, 43-62. Princeton, New Jersey: American School of Classical Studies at Athens.

Perlzweig, J. 1961. *The Athenian Agora VII. Lamps of the Roman Period: First to Seventh Century after Christ*. Princeton, New Jersey: American School of Classical Studies at Athens.

Slane, K. W. 1990. *Corinth XVIII.2: The Sanctuary of Demeter and Kore: The Roman Pottery and Lamps.* Princeton, New Jersey: American School of Classical Studies at Athens.

Slane, K. W. and Sanders, G. 2005. Corinth: Late Roman Horizons. *Hesperia* 74, 243-297.

Slane, K. W. 2008. The End of the Sanctuary of Demeter and Kore on Acrocorinth. *Hesperia* 77, 465-496.

Wiseman, J. 1969. Excavations in Corinth: The Gymnasium Area, 1967-1968. *Hesperia* 38, 64-106.

Wiseman, J. 1970. The Fountain of the Lamps. *Archaeology*, 130-137.

Wiseman, J. 1972. The Gymnasium Area at Corinth, 1969-1970. *Hesperia* 41, 1-42.

Athenian Lamps in the Early Byzantine Period - Export, Import and Imitation

Arja Karivieri

Stockholm University / Institutum Romanum Finlandiae

Abstract: Athenian terracotta lamps dominated the market in Greece from the 3rd to the first half of the 5th century AD. When political and economic changes took place in the society it had a strong impact on the trade networks in the 5th century, whileat the same time the development of the Eastern Roman Empire created new links for the export and import of products. The movement of goods and people was intensified for political, military and religious reasons. Oil lamps were essential for households, for religious services, and also to lighten meeting places, newly built churches and monasteries. As ancient temples were closed and Christianity prevailed, a new market was created for lamp workshops for new important customers: church officials and pilgrims to holy places. This paper analyzes in detail what methods the Athenian lamp producers used in their efforts to keep their trade contacts and market at a time when North African and Asia Minor products became more popular in the Eastern Mediterranean towards the end of the 5th century. Another aspect of the article deals with the imitation of imported lamps during this period.

Keywords: Athens, lamps, terracotta, glass, metal, export, import, imitation, Early Byzantine period.

Athenian terracotta lamps dominated the market in Greece in the 3rd and 4th century AD (Figure 1) and were exported widely in the Eastern Mediterranean area until the late 5th century (Figure 2). After the collapse of the Western Roman Empire in AD 476, the political and economic changes that took place in Late Roman and Early Byzantine society had a strong impact on the trade networks. At the same time, the Eastern Roman Empire continued to develop, creating new links for the export and import of products. The movement of goods and people were intensified for political, military and religious reasons. Oil lamps were essential for the households, but also for religious services in Late Antiquity, not only as offerings in ancient temples and as grave-gifts, but also to lighten the meeting places for new Christian congregations, newly built churches and monasteries. As temples were closed and Christianity prevailed, the building of churches intensified in the 5th century, creating a new market and new important customers for the Athenian lamp workshops, i.e., church officials and pilgrims to holy places.[1]

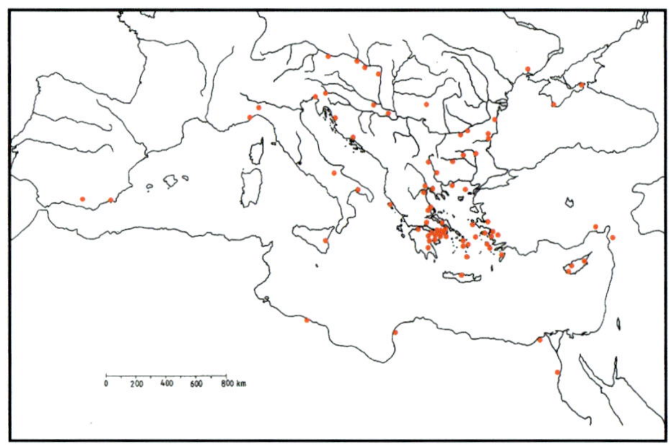

Figure 1. Distribution of Athenian lamps from the 3rd to the late 4th century (Drawing by the author).

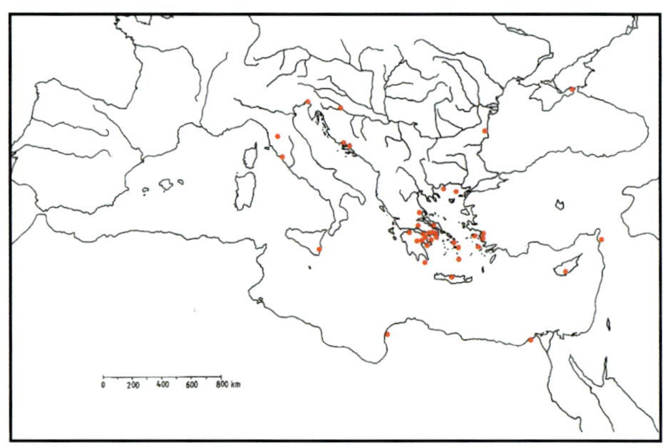

Figure 2. Distribution of Athenian lamps from the 5th to the early 6th century (Drawing by the author).

In this paper I will analyze in detail what methods the Athenian lamp producers used in their efforts to keep their

[1] I wish to take this opportunity to express my congratulations for the organizers of the ILA Round Table in Thessaloniki for giving us the possibility to experience the wonderful hospitality of the Museum of Byzantine Culture, the then-director Mrs. Choulia-Kapelloni for hosting the Round Table in the beautiful auditorium of the Museum, the exhibition 'Light on Light: an illuminating story' and a workshop about Early Christian lamps. During the week we could admire the beautiful results of this initiative. We owe the organization of the Round Table in 2011 to the great efforts of the two national representatives of the ILA in Thessaloniki, Prof. Stella Drougou from the Aristotle University of Thessaloniki and Dr Ioannis Motsianos from the Museum of Byzantine Culture, as well as for Mrs Missailidou-Despotidou and Mrs Demetra Papanikola-Bakirtzi from the Archaeological Institute of Macedonian and Thracian Studies, and Dr Anastasios Antonaras from the Museum of Byzantine Culture, and all their co-workers in Thessaloniki. I also wish to thank Karen Garnett for helping the organizers with practical arrangements and editing of the acts, and Eleni Stoubou-Katsamouri who prepared the leaflet for the abstracts of the Round Table.

trade contacts and markets during the era when North African and Asia Minor products became more popular in the Eastern Mediterranean towards the end of the 5th century. The second part of the paper will discuss the imitation of lamps so common during this period. How were the imitations made and what methods were used? When do the lampmakers change the appearance of the original product and create a new subtype? Does the change in appearance always change the quality of the lamp for worse compared to the original one? In which ways did the lampmakers try to enhance the quality of their copies and what can be said about their successful or less successful results?

We can start with the first aspect, i.e., to study the methods the Athenian lamp producers used in their efforts to keep their trade contacts and markets. The Athenian lampmakers had to make their products for various groups of customers that required different types of distribution networks, i.e., the local market in Attica, the market in Greece,[2] and the export overseas when the lamps were to be distributed via maritime routes.[3] When North African and Asia Minor products gained in popularity in the Eastern Mediterranean towards the end of the 5th century, the Athenian producers adopted the same method they had used in the Late Hellenistic period, when the Knidian lamps became popular -they started to copy the features that had become most popular and also copied complete lamps by the *surmoulage* method. In the second half of the 5th century AD they were forced to do the same, to copy popular features of North African and Asia Minor lamps and even use original imports as matrices for Athenian *surmoulage* copies produced of local clay.

Many interesting details can be discerned in some Athenian lamps that imitate various features of imported lamps. When a *surmoulage* copy did not repeat the details of the original lamp *in toto*, it could be retouched with a sharp tool to give the relief an expression that better resembled the original. At the same time, production and copying of lamps with figurative motifs that were already used in the 3rd century continued in the late 5th and early 6th century. Among these motifs were dolphin, boukranion, two fish, venator with bear, Eros playing double-flute and Poseidon. New mythological motifs were Athena Promachos standing, a motif that was copied from Asia Minor lamps, and Helios.[4] The Theodoulos and Chione workshops reintroduced rays-and-vine lamps that had been so popular in the 3rd and 4th centuries. Other new motifs that were developed in the 4th century are panther, deer and a bird, which could also have been inspired by the figurative motifs in imported lamps.

Athenian lampmakers could either copy details of imported lamps and adapt them to the Athenian overall design or they could copy complete lamps. One method that is attested in the products of the Chione workshop is that the lampmaker copied the top of a North African lamp, combined it with an Athenian base and added the signature of the workshop, i.e., the lampmaker combined a top mold from a North African lamp with a base that was an Athenian adaptation.[5] However, there exists one lamp from the Kerameikos, where Chione has combined an Asia Minor lamp top with a cross monogram in the disk surrounded by a raised edge, with an Athenian base.[6] This practice seems to emphasize that the Chione workshop wished to demonstrate its capabilities in producing imitations of foreign products. At the same time, the same workshop still produced traditional Athenian products, lamps with both figurative motifs and Christian symbols. The other large Attic workshop of the late 5th to early 6th century AD, Soteria, seems to have preferred Asia Minor forms in its production. The majority of lamps from the Chione workshop are unglazed, when all the lamps of Soteria are unglazed and made with well-levigated fine clay.[7]

In the 6th century when products of both Chione and Soteria disappear, the production of anonymous copies of Asia Minor and North African lamps takes over the market. These lamps are generally of poorer quality, unglazed, but sometimes with slip that is flaking and the clay often includes impurities which cause the clay to crack. Thus, the market of terracotta lamps consisted of *surmoulage* copies, copies of Athenian lamps from the 3rd and 4th centuries with some added retouching, and a large group of hybrids whose designs include details from various styles and types, many with relief décor that has been incised in the mold, resulting in a décor in relief.[8] We have to wait until the 7th century before we see Athenian products that have been produced with good quality clay, but which are further developments of the North African lamp types, with well distinguished details in relief.

Now let us look at some 5th and 6th century lamps from the Athenian Agora in detail, first a lamp representing St Peter[9] and a fragment of another one, representing St Paul (Figure 3).[10] The lamp decorated with St Peter carrying the cross in a key-hole shaped disk has a lamp top that imitates North African lamps, with a flat rim decorated with incised stamped symbols that imitate the stamps found on North African products. The figure of St Peter is rendered clumsily and retouched. In comparison, the profile head of St Paul, enclosed in a quadrangular frame and derived from an original North African product, is flanked by floral motifs in relief and shows the fine details of an original product with the characteristic fine clay and thin glaze.

The next two lamps show other interesting features of these late Athenian products.[11] The upper half of the lamp to the right (Figure 4) preserves an image of a bird with a long tail (L 5976), a broad channel to the wick hole and a herringbone branch rim motif in relief on the flat shoulder. The lamp represents a typical Greek copy of a North African lamp from the 6th century. The lamp to the left shows a jeweled cross on the disk (L 5979), a broad channel to the nozzle, and the disk is separated with framing band from the broad, flat rim where clay pearls can be seen between the concentric rings in reliefs, which points to a *surmoulage* copy of a North African original. Both lamps were found in the silt of a major drain, dated to the late 6th and the early 7th century.[12] The lamp with the

[2] Cf. Karivieri 1999; for copies of Athenian lamps in the Corinthia, see also Garnett 1975, 188-195, and Garnett in this volume.

[3] Karivieri 1998.

[4] Karivieri 2001.

[5] Perlzweig 1961, 176, no. 2416, pl. 38; Karivieri 1996, 144.

[6] Karivieri 1996, 189, no. 91, pl. 9.

[7] Karivieri 1996, 135-136.

[8] Karivieri 1996, 58-59.

[9] Perlzweig 1961, 173, no. 2382, pl. 37.

[10] Perlzweig 1961, 99, no. 324, pl. 10.

[11] Camp 1996, 238, pl. 68a.

[12] Camp 1996, 238, note 11.

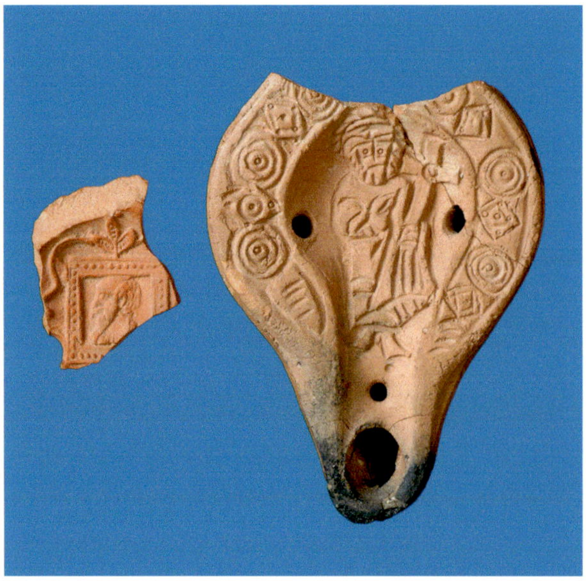

Figure 3. To the left, fragment of a North African lamp, representation of St Paul (L 1153; Perlzweig 1961, no. 324, p. 99, pl. 10). To the right, a lamp decorated with St Peter carrying the cross, the lamp top imitates North African lamps (L 4754; Perlzweig 1961, no. 2382, p. 173, pl. 37) (American School of Classical Studies at Athens: Agora Excavations).

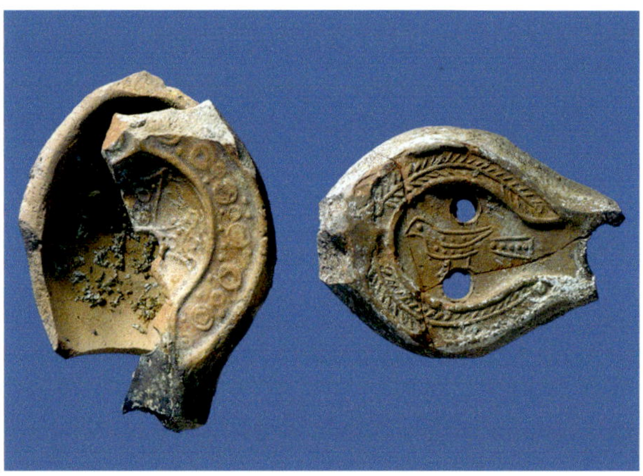

Figure 4. Athenian lamps recovered in the silt of a drain. Lamps with figured disk: a cross (left, L 5979), a bird with long tail (right, L 5976) (American School of Classical Studies at Athens, Agora Excavations).

bird could be somewhat later than the larger *surmoulage* copy that also preserves an Athenian glaze typical of products of the 5th century.

Three other lamps (Figure 5) were found in the same drain[13] and represent typical copies of Asia Minor lamps with globules on the rim, a raised ridge around the wick hole and tongues between the disk and the nozzle. These three lamps (L 5980, L 5978 and L 5977) seem to belong to a later development in the

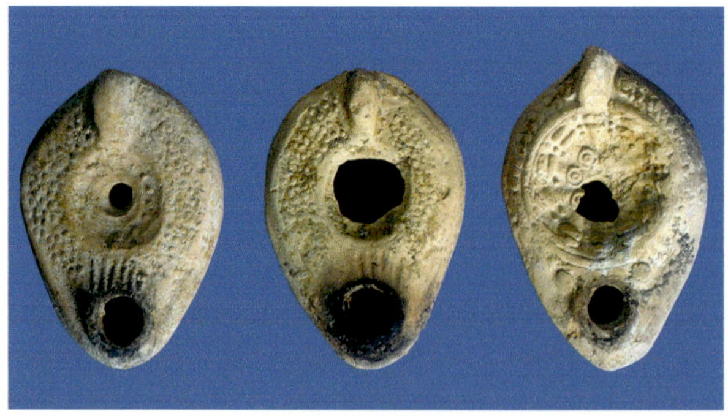

Figure 5. Three lamps recovered in the silt of a drain. Left to right: L 5977, L 5978 and L 5980 (American School of Classical Studies at Athens, Agora Excavations).

copying imported lamps and do not have as good quality as the lamps imitating North African lamps. They could possibly be dated to the second half of the 6th century. However, if we compare the composition of other deposits from the Athenian Agora that can be dated to the same period, i.e., Deposits D 6:1, Q 17:4, M-Q 17-1, O 13-14 and Q 13-14,[14] we can consider that in the Agora deposits imitations of North African lamps have usually better preserved details of decoration than the imitations and adaptations of Asia Minor lamps. It could be that when Asia Minor lamps with raised globuli were copied with the *surmoulage* method with gypsum molds, the copies soon lost the relief when the gypsum mold became worn, resulting in blurry details, especially since the clay that was used for production in the 6th century was no longer well-levigated and included impurities.

These results suggest that the producers of terracotta lamps no longer had the skills or remembered the production techniques used earlier by the Athenian lampmakers and they chose to produce copies and hybrids with simple decoration for everyday use and the local market only. However, other factors have to be taken into account, such as the increasing production and use of glass or bronze lamps by the Christian congregations and in the newly-built churches in Athens.

When glass became a common material in the household, it was used side by side with pottery and metal objects with which glass objects competed as tableware, as containers and grave offerings. The glass objects offer us a possibility to better understand the changes in economy, trade and production of terracotta lamps in Athens after the city had lost its status as the cultural center of Greece but when it was still an important center for the new Christian congregations, for whom the various workshops provided the equipment needed. The so-called Palace of the Giants (Figure 6) in the southeastern corner of the Athenian Agora, that has preserved a destruction horizon from the 6th century, provides several important finds from the 6th century. A group of finds from the palace (Figure 7) shows the character of finds, i.e., imitations of imported lamps, degenerated copies of Athenian products, ampullae, and glass objects.[15]

[13] Camp 1996, 238, note 11.

[14] Cf. list of deposits in Karivieri 1996.

[15] Agora Image 1997.10.0034 and 2012.58.0014 (Neg. 86-151): Three lamps (L 5469, L 5467, L 5453), a glass candlestick (G 576), an ampulla (P 27463) and a glass goblet fragment (G 567) are finds from the

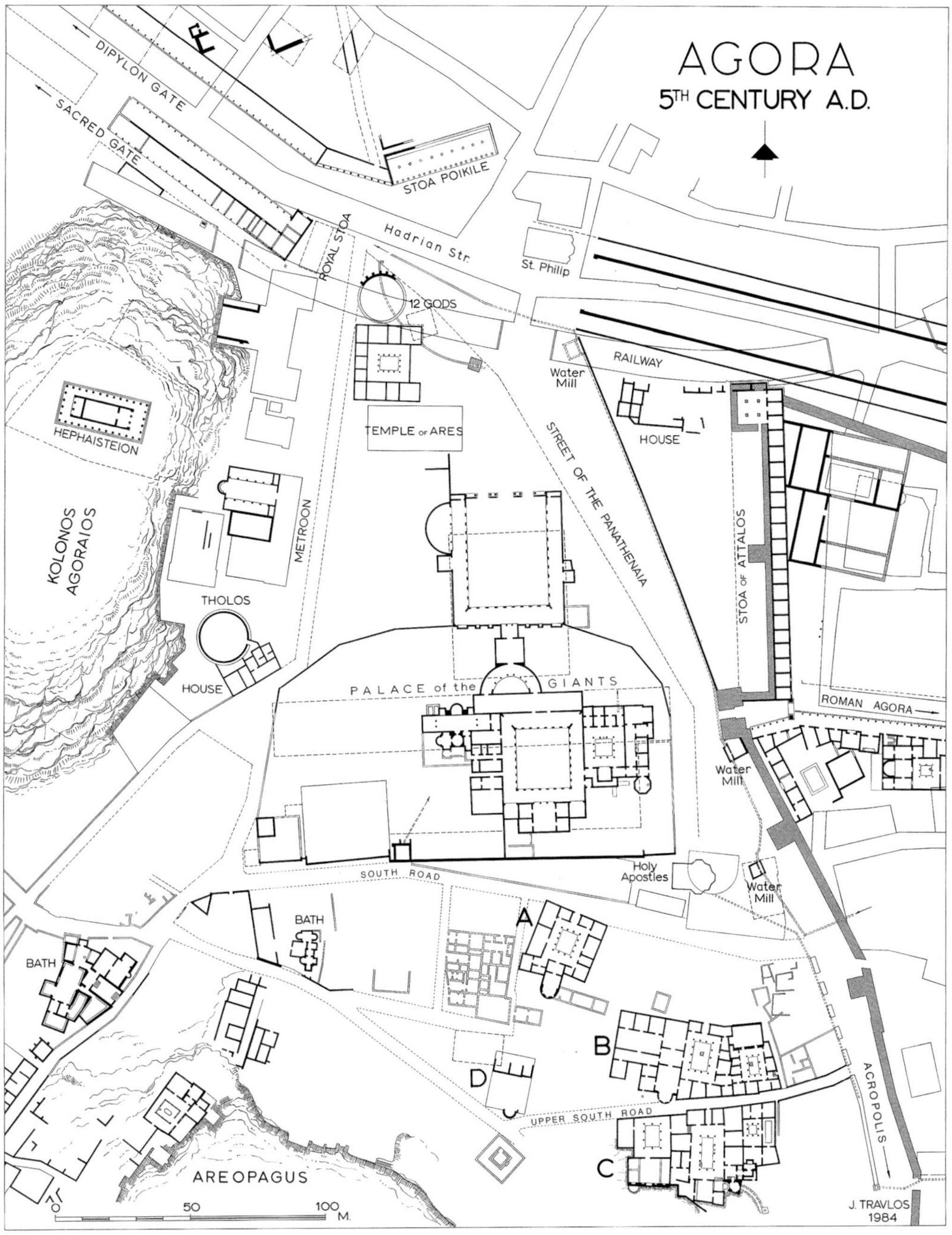

Figure 6. The Athenian Agora in the 5th century AD, plan John Travlos 1984 (American School of Classical Studies at Athens, Agora Excavations).

The glass object G 576 has been interpreted as a candlestick, and it raises an interesting question: how popular were glass lamps in the 6th century? Could they have taken over a part of the market from terracotta lamps?

destruction deposit O 13-14:1 over the floor of the rooms (grid O 13-14) found in 1965. G 576 was published by Weinberg and Stern 2009, ill. 11, pl. 33, no. 374. The three-nozzled lamp L 5467, in the center of the upper row, was erroneously marked as L 5479 in the photo 1997.10.0034, and published in Frantz 1988 as pl. 72 b. I owe this correction to Ms. Sylvie Dumont at the Agora Excavations. L 5467 and L 5479 are both three-nozzled lamps found in the Palace of the Giants, but L 5479 was found in Room B and L 5467 in Room F in the same destruction deposit.

Recent finds of glass lamps and candlesticks for wax candles from the Agora excavations also show that this combination is an aspect of the lighting trends of the time that has to be taken into account. Glass lamps as well as metal lamps[16] (Figure 8) became more popular in official buildings and especially for lighting of Christian churches where glass lamps provided more light and an atmosphere better suited for grandiose church architecture. This fact is attested by the description of Hagia Sophia written by Paul the Silentiary, where a long

[16] Camp 2007, B 2150, p. 635-636, fig. 6: bronze bowl, hanging from three chains, from a context of 5th-6th century, represents a possible example for a candleholder made of bronze.

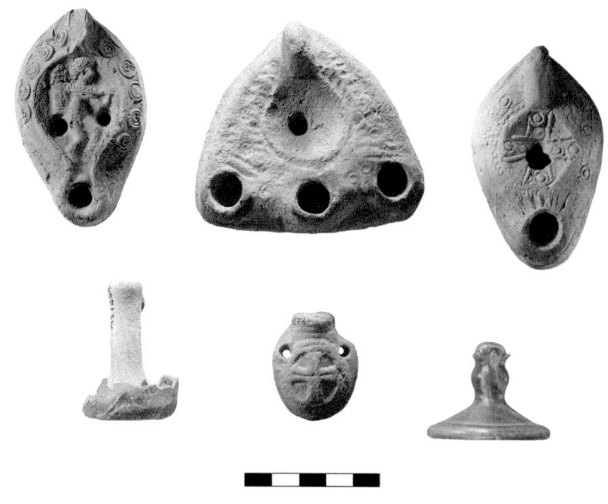

Figure 7. Three lamps (L 5469, L 5467, L 5453), a glass candlestick (G 576), an ampulla (P 27463) and a glass goblet fragment (G 567) illustrating the deposit over the floor of the rooms (O 13-14:1) found 1965 in the Palace of the Giants. (American School of Classical Studies at Athens: Agora Excavations).

passage describes the glass and metal lamps and chandeliers that illuminated the great church of Justinian. In more humble surroundings such as households wax candles made of beeswax gained in popularity. Athens had been famous for its honey during antiquity and was certainly a place where wax candles could gain in popularity, be used in private dwellings and as votive offerings in churches. Furthermore, it was easier and quicker to restart the production of wax candles following periods of unrest than it was to build new kilns and restart the production of pottery and terracotta lamps.

All in all, various factors have to be taken into account when we study evidence for the decline of Athenian terracotta lamp production in the 6th century. The answer is not always as easy and uncomplicated as it may seem.

Bibliography

Camp, J. McK. 1996. Excavations in the Athenian Agora: 1994 and 1995. *Hesperia* 65, 231-261.

Camp, J. McK. 2007. Excavations in the Athenian Agora: 2002-2007. *Hesperia* 76, 627-663.

Frantz, A. 1988. *The Athenian Agora 24: Late Antiquity: A.D. 267-700.* Princeton (NJ): American School of Classical Studies at Athens.

Garnett, K. S. 1975. Late Roman Corinthian Lamps from the Fountain of the Lamps. *Hesperia* 44, 173-206.

Karivieri, A. 1996. *The Athenian Lamp Industry in Late Antiquity,* Papers and Monographs of the Finnish Institute at Athens, 5. Helsinki: Finnish Institute at Athens.

Karivieri, A. 1998. The Athenian Lamp Industry and Lamp Trade from the Fourth to the Sixth Century after Christ, in N. Cambi and E. Marin (eds) *Acta XIII Congressus Internationalis Archaeologiae Christianae (Split-Porec 1994),*

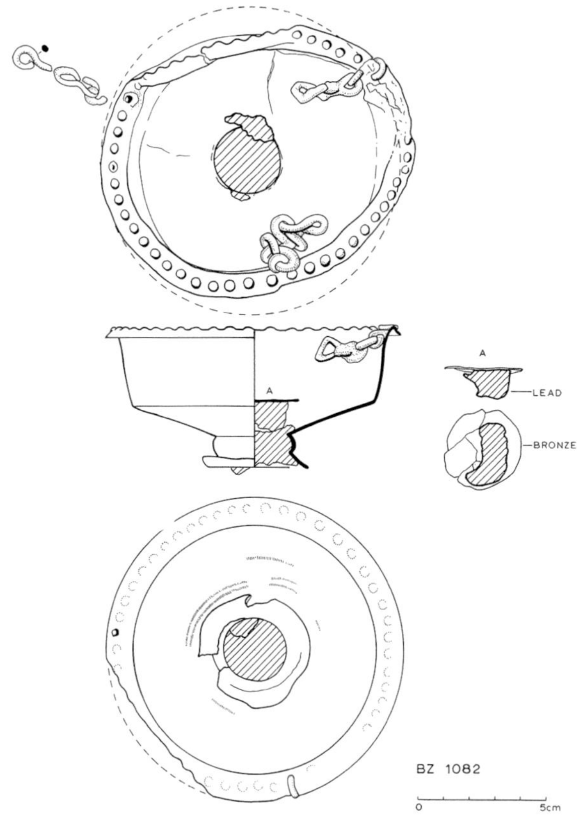

Figure 8. Bronze bowl hanging from three chains (B 2150; Camp 2007, 635, fig. 6). (American School of Classical Studies at Athens, Agora Excavations).

vol. III (Studi di Antichità cristiana pubblicati a cura del Pontificio Istituto di Archeologia Cristiana, 54), 421–428. Split: Arheoloski muzej; Città del Vaticano: Pontificio Istituto di archeologia cristiana.

Karivieri, A. 1999. Statistical Aspects on Lamp Distribution: Athens-Corinthia 200-550 AD, in *Akten des Internationalen Symposions Die Ausgrabungen im Kerameikos, Bilanz und Perspektiven: Athen, 27. - 31. Januar 1999*, 269-281. Mainz am Rhein: Zabern.

Karivieri, A. 2001. Mythological Subjects on Late Roman lamps and the Persistence of Classical Tradition, in J. Fleischer, J. Lund and M. Nielsen (eds) *Late Antiquity - Art in Context* (Acta Hyperborea 8), 179-198. Copenhagen: Museum Tusculanum.

Perlzweig, J. 1961. *Lamps of the Roman Period: First to Seventh Century after Christ. The Athenian Agora 7.* Princeton (NJ): American School of Classical Studies at Athens.

Weinberg, G. D. and Stern, E. M. 2009. *Vessel Glass.* Princeton, (NJ): American School of Classical Studies at Athens.

Aspects of Typology, Chronology and Iconography in the Regional Lamp Production in the Area of Present-day Israel (4th-7th centuries AD)

Renate Rosenthal-Heginbottom
Independent Scholar, Berlin, Germany

Abstract: In the Early Byzantine period the local manufacture of clay lamps is predominant with imports not exceeding 5%. With a tendency to local, mostly unidentified workshops, and a greater uniformity in the lamp development when compared to the Roman period, three general observations can be made: firstly, the majority are mold-made with high relief decoration; secondly, the morphological feature is the large central fill hole, resulting in placing the decoration on the shoulder and the wide nozzle; and thirdly, local groups permit religious and ethnic attributions. This paper will focus on lamps from domestic and funerary contexts rather than on those in museums and private collections, many of which were acquired on the antiquities market. Furthermore, the typology will be based on shape and decoration and not on site names referring to the find-spot, but seldom to the actual workshop. Several groups and their variants evolved. The fourth-century small ovoid lamps with a large central fill hole are derivatives of the Roman disk lamps, though mostly without a decorated disk. The shoulder decoration is simple and uniform. The lamps with a bow-shaped nozzle occur in small and large versions; they constitute a major group with a wide distribution, their variations being long-lived. The shoulder and the nozzle are decorated with manifold geometric and floral patterns, sometimes with animals and amphorae; all arranged in a manner termed 'Oriental style'. The *polilychnoi* with two to seven nozzles can be adorned with fishes or architectural elements. The other dominant group are the slipper lamps with three sub-types: small and large lamps, lamps with high handle and wheel-made lamps. Altogether they are austere in style and express little originality. The first sub-type is decorated with a radial pattern on the shoulder and a 'palm-branch' or cross on the nozzle. Lamps inscribed in Greek with religious formulae are also common. Lamps with high handle adorned with crosses, heads or geometric patterns are rare on sites in Israel and influenced by Syrian prototypes; it is not clear whether they are local products or imported. The wheel-made lamps have a circular reservoir and a projecting nozzle, with a gradual upward slope. Two distinct northern groups are lamps with low relief decoration and ovoid lamps with a shallow channel on the nozzle. Lastly, there are lamps which can be attributed to Christians (with symbols such as the cross, dove, eagle, pheasant, fish), Jews (the seven-branched candlestick) and Samaritans (a well-defined regional group, a few with Samaritan inscriptions). Jewish and Christian motifs and biblical scenes are found on North African lamps, the main group of imported lamps. In a brief summary the significance of lighting in domestic and public life, in mortuary and religious practice as well as consumer behaviour will be considered.

Keywords: Beit Nattif, Baysān theatre workshop, mold-made, wheel-made, bow-shaped nozzle, channel-nozzle lamps, multi-nozzled, slipper lamps, inscribed lamps, stamped decoration.

Introduction[1]

In the Byzantine period the local manufacture of clay lamps is predominant with imports being negligible. While there is great variation with regard to shapes and decorations, three general observations can be made: first, the majority of lamps are mold-made with high relief decoration; second, the dominant morphological feature is a large central fill hole with the decoration placed on the shoulder, the wide nozzle or sometimes on a small disk; third, local distribution permits religious and ethnic attributions and is evidence for production and consumption within a restricted market. The geographical differentiation is according to south and north –the south with Jerusalem as center and the north with Nysa-Scythopolis as center. The division according to ethnic and religious borders is the following: Christians live in Jerusalem and its surroundings, the Jews having been expelled. Jews are settled in southern Judea and Upper Galilee and the Samaritans inhabit the Samaria district and the area along the coast from Tel Aviv to Haifa. The pagan population did not live in a defined region, but mixed with others. Their role in culture was gradually declining in the Byzantine period. This division is reflected in the material culture, especially lamps, figurines, and mosaics. The lamps are classed by shape and decoration and not by typology or site name, since the latter refers generally to the find-spot, but seldom to the actual site of manufacture. Intentionally no typology is presented for several reasons. First, it has become custom to publish a lamp corpus from a well-documented site by types, creating an increasing number of site typologies but loosing track

[1] The overview does not pretend to be a comprehensive study of the lamp development; there will be neither a typological presentation nor a full reference list. However, as far as possible the illustrations are taken from recent excavation reports where additional information can be found. For a general account on Byzantine lamps, see Magness 1993, 249-259. Some references to collections acquired from local dealers are also included, see Rosenthal and Sivan 1978; Israeli and Avida 1988. The report on the lamps from the Bet Shean excavations should be consulted for updated bibliography, see Hadad 2002. For Umayyad period lamps in the Byzantine tradition (early channel-nozzle type), see the finds from the Baysān (Nysa-Scythopolis) theatre workshop, for which the *terminus ante quem* is the earthquake in the year AD 749; Bar-Nathan 2011, 298-299, 301-303.

of general trends;[2] second, regional groups are classified by types, not according to morphological features, but by ethnic and religious considerations.[3] The dilemma is that every time different criteria are employed for defining a type.

This article comprises three sections; the first section presents the predominant shapes, the second section deals with regional shapes and rare forms, and in the third section the iconography is addressed. So far, few workshops have been identified.[4]

Section 1: Predominant types

1.1 Small ovoid lamps with large central fill hole (Figure 1)

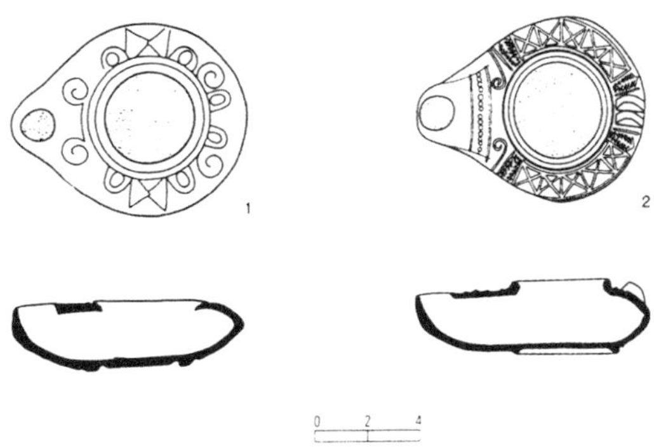

Figure 1. Ovoid lamps from Jerusalem (Shapira and Peleg 2003, 106, pl. I.17:1; Peleg-Barkat and Adler 2011, 251, fig. 14.3:5).

By shape these lamps are derivatives of the Roman disk lamps, though without the plain or decorated disks. The shoulder decoration is simple and uniform, mainly radial designs. Herringbone patterns, circles, triangles and lozenges, rosettes; a row of masks, even antithetic birds and fishes are also found.[5] A variant has a handle which constitutes part of the body. The type seems to have evolved in the south, found already in burials of the third century, but is prevalent in the 4th and 5th centuries both in the south and in the north.[6]

1.2 Lamps with bow-shaped nozzle (Figure 2)

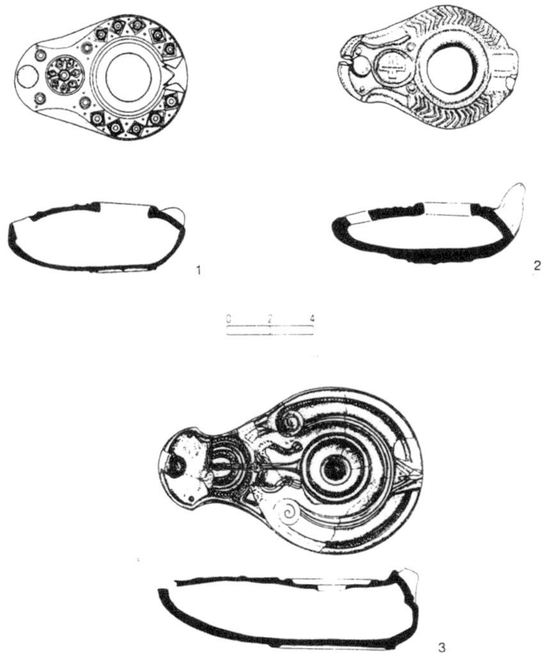

Figure 2. Lamps with bow-shaped nozzle from Jerusalem (Shapira and Peleg 2003, pl. I.17:2-3), Beth Guvrin (Magness 2008, 143, fig. 5.9:1) and Hammat Gader (Coen-Uzzielli 1997, 331, pl. III:2).

Lamps with bow-shaped nozzles occur in small and large versions and constitute a major group with sub-types and a wide distribution, their variants being long-lived. The earliest occurrence is documented in Beit Nattif to the west of Jerusalem, where cisterns filled with the refuse of a potter's workshop including molds have been unearthed, dated to the 4th-5th centuries AD.[7] The shoulder and the nozzle are decorated with manifold geometric and floral patterns, animals and amphorae, menorah and cross. It is a concept characterized by the filling of the entire surface with elements of decoration. The latest variation of lamp with a bow-shaped nozzle is found in northern Palestine and Transjordan and is common in the 5th and early 6th centuries AD. Its much more sparsely set decoration is geometric and vegetal, including rare stylized human faces.[8]

1.3 Small and large slipper lamps, also called candlestick lamps (Figure 3)

Altogether these lamps are austere in style and express little originality. They are decorated with a radial pattern on the shoulder and a palm-branch or cross on the nozzle. A very popular type in the south and north, it starts to appear in the 5th century, is most widespread in the 6th and continues into Early Islamic times, possibly to the end of the Umayyad period (AD 750). Lamps inscribed in Greek with religious formulas are quite common. The most recurrent phrase is: ΦΩΣ ΧΡΙCΤΟΥ ΦΕΝΙ ΠΑΣΙΝ – Φῶς Χριστοῦ φαίν(ει) πᾶσιν – the light of Christ enlightens all. There are abbreviations and sacred letters as well as unintelligible letter combinations.

[2] Hadad 2002 for Nysa-Scythopolis.

[3] Sussman 1989 for 'northern stamped oil lamps'.

[4] Beit Nattif /Sections 1.1-1.2: Baramki 1936, pl. XII; Israeli and Avida 1988, 114; Caesarea / Section 2.5: Sussman 1980; Israeli and Avida 1988, 98; Ḥorbat 'Uẓa / Section 2.2: Avshalom-Gorni 2009, 75-76; Nysa-Scythopolis / Section 2.8: Hadad 2002, 127-129; a clay mold acquired on the antiquities market / Section 1.3: Rosenthal and Sivan 1978, 165, no. 684.

[5] Rosenthal and Sivan 1978, nos. 405, 410; Vincenz 2001, 109, 114, fig. 3.10:14.

[6] Rosenthal and Sivan 1978, 99-103; Magness 1993, 249-250; Ben-Arieh and Coen-Uzzielli 1996, 88, fig. 4.12:1; Shapira and Peleg 2003, 104; Hadad 2002, 24-26, Type 15; Magness 2008, 130; molds at Beit Nattif.

[7] Baramki 1936; Rosenthal and Sivan 1978, 104-108; Ben-Arieh and Coen-Uzzielli 1996, 87-89, figs 4.11, 4.12:2-5; Magness 2008, 130.

[8] Hadad 2002, 26-35, Type 17; 53-61, Types 21-22.

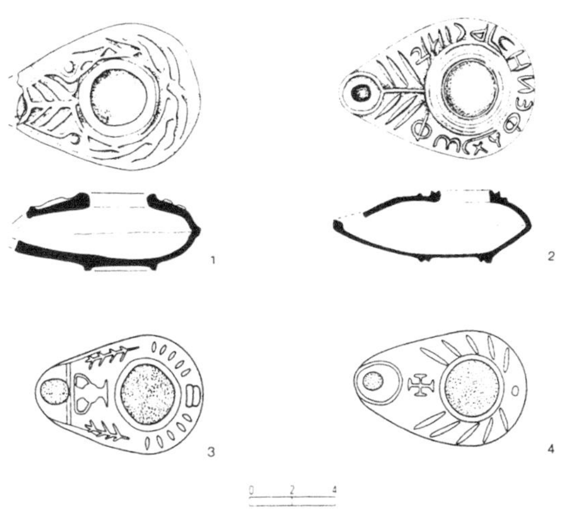

Figure 3. Slipper lamps from Jerusalem (Shapira and Peleg 2003, 106, pl. I.17:4-5; 188-189, pl. II.4:4; Vincenz 2011, 107, fig. 3.9:8).

1.4 Lamps with high handle (Figure 4)

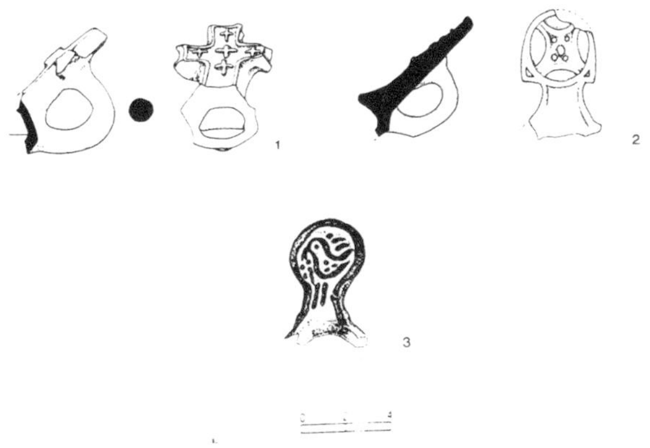

Figure 4. Lamps with high handle from Jerusalem (Vincenz 2011, 87, fig. 3.1:11; Shapira and Peleg 2003, 106, pl. I.17:9) and Ḥorbat 'Uẓa (Avshalom-Gorni 2009, 77, fig. 2.49:8).

Such lamps do not constitute a well-defined type. With a southern and northern distribution they are dated to the 5th and 6th centuries AD. The handle can be cross-shaped or decorated with a cross.[9]

1.5 Wheel-made lamps (Figure 5)

While the majority of the lamps in the Byzantine period are mold-made, a single boot-shaped type is wheel-made (i.e., the reservoir, the nozzle hand-made). These plain lamps occur in two variants, with either a smooth or a ribbed body. The reservoir is circular with a projecting nozzle, generally with a gradual upward slope and a loop handle at the rear.[10] More widespread in the south, including the Negev, the lamps were used by both Jewish and Christian populations from the 6th to the late 7th centuries; for the north see Nysa-Scythopolis.[11] They are also found in Transjordan and Cyprus. Singular is a circular lamp with twelve nozzles from the Gaza area, which was either placed on a metal plate suspended from the ceiling or upon the chalice of a candelabrum.[12] Once considered to reveal Mesopotamian influence[13] I no longer concur with this attribution, since there is no evidence whatsoever for trading in objects of daily use with the Sassanid empire.

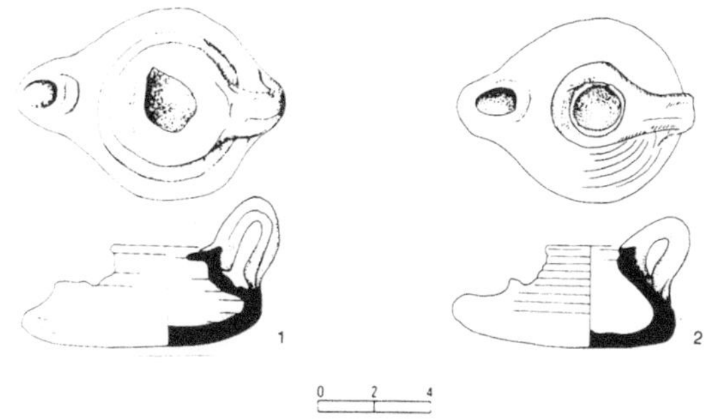

Figure 5. Wheel-made lamps from the monastery at Khirbet Jemameh (Gophna and Feig 1993, 105, fig. 15:13-14).

1.6 Early channel-nozzle lamps (Figure 6)

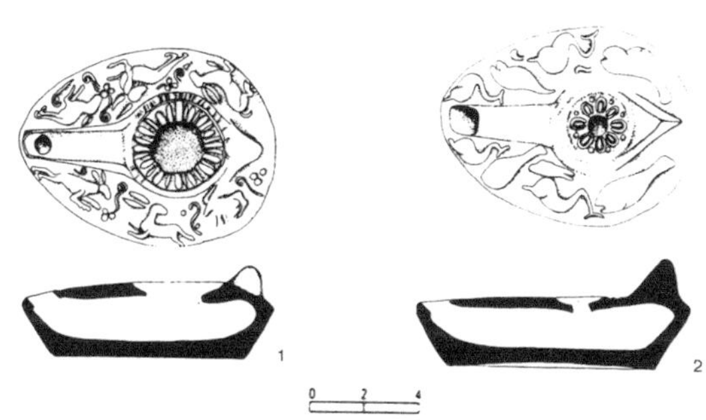

Figure 6. Early channel-nozzle lamps from Khirbet el Shubeika (Tatcher 2002, 280, fig. 5:10 and 281, fig. 6:13).

Like the mold-made slipper lamps and the wheel-made lamps this type begins in the Byzantine period and continues into Islamic times, thus verifying the assumption that historical upheavals together with a change in religious belief do not cause an immediate change in material culture. Of ovoid shape the reservoir is bi-conical; the distinct features are a channel between the disk and fill hole and a low knob handle. The decoration comprises floral, zoomorphic, and geometric motifs; it is placed on the shoulder and occasionally on the disk and on the channel.[14] A great variety of lamps came to

[9] Magness 1993, 254; Hadad 2002, 64-66, Types 26-27.

[10] Gihon 1993, 243-244, pl. 45:14a-19; Sussman 2004, 106*-107*, fig. 5:2-3. [Asterisks indicate English section].

[11] Hadad 2002, 72-73, Type 31.

[12] Rahmani 1983, 221, fig. 1.1, pl. 26:C.

[13] Rosenthal and Sivan 1978, 122-123; Magness 2008, 132 'Persian' lamp.

[14] Rosenthal and Sivan 1978, 123; Magness 1993, 255-258; Hadad 2002, 70-72, Type 30; Magness 2008, 131; Umayyad workshop products: Hadad 2002, 78-82, Type 35; Bar-Nathan 2011, 298-299, 301-303.

light in two burial caves at Khirbet el Shubeika in western Galilee, together with pear-shaped lamps with stamped decoration[15] and early channel-nozzle type lamps of the Early Islamic period. The main period of use of the caves was during the 6th century.[16] The shape is not consistent; there are lamps with small or large fill holes and with either a flat base or base-ring. The morphological diversity points to different workshop traditions; whether it further implies a chronological difference is not clear. Beginning in the late 6th century such lamps were still produced into the first half of the 8th century. They are more common in the north than in the south and are also found in the Lebanon and on Cyprus.

Section 2: Regional types and rare forms

2.1 Bi-lanceolate lamps (Figure 7)

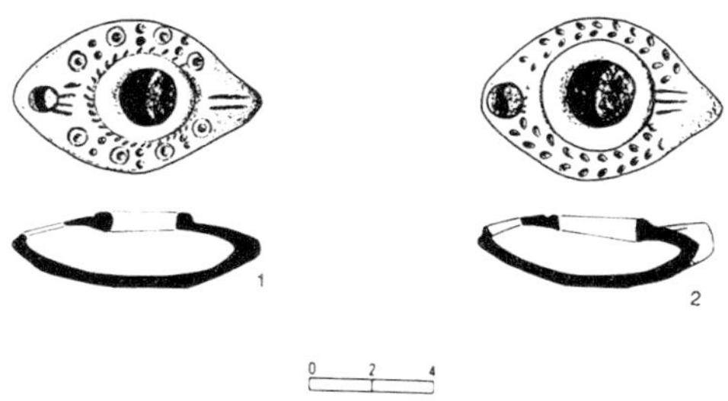

Figure 7. Bi-lanceolate lamps from Hammat Gader (Coen-Uzzielli 1997, 330, pl. II:3, 6).

These small lamps have a flat oblong body, a nozzle incorporated in it and a pointed handle, projecting horizontally and not higher than the reservoir. A large fill hole is surrounded by a prominent ridge. The impressed -stamped or incised- decoration comprises herringbone patterns, lozenges, circles, dots, and ladder decoration.[17] Restricted to the north, the lamps were used from the late 3rd to the early 5th centuries.

2.2 Lamps with circular reservoir and elongated nozzle (Figure 8)

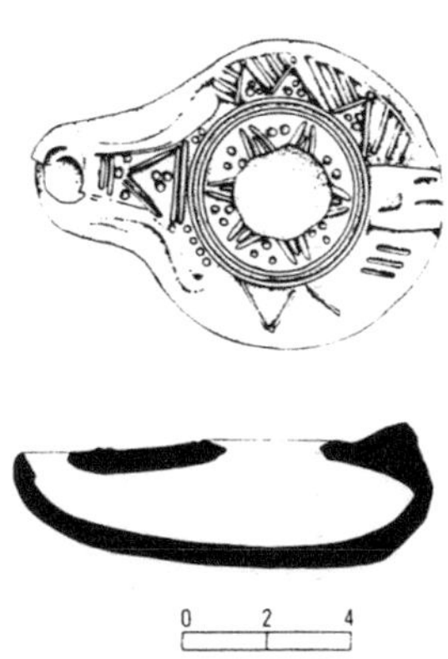

Figure 8. Lamp with circular reservoir and elongated nozzle from Ḥorbat 'Uẓa (Avshalom-Gorni 2009, 76, fig. 2.48:3).

There is a narrow disk and a small central fill hole as well as a high tongue-shaped handle or a low lug handle. The geometric and floral decoration covers the shoulder, disk and nozzle, similar to the following group. At Ḥorbat 'Uẓa five lamps made from the same mold were unearthed which suggests local production. The lamps were mainly found in loci attributed to Stratum 8, dated to 340-410 AD.[18]

2.3 Lamps with 'crowded ornamentation' (Figure 9)

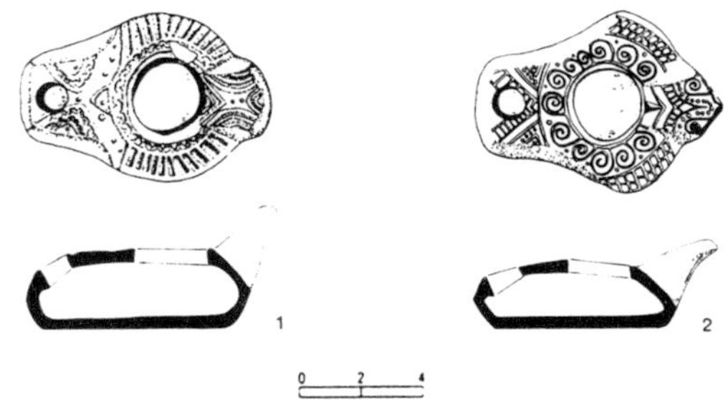

Figure 9. Lamps with 'crowded ornamentation' from Hammat Gader (Coen-Uzzielli 1997, 330, pl. V:4-5).

Such lamps form the predominant type at Nysa-Scythopolis and are widespread in the Bet Shean Valley and Samaria and rare on other sites in the north.[19] The geometric decoration is diversified and covers the entire upper part of the lamp, including the differently shaped handles The size of the lamps vary from 8 to 14 cm. Occasional designs include the seven-branched candlestick and cross, a stylized human face and amphora. Restricted to the 5th century, their use was discontinued in the early 6th for a reason not yet clear.

2.4 Lamps in the Beit Nattif tradition (Figure 10)

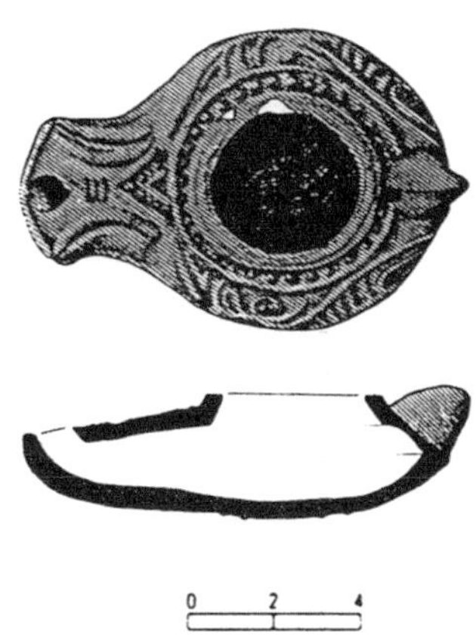

Figure 10. Lamp in the Beit Nattif tradition from Hammat Gader (Coen-Uzzielli 1997, 331, pl. III:4).

[15] See Section 2.6.
[16] Tatcher 2002, *186, 280-283, figs 5-8:4.
[17] Rosenthal and Sivan 1978, 111-112; Coen-Uzzielli 1997, 320-322.
[18] Avshalom-Gorni 2009, 5, 75-76.
[19] Hadad 2002, 37-50, Type 19.

At Nysa-Scythopolis there are two groups described as 'imitation of Beit Nattif lamps'.[20] One group has a bow-shaped nozzle and a conical handle. The large fill hole is surrounded by ridges and the decoration covers the shoulder and sometimes the nozzle. Most common is the geometric and vegetal ornamentation; stylized human faces and birds are rare.[21] The other group has a rounded or a straight-edged broad nozzle decorated with amphora or menorah or wheel. The shoulder has mostly a radial or herringbone pattern, sometimes a scroll with clusters of grapes or a guilloche pattern.[22] Both groups are restricted to the north and Transjordan and are dated to the 5th and early 6th centuries.

2.5 Lamps with decorated disk (Figure 11)

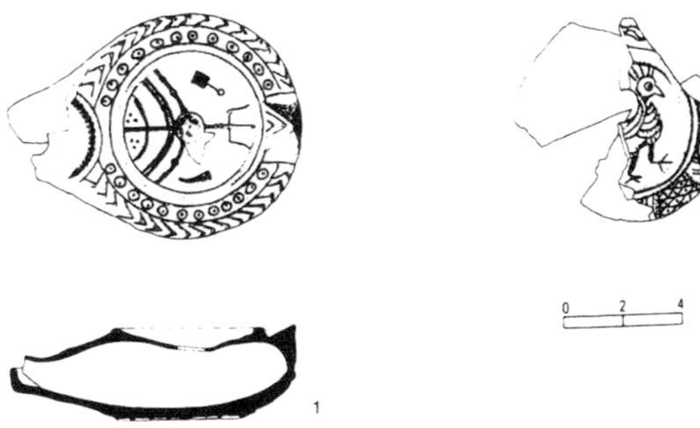

Figure 11. Lamps with decorated disk from Bet Guvrin (Magness 2008, 135, fig. 5.2:2-3).

A fairly uncommon group found in small numbers in the south and north combines the Roman circular shallow reservoir with a large disk and the Byzantine features of a wide nozzle and a pyramidal handle.[23] Decorated with menorah and cross, these lamps were used by both Jews and Christians and are found in the south and north.[24] The depiction of a temple façade is exceptional.[25] In the synagogue of En Gedi on the Dead Sea they occur in a stratum dated *c.* 450-550 AD. A production center existed at Caesarea.[26]

2.6 Lamps with stamped decoration (Figure 12)

They are part of the 'northern oil lamps' for which a typology has been suggested by Sussman (1989). The most common lamps have a pear-shaped body and a low handle, a disk and often a channel, thus they could be a forerunner of the early channel-nozzle lamps.[27] Crudely executed, the impressed geometric and floral designs and herringbone patterns are carelessly applied. There are also undecorated lamps. Production centers were in Galilee and existed from the late 3rd to the early 5th centuries.[28] The date is corroborated by a group of lamps from shops in a street at Paneas; the shops

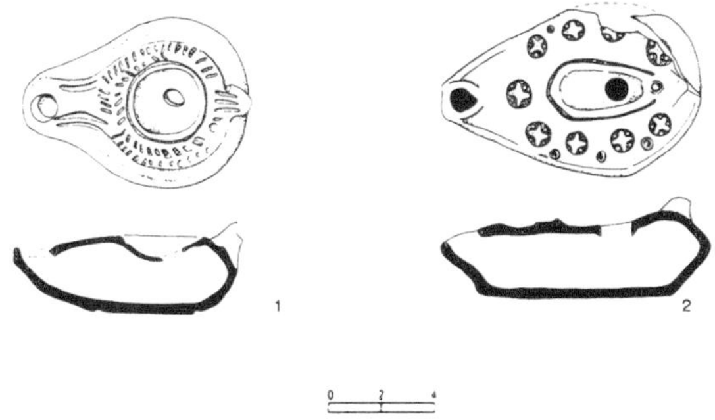

Figure 12. Lamps with stamped decoration from Paneas (Israeli 2008, 131, fig. 8.1:1) and Ḥorbat 'Uẓa (Avshalom-Gorni 2009, 77, fig. 2.49:1).

existed until the early 5th century.[29] A sixth-century date is suggested by the lamps from the burial caves at Khirbet el Shubeika.[30]

2.7 Samaritan lamps (Figures 13, 14)

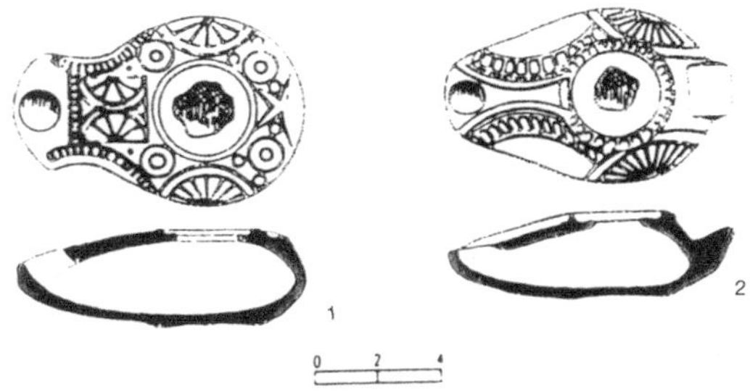

Figure 13. Samaritan lamps from Samaria (Crowfoot 1957, 375, fig. 89:2-3).

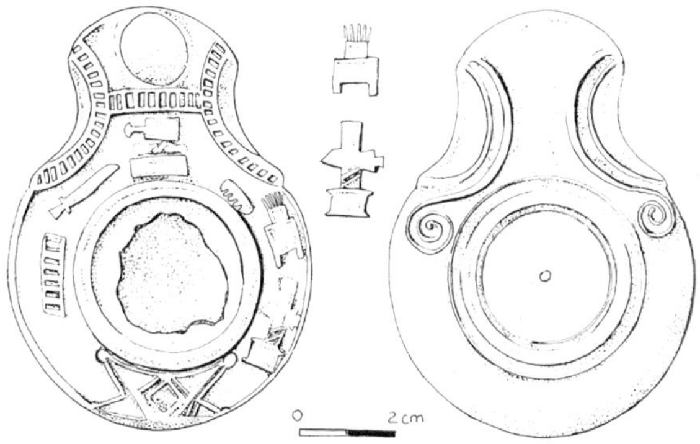

Figure 14. Samaritan lamp with the Sacrifice of Isaac (Sussman 1998, 184, fig. 1).Figure 14. Samaritan lamp with the Sacrifice of Isaac (Sussman 1998, 184, fig. 1).

The Samaritans occupied the hilly Samaria district and the coastal strip from Tel Aviv to Haifa. The lamp type defined as Samaritan forms a distinct group and is based on the rare use of the Samaritan script, on the uniform linear style and

[20] See Section 1.2.
[21] Hadad 2002, 56-61, Type 22.
[22] Hadad 2002, 53-56, Type 21.
[23] Hadad 2002, 61, Type 23; Magness 2008, fig. 5.2:2-3.
[24] Siegelmann 1992, 65*, fig. 4:3.
[25] Israeli and Avida 1988, 112, no. 321; Siegelmann 1992, 65*, fig. 4:1.
[26] Sussman 1980; Israeli and Avida 1988, 98.
[27] See Section 1.6.
[28] Sussman 1989, 42.
[29] Israeli 2008, 138.
[30] Tatcher 2002, *186, 277-279, figs 2-4.

designs and on the abundance of lamps in Samaria and nearby sites as well as in coastal sites.[31] The earliest lamps date from the late 3rd and early 4th centuries and sub-types and derivatives continued until the Samaritan uprisings in the 6th century at least, when the Samaritan faith was prohibited and the population persecuted and diminished.

2.8 Ovoid lamps with high curved handles (Figure 15)

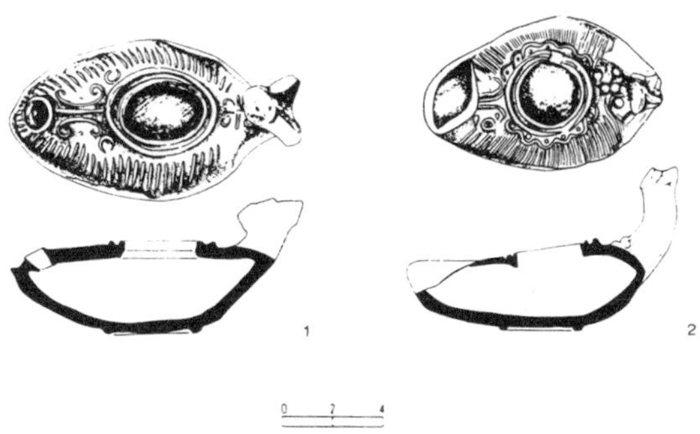

Figure 15. Ovoid lamps with high handle from Hammat Gader (Coen-Uzzielli 1997, 334, pl. VI:5-6).

The characteristic feature is the handle ending in hand-formed zoomorphic heads. Produced at Jerash (Gerasa) where several molds have been found, the lamps are given that name.[32] They are common in Jordan and the Hauran, but rare in Israel, with two molds unearthed at Nysa-Scythopolis.[33] At Nysa-Scythopolis they are dated to about AD 600-750, while in Jordan and the Hauran they continue to be used in the second half of the 8th and the beginning of the 9th centuries.[34]

2.9 Multi-nozzled lamps

Mold-made multi-burners with two to nine nozzles occur in several types and in different sizes.[35] The lamps have a loop handle or a solid handle and either a large fill hole surrounded by a ridge or a smaller fill hole in the center of a narrow decorated disk and a ridge separating shoulder and disk. The shoulder and the upper part of the wide nozzle are always elaborately decorated, the most common motif being the arcade.[36] Circular wheel-made lamps with nine and fifteen nozzles are rare finds.[37]

Section 3: Iconography

After presenting the lamps well documented in archaeological excavations in both domestic and funerary contexts, let us look at the iconography. The predominant decorative elements are geometric and floral designs applied to shoulder, nozzle and handle in never-ending variations, often covering the entire surface. This popular linear style is described by M. Avi-Yonah as "Oriental preference for an 'endless' pattern" which has a long tradition in the visual arts of the Levant.[38] Animals, often in an antithetic position (cocks, peacocks, birds, fishes, lions, horses) are fairly common on lamps and are no ethnic or religious marker. The depiction of architectural structures (temple façade, arcades) and objects (amphorae, knives and daggers) is scarce; stylized representations of human heads are exceptional.[39] Lamps with a specific iconographic subject and religious or ethnic connotation are comparatively rare (menorah, cross, Sacrifice of Isaac, Mount Gerizim). Yet since such lamps were and are always desired objects of dealers and collectors, their number is more prevalent in publications of collections than in domestic and funerary contexts. To a certain degree the subjects depicted on lamps satisfied a need for self-identification; they may have symbolic and/or apotropaic significance. Their interpretation should be viewed in a wider context of the visual arts, considering mosaics and paintings, sarcophagi and lead coffins, gems and coins, clay and metal vessels. For a thorough treatment of motifs of decoration and their interpretation L. Y. Rahmani's study on the lead coffins can be consulted;[40] since the majority of lamps bear isolated motifs their interpretation is largely speculative.

Menorah and cross can indicate Jewish, Samaritan and Christian consumers (Figure 2). The palm-branch is interpreted as the tree of life, and when combined with the two steps[41] it is thought to refer to the rite of baptism, the act of entering the baptismal font by stairs on one side and leaving it from the other as a new-born human being. The phrase 'the light shines' is associated with the Church of the Holy Sepulchre, often represented on ampullae and eulogia tokens.[42]

The depiction of fishes is relatively rare. Two multi-nozzled lamps, by fabric Palestinian products, were acquired on the antiquities market in Jerusalem in 1937.[43] With the fishes alone it is not possible to decide whether a genre subject or a Christian symbol is intended. The plastic lamp in the shape of a fish has a cross on one side of the body and a Christogram on the other; however, even though known in more than a dozen examples they are fakes.[44] Four fishes were placed on the shoulder of ovoid lamps with large fill hole[45] and on an early channel-nozzle lamp.[46]

A biblical theme shared by Jews, Samaritans, and Christians is the Sacrifice of Isaac. On the nozzle of a Samaritan lamp there is a stone altar with an elongated human figure (note the arm) and a hand in rectangular form without fingers representing God. On the right shoulder the ram, another altar, and on the left a dagger and two unidentified objects can be made out

[31] Sussman 1978; Israeli and Avida 1988, 135-141; Hadad 2002, 25-37, Type 18 and 61-64, Type 24.
[32] Scholl 1991.
[33] Hadad 2002, 127, nos. 500-501.
[34] Rosenthal and Sivan 1978, 139; Hadad 2002, 68-71, Type 29; Bar-Nathan 2011, 300.
[35] Rosenthal and Sivan 1978, nos. 442-447, 471-472; Israeli and Avida 1988, nos. 305-306, 330, 364-367; for Samaritan lamps: Sussman 1978, pl. 45; Rosenthal and Sivan 1978, nos. 520-521, 525.
[36] Rosenthal and Sivan 1978, 108.
[37] Rahmani 1983, 221, fig. 1:1; Ayalon 1996, figs 1-3; for a lamp with six nozzles from the antiquities market see Rosenthal and Sivan 1978, 128, no. 529.
[38] Avi-Yonah 1944, 112.
[39] For Section 2.4 see Hadad 2002, nos. 260-261, 265; for Section 2.6 see Sussman 1964.
[40] Rahmani 1999, 16-72.
[41] Figure 3:1-2; the motif is termed "stylized flight-of-steps design" by Loffreda 2001, 47-50
[42] Loffreda 2001, 22-28.
[43] Rosenthal and Sivan 1978, 109-110, nos. 445, 447.
[44] Loffreda 2001, 52, fig. 100; Buora and Ristow 2014.
[45] Israeli and Avida 1988, 106-107, nos. 298-302.
[46] Shapira and Peleg 2003, 188-189, pl. II.4:10.

(Figure 14). The scene is depicted in the wall paintings of the synagogue of Dura Europos and of the Christian chapel of el-Bagawat in Lower Egypt[47] and on North African lamps.[48] Jews and Samaritans share the same religious heritage, thus on lamps and on mosaic floors we find motifs such as a seven-branched candlestick, amphora, the Torah shrine, the temple façade, a harp, and knives and daggers as well as Mount Gerizim -the Samaritan's holy mountain- and short texts in the Samaritan script.[49]

North African lamps form the main group of imported lamps, yet they represent only a minor fraction of the lamps unearthed.[50] Sussman's statement that such lamps 'are found in the country in abundance' is not corroborated by finds.[51] I do not believe that the reason is economic. Much more likely, imported lamps were not considered superior to local products. Fine tableware was imported on a large scale, beginning with African Red Slip Ware around AD 300[52] and continuing with Cypriot and Phocaean Red Slip Wares. Thus trading networks existed and could have been utilized for importing lamps, if wished.

Conclusions

The significance of lighting in domestic and public life, in mortuary and religious practice is attested by a wealth of archaeological data. Lamps are objects of everyday use, mass-produced, and often without artistic quality; therefore, it is frequently assumed that pictorial representations are merely decorative and that lamps were bought casually and not purposely. On the other hand, religious symbols are often interpreted as being used by members of the faith they stand for. However, the determination should not be based on a single lamp but on the total assemblage of finds, and even then the interpretation is not equivocal, as evidence from the south cemetery of Bet Guvrin indicates.[53] The decoration of lamps with menorah and cross can be taken as a marker for a change from a Jewish to a Christian burial ground. It could mean that Jewish families converted to Christianity or that a replacement of population was carried out. Yet, the majority of the lamps were found together with those adorned with non-specific floral and geometric designs; a few bore so-called pagan motifs such as a bird, cock or fish -and even these are not entirely neutral.[54] For Bet Guvrin historical sources permit us to draw the conclusions that during the 2nd and 3rd centuries it was a Roman urban center with a pagan population and some Jewish inhabitants. Already in AD 325 the city had a bishop, suggesting that conversion had taken place. Whether there was still a Jewish population in the 6th century Christian town is unknown for lack of archaeological and historical evidence.[55] Notwithstanding the cautious consideration of evidence, there is still the element of consumer behavior: objects are acquired because of availability, preference or cost.

The inscribed lamps form a special class. They are considered eulogies or sacred objects used in daily life and in funerary rites and acquired by pilgrims visiting the Christian sites.[56] Did the owner / user read the inscription or was it sufficient to know the formulas by heart? In a society in which the average person could not read or write, pictorial images assume a spiritual language and message that is no longer evident to the modern spectator. Private possessions are a matter of personal taste, whether in ancient or modern times. When, however, lamps with a specific subject occur in certain quantities, they also reflect the religious and profane beliefs and practices of the multi-cultural Near Eastern population.

Bibliography

Ayalon, E. 1996. A Rare Multi-nozzled Byzantine Lamp from Kiludiya. *'Atiqot* 29, 63-67.

Avi-Yonah, M. 1944. Oriental Elements in the Art of Palestine in the Roman and Byzantine Periods. *Quarterly of the Department of Antiquities of Palestine* 10, 105-151.

Avshalom-Gorni, D. 2009. The Pottery Workshops from Strata 9-6; The Oil Lamps from Strata 9-6, in N. Getzov, D. Avshalom-Gorni, Y. Gorin-Rosen, E. J. Stern, D. Syon, and A. Tatcher, *Ḥorbat 'Uẓa. The 1991 Excavations. Volume II: The Late Periods* (Israel Antiquity Authority Reports 42), 23-27; 75-78. Jerusalem: Israel Antiquities Authority. Avshalom-Gorni, D. 2009. The Pottery Workshops from Strata 9-6; The Oil Lamps from Strata 9-6, in N. Getzov, D.

Bar-Nathan, R. 2011. The Pottery Corpus, in R. Bar-Nathan and W. Atrash, *Baysān. The Theater Pottery Workshop* (Israel Antiquity Authority Reports 48), 229-343. Jerusalem: Israel Antiquity Authority.

Baramki, D. C. 1936. Two Roman Cistern at Beit Nattīf. *Quarterly of the Department of Antiquities of Palestine* 5, 3-10.

Ben-Arieh, R. and Coen-Uzzielli, T. 1996. The Pottery, in G. Avni and Z. Greenhut, *The Akeldama Tombs. Three Burial Caves in the Kidron Valley, Jerusalem* (Israel Antiquity Authority Reports, no. 1), 73-93. Jerusalem: Israel Antiquity Authority.

Calderon, R. 2000. Roman and Byzantine Pottery, in Y. Hirschfeld, *Ramat Hanadiv Excavations. Final Report of the 1984-1998 Seasons*, 91-165. Jerusalem: The Israel Exploration Society.

Buora, M. and Ristow, S. 2014. Tonlampen in Fischform in frühchristlicher Zeit? *Bonner Jahrbücher* 214, 229–241.

Coen-Uzzielli, T. 1997. The Oil Lamps, in Y. Hirschfeld, *The Roman Baths at Hamat Gader. Final Report*, 319-346. Jerusalem, The Israel Exploration Society.

Crowfoot, G. M. 1957. Lamps, and an Early Stone Lamp Holder, in J. W. Crowfoot, G. M. Crowfoot and K. Kenyon, *The Objects from Samaria, Samaria-Sebaste Volume III*, 365-378. London: Palestine Exploration Fund.

Ennabli, A. 1976. *Lampes chrétiennes de Tunésie (Musées du Bardo et de Carthage)*. Paris: Centre National de la Recherche Scientifique.

Gichon, M. 1993. *En Boqeq. Ausgrabungen einer Oase am Toten Meer. Band I: Geographie und Geschichte der Oase. Das spätrömisch-byzantinische Kastell.* Mainz: Philipp von Zabern.

[47] Sussman1998, 187-188, figs 5-6.
[48] Ennabli 1976, 43, pl.1:14-15.
[49] Sussman 1978, 242-243; 1998; Naveh 1998, 99-100.
[50] En Boqeq: Gihon 1993, 245, pl. 45:22; Jerusalem: Shapira and Peleg 2003, 105-107, pl. 1.17:17; Ashkelon: Johnson 2008, 128, 131, no. 388; Caesarea: Vine and Hartelius 1986, 374-377, figs 7-8; Geva (Mishmar Ha-'Emeq): Sussman 1988, 94, no. 28; Ramat Hanadiv: Calderon 2000, 155, fig. 46; Bet Shean: Haddad 2002, 133-135.
[51] Sussman 1989, 42.
[52] Hayes 1972, 68-73, Form 50.
[53] Magness 2008, 132-134.
[54] Magness 2008, 135, fig. 5.2:2.
[55] Magness 2008, 134.
[56] Loffreda 2001, 52-54.

Gophna, R. and Feig, N. 1993. A Byzantine Monastery at Kh. Jemameh, *'Atiqot* 22, 97-108.

Hadad, S. 2002. *The Oil Lamps from the Hebrew University Excavations at Bet Shean* (Excavations at Bet Shean Volume 1, Qedem Reports 4). Jerusalem: The Institute of Archaeology; The Hebrew University of Jerusalem.

Hayes, J. W. 1972. *Late Roman Pottery*. London: W. Heffer and Sons Ltd. Cambridge.

Israeli, S. 2008. Area F: The Lamps from the Shops, in V. Tzaferis and S. Israeli, *Paneas Volume I: The Roman to Early Islamic Periods, Excavations in Areas A, B, E, F, G and H* (Israel Antiquity Authority Reports 37), 129-139. Jerusalem: Israel Antiquities Authority.

Israeli, Y. and Avida, U. 1988. *Oil Lamps from Eretz Israel. The Louis and Carmen Warsaw Collection at the Israel Museum, Jerusalem*. Jerusalem: The Israel Museum.

Johnson, B. L. 2008. *Ashkelon 2: Imported Pottery of the Roman and Late Roman Periods*. Winona Lake, IN: Harvard Semitic Museum Publications.

Loffreda, S. 2001. *Light and Life. Ancient Christian Oil Lamps from the Holy Land* (Studium Biblicum Franciscanum Museum 13). Jerusalem: Franciscan Printing Press.

Magness, J. 1993. *Jerusalem Ceramic Chronology circa 200-800 CE*. Sheffield: Sheffield Academic Press.

Magness, J. 2008. The Oil Lamps from the South Cemetery, in G. Avni, U. Dahari and A. Kloner, *The Necropolis of Bet Guvrin-Eleutheropolis* (Israel Antiquity Authority Reports 36), 121-178. Jerusalem: Israel Antiquities Authority.

Naveh, J. 1998. Scripts and Inscriptions in Ancient Samaria. *Israel Exploration Journal* 48, 91-100.

Peleg-Barakat, O. and Adler, N. 2011. The Pottery Assemblage from the Structure in Area I, in E. Mazar, *The Temple Mount Excavations in Jerusalem 1968-1978 Directed by Benjamin Mazar. Final Reports Volume IV: The Tenth Legion in Aelia Capitolina* (Qedem 52), 245-252. Jerusalem: The Institute of Archaeology; The Hebrew University of Jerusalem.

Rahmani, L. Y. 1983. Finds from a Sixth to Seventh Centuries Site near Gaza: II. Pottery and Stone Objects. *Israel Exploration Journal* 33, 219-230.

Rahmani, L. Y. 1999. *A Catalogue of Roman and Byzantine Lead Coffins from Israel*. Jerusalem: Israel Antiquities Authority.

Rosenthal, R. and Sivan, R. 1978. *Ancient Lamps in the Schloessinger Collection* (Qedem 8). Jerusalem: The Institute of Archaeology; The Hebrew University of Jerusalem.

Scholl, T. 1991. The Chronology of Jerash Lamps: Umayyad Period. *Archeologia* 42, 65-84.

Shapira, L. and Peleg, O. 2003. Pottery Lamps of the Byzantine Period from Area XV. Byzantine and Early Islamic Pottery Lamps from the 'House of the Menorot' in Area VI, in E. Mazar, *The Temple Mount Excavations in Jerusalem 1968-1978 Directed by Benjamin Mazar. Final Reports Volume II: The Byzantine and Early Islamic Periods* (Qedem 43), 104-108, 187-190. Jerusalem: The Institute of Archaeology; The Hebrew University of Jerusalem.

Siegelmann, A. 1992. Roman and Byzantine Remains in the Northern Coastal Plain. *'Atiqot* 21, 63*-67*, 178 (Hebrew with English summary).

Sussman, V. 1964. Representation of a Human Head on a Lamp. *Israel Exploration Journal* 14, 99-100.

Sussman, V. 1978. Samaritan Lamps of the Third-Fourth Centuries A.D. *Israel Exploration Journal* 28, 238-250.

Sussman, V. 1980. Moulds for Lamps and Figurines from a Caesarea Workshop. *'Atiqot* (English Series) 14, 76-79.

Sussman, V. 1988. Lamps from Geva (Abu-Shusha), in B. Mazar (ed.) *Geva. Archaeological Discoveries at Tel Abu-Shusha, Mishmar Ha-'Emeq*, 89-119. [Jerusalem]: The Israel Exploration Society; Hakibbutz Hameuchad Publishing House Ltd. (Hebrew).

Sussman, V. 1989. Northern Stamped Oil Lamps and Their Typology. *Michmanim* 4, 22-58.

Sussman, V. 1998. The Binding of Isaac as Depicted on a Samaritan Lamp. *Israel Exploration Journal* 48, 183-189.

Sussman, V. 2004. Ḥorbat Rimmon: Pottery and Small Finds. *'Atiqot* 46, 99*-112*.

Tatcher, A. 2002. Excavations at Khirbet el-Shubeika 1991, 1993. The Lamps, in Z. Gal (ed.) *Eretz Zafon. Studies in Galilean Archaeology*, 266-288. Jerusalem : Israël Antiquities Authority.

Vincenz, A. de 2011. The Pottery Assemblage from the Bathhouse, in E. Mazar, *The Temple Mount Excavations in Jerusalem 1968-1978 Directed by Benjamin Mazar. Final Reports Volume IV: The Tenth Legion in Aelia Capitolina* (Qedem 52), 85-117. Jerusalem: The Institute of Archaeology; The Hebrew University of Jerusalem.

Vine, K. and Hartelius, G. 1986. Ceramic Lamps from the Hippodrome of Caesarea Maritima – 1974, in L. T. Geraty and L. G. Herr (eds) *The Archaeology of Jordan and Other Studies Presented to Siegfried H. Horn*, 365-426. Berrien Springs, MI: Andrews University Press.

Argos and its Lamps from the Late 4th to 7th Century AD

Lambrini Koutoussaki
University of Fribourg, Switzerland

Abstract: Over 5000 lamps were found in Argos during the French excavations (Ecole Française d'Athènes) from 1972 to 1991 in the Agora and the Thermes area. About 10% of them belong to the 5th, 6th and 7th century AD, an important period of historical evolution for the city. The most of them have the characteristics of lamps, which we call today 'Christian' Lamps. Argos was, without doubt, one of the main production centers in the region and a lot of *officines* were specialized in the production of copies. We find some common types, but also those with a very local character. Their form is particular and seems to have been produced for only a short period.

Keywords: Argos, lamps, Christogram, *officines.*

Argos, the most ancient city of Europe according to a large sign at the entrance to the city, has experienced historical events typical of most cities of great antiquity. But even though Argos was conquered and destroyed many times, its territory was occupied without interruption. Systematic excavations over the last 40 years have provided clear evidence that the Argive population persistently continued to produce, to import and to export. In brief, they continued to have a normal life. Lamps found during the excavations prove without a doubt that their production was never interrupted. This thesis is quite logical, because of the importance of the object: human beings cannot live without light!

For the epoch with which we are concerned, the Early Byzantine Period between the late 4th and the 7th centuries AD, the evidence is clear: Argos passed from one Empire to another! The domination of pagan Rome is replaced with that of Christian Byzantium. The transformations made during this era in the public areas (Figures 1-2), such as in the Argive Agora and its environs, are still very visible today.

From the reign of Constantine the Great at the beginning of the 4th century AD onwards, paganism became less important as Christianity expanded. In the year AD 392 Theodosius I ordered the closure of all the temples consecrated to the pagan religions. A few years later, in 396, Alaric invaded and destroyed a large number of Greek regions and cities,

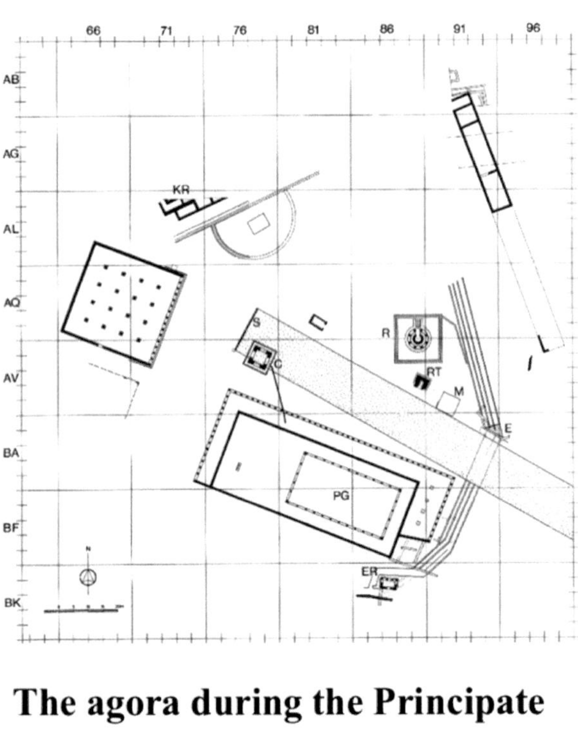

Figure 1.

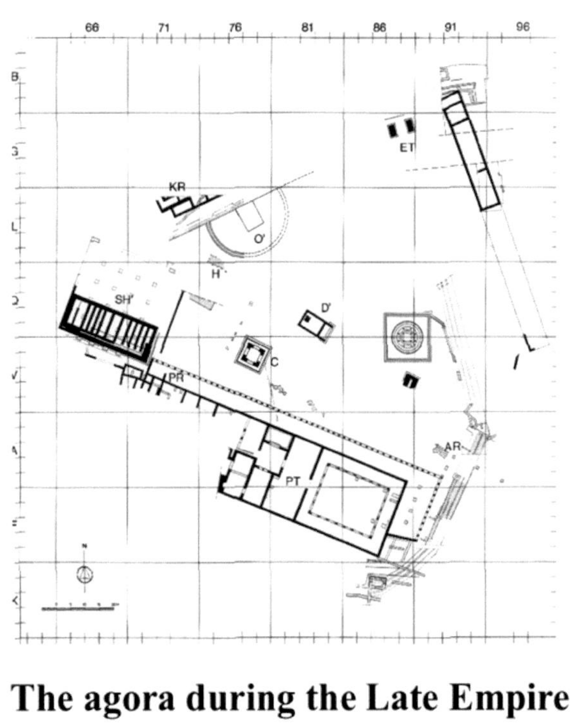

Figure 2.

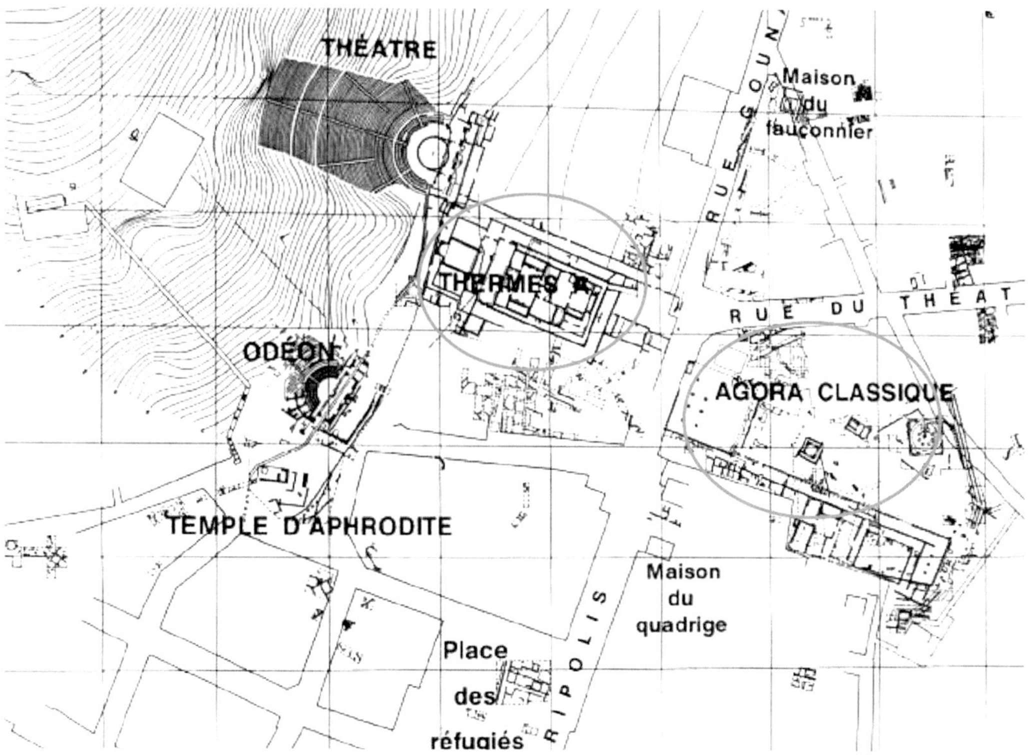

Figure 3. Plan of Argos indicating excavation areas of the Agora and of the Roman Bath.

including Argos. It is a paradox that the Christian community was indeed not worried about this dramatic situation! The destruction of all the buildings which belonged to the pagan religion was God's wish. The excavations have pointed out that many pagan sanctuaries and public sites were abandoned at this time. Excavators uncovered a well from the Agora area, which can be dated to the transition from the 4th to the 5th century AD. The well contained much evidence of destruction, such as disfigured statues, as well as a large quantity of common everyday ceramics and about 100 lamps of local production.

All these very important changes provide evidence of the intense expansion of the new religion. The willpower of its followers constituted the basis for this fast expansion. During its first centuries (1st to 4th century AD), the Christian community acted as a closed ensemble and this is obvious in the archaeological record of Argos. Once the new religion became officially sanctioned as the imperial religion, the urban landscape changed: the Agora, until then the center of public life, became less influential. Many private houses were built into or over the ruins of ancient pagan buildings, and there is a relative impoverishment of this sector of the city. But at the same time, in the South-East area of the city, many luxury houses and other buildings were constructed (Figure 3). This fact shows very clearly the new dynamism of the inhabitants. The South-East area became the new administrative and religious center of Argos. After the so-called 'victory of Christianity', Argos seems to be forgotten for a while (late 4th to 5th Century AD). In the 5th century AD Argos became a Bishop's seat! Many churches date from the late 5th to the beginning of the 6th century AD. For example, one church was constructed on the Aspis Hill, and another was built over the ruins of the Temple of Apollon Pythaeus. A large percentage of the decoration of these churches has been preserved and restored. Some of the churches are still in current use. A number of cemeteries from the same period were excavated. They were found near the sacred Christian places, and were also located in areas around the town where there were earlier Roman cemeteries, which were then reused by the Christians. For that reason, it is not so easy for the excavators to distinguish the pagan tombs from those of the Christians. Apparently at this time the Christians still tolerated the pagan burial practices. Archaeological remains dating from the reign of Justinian in the middle of the 6th century AD have shown that a lot of the urban areas were simply abandoned, and the Christian expansion occurred in adjoining areas. But we can also observe that other quarters of the ancient city were reorganised. We know nothing about the reasons for this reorganisation. But we can safely say that the new Christian society changed very fast and the wealth of the community came into the hands of the main local authorities who had the power! The historians are not always in agreement about this fact, and Argos was not the only city where these transitions are apparent in the archaeological record.

The earthquakes of AD 522 and 551 and the plagues of the 6th century surely played a vital role in these transitions. But a most important danger was the invasion by enemies coming from the North. The years 584 or 585 may be the time of the Slav incursion: the ceramics and also the place names prove that an important part of this population moved into the Argolid and settled themselves in the city of Argos and its environs.

The next period, from the 7th to 9th century, is known as the 'Argive Dark Ages'. This phenomenon is observed in all areas

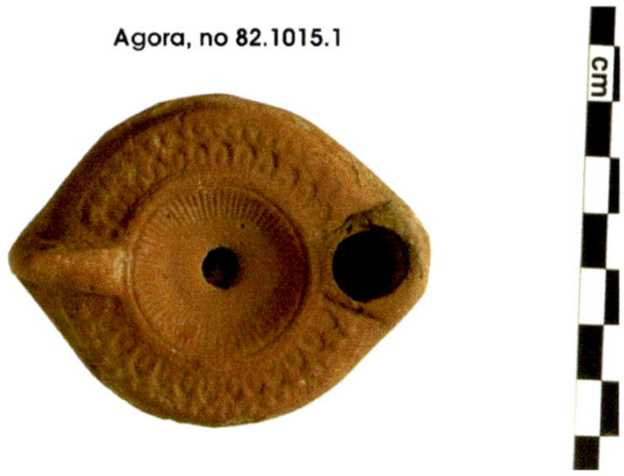

Figure 4.

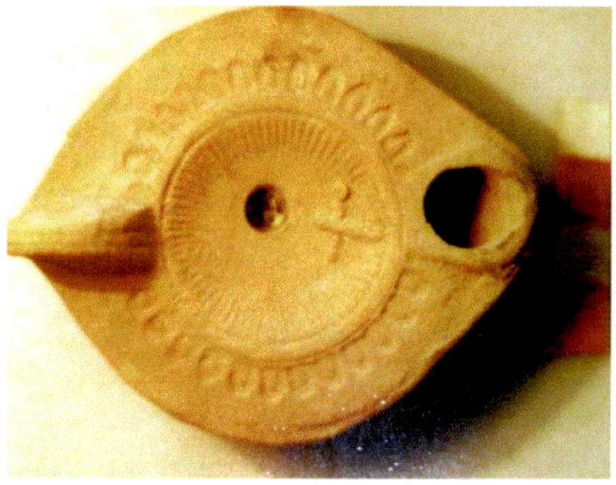

Figure 5.

of the Peloponnese. For Argos particularly, we are not yet able to refine our dating for this period. Results from excavations concluded that settlements of this period were mostly poor.

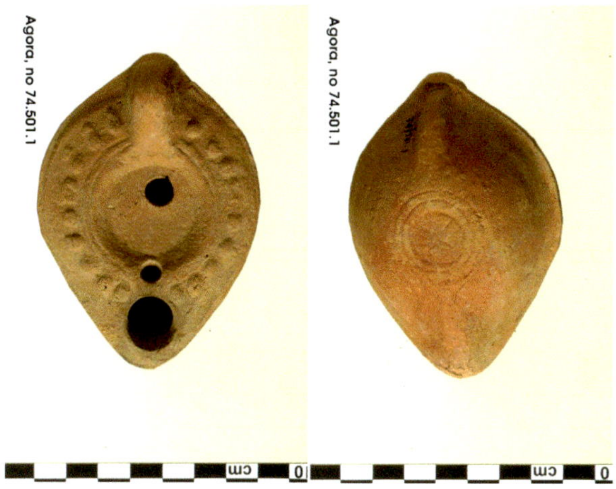

Figures 6 and 7.

The first centuries of the Christian community were difficult but also vital for establishing continuity. Motifs and stylistic customs in the artistic sector changed more slowly than in other sectors. The 'old' Greco-Roman pagan modes were initially continued by the Christian artists. But thereafter we can speak of the birth of Christian art. In Argos, it seems that many lamp factories were specialising in both pagan and Christian styles and used pagan decorations in combination with new ones. Initially they just added symbols, which became the Christian symbols par excellence, such as the cross, the Christogram or the Constantinian monogram, the fish, the palm tree, etc. Within the context of ceramic products, lamps constitute a key group of objects where we can closely observe this 'new' mode.

In the following examples found during the French excavations in the Agora and the Roman Baths of Argos, I would like to illustrate this passage from the designs of the pagan era to those of the Christian era. All the lamps date from the end of 4th to the 7th century AD and constitute an ensemble of very common Corinthian and Argive types, manufactured in local workshops. Their quality was mostly good and they were probably produced for the local market. Many of the workshops were located in the South-East area outside of the Agora.

The first examples (Figures 4-5) are copies of lamps, the originals of which belong to the Corinthian Broneer Type XXVII a.[1] Lamp 244 (Figure 5) has a little cross on the disk, which was added by hand. There is no signature and the number of examples found with this variation is insignificant in comparison with the number of finds of the main type, which is without a cross.

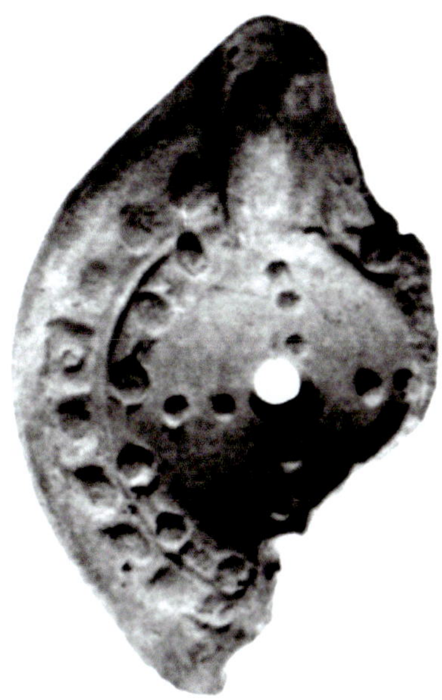

Figure 8.

The next lamps[2] represent local types, and date from the end of 4th into the first half of the 5th century AD (Figures 6-8). The majority of examples of these lamps have no decoration on the disk. Our example has a representation of a Greek letters *iota* and *chi*, the capital initials *I* and *X* of 'Jesus Christ'

[1] Koutoussaki 2008, nos. 243 and 244.
[2] Koutoussaki 2008, nos. 253 and 374.

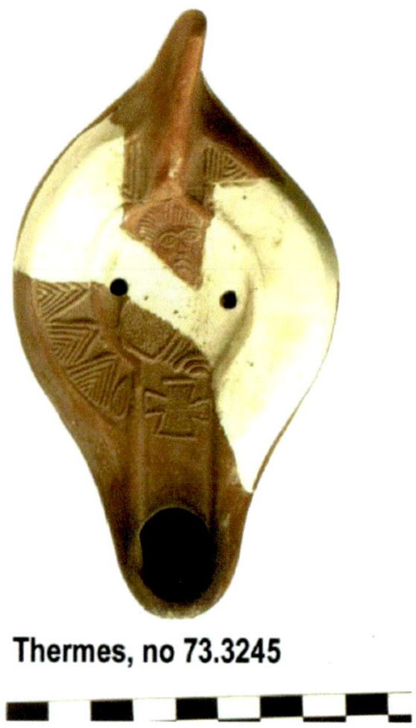

Figure 9.

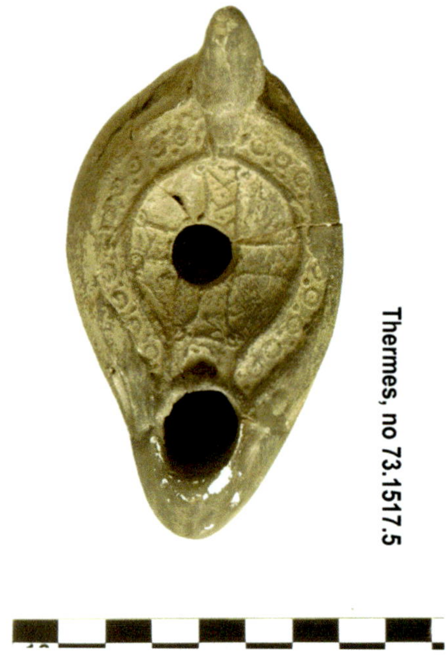

Figure 11.

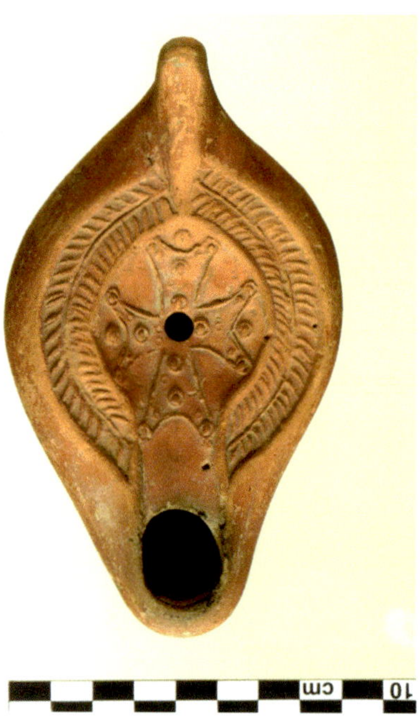

Figure 10.

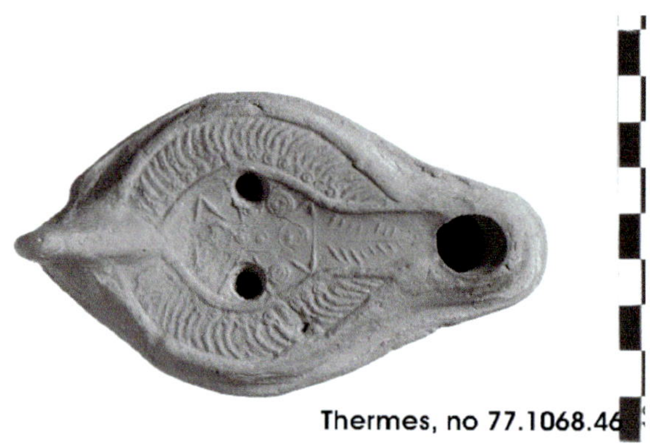

Figure 12.

in Greek, on the base. The latest examples of this type are decorated with a cross on the disk. Despite everything, we can not be sure that this addition is absolute evidence of the new religion.

The appearance of Christian lamps, in particular North-African style lamps, during the 5th century AD, confirms that the Christians in Argos were becoming a very important presence. Christian motifs and subjects from the bible became very popular (Figure 9). A great number of North African style lamps clearly show this popularity.

The symbol most commonly represented is definitely the cross. It appears in various forms: Greek, Byzantine and Maltese crosses are the most common (Figures 10-12). The Christogram is also frequently represented (Figures 13-14). Interestingly, a number of variations were unique and typical of Argive creations, which may provide concrete proof for contemporary local workshops. Another important feature is the appearance of marks on the base. Some of these marks have also been found on lamps in Corinth or Athens (Figures 15-16). It seems that some workshops in the Argolid had significant expertise in lamp production.

From the end of the 5th into the 7th century AD there are a great number of imitations and copies of imported lamps. One type, originally from Asia Minor (Broneer Type XXIX), is well represented in Argos (Figures 17-18). The decoration is rather simple. The signature on the base is complete but illegible. Broneer proposed the reading QAUMASIS. Bovon has read the letters as AILOUROU and proposed that it is probably

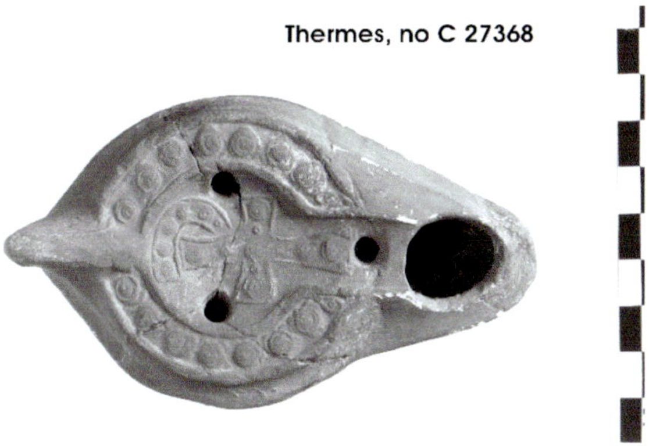

Figure 13.

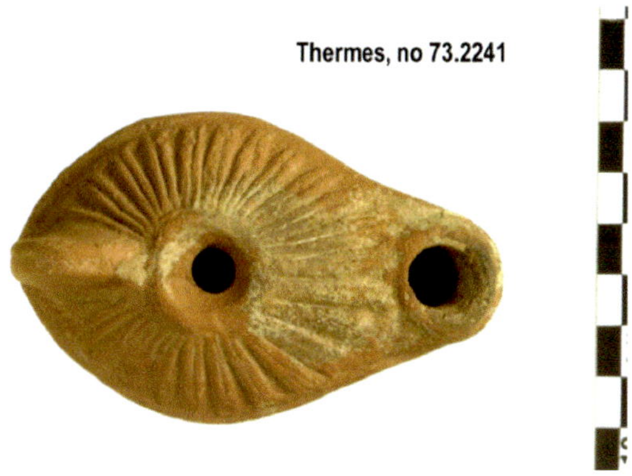

Figure 17.

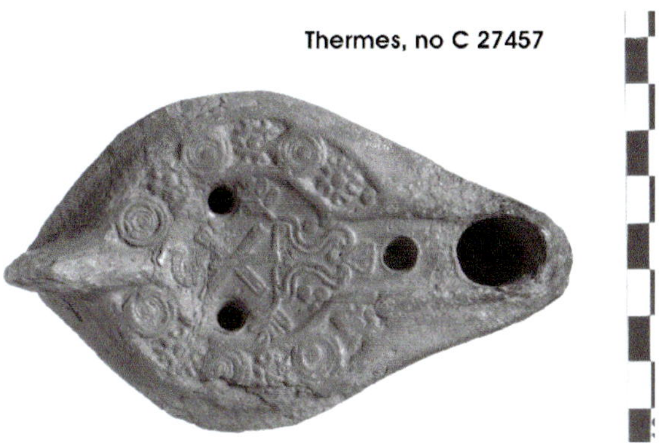

Figure 14.

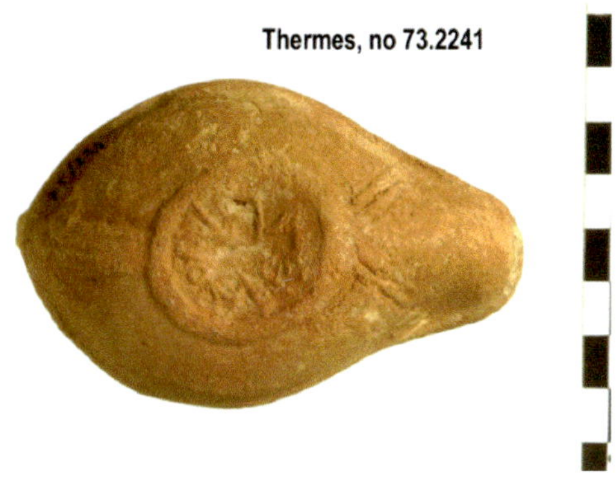

Figure 18.

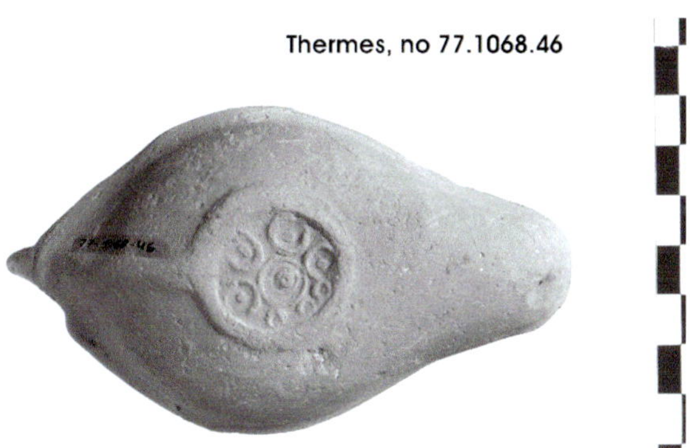

Figure 15.

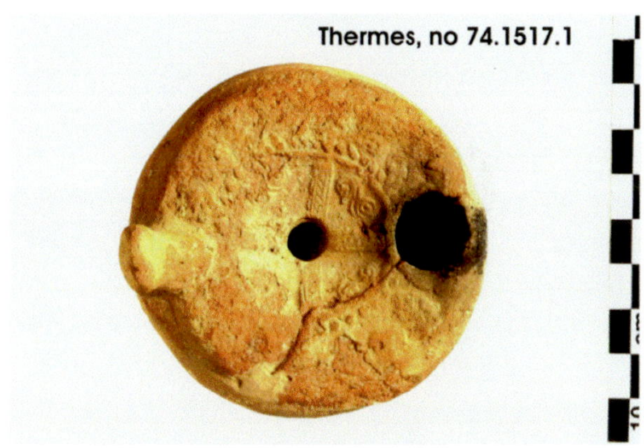

Figure 19.

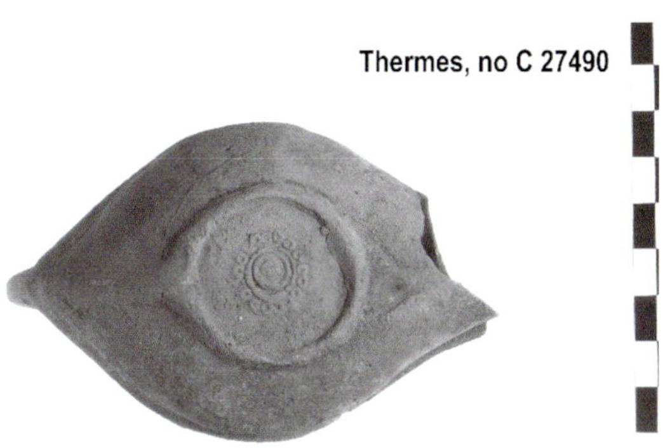

Figure 16.

a votive signature and not that of a workshop. Unfortunately, there is no other evidence of this signature in Argos, which could help provide an explanation for any connection with a local workshop or family. It seems that this particular series of lamps was produced for a short time, and maybe for a specific reason.

The last type of lamp presented in this paper belongs also to a particular type, classified by Broneer under Type XXXII (Figures 19-21). The round shape of these lamps is very striking and the orange-buff clay used in many of the examples is very

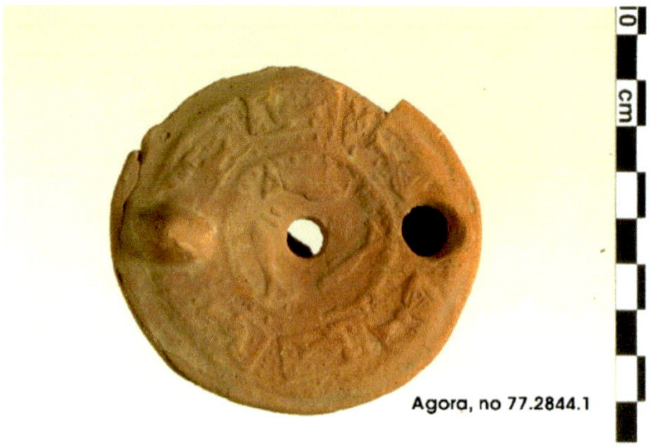

Figure 20.

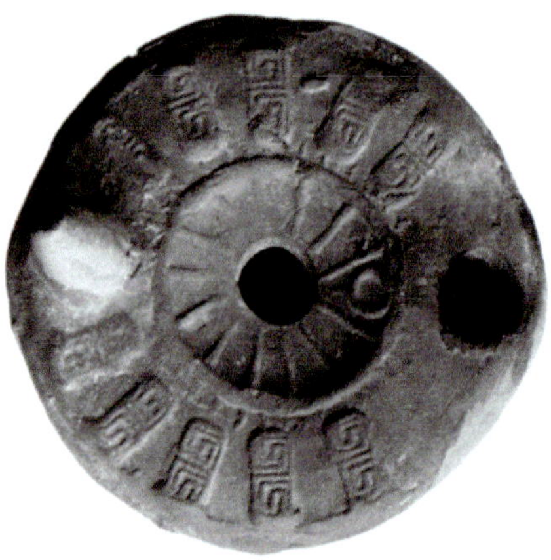

Figure 21.

fine. At the time of his writing, Broneer[3] thought that Corinth was the only site where this type of lamp had been found in the Peloponnese. But the examples found in the Thermes and in the Agora of Argos are most significant. It is possible that the type was produced somewhere in the region, if not in the city itself, during the 6th and the beginning of the 7th century AD. These lamps have a round shape with the nozzle integrated into the rim. The base is mostly plain within a ring and there is no lampmaker's signature. The disk and rim motifs are varied: we can find not only 'pagan' motifs such as Greek meanders, spirals and volutes, but also Christian motifs such as fishes, crosses and birds. This type represents some of the last traces of the local and regional lamp production before the arrival of the 'Argive Dark Ages'.[4] To date there is no evidence suggesting that there were any major ceramic workshops or any other production within the city area during these late years when the Argive population had to survive in very difficult conditions.

In conclusion, Argos was, without doubt, one of the main centers of lamp production in the region during the period from the 4th to the early 7th century AD. Many *officines* specialized in the production of copies, but they also developed some new shapes and types.

Bibliography

Broneer, O. 1930. *Corinth. IV.2: Terracotta Lamps* (Results of Excavations Conducted by the American School of Classical Studies at Athens). Cambridge, Mass: Harvard University Press.

Bovon, A. 1966. *Lampes d'Argos*. Paris: J. Vrin.

Piérart, M. and Touchais, G. 1996. *Argos. Une ville grecque de 6000 ans.* Paris: Editions du Centre National de la Recherche Scientifique.

Oikonomou-Laniado, A. 2003. *Argos paléochrétienne: contribution à l'étude du Péloponnèse Byzantin.* Oxford: Archaeopress, British Archaeological Reports, International Series 1173.

Koutoussaki, L. 2008. Lampes d'Argos. Les lampes en terre cuite découvertes dans l'agora et les thermes. Campagnes de fouilles 1972-1991. Unpublished PhD dissertation, University of Fribourg.

[3] Broneer 1930, 120-121.

[4] Broneer 1930, 120-121.

A Three-Handled, Calyx-Shaped Glass Lamp from Thessaloniki and its Archaeological Context

Anastassios Ch. Antonaras
Thessaloniki, Museum of Byzantine Culture

Abstract: In the excavations conducted in 1960 a very rare form of glass lamp was found in a sarcophagus excavated in the eastern necropolis of Thessaloniki, very close to the city walls. It is a relatively large, calyx-shaped, three-handled vase and its shape is reminiscent of the mosque lamps of the second millennium AD. The finds from the sarcophagus along with those from the neighboring burials are published here in order to ascertain the date of the glass lamp. These include forms, well-known in late 3rd-early 4th century Thessaloniki, of clay and glass vessels, jewels and clothing implements. Along with a few coins, these forms are compared to finds from other parts of the Roman world and they are presented in the context of their contemporaneous findings.

Keywords: Thessaloniki, Roman necropolis, glassware, glass lamps, sarcophagi, clay vessels, jewelry, inscription.

This paper concentrates on the presentation of an excavation hitherto unpublished that yielded among its finds an intact calyx-shaped glass lamp, already published by the author in Lychnological Acts II.[1] Since it is a form very rarely found and because of its resemblance to medieval mosque lamps, its dating could easily be set in doubt. Therefore, it seems useful to thoroughly present this find within its archaeological context.[2]

The glass lamp (ΜΘ 3759) (Figure 1) has a fire-polished rim, a wide funnel-shaped mouth, a calyx-shaped body with a flat bottom, and the low conical ring base. There are three vertically positioned, round in cross-section handles around the rim. A fine glass thread is wound three times around the lower part of the neck just below the bases of the handles. The vessel is made of clear colorless glass with a greenish tinge. The lamp is free-blown, and the base is formed by pressing the lower part with pincers. A rod was then affixed to the base, the moil was cut off, the rim was widened and then fire-polished, and the handles were affixed. Finally, while the vessel was being turned horizontally, a fine thread of glass was applied around the base of the mouth. Lamps of this type were used suspended from *polyangistra* but in rare cases they could have been also used as free-standing lighting devices as their base offers a quite stable footing.[3]

This form seems to be morphologically related to or to be a descendant of the bell-shaped vessels usually standing on a stemmed base, Clasina Isings' typology form 36, specifically 36a or 36c, that are dated to the 3rd-4th century.[4] It seems that this form evolved already before the 7th century into a type of glass lamp with similar body shape, a stemmed conical base, and equipped with three long-tailed handles placed around the rim.[5] Probably from that form evolved a type of lamps known in the Arab world in the 9th-10th century; a type with wide conical body and three small handles around the rim that was standing on an attached, solid and beaded, stemmed conical base.[6] The form under discussion though, appears to follow a different evolutionary line, and is possibly connected with forms/shapes preserved only in early Christian silver examples.[7] These are vessels found in several 6th century hoards in the eastern Mediterranean that share the same low ring base, and an essentially calyx-shaped body, which is quite shorter though compared to our the glass example, and they have a far taller and wider funnel mouth. Mosque lamps possibly evolved from these silver vessels, and have their handles now placed on the shoulders and no longer on the rim. This Mosque lamp form first appeared at the end

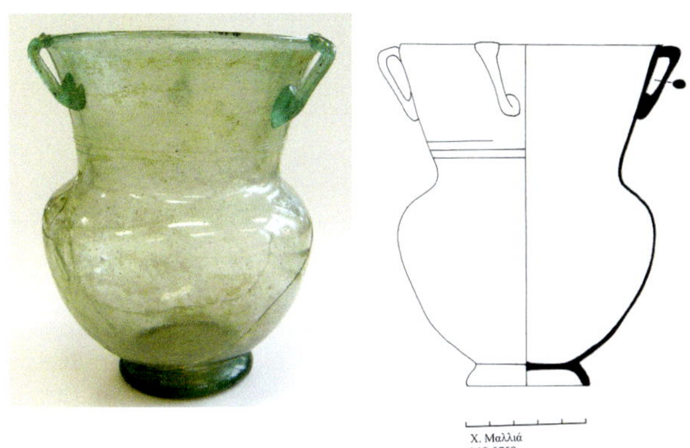

Figure 1. Photo and Drawing of Lamp ΜΘ 3759. © Archaeological Museum of Thessaloniki. Photo by A. Antonaras.

---

[1] Antonaras 2008, 23-30.

[2] I would like to thank Dr P. Adam-Veleni, the director of the Archaeological Museum of Thessaloniki, for giving me the opportunity to conduct this study.

[3] On use of lamps in Byzantium see Motsianos 2011, 205-260.

[4] Isings 1957, 50-52, form 36; Harden *et al.* 1987, 123, no. 55; Whitehouse 2001, nos. 674, 782.

[5] Steckner 1988, 257-270. Several examples of this form were also found at Pyrgi on Crete, Coutsinas in preparation.

[6] Carboni and Whitehouse 2001, 76, no. 6.

[7] Similar shapes with the same low base but with wider rim appear among 6th century silver lamps, which were found in several hoards from the eastern Mediterranean. Newman and Lie 1992, 82-83, fig. 37.1, 37.2; Mundell-Mango 1992, 137, fig. 18; Snow 1992, 202, fig. 2.

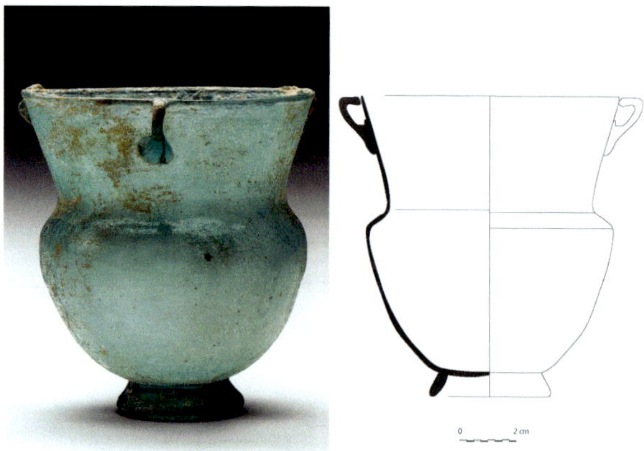

Figure 2. Photo and Drawing of Lamp inv. no. AA 2004-222. © Musée d'art et d'histoire, Ville de Genève. Photo: Musée d'art et d'histoire, Ville de Genève.

of the first millennium and their use culminated during 13th and 14th century. Examples from the last centuries were almost always supplemented with enamelled decoration.[8]

Concerning direct glass comparanda, an analogous find, of the rim and upper body of a colorless example, was spotted in the unpublished material of a salvage excavation in the ditch of the western city walls of Thessaloniki. The find has no firm dating, but it, along with the rest of the glass and clay finds of the site, can be generally dated in the Late Antique period.[9] An identical intact example of unknown provenance and date, originally from the Janet Zakos (Genève) collection, is kept in the Art and History Museum of Geneva[10] (AA 2004-222) (Figure 2).

The excavation, in which the glass lamp was found (Figure 3) is not noted in the *Archaeologicon Deltion* or any other archaeological journal of the time and the only thing that ties together our glass lamp with the remaining of the movable finds are the short provenance notes in the register of the Archaeological Museum of Thessaloniki. In them mention is made that they were found on 23 and 25 April in 1960, at 44, Vasilissis Sofias street, which nowadays is named Ethnikis Amynis. This site is situated extremely close to the eastern city walls and is also very close to the main eastern gate, leading to the city's main street. All in all this location was a very important part of the city's necropolis and based on its

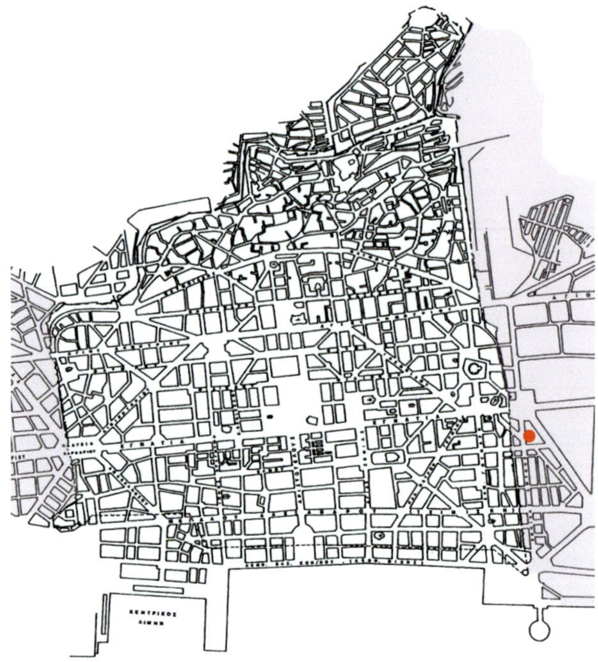

Figure 3. Thessaloniki, city plan. Red dot on the find spot of the glass lamp.

proximity to the walls, a very old part of it as well. It should be noted that the use of the eastern necropolis ceases in the 7th century, along with the practice of extramural funerals in general, probably a result of the barbaric invasions and the consequential fear that the graves would be desecrated by the invaders.[11]

In the register of the Archaeological Museum of Thessaloniki we can read that the glass lamp was found in the third sarcophagus of the excavation, which leads to the conclusion that at least three sarcophagi were found within the plot. Nevertheless no information connecting the site with any of the extant Roman sarcophagi is preserved. What is known though is that all 290 sarcophagi found in the city's necropolises are dated between the 2nd and the 3rd century, a time frame that presents a *terminus post quem* for our find.[12]

A working hypothesis on the whereabouts of the excavation can be made. According to the archives of the 11th Archaeological District, nowadays the 16th Ephorate of Prehistoric and Classical Antiquities, three heavy sarcophagi were found at Vasilissis Sofias street (no number is given)near the Sintrivani square. It must be noted that #44 is the second property south of this square. They were removed from the site sometime before 21 July 1957, according to report no. 432/21-6-1957 of the Ephor of Antiquities Charalampos Makaronas to the Directorate of Antiquities, and were left in the public garden across the street in front of the site where the new archaeological museum was to be built.[13] The Museum

[8] Kröger 1995, 182, no. 235, find from Nishapur, 10-11th century; Saldern 1980, 174, no. 176, probably from Iran, 8-9th century; Goldstein 2005, 82, nos. 80 and 85, no. 82 probably from Iran, 8-10th century; Saqqarah, Crowfoot and Harden 1931, 205, pls. 30, 46; Saint-Nicolas in Myra, 11-12th century Olcay 2001, 86, fig. 6i; Caesarea Maritima late 10-11th century; Pollak 2000, 240-241, fig. 6.8; Serçe Limani *c.* 1025, Bass *et al.* 2009, 372-373. For 13-14th centuries examples see Carboni and Whitehouse 2001, 227-238, nos. 113-118; Goldstein 2005, 286-87, no. 311 from Egypt, *c.* 1385.

[9] For the excavation see Marki 1999, 579. For the clay finds of the site see Tzevreni 2011, 39-40.

[10] That was reported to me by the colleague and friend Marie-Dominique Nenna. For this information I warmly thank her from this position as well. I also thank the museum's curator Marielle Martiniani-Reber, and the Museum's photographic service for rendering the object's photos.

[11] Marki 2006, 47, 49-61; Makropoulou 2006, passim; Makropoulou 2007, 475.

[12] Stefanidou-Tiveriou 2010, 151-188, esp. 152.

[13] Daux 1958, 760-761, figs 9-10 (registered as ΜΘ 5684, ΜΘ 5685, and probably ΜΘ 5686).

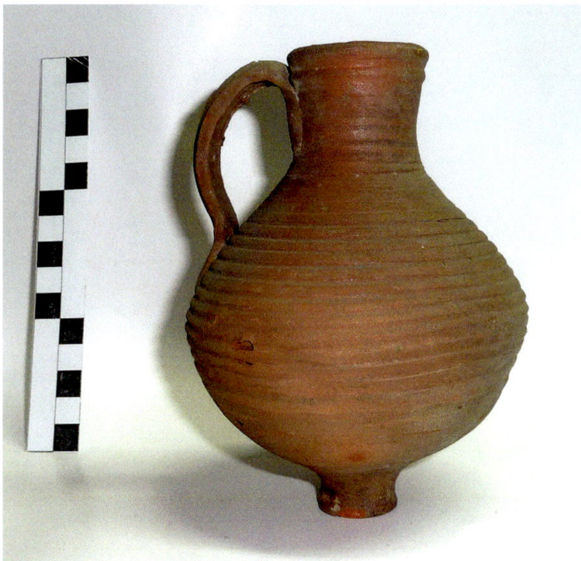

Figure 4. Jug MΘ 5266. © Archaeological Museum of Thessaloniki. Photo by A. Antonaras.

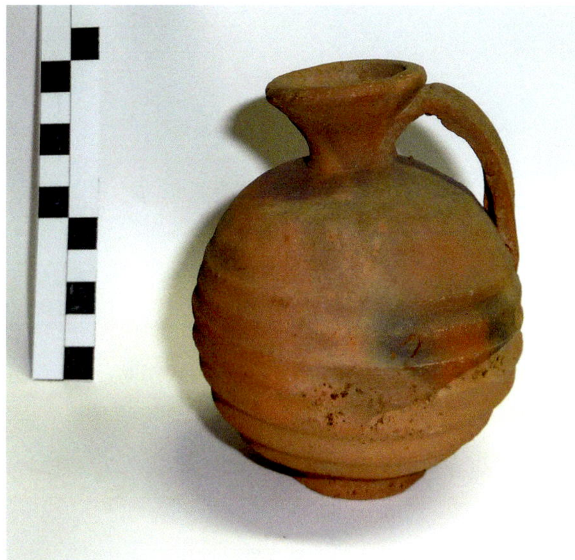

Figure 5. Jug MΘ 5262. © Archaeological Museum of Thessaloniki. Photo by A. Antonaras.

was inaugurated on 27 October 1962, but construction work begun much earlier, probably in 1960.[14] All movable finds from Vasilissis Sofias 44, bear as find dates 23 and 25 April 1960, although there is evidence that the building on the site of the excavation was constructed earlier and it was already inhabited by 1959.[15] So, it is quite possible that the three sarcophagi from Vasilissis Sofias street, near by the Sintrivani square, were indeed from the plot on #44 and that they were left in the public garden intact and sealed until construction of the new museum started, when they were opened and their contents were registered with find dates as the dates of the final opening of the sarcophagi.[16]

Apart of this hypothetical part the hard evidence that dates our glass lamp are the registered movable finds that were indeed associated with it according to the Museum's register.

All clay vessels that were unearthed in the site belong to forms of vessels widely used in Thessaloniki's necropolises, and can be dated with certainty in the late 3rd century and mainly in the 4th century. Almost all of them are made with light yellowish clay and only two of them bear a coating. Eight of them are jugs, one spindle-shaped (MΘ 5266), one globular (MΘ 5262), one pear-shaped (MΘ 5269), one bulbous (MΘ 5334) and four cylindrical ones (MΘ 5264, 5294, 5295, 5296, (Figures 4-8).[17] There is also one globular cup (MΘ 5263) and three unguentaria, two spherical (MΘ 5265, 5268) and one spindle-shaped (MΘ 5267) (Figures 9-11).[18]

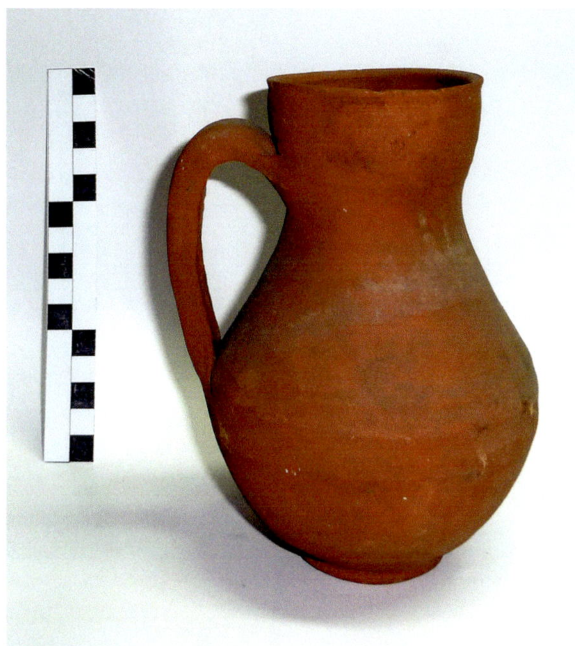

Figure 6. Jug MΘ 5269. © Archaeological Museum of Thessaloniki. Photo by A. Antonaras.

[14] The document that arranged the size and the exact shape of the museum's plot was issued already on 23 September 1959.

[15] Personal communication with inhabitant of the building, Mrs. Kaliampetsou, September 2011.

[16] I thank wholeheartedly Dr Demetra Aktseli who very kindly helped me through the maze of the archives of Archaeological Service of northern Greece. This whole hypothesis of the opening of the sarcophagi at a later date is corroborated by the fact that there is at least one other sarcophagus that was excavated in the 60's and was kept sealed for several decades in the front yard of the Archaeological Museum of Thessaloniki.

[17] For an overview of the Late Roman and Early Christian clay finds from Thessaloniki see Papanikola-Bakirtzi 2010, 263-298; Petsas 1966, 335, pls 348a, 349a-d; Nalpantis 2003, 119-123; Makropoulou 2007, 257-270; Marki 2006, 206, pls 68-69; Makropoulou, 'Grave Finds'; Papanikola-Bakirtzi 2002, no. 735 [E. Pelekanidou].

[18] MΘ 5334: Petsas 1966, 334-339, pl. 353η, in the middle; Pazaras 1997, 484-5, fig. 19 from a grave used in the Early Christian period until the 6th century; Tzanavari 2010, 453-454, fig. 5, 13, draw. 5, 4th century. MΘ5262: Petsas 1969, 304, pl. 314γ, from N. Michaniona, 4th

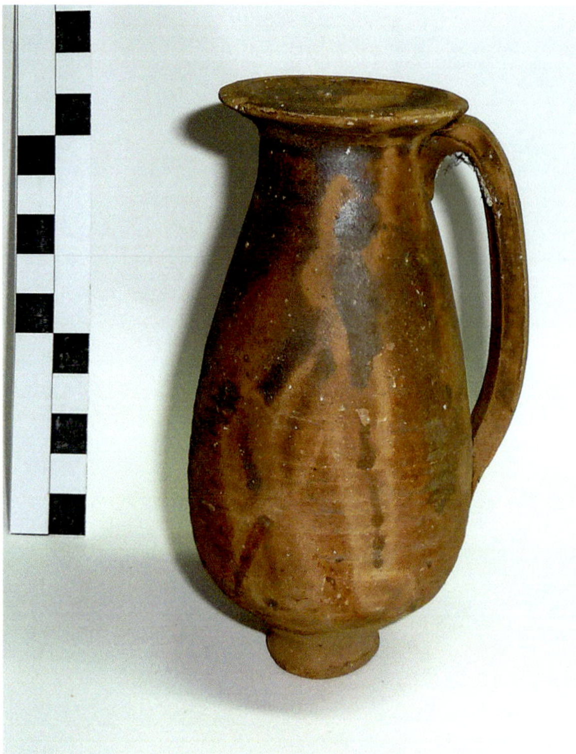

Figure 7. Jug MΘ 5334. © Archaeological Museum of Thessaloniki. Photo by A. Antonaras.

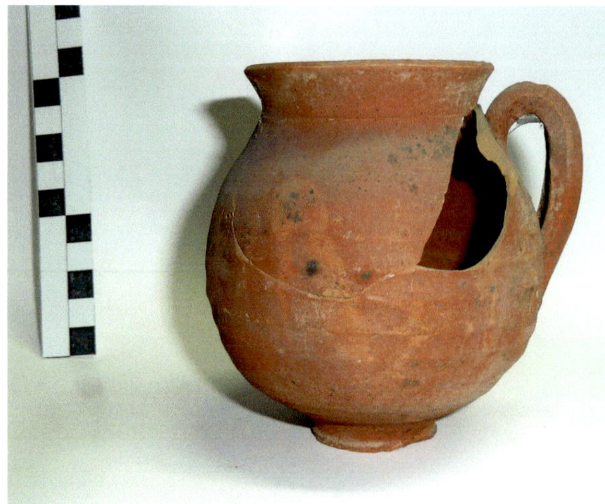

Figure 9. Cup MΘ 5263. © Archaeological Museum of Thessaloniki. Photo by A. Antonaras.

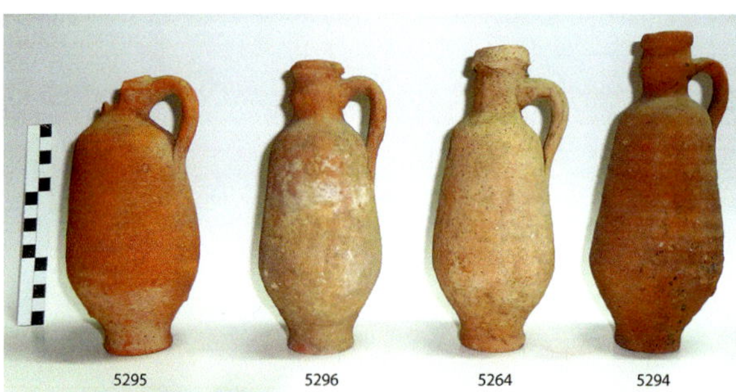

Figure 8. Jug MΘ 5295, 5296, 5264, 5294. © Archaeological Museum of Thessaloniki. Photo by A. Antonaras.

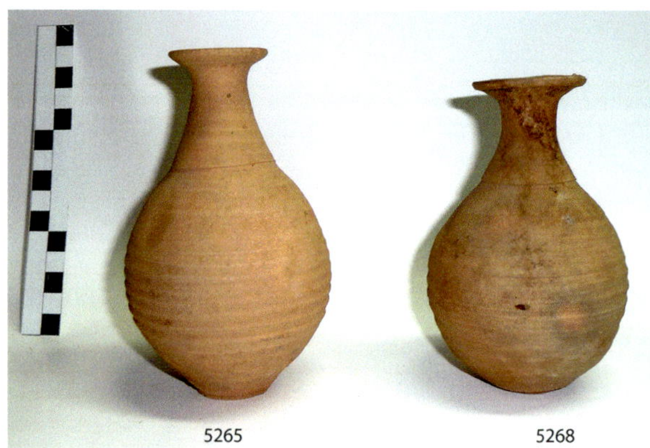

Figure 10. Unguentaria MΘ 5265, 5268. © Archaeological Museum of Thessaloniki. Photo by A. Antonaras.

century; Nalpantis 2003, 33, no. BK 4506/9, 4th century; Panti 2010, 476, fig. 1.ε, first half of the 4th century. MΘ 5265 and MΘ 5268: Petsas 1966, 334-339, pl. 353η, at the right; Petsas 1967, 391, pl. 296γ , at the right, dated in the 3rd century; Petsas 1969, 304, pl. 314β, from N. Michaniona, Thessaloniki, dated in the 4th c; Pazaras 1981, 375, pl. 1β vessel no BK 2469, mid 3rd century. Also published in *Thessaloniki History and Art*, 51, 53, pl. II.11.6; Nalpantis 1992, 316, fig. 16, nt. 20 [BK 4532/2], 4th century; Pelekanidou 1993, 376, fig. 10 from late Roman graves, dated before the 5th century; Nalpantis 2003, on the form generally p. 123 where it is dated in the first half of the 4th century, also examples are presented at p. 32 BK 4506/7, 4th century, p. 83 BK 4506/37, early 4th century, p. 45 BK 4506/15 3rd-4th century, p. 47 BK 4506/16, second half of 4th century, p. 58 BK 4506/22 3rd-4th century; Makropoulou 2007, six examples: BK 4544/24, p. 215 pl. 63, first half of the 4th century; BK 4500/8, p. 29, pl. 63, first half of the 4th century; BK 4500/6, p. 35, pl. 63, first half of the 4th century; BK 4500/25, p. 52, pl. 63, 4th century; BK 4500/3, p. 33, pl. 63, first half of the 4th century; BK 4500/29, p. 53, pl. 63, first half of the 4th century; Pazaras 2009, 129, unguentarium form 2, from the second quarter to the end of the 4th century; Tzanavari 2010, 453, fig. 14, probably 4th century; Panti 2010, 478, no. 23, fig. 2.α, late 3rd to first half of the 4th century, p. 478, no. 25, fig. 3.γ third quarter of the 4th century. MΘ 5266: Petsas 1969, 304, pl. 314 γ, from New Michaniona, 4th century; Giouri 1970, 378, pl. 316ε; Pazaras 1981, 374, pl. 1β, vessel no. BK 2471, 5th century. Also published in *Thessaloniki History and Art*, 51, 53, pl. II.11.7; Makropoulou 1991, 259, 261, figs 1, 10, 18, 4th century; Nalpantis 1992, 316, fig. 19, nt. 20 [BK 4532/1], 4th century; Nalpantis 2003, on the form generally p. 120 where is dated between the last quarter of the 3rd and the first half of the 4th century; also somewhat slimmer examples p. 29, BK 4506/4, BK 4506/5, BK 4506/6, late 3rd- mid 4th century, p. 34, BK 4506/11, 4th century, p. 66 BK 4506/29, early 4th century, p. 79, BK 4506/34, last quarter of the 3rd century; Makropoulou 2007, two examples: BK 4500/1, p. 32, pl. 63, late 3rd to second half of the 4th century, BK 4500/12, p. 42, pl. 63, late 3rd to second half of the 4th century; Pazaras 2009, 126, Jug form 10, 4th century; Tzanavari 2010, 450-451, fig. 9, dated in the late 3rd to early 4th century according to similar finds from Thessaloniki; Graekos 2010, 432-433, fig. 1, 2 late product of the 'urnette a collarino' workshop third quarter of the 3rd century, fig. 11 slimmer late third-century version like Nalpantis' 'local' Thesalonian examples; Panti

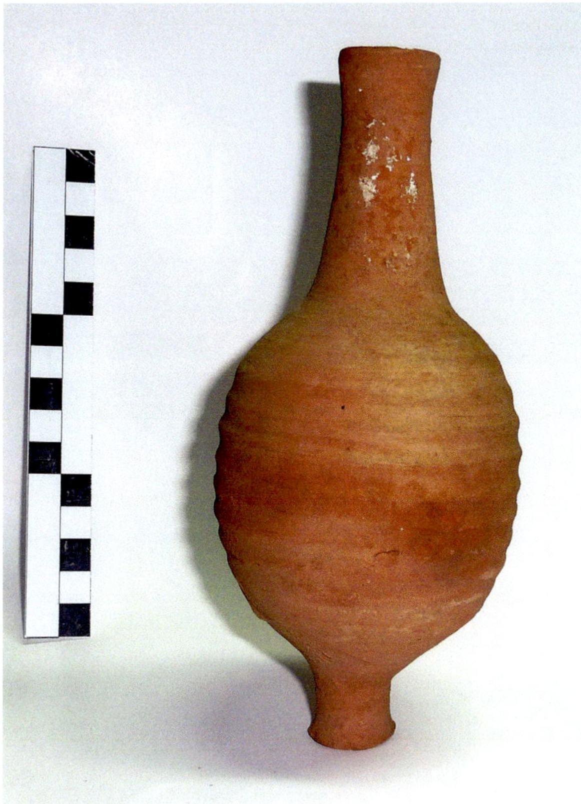

Figure 11. Unguentarium MΘ 5267. © Archaeological Museum of Thessaloniki. Photo by A. Antonaras.

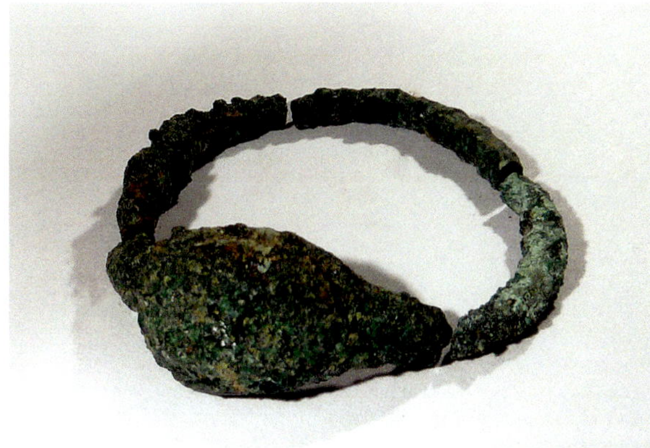

Figure 12. Bracelet MΘ 3020. © Archaeological Museum of Thessaloniki. Photo by A. Antonaras.

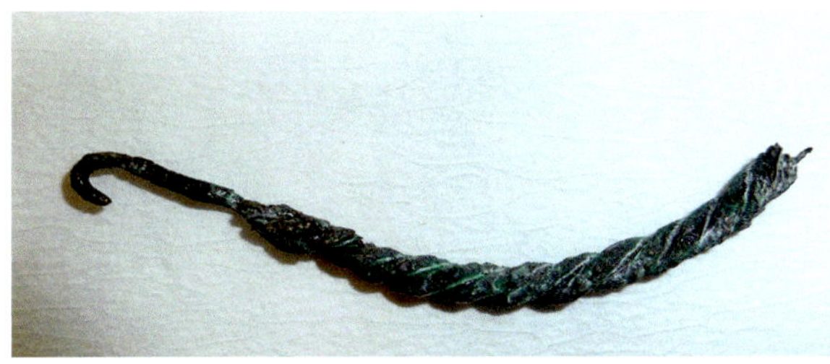

Figure 13. Bracelet or *torque* MΘ 3021. © Archaeological Museum of Thessaloniki. Photo by A. Antonaras.

Metal finds from the site include a plain, cast bronze bracelet with thick, open ends, squeezed to fit a really small wrist (MΘ 3020, Figure 12), [19] a partly preserved bronze object made of two twisted wires, either a bracelet corresponding to a wider wrist or possibly a *torque*, i.e., a chocker (MΘ 3021, Figure 13)[20] and two small bronze rings, with low, oval bezels, partly and badly preserved (Figure 14-15). The bezel of the first ring is undecorated. The bezel of the second ring (MΘ 25905) bears a standing, probably female, figure rendered in intaglio. Two more, bronze items, associated with clothing were discovered. The first is a belt buckle with an oval body made of a folded plate, decorated at the fringes with clusters of triple incised grooves and a compressed, oval, omega-shaped ring.[21] The buckle was obviously gilded, and a golden patina is preserved in some of the grooves at the periphery of the body (Figure 16). The second (MΘ 25906) is a fragmentarily preserved, heart-shaped, inscribed plate, probably a decorative ending of a belt. At the narrow end of the plate are partly distinguishable two holes for the rivets that kept it attached to the belt. At the wider end, each curved part is defined by a simple incised volute.[22] A Christogram is preserved at the center of the upper part, and below it is an inscription in three lines: MARTINE/.

2010, 476-467, no. 9, fig. 2.α, late 3rd to first half of the 4th century. MΘ 5263: Petsas 1967, 391, pl. 296β, at the center, dated *c.* 200; Petsas 1967a, 393, pl. 300γ, four late Roman examples; Giouri 1970, 378, pl. 316στ (two vessels at the left); Nalpantis 2003, vessel no. BK 4506/36, p. 79, last quarter of the 3rd century; Makropoulou 2007, vessel no BK 4500/21, p. 50, pl. 64, 4th century; Kallintzi and Chryssaphi 2010, vessel no MA 7437, p. 391, 399, fig. 3α jug form 1, dated according to similar finds between the 1st and the 3rd century; cf. Malama and Darakis 2008, four examples: MA 6470, 6496, 6497, 6773, p. 403-405, pl. 64, mid 3rd to 4th century. MΘ 5269: cf. Makropoulou 2007, three examples: BK 4501/18, p. 82, 84, pl. 10. 4th century; BK 4499/8, p. 101, pl. 64 4th century; BK 4500/27, p. 52, pls 13, 63, 4th century. Also, cf. Chrysostomou 2010, 510, no. 38, fig. 19 from a grave with early Christian coins (no dates provided) from Apsalos, Almopia. MΘ 5294 5295 5296: Makropoulou 2007, four examples: BK 4544/30, p. 217 pl. 15, first half of the 4th century; BK 4524/5, p. 243, pl. 15 first half of the 4th century; BK 4597/3, p. 332 pl. 15 4th century; BK 4597/39, p. 305-306 pl. 15, first half of the 4th century; Graekos 2010, 435-436, fig. 20 a late Roman find; Panti 2010, 477, no. 13, fig. 2γ, 3rd-first half of the 4th century. MΘ 5267: Petsas 1967, 391, pl. 296γ , at the left, dated in the 3rd century; Petsas 1969, p. 304, pl. 314β, from Nea Michaniona 4th; Makropoulou 1991, 259, fig. 12. Also published in Makropoulou 2007, 44, pl. 63, vessel no. BK 4500/16, dated first half of the 4th century; Tzanavari 2010, 453-54, fig. 4, 15, 4th century according to similar finds; Pazaras 1997, 485-486, fig. 20, 4th century. Also published in Pazaras 2009, 128, unguentaria form 1, late 3rd to after the middle of the 4th century.

[19] Makropoulou 2007, BKO 269/1, BKO 269/7, p. 165, pl. 72, both from a 4th century grave.

[20] Cf. Malama and Darakis 2008, 107, 428, pl. 79: MA 6440 from a grave dated to the first half of the 3rd century. Also, Kourkoutidou-Nikolaidou 1997, 141, fig. 62 lower left for a third-fourth-century example from Thessaloniki; Pologiorgi 1998, 77, no. MΩ 445, Roman period.

[21] For identical examples excavated in the western necropolis of Thessaloniki, dated to the end of the 3rd and into the 4th centuries, see Makropoulou 2007, 462-463, pl. 35 (BKO 219/4α 219/9α).

[22] For a relevant example from Thessaloniki, dated from the end of the 3rd to the 4th century, see Makropoulou 2007, 463, pl. 35 (BKO 268/3).

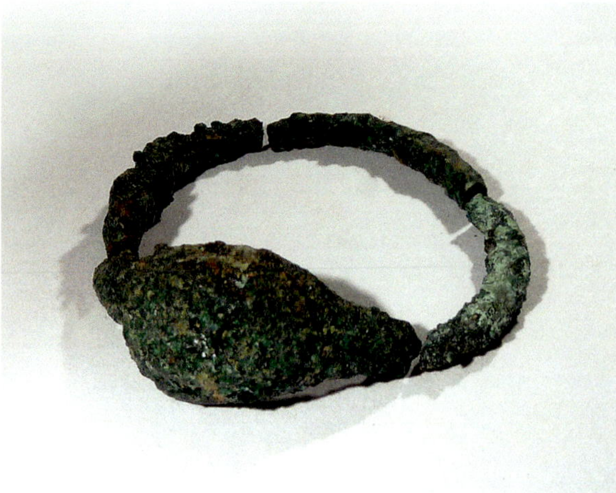

Figure 14. Ring. © Archaeological Museum of Thessaloniki. Photo by A. Antonaras.

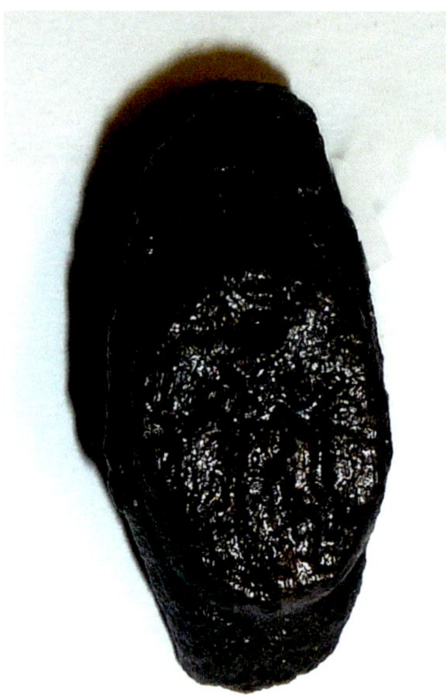

Figure 15. Ring MΘ 25905. © Archaeological Museum of Thessaloniki. Photo by A. Antonaras.

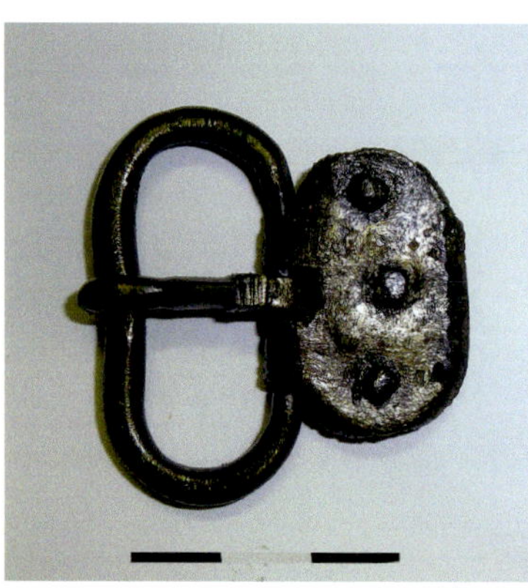

Figure 16. Buckle unregistered. © Archaeological Museum of Thessaloniki. Photo by A. Antonaras.

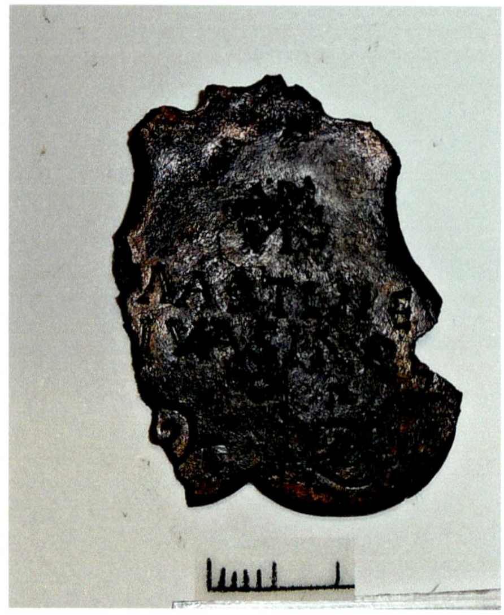

Figure 17. Belt finial MΘ 25906. © Archaeological Museum of Thessaloniki. Photo by A. Antonaras.

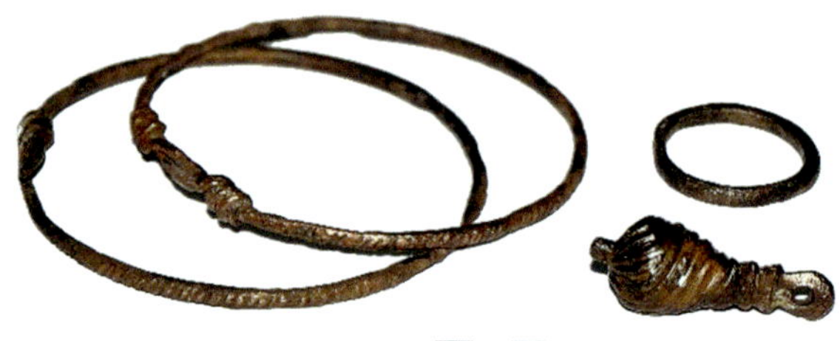

Figure 18. Jewelry BKO 5. © Museum of Byzantine Culture. Photo by A. Antonaras.

IVASIND./O, transcribed as *Martine, vivas in Deo*[23] (Figure 17). Four more of the bronze findings from the third sarcophagus are kept in the Museum of Byzantine Culture (BKO 5, Figure 18).[24] Two identical bracelets were made of a fine, flattened wire. The ends of the wire are twisted forming a loop, each one of them set within the other end allowing for an easy adjustment of the bracelet's size.[25] The outer side of their body is decorated with twisted, parallel incisions. A simple ring and a twisted bulbous ornament or pendant conclude this group.[26] In addition, one cylindrical, bronze pyxis (MΘ 29881) found in the third sarcophagus belongs to a pyxis type known from other fourth-century burials in Thessaloniki (Figure 19).[27]

[23] A similar object is exhibited at the Metropolitan Museum of Art (1993.166), which was also dated to the 4th century. On this type of inscription see DACL XV, 3140 ff.

[24] According to their tag they were unearthed in the third sarcophagus of Zervas' plot on Angelaki Street, 25 April 1960 and they are registered under no. BKO 5 in the Museum of Byzantine Culture.

[25] Three similar bracelets are exhibited in the Museum of Byzantine Culture, finds from a 3rd or 4th-century grave in the western necropolis of Thessaloniki (BKO 228/36-228/36).

[26] Similar finds are kept in the Museum of Byzantine Culture, from 4th century burials in the eastern necropolis of Thessaloniki (BKO 7, BKO 8, BKO 10 and esp. BKO 11).

[27] Kourkoutidou-Nikolaidou 1997, 136. Nalpantis 2003, 139, pl. 48

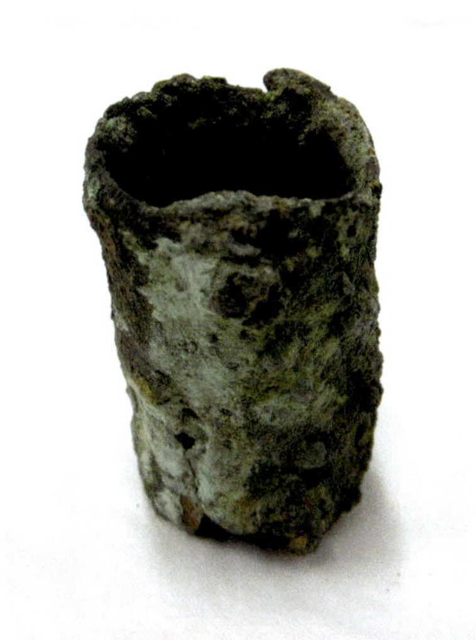

Figure 19. Pyxis MΘ 29881. © Archaeological Museum of Thessaloniki. Photo by A. Antonaras.

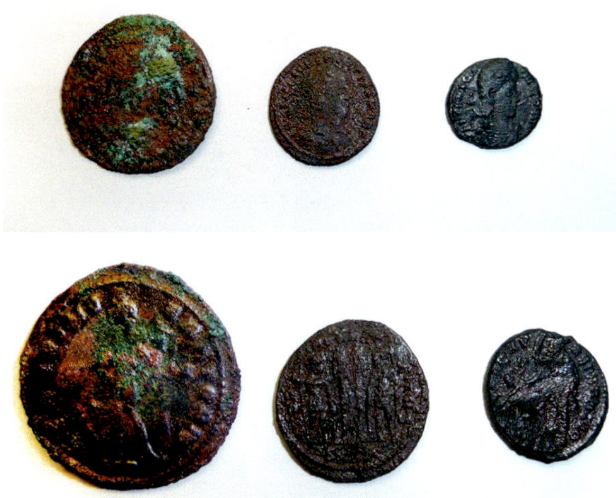

Figure 20. Coins. © Archaeological Museum of Thessaloniki. Photo by A. Antonaras.

Five bronze Roman coins are also found in the site, two of them are illegible and the other three are dated in the 4th century (Figure 20).[28] None of the coins are clearly associated with the sarcophagus which contained the glass lamp. Finally some small iron hobnails, apparently from the shoes of one of the deceased, were also preserved (Figure 21).[29] A few glass

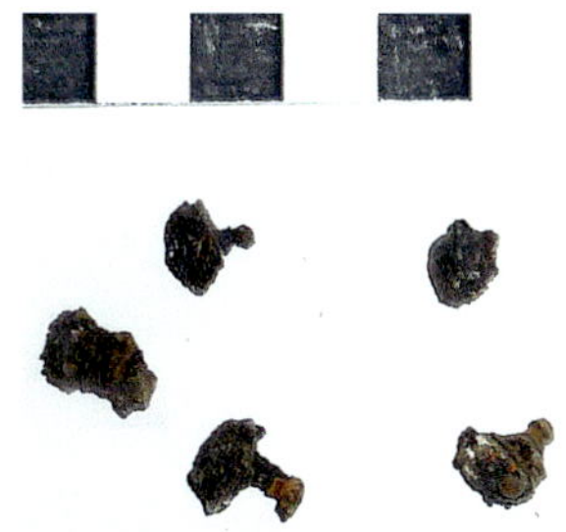

Figure 21. Hobnails. © Archaeological Museum of Thessaloniki. Photo by A. Antonaras.

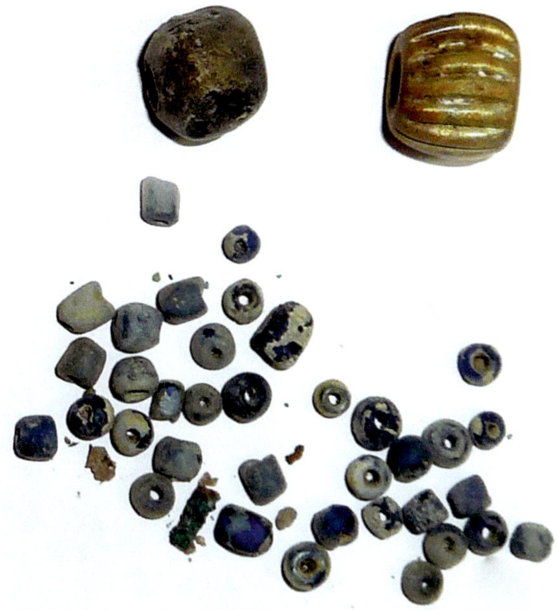

Figure 22. Beads. © Archaeological Museum of Thessaloniki. Photo by A. Antonaras.

beads: 36 smooth and round, dark blue ones, and two dark-colored, round, ribbed ones, all of well-known Late Antique types, were also collected from the graves (MΘ 25903, 25904, 25910, 25915, Figures 22-23).[30]

upper part, finds dated to the first half of the 4th century. Makropoulou 2007, 72-73, 78, 88, pl. 38, two pyxides from a burial complex dated to the second half of the 4th century.

[28] A coin struck at Cyzicus in 309-310 of the type Genio Imperatoris, see. RIC 6, 588, nos. 56-57 or p. 589, no. 63. A coin of 330-335 of the type Gloria Exercitus, two solders, two standards, cf. LRBC I 835 ff. And one coin of Constantine II from 351-361, of the third subtype of the Fallen Horse type (FH III), cf. LRBC II, 1681 ff. I thank for the identifications of the coins prof. Dr Michael Ierardi, Bridgewater University.

[29] For a general overview on Roman footwear see Goldman 2001, and especially on nails preserved at the sides of slippers with wooden soles see Goldman 2001, p. 144 which might explain the small number of nails found in some graves, as in our case, as opposed to finds of greater numbers of nails, most probably belonging to hobnailed nail *caligae*. For finds from the Macedonian region see: Chrysostomou 2000, 291, no. 825, pl. 329, p. 295, no. 839 ,draw. 119, pl. 333, p. 539-40 on finds from Edessa, with further bibliography on the matter; Malama and Darakis 2008, 435, more than 800 shoe-nails were collected from seventeen burials from Nea Kerdylia, Serres, dated mainly in the early 3rd century with further bibliography. Other finds from Thessaloniki include: Petsas 1970, 349, late 3rd century; Papanikola-Bakirtzi 2002, 113, cat. no. 106 [D. Makropoulou]; Nalpantis 2003, 140-141, pls 49, 50, 303 shoe nails were collected from five burials dated in the 3rd and 4th centuries; Makropoulou 2007, 388-389, 466 nails from 13 graves; Pazaras 2009, 149, 179 nails from three graves from Limori Epanomi, 4th according to the graves' inventory.

[30] On ribbed/melon beads see Malama and Darakis 2008, three sets: 116-117, pl. 83: MA 6465 from a late 2nd to early 3rd century grave;

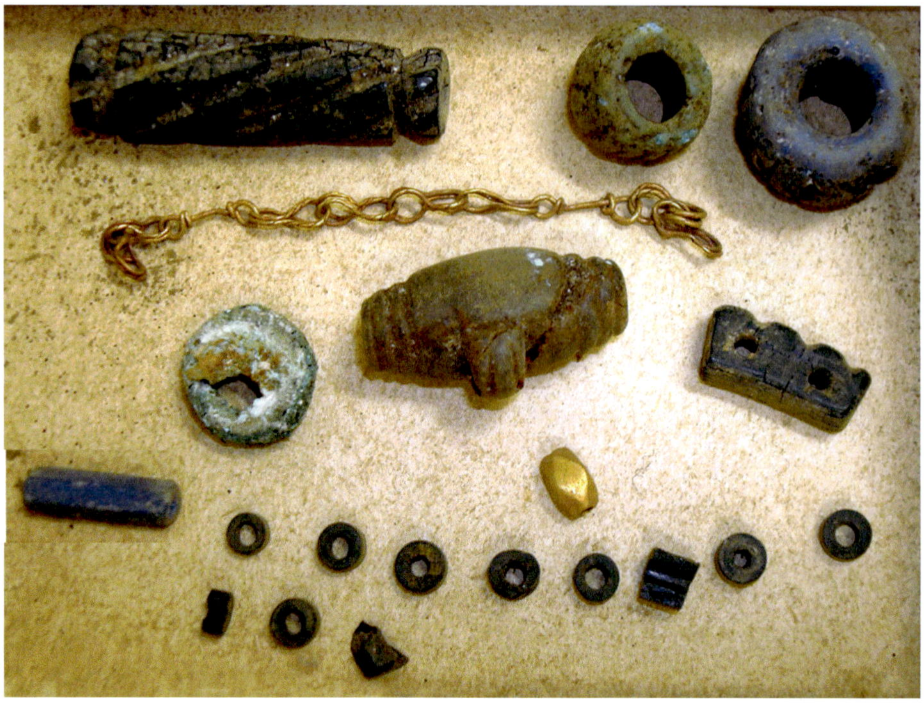

Figure 23. Beads MΘ 25903, 25904, 25910, 25915. © Archaeological Museum of Thessaloniki. Photo by A. Antonaras.

A dark green stone palette for cosmetics was also found in one of the graves of the site. Its upper and lower surfaces are totally smooth, while the sides are oblique and only very roughly carved (Figure 24). This detail indicates that the stone palette was originally placed in a frame of wood or of some other organic material that has totally perished leaving no traces in the excavation, or that it originally was part of a metal box with medical tools, out of which nothing else was placed in the grave, or survived the decay. Similar stone palettes are known from the Ancient and Roman world but they do not occur in Byzantine layers.[31]

The excavation yielded a partly preserved, two-sided wooden comb (Figure 25) decorated with a simple, wavy, geometrical engraving at its middle part (MΘ 3022, Figure 26).[32] Finally, among the findings of the sarcophagus are included two braids of human hair (*c.* 20 cm long), indicating the sex of the owner, seemingly made of human hair, that are quite well preserved. Because of their good state of preservation, which

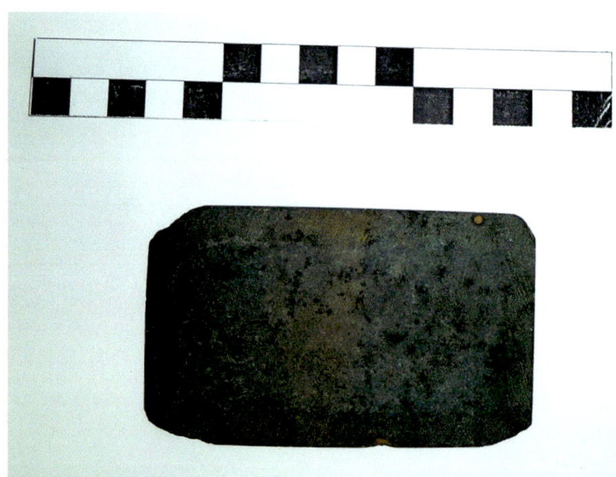

Figure 24. Palette. © Archaeological Museum of Thessaloniki. Photo by A. Antonaras.

p. 145-146, pl. 83: MA 6559 from a grave of the first half of the 3rd century; p. 152, pl. 83: MA 6574 from a grave of the first half of the 3rd century; and Antonaras 2010, 319-323, where a short overview of the bead forms excavated in Thessaloniki is presented. Also Arveiller-Dulong, Nenna 2011, 192, nos. 241-242 for Late Roman examples. On smooth ones see Malama and Darakis 2008, one set with smooth beads: p. 226-27, pl. 84, MA 6796, a late 3rd to early 4th century grave; Antonaras 2010, op.cit.; and Ignatiadou 2010 [K. Sismanidis] 345-348, nos. 365, 370, 371 dated to the early 2nd century.

[31] For relevant finds from central Macedonia see Malama and Darakis 2008: 247, 436 pl. 86: MA 6838 from a mid-4th century grave; p. 124, 247, pl. 86, MA 6488, from a late 2nd-early 3rd century grave; Chrysostomou 2000, 208, no. 497, pl. 212 and p. 293, pl. 330. For their use in medical boxes, see Künzl 1983, 170-171, 222-223; Boyer and Mourey, 1990, 234, note 45 with many relevant finds from medical boxes.

[32] For 12 examples from the Roman military camp at Vechten, Netherlands and generally on Roman wooden double-sided combs, mainly made of boxwood, and their use equally by women and men, or at least soldiers, see Derks and Vos 2010, 53-77 where a rich bibliography on analogous finds is noted. For ten first-century AD almost identical examples found in Qumran in the Judean desert, see Magness 2002, figs 56, 57. For four new finds from Karanis in Fayum see Derks and Vos 2010, 68, note 13. For Roman and Late Antique bone combs from the Balkan hinterland and central Europe see Petković 1995, passim and especially 21-28, where the use of these bone implements is dated between the 4th and 6th centuries. For Roman one-piece double-sided bone combs see Derks and Vos 2010, 56, 67, note 5.

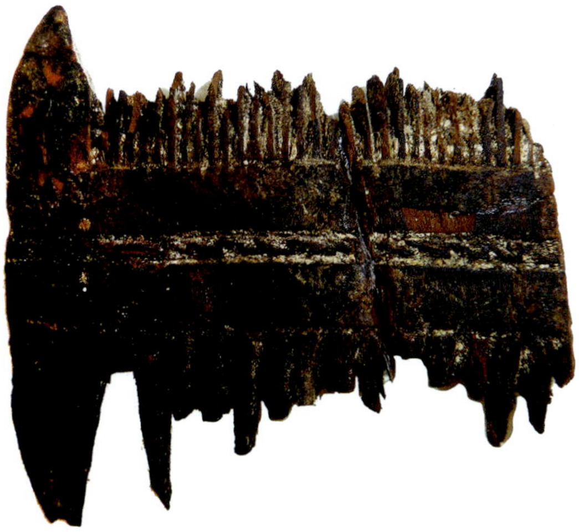

Figure 25. Comb MΘ 3022. © Archaeological Museum of Thessaloniki. Photo by A. Antonaras.

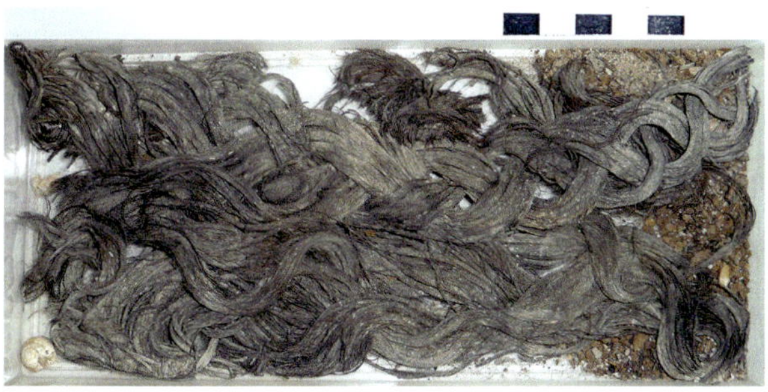

Figure 26. Braided hair. © Archaeological Museum of Thessaloniki. Photo by A. Antonaras.

is quite unusual for Thessaloniki's humid conditions, it could be assumed that they may present some kind of wig that had already been embalmed or chemically processed before its final placement in the sarcophagus. The hypothesis that it was a wig is corroborated by the fact that some threads are interconnected with the hair and a small part of a possibly linen (?) fabric is preserved attached with a thread or sewn on one of the braids.[33]

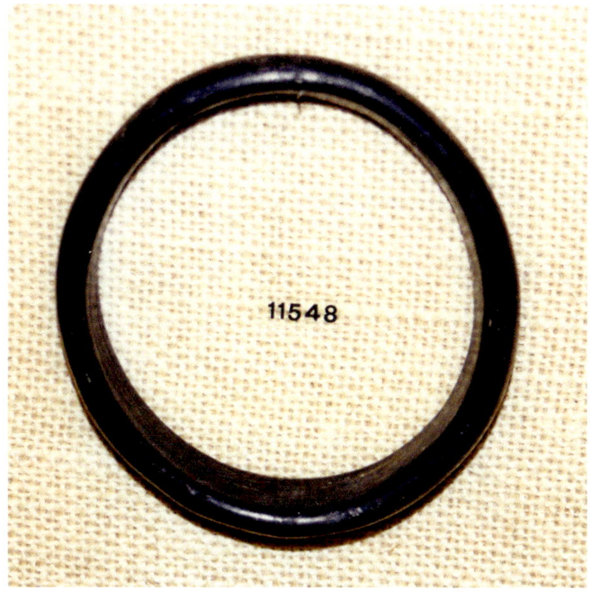

Figure 27. Bracelet MΘ 11548. © Archaeological Museum of Thessaloniki. Photo by A. Antonaras.

The dating of the glass lamp is assisted, by the third sarcophagus' contents that comprises a bronze pyxis, two bronze bracelets, one ring, one bulbous ornament, and a necklace with jet, gold and glass beads on a golden chain. All of these indicate that a woman was buried in the sarcophagus. In addition, a glass bracelet and a spherical glass vessel of the Isings 101 type were also found in it, which also clearly suggest a burial date somewhere in the late third or in the fourth century. The bracelet (MΘ 11548, Figure 27) is made of dark blue, seemingly black glass, and is seamless. The use of glass bracelets in the eastern Mediterranean cannot be dated before the mid-3rd century AD and indicate a date in Late Antiquity (Figure 28).[34] Glass flasks of Isings form 101, also known as Antonaras' form 51a, are very widespread and they can be dated between the late 3rd and the 5th century, as several better dated examples from Thessaloniki prove.[35]

In conclusion it is clear that the glass lamp from Vas. Sofias 44 sarcophagus no. 3 can quite certainly be ascribed to the

[33] Similarly, almost identical finds of blond braids have been preserved in a lead coffin which was placed within a marble sarcophagus excavated in 1962 at Anapafseos street, at the eastern necropolis. For a gold-weft, purple cloth that was found in the same sarcophagus reg. no. 18218 see Vokotopoulou 1996, 88 and fig. on p. 89. Other finds of human braids and wigs are preserved in burials in different parts of the empire: Britain, York: Allason-Jones 1989, 133-134; France, Les Martres-de-Veyre: Audollent 1923, 275-328, esp. 284, pl. 8; Egypt: Walker and Bierbrier 1997, 208-9, no. 302; Masada, Judaea: Yadin 1966, 54, col. pl. p. 56, pl. p. 196; Zias 1998; Rumania, Tomis/Constantia: Canarache 1967, 40. I thank Dr Katerina Tzanavari for informing me on the find from Tomis. On Roman female hairstyles and the use of wigs see Bartman 2001, passim and esp. 7, 10, 14-15, where all prior bibliography is cited. On references/citations of ancient Greek and Latin sources concerning hair-dyeing and wig-wearing, see Reinach 1896, 1452-1453; Bartman 2001, 14. For a second-century witness on the wide use of wigs and the castigation of their use on the heads of the deceased on behalf of the Christians see Clement, *Paedagogus*, 3.11.63.

[34] The glass bracelet (MΘ 11548) is presented in *Thessaloniki from prehistoric to Christian era* 1986, 132-33, pl. 133α and published in Ignatiadou 2010 [Ai. Tzanavari] 297, no. 272 2nd-3rd century. Glass bracelets in the Mediterranean region spread from about the mid-3rd century onwards, see Spear 1988, 51 and Antonaras 2006, 423-434. Also see Arveiller-Dulong and Nenna 2011, 249-260, nos. 350-371. For three other glass bracelets from Thessaloniki dated in the Roman period see *Thessaloniki from prehistoric to Christian era* 1986, 132-133, pl. 133; *6000 years of Tradition* 1997, 150, no. 149 [D. Ignatiadou]; Ignatiadou 2010[Ai. Tzanavari] 297-298, nos. 273-275. For two more examples from Thessaloniki see Makropoulou 1997, 65, pl. 3. Also see Makropoulou 2007, 105-106, pl. 36 BYM 43/1, probably second half of 4th century and p. 242, pl. 36 BYM 58/1, probably second half of 4th century. Another Late Roman example from the eastern necropolis of Thessaloniki is mentioned in Pelekanidou 1993, 378. For an overview on this subject and on finds from Thessaloniki see Antonaras 2010, 323, with prior bibliography.

[35] Antonaras 2009, 192-196, form 51a.

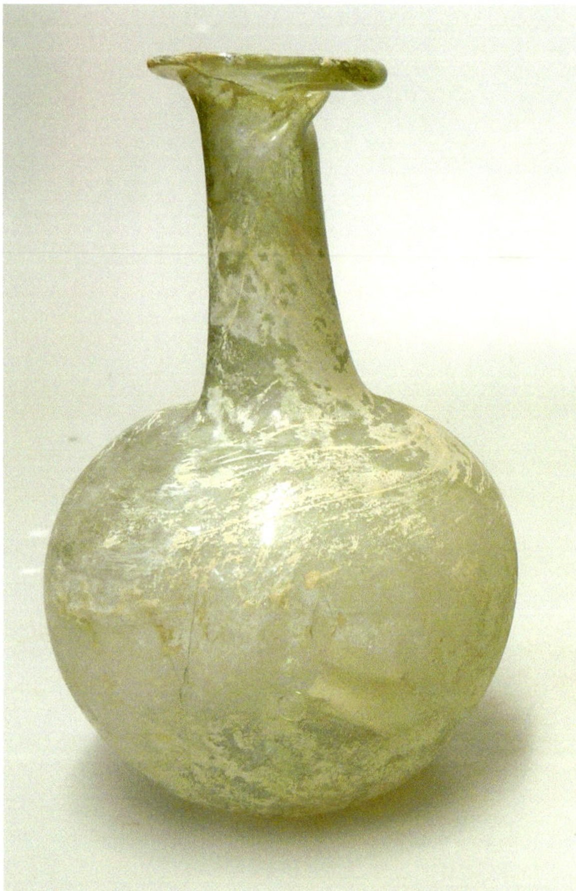

Figure 28. Bottle, ΜΘ 3754. © Archaeological Museum of Thessaloniki. Photo by A. Antonaras.

period between the end of the 3rd and the first half of the 4th century, a dating also attested by the use of the necropolis, the sarcophagus itself and the remaining of the burial inventory items of clay, glass, metal and stone.

Bibliography

6000 years of Tradition. Greek Jewellery, 1997. Exhibition catalogue, Villa Bianca, Thessaloniki 1997. Athens: Archaeological Receipts Fund.

Allason-Jones, L. 1989. *Women in Roman Britain*. London: British Museum Publications.

Antonaras, A. 2006. Γυάλινα μεσοβυζαντινά βραχιόλια. Συμβολή σε θέματα διάδοσης, παραγωγής, τυπολογίας και χρήσης (Middle byzantine glass bracelets. Contribution to issues of distribution, production, typology and use). *Δελτίον Χριστιανικής Εταιρείας*, period Δ΄, vol. ΚΖ, 423-434.

Antonaras, A. 2008. Glass Lamps of the Roman and Early Christian Periods. Evidence from the Thessaloniki Area, in C. A. Roman and N. Gudea (eds) *Lychnological Acts 2. Acts of the 2nd International Congress on Ancient and Middle Age Lighting Devices, Zalău-Cluj-Napoca, 13th-18th of May 2006. Trade and Local Production of Lamps from the Prehistory until the Middle Age*, 23-30. Cluj-Napoca: Editura Mega.

Antonaras, A. 2010. Glassware in Late Antique Thessalonikē (Third to seventh centuries C.E.), in L. Nasrallah, Ch. Bakirtzis and S. Friesen (eds), *From Roman to Early Christian Thessalonikē, Studies in Religion and Archaeology*, 299-332. Cambridge Mass.: Harvard University Press.

Arveiller-Dulong, V. and Nenna, M. D. 2011. *Les verres antiques du Musée du Louvre, Parures, instruments et éléments d'incrustation*. Paris: Musée du Louvre Éditions.

Audollent, A. 1923. Les tombes gallo-romaines a inhumation des Martres-de-Veyre (Puy-de-Dome). *Memoires a l'Académie des Inscriptions et Belles-Lettes* 13, 275-328.

Bartman, E. 2001. Hair and the Artifice of Roman Female Adornment. *American Journal of Archaeology* 105, 1, 1-25.

Bass *et al.* 2009. Bass, G. F., Brill, R., Lledó, B. and Matthews, S. D., *Serçe Limanı vol 2: The Glass of an Eleventh-century Shipwreck*. Austin, Texas: University of Texas Press; Chesham: Combined Academic.

Boyd, S. and M. Mundell-Mango (eds) 1986. *Ecclesiastical Silver Plate in Sixth-Century Byzantium*. Papers of the Symposium held May 16-18, 1986 at The Walters Art Gallery, Baltimore and Dumbarton Oaks, Washington, D.C. Washington, D.C.: Dumbarton Oaks Research Library and Collection.

Boyer, R. and Mourey, W. 1990. Le matériel d'oculiste. *Gallia* 47, 224-235.

Canarache, V. 1967. *Le Musée d'Archéologie Constantza*.

Carboni, S. and Whitehouse, D. 2001. S. Carboni, D. Whitehouse with contributions of R. H. Brill and W. Gudenrath, *Glass of the Sultans*, The Metropolitan Museum of Art, New York. The Corning Museum of Glass, Corning, Benaki Museum, Athens. New York: The Metropolitan Museum of Art; New Haven; London: Yale University Press.

Chrysostomou, A. 2000. Αρχαία Έδεσσα: Τα νεκροταφεία. Unpublished PhD dissertation, Aristotle University of Thessaloniki.

Chrysostomou, A. 2010. Κεραμική Ύστερης Αρχαιότητας από την Έδεσσα και την περιοχή της Αλμωπίας (Late Antique pottery from Edessa and Almopiain the prefecture of Pella), in D. Papanikola-Bakirtzis and D. Koussoulakou (eds) *Κεραμική της Ύστερης Αρχαιότητας από τον Ελλαδικό χώρο, Επιστημονική Συνάντηση, Θεσσαλονίκη 12-16 Νοεμβρίου 2006*, 505-519. Thessaloniki: Archaeological Institute of Macedonian and Thracian Studies (in Greek with English abstract).

Coutsinas, N. (in print). Early Byzantine Glass from Eleutherna Pyrgi (Sector II). Πρακτικά διεθνούς συνεδρίου *'History, Technology and Conservation of Ancient Metal, Glasses and Enamels', Αθήνα, 16-19 Νοεμβρίου 2011.*

Crowfoot, G. M. and Harden, D. B. 1931. Early Byzantine and Later Glass Lamps. *Journal of Egyptian Archaeology* 17, 196-208.

Daux, G. 1958. Chronique de fouilles. *Bulletin de correspondance hellénique* 82, 644-830.

Derks, T. and Vos, W. 2010. Wooden combs from the Roman fort at Vechten: the bodily appearance of soldiers. *Journal of Archaeology in the Low Countries* 2-2 (November 2010), 53-77.

Giouri, E. 1970. Χαλκιδική, Νέα Καλλικράτεια. *Αρχαιολογικόν Δελτίον* 25 Β2-Χρονικά, 378.

Goldman, N. 2001. Roman Footwear, in J. L. Sebesta and L. Bonfante (eds) *The World of the Roman Costume*, 101-130. Madison: University of Wisconsin Press.

Goldstein, S. M. 2005. S. M. Goldstein with contributions by J. M. Rogers, M. Gibson and J. Kroger, *Glass: From Sassanian antecedents to European imitations*,

Nasser D. Khalili Collection of Islamic Art vol. 15. London: The Nour Foundation, in association with Azimuth Editions.

Graekos, I. 2010. Κλειστά σύνολα κεραμικής από τα νεκροταφεία της ΄Υστερης Αρχαιότητας στη Νέα Καλλικράτεια Χαλκιδικής (Closed pottery assemblages from cemeteries dating from Late Antiquity at Nea Kallikrateia, Chalkidiki), in D. Papanikola-Bakirtzis and D. Koussoulakou (eds) *Κεραμική της ΄Υστερης Αρχαιότητας από τον Ελλαδικό χώρο, Επιστημονική Συνάντηση, Θεσσαλονίκη 12-16 Νοεμβρίου 2006*, 429-443. Thessaloniki: Archaeological Institute of Macedonian and Thracian Studies (in Greek with English abstract).

Harden *et al.* 1987: Harden, D. B., Hellenkemper, H., Painter, K. and Whitehouse, D., *Glass of Caesars*. Milan: Olivetti.

Ignatiadou, D. (ed.) 2010. *Glass Cosmos,* exhibition catalogue, Archaeological Museum Thessaloniki 2010. Thessaloniki: The Archaeological Museum of Thessaloniki.

Isings, C. 1957. *Roman Glass from dated finds*. Groningen / Djakarta: J. B. Wolters.

Kallintzi, K. and Chryssaphi, M. 2010. Κεραμική της ΄Υστερης Αρχαιότητας από τα Άβδηρα (Late Antique Pottery from Abdera), in D. Papanikola-Bakirtzis and D. Koussoulakou, eds) *Κεραμική της ΄Υστερης Αρχαιότητας από τον Ελλαδικό χώρο, Επιστημονική Συνάντηση, Θεσσαλονίκη 12-16 Νοεμβρίου 2006*, 386-401. Θεσσαλονίκη (in Greek with English abstract).

Kourkoutidou-Nikolaidou, E. 1997. From the Elysian Fields to the Christian paradise, in L. Webster and M. Brown (eds) *The Transformation of the Roman World AD 400-900*, 128-142. London: British Museum Press.

Kröger, J. 1995. *NISHAPUR Glass of the Early Islamic Period*. New York: The Metropolitan Museum of Art.

Künzl, E. 1983. *Medizinische Instrumente aus Sepulkralfunden der Römischen Kaiserzeit*. Köln: Rheinland Verlag; Bonn: In Kommission bei R. Habelt.

Magness, J. 2002. *The Archaeology of Qumran and the Dead Sea Scrolls*. Grand Rapids, MI: Eerdmans.

Makropoulou, D. 1991. Δυτικό νεκροταφείο Θεσσαλονίκης: αρχαιολογικές έρευνες στην οδό Λαγκαδά. *Το Αρχαιολογικό Έργο στη Μακεδονία και Θράκη* 5, 257-270.

Makropoulou, D. 1997. Κοσμήματα από τον 3ο έως τον 6ο αι. μ.Χ. από ανασκαφές της Θεσσαλονίκης. Συμβολή στη μελέτη της παλαιοχριστιανικής αργυροχρυσοχοΐας. *Θεσσαλονικέων Πόλις* 3, 56-69.

Makropoulou, D. 2006. Grave Finds and Burial Practices in Thessaloniki (4th-15th century). *Proceedings of the 21st International Congress of Byzantine Studies*, London 2006. Online http://www.byzantinecongress.org.uk/paper/II/II.4_Makropoulou.pdf visited 22 January 2010.

Makropoulou, D. 2007. Τάφοι και ταφές από το δυτικό νεκροταφείο της Θεσσαλονίκης (β' μισό 3ου αιώνα - 6ος αιώνας μ.Χ.). Δομικά χαρακτηριστικά - Ταφικές πρακτικές Κινητά ευρήματα. Unpublished PhD dissertation, University of Athens.

Malama, P. and Darakis, K. 2008. Νεκροταφείο Ρωμαΐκών χρόνων στα Νέα Κερδύλλια Σερρών (Cemetery of the Roman Period at Nea Kerdyllia, Serres, Greece). Θεσσαλονίκη: ΙΗ΄ Εφορεία Προϊστορικών και Κλασικών Αρχαιοτήτων – Αρχαιολογικό Ινστιτούτο Μακεδονικών και Θρακικών Σπουδών.

Marki, E. 1999. Συμβολή των οδών Ειρήνης και Λαγκαδά. *Αρχαιολογικό Δελτίο* 54, B2, 579.

Marki, E. 2006. Η νεκρόπολη της Θεσσαλονίκης στους υστερορωμαϊκούς και παλαιοχριστιανικούς χρόνους (μέσα του 3ου έως μέσα του 8ου αι. μ.Χ.). Athens: Archaeological Receipts Fund.

Motsianos, I. 2011. Φως ιλαρόν: ο τεχνητός φωτισμός στο Βυζάντιο (Joyful light: the artificial lighting in Byzantium), Unpublished PhD dissertation, Volos, University of Thessaly. http://hdl.handle.net/10442/hedi/31739

Mundell-Mango, M. 1992. The Monetary Value of Silver Revetments and Objects Belonging to Churches A.D. 300-700, in S. Boyd and M. Mundell-Mango (eds) 1986. *Ecclesiastical Silver Plate in Sixth-Century Byzantium*. Papers of the Symposium held May 16-18, 1986 at The Walters Art Gallery, Baltimore and Dumbarton Oaks, 123-136. Washington, D.C.: Dumbarton Oaks Research Library and Collection Washington.

Nalpantis, D. 1992. Νεότερα ευρήματα από σωστικές ανασκαφές στη Θεσσαλονίκη. *Το Αρχαιολογικό Έργο στη Μακεδονία και Θράκη* 6, 311-326.

Nalpantis, D. 2003. *Ανασκαφή στο οικόπεδο του Μουσείου Βυζαντινού Πολιτισμού. Ταφές και ευρήματα*. Athens: Archaeological Receipts Fund.

Newman, R. and Lie, H. 1992. Technical Examination and Conservation of Object, in S. Boyd and M. Mundell-Mango (eds) 1986. *Ecclesiastical Silver Plate in Sixth-Century Byzantium*. Papers of the Symposium held May 16-18, 1986 at The Walters Art Gallery, Baltimore and Dumbarton Oaks, 77-88. Washington, D.C. Washington, D.C., Dumbarton Oaks Research Library and Collection Washington.

Olcay, Y. 2001. Lighting Methods in the Byzantine Period and Findings of Glass Lamps in Antatolia. *Journal of Glass Studies* 43, 77-87.

Panti, A. 2010. Κεραμική της ΄Υστερης Αρχαιότητας από το ανατολικό νεκροταφείο Θεσσαλονίκης (Late antique pottery dating from the eastern cemetery of Thessaloniki), in D. Papanikola-Bakirtzis and D. Koussoulakou (eds) *Κεραμική της ΄Υστερης Αρχαιότητας από τον Ελλαδικό χώρο, Επιστημονική Συνάντηση, Θεσσαλονίκη 12-16 Νοεμβρίου 2006*, 466-485. Θεσσαλονίκη (in Greek with English abstract).

Papanikola-Bakirtzi, D. (ed.) 2002. *Everyday Life in Byzantium*, exhibition catalogue Thessaloniki, White Tower, 2001. Athens: Kapon Editions.

Papanikola–Bakirtzi, D. 2010. Ceramics in Late Antique Thessalonike, in L. Nasrallah, Ch. Bakirtzis and S. Friesen (eds) *From Roman to Early Christian Thessalonikē, Studies in Religion and Archaeology*, 263-298. Cambridge, Mass.: Harvard Theological Studies, Harvard Divinity School.

Papanikola-Bakirtzis, D. and Koussoulakou, D. (eds) *Κεραμική της ΄Υστερης Αρχαιότητας από τον Ελλαδικό χώρο, Επιστημονική Συνάντηση, Θεσσαλονίκη 12-16 Νοεμβρίου 2006*, Πρακτικά, β τόμος. Thessaloniki: Archaeological Institute of Macedonian and Thracian Studies.

Pazaras, Th. 1981. Δύο παλαιοχριστιανικοί τάφοι από το δυτικό νεκροταφείο της Θεσσαλονίκης. *Μακεδονικά* 21, 373-389.

Pazaras, Th. 1997. Ανασκαφή στη θέση Λιμόρι Επανωμής (1997). *Το Αρχαιολογικό Έργο στη Μακεδονία και Θράκη* 11, 481-94.

Pazaras, Th. 2009. *Ανασκαφικές έρευνες στην περιοχή της Επανομής Θεσσαλονίκης. Το νεκροταφείο στο Λιμόρι και η παλαιοχριστιανική βασιλική στο Μπγιαδούδι*. Thessaloniki: Centre of Byzantine Studies.

Pelekanidou, E. 1993. Νέα ευρήματα στο ανατολικό νεκροταφείο της Θεσσαλονίκης. *Το Αρχαιολογικό Έργο στη Μακεδονία και Θράκη* 7, 373-387.

Petković, S. 1995. *Rimski predmeti od kosti i roga sa teritorije Gornje Mezije*, Belgrade, Arheološki Institut.

Petsas, F. 1966. Ανασκαφή Πανεπιστημιουπόλεως Θεσσαλονίκης. *Αρχαιολογικόν Δελτίον* 21, B2 Χρονικά, 334-339.

Petsas, F. 1967. Ανασκαφή Πανεπιστημιουπόλεως Θεσσαλονίκης. *Αρχαιολογικόν Δελτίον* 22, B2 Χρονικά, 391-393.

Petsas, F. 1967a. Ανασκαφή Οικοπέδου Σ. Γεωργιάδου. *Αρχαιολογικόν Δελτίον* 22, B2 Χρονικά, 393-395.

Petsas, F. 1969. Θεσσαλονίκης, Νέα Μηχανιώνα. *Αρχαιολογικόν Δελτίον* 24, B2 Χρονικά, 304.

Petsas, F. 1970. Αρχαιότητες και μνημεία της κεντρικής Μακεδονίας. *Αρχαιολογικόν Δελτίον* 25, B2 Χρονικά, 347-361.

Pollak, R. 2000. Glass Vessels of a Fatimid Hoard from Caesarea Maritima. *Annales du 14e congrès de l'association internationale pour l'histoire du verre, Italia-Venezia-Milano*, 238–242. Lochem: Association internationale pour l'histoire du verre.

Pologiorgi, M. 1998: *Μνημεία του δυτικού νεκροταφείου του Ωρωπού. Οικόπεδο Οργανισμού Σχολικών Κτιρίων*. Athens: Archaeological Receipts Fund.

Reinach, S. 1896. Galerus, in Ch. Daremberg and E. Saglio, *Le Dictionnaire des Antiquités Grecques et Romaines, vol.* 2 (2), cols 1452-3.

von Saldern, A. 1980. Glass 500 B.C. to A.D. 1900: the Hans Cohn collection, Los Angeles/Cal. = Glas von der Antike bis zum Jugendstil: Sammlung Hans Cohn, Los Angeles/Cal., c1980. Mainz on Rhine: Philipp von Zabern.

Snow, C. 1992. From Ingot to Object: Fabrication Techniques Used in the Manufacture of the Hama Silver, in S. Boyd and M. Mundell-Mango (eds) 1986. *Ecclesiastical Silver Plate in Sixth-Century Byzantium.* Papers of the Symposium held May 16-18, 1986 at The Walters Art Gallery, Baltimore and Dumbarton Oaks, 197-202. Washington, D.C.: Dumbarton Oaks Research Library and Collection Washington.

Spear, M. 1988. The Pre-Islamic Glass Bracelets of Palestine. *Journal of Glass Studies* 30, 51-62.

Steckner, C. 1988. Phanokantharoi und Kylikeia, Dionysische Lichtgefässe in architektonischem Kontext. *Annalles de l'Association Internationale pour l'Histoire du Verre* 11, 257-270.

Stefanidou-Tiveriou, T. 2010. Social Status and Family Origin in the Sarcophagi of Thessalonike, in L. Nasrallah, Ch. Bakirtzis and S. Friesen (eds) *From Roman to Early Christian Thessalonikē, Studies in Religion and Archaeology*, 151-188. Cambridge, Mass.: Harvard Theological Studies, Harvard Divinity School.

Thessaloniki from prehistoric to Christian era (Θεσσαλονίκη από τα προϊστορικά μέχρι τα χριστιανικά χρόνια) 1986. Exhibition Catalogue, Archaeological Museum of Thessaloniki. Athens: Archaeological Receipts Fund.

Thessaloniki History and Art (Θεσσαλονίκη Ιστορία και Τέχνη) 1986. Exhibition Catalogue, White Tower, Thessaloniki. Athens: Archaeological Receipts Fund.

Tzanavari, K. 2010. Ταφικά σύνολα κεραμικής των όψιμων αυτοκρατορικών χρόνων από την αρχαία Λητή (Funerary assemblages of late imperial pottery from Lete), in D. Papanikola-Bakirtzis and D. Koussoulakou (eds) *Κεραμική της Ύστερης Αρχαιότητας από τον Ελλαδικό χώρο, Επιστημονική Συνάντηση, Θεσσαλονίκη 12-16 Νοεμβρίου 2006*, 444-465. Thessaloniki: Archaeological Institute of Macedonian and Thracian Studies (in Greek with English abstract).

Tzevreni, St. 2011. Clay lamps from the ditch of the Western city walls of Thessaloniki (Τα λυχνάρια από την τάφρο των δυτικών τειχών της Θεσσαλονίκης) in the book of abstracts of the 4th International ILA Round-Table, Lighting in Byzantium, Thessaloniki, October, 11–14, 2011, 39-40.

Vokotopoulou, I. 1996. *Οδηγός Αρχαιολογικού Μουσείου Θεσσαλονίκης*. Athens: Kapon Editions.

Walker, S. and Bierbrier, M. 1997. *Ancient Faces: Mummy Portraits from Roman Egypt*. London: British Museum Press.

Whitehouse, D. 2001. *Roman Glass in the Corning Museum of Glass, volume Two*. Corning, New York: Corning Museum of Glass.

Yadin, Y. 1966. *Masada: Herod's Fortresss and the Zealots' Last Stand*. New York: Random House.

Zias, J. 1998. Whose Bones?. *Biblical Archaeology Review* 24, 40-5, 64-6.

Some Remarks on a Category of Wheel-Made Lamps: A ‘Koinè’ in the 6th and the 7th Centuries AD?

Anastasia G. Yangaki

Senior Researcher, Institute of Historical Research, National Hellenic Research Foundation, Athens, Greece

Abstract: This paper explores one category of wheel-made lamp, which is characterized by an ovoid body, an oblong nozzle and a cylindrical neck. Based on the diffusion of its six sub-types and the range of the production of each, it is apparent that such wheel-made lamps were common during the 6th and the 7th centuries AD in regions of the eastern Mediterranean. It is further indicated that one of these sub-types, which is common in the Aegean region, constitutes a ‘koinè’ in this area. This category of lamp is one of the missing links between the wheel-made lamps of the Roman and Middle Byzantine periods.

Keywords: wheel-made lamps, typology, distribution, 6th-7th centuries AD, Aegean region, eastern Mediterranean.

Introduction

Regarding the mold-made lamps of Late Antiquity, analytical typologies containing further sub-types and featuring major production centers, have been proposed.[1] As for wheel-made lamps, their wide dissemination during Classical, Hellenistic and Roman times suggests their uninterrupted use down through the centuries. Furthermore, researchers have noted that the continuous presence of these lamps should be taken as given for the Late Antique period as well, even if mold-made lamps came to the fore during this period.[2] In fact, only a few years ago have researchers identified the basic categories of wheel-made lamps in use in the eastern Mediterranean in the Late Antique and Byzantine periods.[3] It should be borne in mind that the lack of a high degree of formal standardization in the technique of their manufacture does not offer the advantage of a faithful reproduction of a prototype, an advantage offered by the use of a mold. Most of these categories can be subdivided into numerous sub-types (or preferably ‘groups’). We have sufficient evidence to form a quite detailed picture, particularly for the 6th and the 7th centuries AD. In fact, a typological classification for the lamps of the latter period has already been proposed by the author. Most of these wheel-made lamps are closed in shape, with circular or oblong bodies with or without a neck and most have a handle.[4]

Wheel-made lamps with closed, ovoid body, long nozzle and cylindrical neck

The main aim of this study is to comment on a specific category of these wheel-made lamps (type 2 in the afore-mentioned typology), which represents a rather interesting case study that may shed light on the on-going debate on the use of wheel-made lamps in parallel with the use and the predominance of the mold-made lamps.

More particularly, this category of wheel-made lamp is distinguished from the others by the closed, ovoid form of the body, which elongates to a lesser or greater degree towards a long, projecting nozzle which is tubular and, in most cases, gradually upward sloping. The base is flat to convex. The walls are either concave or conical, and there is sometimes a ridge at the junction with the neck or nozzle. The neck is cylindrical, tall (to a greater or lesser degree) and of varying widths. Surrounding the central hole of the reservoir, it combines with the body to make the lamp rather pear-shaped. The end of the nozzle’s rim can be flat, angular or rounded, with variations of each. The category generally features a loop handle attached directly to the neck and the body, opposite the nozzle; in some cases, the handles are higher than the rim. Several variations exist with regard to the formation of the neck and body. The form of the lamp makes it clear that its body and neck were manufactured by the potter on the wheel;[5] in fact, this is further corroborated by the form of the first sub-type, quite common on the island of Crete (Figure 1), which clearly corresponds, without the nozzle, to small flagons (Figure 2) commonly produced by local workshops. The nozzle and handle were attached to the reservoir in a second phase.[6] That the nozzle was attached to the main body of the lamp is obvious from broken specimens, and finger marks resulting from the pressure of attaching these parts to the main body of the lamp are sometimes visible at the junction of the reservoir with the nozzle and handle. Based on the formation of these lamps’ walls, neck and nozzle, a finer classification has already been proposed which divides this type into six sub-types, or ‘groups’,[7] since, given the lack

[1] To note just a few of these typologies: Broneer 1930; Perlzweig 1961; Deneauve 1969; Broneer 1977; Anselmino, Pavolini 1981, 184-207; Bussière 2000; Bonifay 2004, 312-430.

[2] Orssaud, Sodini 1997, 63; Yangaki 2005, 232-233; Motsianos 2010, 272-273.

[3] For some of the typologies on wheel-made lamps: Orssaud, Sodini 1997, 63-71; Motsianos 2005, 249-250; Yangaki 2005, 232-241.

[4] For a more detailed presentation of the main categories of the lamps of these centuries and a preliminary classification: Yangaki 2005, 232-235.

[5] On manufacturing details of wheel-made lamps see, in addition, Motsianos 2005, 248-250.

[6] See also the remarks by D. G. Barrett (Barrett 2008, 105).

[7] Since a detailed presentation of these sub-types exists (Yangaki 2005, 233-234), only the basic elements of each shall be presented here. Regarding the basic criteria used to classify wheel-made lamps

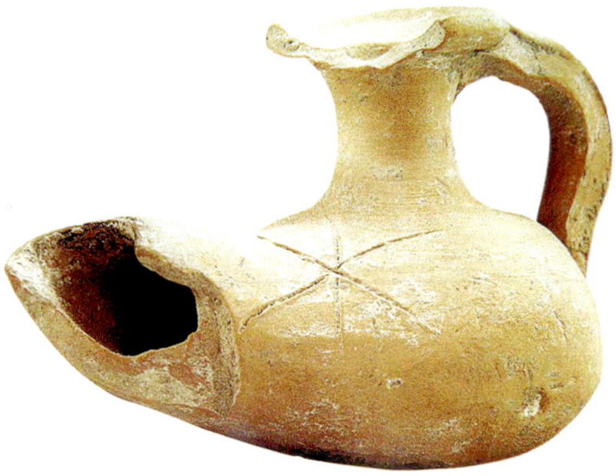

Figure 1. Wheel-made lamp with an ovoid, elongated body, oblong nozzle and cylindrical neck (first sub-type) from Eleutherna, Crete [Inv. no. 19192; photo: Stampolidis 2004, 207, no. 152 (P. Themelis)].

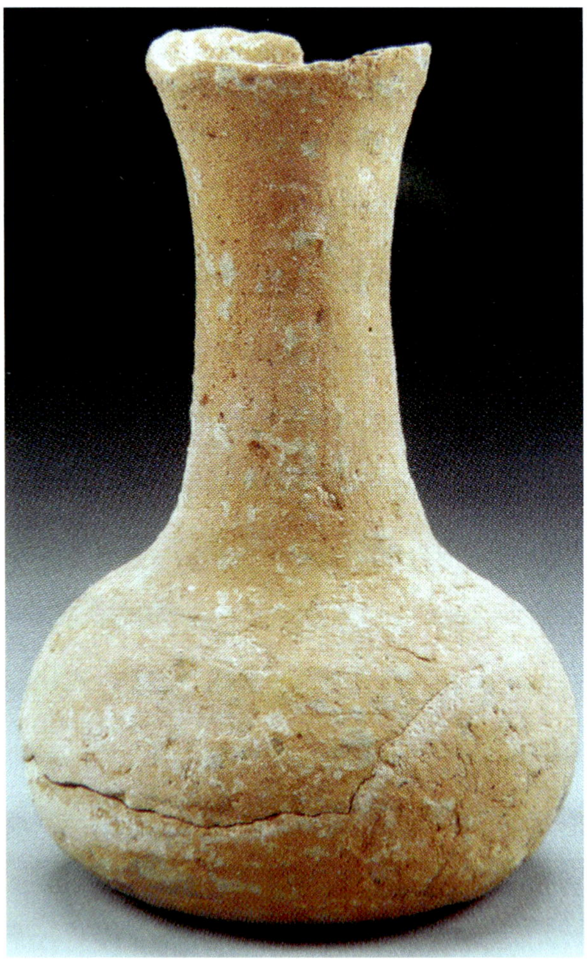

Figure 2. Small clay flagon from Eleutherna, Crete [Inv. no. 15738; photo: Stampolidis 2004, 196, no. 116 (P. Themelis)].

of standardization already mentioned, these should not be considered as strictly fixed. A brief note on each is essential to familiarize the reader with the material and –above all– since a comprehensive presentation of old data combined with references to the latest evidence from publications and to specific maps[8] charting the diffusion of each sub-type is crucial for the argument that follows.

Thus, the first sub-type is comprised of lamps mainly characterized by a long, narrow neck. They have a flat base and concave walls which converge at the neck. The rim is either flat or out-turned, with various variations as to form. The loop handle starts from the rim or the neck and ends on the body of the lamp (Figure 1, 3a-b). This type was most common during the 6th and 7th centuries AD on the island of Crete, where (see map, Figure 4), it has been found at many sites, sometimes in significant numbers.[9] Lamps of this form have been attributed to various Cretan workshops, including the workshops of Gortyn. The surface of the bodies of many of these lamps display three, deep incisions that create a star-like motif. This motif is believed to represent the production of a specific workshop[10] and occurs, besides Gortyn, on lamps of the same form found not only in Matala, Simi and Eleutherna, but also in Kenchreai in the Peloponnese,[11] implying that they probably came from the same workshop.[12] It should be noted that a less refined version of a similar motif has been found on a lamp with the same shape found on the island of Thera.[13] The use of analytical techniques on the fabric of this lamp could verify its Cretan provenance. This sub-type mainly dates from the mid-6th and 7th centuries AD. However, given that various sub-forms of this 'group' occur in Gortyn in later layers of the excavation[14] dated to the 8th and probably the early 9th century, while glazed examples have come to light in Sabratha,[15] for which a date in the 9th century has been proposed, a later use for the first sub-type seems plausible.[16] Based on its distribution (Figure 4), it is an eastern Mediterranean regional form which was particularly well-diffused on Crete.

The second sub-type is characterized by a concave body or ridged walls, a short and more or less wide neck, and a long,

in particular, and the difficulties of establishing a firm typological classification, see the related analytical comments by C. A. Roman (Roman 2000, 100-101), with the related bibliography.

[8] For the maps that follow, only data which provide secure and reliable indications about the provenance of the lamps have been used. The information on the maps derives, amongst others, from the bibliography presented in notes 9, 11, 13, 15 (for the first sub-type), notes 17-18 (for the second sub-type), notes 19-21 (for the third sub-type), notes 22-23 (for the fourth sub-type) and notes 24, 26-27 (for the fifth and sixth sub-type).

[9] For all these references: Yangaki 2005, 233 and notes 1437-1438; see also: Baldini, Parello 2001, 167-171, pl. XXXI, c-g. In addition, see: Bonetto *et al.* 2010, 902 fig. 3; Albertocchi 2011, 111; Fiolitaki 2011, 100-101, fig. 40; Xanthopoulou 2015, 588-589, fig. 3, δ; Sythiakaki 2018, 327, nos. 231-236.

[10] Baldini, Parello 2001, 168-169.

[11] Matala: Vallianou 1988, pl. 339, a, c, d; Simi Viannou: Lebessi 1992, 215, pl. 91b; Eleutherna: Stampolidis 2004, 207, no. 152 (P. Themelis); Kenchreai: Williams 1981, 87, no. 466, pl. 21, 466.

[12] However, other opinions have been expressed as to the meaning of the star-like motif; see, as an example: Chatzi-Vallianou 1995, 1031.

[13] Gerousi 2010, 231 fig. 34; Gerousi-Bendermacher 2013, 170, no. 248.

[14] For details of the various sub-divisions: Baldini, Parello 2001, 169-171, 182.

[15] Joly 1974, 58, no. 1329, 205, no. 1329, pl. LVIII, 1329.

[16] See also Yangaki 2016, 223.

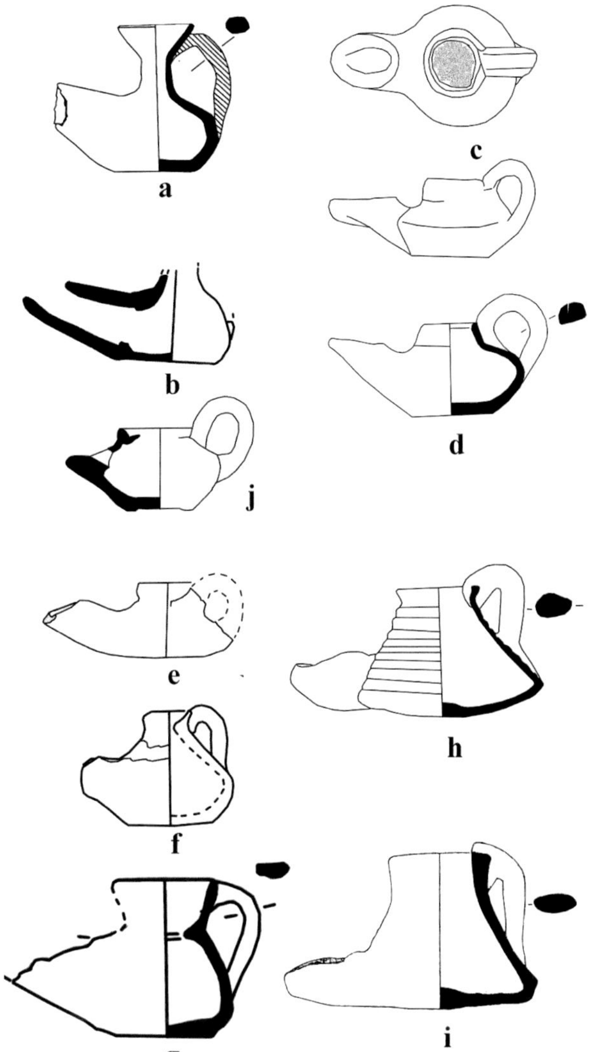

Figure 3. Main sub-types of the wheel-made lamp with ovoid body, oblong nozzle and cylindrical neck: a-b: first sub-type, c-d: second sub-type, e-g: third sub-type, h: fifth sub-type, i: sixth sub-type, j: fourth sub-type (all drawings by the author: a: after Baldini, Parello 2001, pl. XXXIII, e; b: Yangaki 2005, fig. 67, e; c: after Kuzmanov 1992, pl. 48, 34; d: after Bailey 1988, pl. 125, Q 3339; e: after Bailey 1988, pl. 56, Q 2275 MLA; f: after Bailey 1988, pl. 56, Q 2274 MLA; g: after Baldini, Parello 2001, pl. XXXIV, b; h: after Hadad 2002, 73, no. 317; i: after Hadad 2002, 73, no. 319; j: after Prokopiou 1995, 262 fig. 5. β).

projecting nozzle (Figure 3c-d). It dates from the 6th and 7th centuries AD. Based on recently published evidence which supplements existing examples and is illustrated on the respective map (Figure 5), this sub-type occurs particularly in the Aegean and adjacent regions, and was most densely distributed at the south-west end of Asia Minor, particularly in sites in Lycia and on the neighbouring islands of the Dodecanese.[17] Isolated pieces have been published from Palestine, Egypt, Algeria and the Balkans.[18]

The third sub-type is characterized by concave or ridged walls, a neck which is either very low or high and giving the impression of a funnel around the central fill hole since the diameter of the fill hole is much smaller than the diameter of the neck, because of the existence of a band which creates a form of collar Variations on this general form exist in terms of where this 'collar' is attached – around the hole, higher up, around the neck or to the rim (Figure 3, e-g). The nozzle is oblong. In Gortyn, this form has been dated to between the mid-6th and mid-7th centuries AD,[19] a date that coincides with that proposed for similar lamps from Egypt.[20] Lamps of this 'group' occur in Crete, Nea Anchialos, the Balkans, Egypt and Palestine[21] (Figure 6), but are not as common as those of the previous two; more numerous finds are mainly restricted to Gortyn.

The fourth sub-type has a shortish neck and resembles the previous type in featuring a similar collar between the neck and the fill hole. It differs mainly in the form of the nozzle (Figure 3, j), which is short and formed in most cases out of the lamp wall, and occasionally with the addition of a piece of clay but also in the form of the shortish neck and has several variations. This type dates from between the 6th and 9th centuries.[22] The few examples of the type are mostly from Cyprus, though some have also come to light in Libya (Figure 7).[23]

[17] For additional detail on Aegina, Corinth, Thasos, Thrace, Kalymnos, Kos, Xanthos, Limyra, Crete and Cyprus, see the related, analytical references brought together in Yangaki 2005, 233, note 1439. Additionally, in relation to Rhodes: Bairami 2010, 267, fig. 3, ΠΧ1857; Katsioti 2017, 548-550, nos. WM 1-WM 2; for Thera: Gerousi 2010, 226, 231, fig. 33; Gerousi-Bendermacher 2013, 168-171, nos. 239, 241-247, 249-254; for Xanthos and Limyra, Lycia, see respectively: Grünewald 1984, 61, fig. 31, 1-4; Des Courtils 2001, 235, fig. 18 and Vroom 2004, fig. 8; for Olympos, Lycia: Olcay Uçkan *et al.* 2007, 132, fig. 1; Öztaşkin, Öztaşkin 2012, 346, fig. 9, a; for Dereağzi, Lycia: Gregory 1993, 137, pl. 26, 13; for Alicante on the Iberian peninsula: Reynolds 1995, 198, appendix C.1, fig. 134.

[18] For all the respective references: Yangaki 2005, 233, and note 1439. For additional references: Kassos: Giannikouri, Zervaki 2009, 29-30, fig. 42 (lamp center right); Cyprus: Oziol 1977, 45-46, nos. 96-98, pl. 7, 96-98 (for the arguments relating to the date of these lamps, see analytically: Yangaki 2005, 233, note 1439 [with other related references]); Egypt: Petrie 1905, 13 pl. LXIX, 83; Bailey 1996, 133 Q 2277 bis MLA, pl. 174, Q 2277 bis; Bulgaria: Kuzmanov 2002, 249 type XV, 261 fig. 10, 88.

[19] This is type II, 'ad alto collo': Baldini, Parello 2001, 182.

[20] Bailey 1988, 274, Q2270 MLA-Q2275 MLA. For similar lamps of Egyptian provenance: Lyon-Caen 2005, 225-226, pl. 105, fig. 4; Chrzanovski 2013, 188, 225 nos. 41-42.

[21] For a detailed presentation of its distribution on Crete (Gortyn, Matala, Kommos) and the related bibliography: Yangaki 2005, 234 and note 1440 (with all the respective references). For Itanos: Xanthopoulou 2015, 588-589 fig. 3, ε. For Nea Anchialos: Lazaridis 1965, 19, pl. 18a. For Egypt, see the previous note. For Palestine: Macalister 1912, pl. CI (a), no. 11; Rosenthal, Sivan 1978, 122-123, no. 508. For the Balkans, see examples from Sirmium: Rubright 1973, 53-55, 65-71, pls XV-XIX, pl. XXIV, 154, 159, 236. See also an example from the Rivel collection (Bussière, Rivel 2012, 298-299, no. 356) and an example from the Bouvier collection (Chrzanovski 2013, 188, 225, no. 40).

[22] Oziol, Pouilloux 1969, 25, fig. 19, pl. XI, 476-477; Oziol 1980, 399; Joly 1974, 57.

[23] Cyprus: Salamis: Oziol, Pouilloux 1969, 25, fig. 19, 115, no. 476, pl. XI, no. 476; Oziol 1980, 399, fig. 6; Amathus: Prokopiou 1995, 262, fig. 5. α1-α4, β, 264, with several variants; Kourion: Prokopiou 1995, 265; Libya, Sabratha: Joly 1974, 57, fig. 4, no. 1328, pl. LVIII, 1328. See also

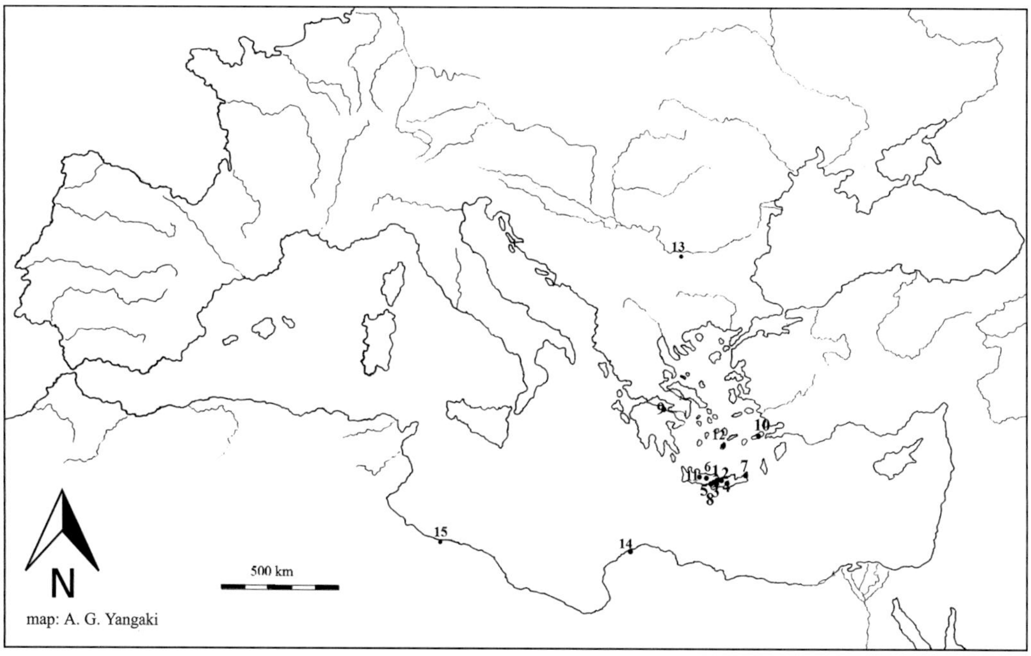

Figure 4. Map with the distribution of the first sub-type: 1. Knossos, 2. Amnisos, 3. Gortyn, 4. Simi Viannou, 5. Matala, 6. Eleutherna, 7. Itanos, 8. Mitropolis, 9. Kenchreai, 10. Kos, 11. Vafes. 12. Thera, 13. Sadovec, 14. Ptolemais, 15. Sabratha (schematic map by the author).

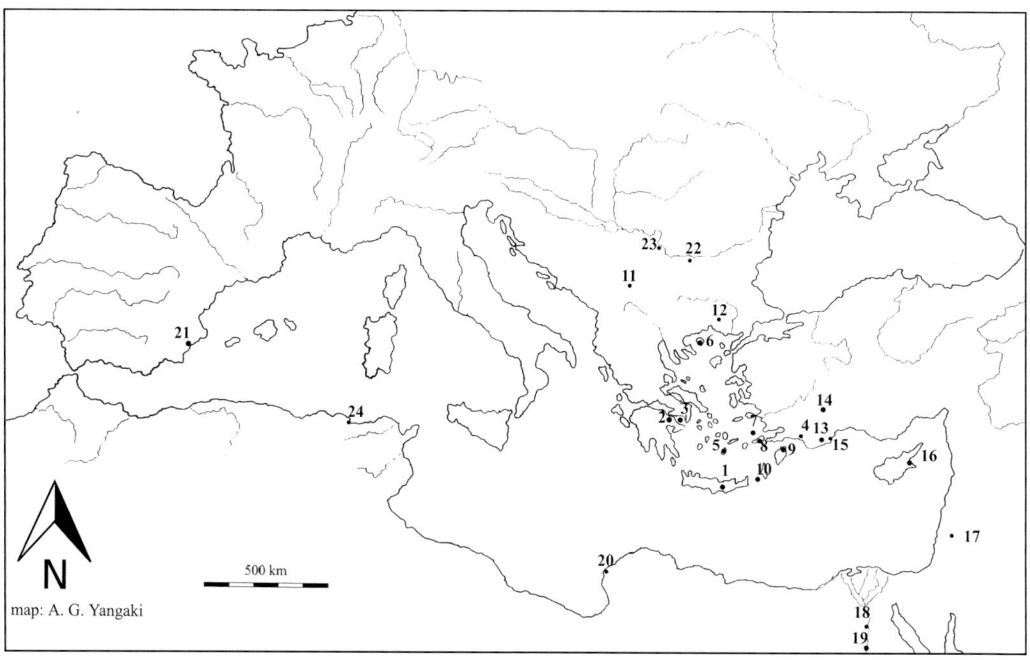

Figure 5. Map with the distribution of the second sub-type: 1. Gortyn, 2. Corinth, 3. Aegina, 4. Xanthos, 5. Thera, 6. Thasos, 7. Kalymnos, 8. Kos, 9. Rhodes, 10. Kassos, 11. Caričin Grad, 12. Arisvi, 13. Limyra, 14. Dereağzi, 15. Olympus, 16. Salamis, 17. Ain Shems, 18. Ehnasya, 19. Asyut, 20. Benghazi, 21. Alicante, 22. Sadovec, 23. Ratiaria, 24. Annaba (schematic map by the author).

The last two groups are comprised of lamps with flat or convex bases, oblique walls with a ridge at the point of transition to the base, a short and broad funnel neck, and an oblong nozzle attached to the reservoir at a rather low point on the body (in comparison with the other sub-types). Part of the loop handle is higher than the rim (Figure 3, h-i). These two sub-types are differentiated from one another by the existence in the

some similar lamps from the Anawati collection (Djuric 1995, nos. C339-C344).

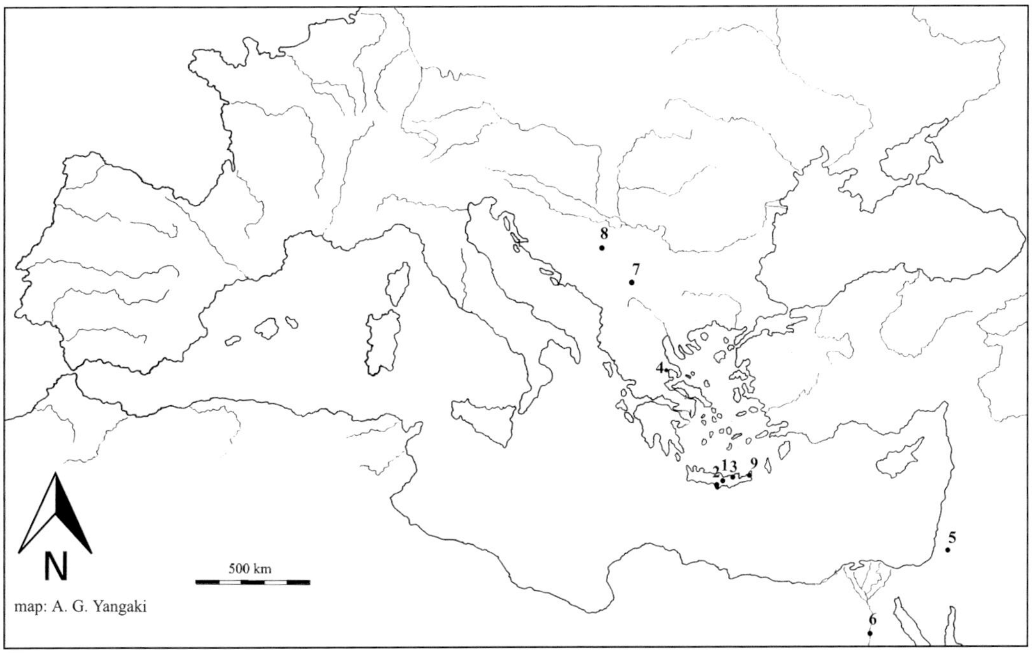

Figure 6. Map with the distribution of the third sub-type: 1. Gortyn, 2. Matala, 3. Kommos, 4. Nea Anchialos, 5. Gezer, 6. Asyut (area), 7. Caričin Grad, 8. Sirmium, 9. Itanos (schematic map by the author).

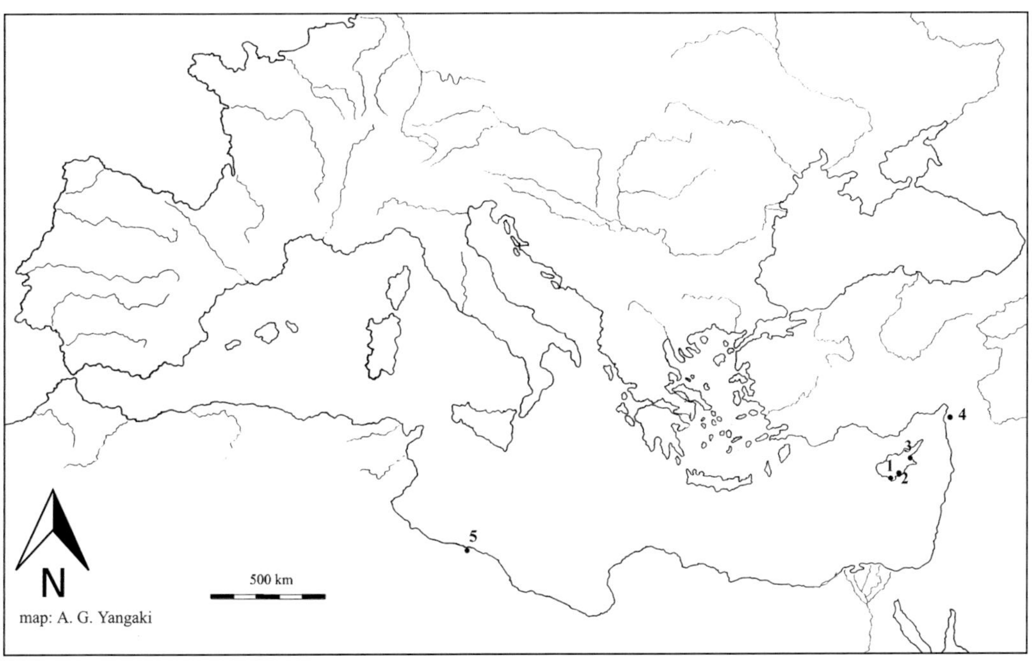

Figure 7. Map with the distribution of the fourth sub-type: 1. Amathus, 2. Kourion, 3. Salamis, 4. Antioch, 5. Sabratha (schematic map by the author).

fifth sub-type of particularly deep grooves around the body (Figure 3, h); it is known as a 'boot-shaped lamp'.[24] Given the distribution of these lamps (Figure 8), which usually come to light together in archaeological contexts,[25] they are restricted mainly to southern Palestine, to many sites in the Negev desert and to Jordan.[26] Outside Palestine and neighbouring

[24] Macalister 1912, pl. LXXII, no. 11; Colt 1962, 63, no. 23; Kennedy 1961-1963, 105, no. 630, pl. XXV, 630; Chrzanovski 2013, 188, 225, no. 39. For this type and its various designations see Bussière, Rivel 2012, 298, no. 357.

[25] Rosenthal, Sivan 1978, 122.

[26] For its distribution: Rosenthal, Sivan 1978, 123; Rahmani 1983, 219,

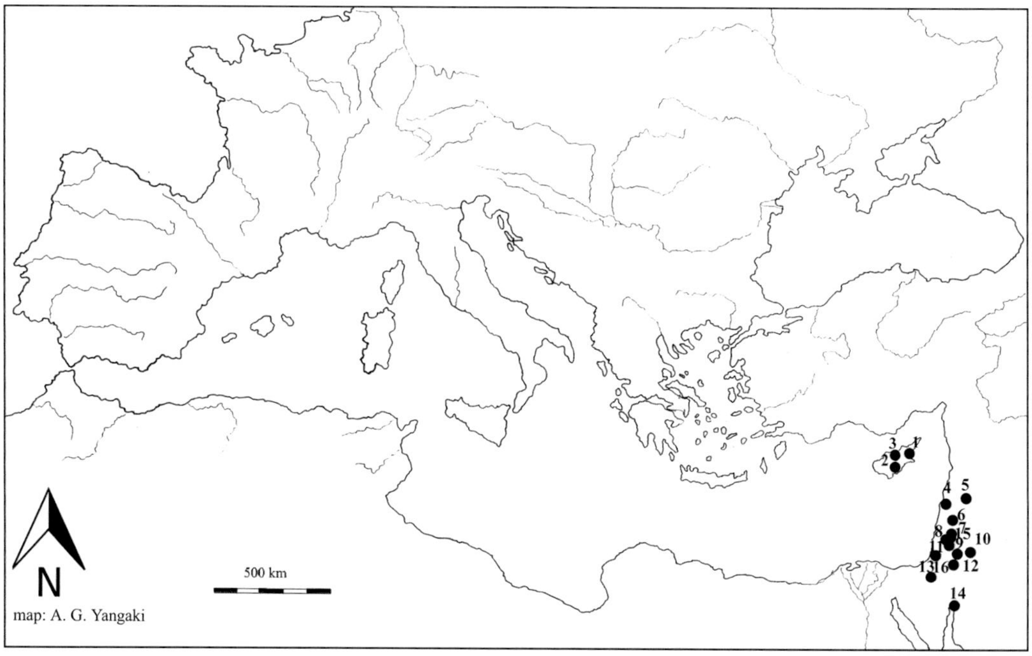

Figure 8. Map with the distribution of the fifth and the sixth sub-types: 1. Salamis, 2. Amathus, 3. Lefka, 4. Casesarea, 5. Beth Shean*, 6. Samaria, 7. Tell en-Nasbeh*, 8. Gezer*, 9. Siyar al-Ghanam (Bethlehem), 10. Mount Nebo, 11. Tel Melfasim (near Gaza), 12. Mampsis*, 13. Nessana*, 14. Aqaba, 15. Jerusalem – Ramat Rahel, 16. Masada (*: besides the fifth type, also occurrence of the sixth; schematic map by the author).

areas, 'boot-shaped' lamps have come to light in Cyprus, while similar lamps are held in the collections of the museums of Florence and Warsaw.[27] Earlier publications presented these sub-types as Hellenistic, while other pieces were dated to the 3rd and 4th centuries AD.[28] However, the latest excavated material would suggest that these lamps came into use in the 5th century, and were most widespread during the 6th, 7th and 8th centuries AD.[29] It has been suggested that the sixth sub-type may date from a little earlier than the fifth century,[30] even if they appear simultaneously in some areas.[31]

Discussion

The above would indicate that this category of wheel-made lamp is generally typical of the 6th and 7th centuries AD, since this was the period in which it was most widespread. Examples of lamps of a similar, general form –many of them glazed– from the 4th and first half of the 5th centuries are encountered in regions of the Roman provinces of Pannonia, Dacia (mostly Dacia Ripensis) and Moesia Prima.[32] Nevertheless, these earlier forms have mostly a small, vertical stub handle. However, no similar, earlier examples have yet been found in regions of the eastern Mediterranean, particularly in and around the Aegean region, where most of the sub-groups of the category were most diffused.

As for lamps occurring in these areas which could be considered as prototypes for this category of wheel-made lamps, the wheel-made clay lamps of the Classical and –above all– Hellenistic eras had a similar shape to these later lamps. For example, Hellenistic clay lamps from Athens and Sparta[33] display similarities with lamps from sub-type 2, while lamps from Athens and Atalanti[34] from between the 1st century BC and the 1st century AD, and lamps from Dimitrias,[35] reveal considerable similarities with lamps from sub-type 3. It is important to note, however, that the specific category of wheel-made lamp –and particularly sub-types 2 and 3– has much in common with the type 3 copper or copper-alloy metal lamps from the classification proposed by M. Xanthopoulou.[36]

221; Hadad 2002, 72 and, more recently: Yangaki 2005, 234, note 1448 (with a detailed presentation of the related bibliography); Bussière, Rivel 2012, 298-299, no. 357 (with additional examples); Chrzanovski 2013, 188.

[27] Cyprus: Oziol 1977, 288-289, no. 883, pl. 48, no. 883; Salamis: Oziol, Pouilloux 1969, 115, no. 475, pl. XI, 475, pl. XX, 475; Oziol 1980, 399-400, fig. 7; Amathus: Prokopiou 1995, 263, fig. 6, δ-ε, 264; Lefka: Oziol 1977, 289, no. 884, pl. 48, 884; Florence: Michelucci 1975, 115-116, type XXXVII, nos. 410-411, pl. XXIV, 410, pl. XXXIII, no. 411; Warsaw: Bernhard 1955, 345, pl. CXVII, no. 384 bis.

[28] Oziol 1977, 288, no. 883; Rosenthal, Sivan 1978, 122-123.

[29] For a more complete presentation of these two sub-types and a revision of the date of the fifth sub-type, in particular: Rahmani 1983, 219, 221 (with the related bibliography); Hadad 2002, 72 and, more recently: Barrett 2008, 105; see as well Bussière, Rivel 2012, 298; Chrzanovsi 2013, 188.

[30] Rahmani 1983, 221.

[31] Rosenthal, Sivan 1978, 122.

[32] For a detailed presentation of these examples: Rubright 1973, 53-55, 65-71, pls XV-XIX, pl. XXIV, 154, 159, 236 (Sirmium in Pannonia); Benea 1990, 151 type XIV, fig. 20, 131 a-b (Tibiscum in Dacia); Kuzmanov 1992, 227, type 9, pl. 48, 34-35 (Sadovec in Dacia Ripensis); Cvjetićanin 2006, 100, fig. 6, LRGL 3, 104-105 (several sites in Dacia Ripensis and Moesia Prima).

[33] Athens: Howland 1958, 115-116 types 36A-36B, pl. 17, 487, pl. 18, 488, 490, pl. 44, 487, 488, 490. Sparta: Zavvou 1999, 164-165, fig. 10. See also some examples of Hellenistic lamps from Bosporus (Zhuravlev 2012, 365-371, figs 5-8, 11). For the main types of wheel-made lamps of the Archaic Greek and the Hellenistic periods see Sussman 2009, 9-27.

[34] Athens: Howland 1958, 121, type 37C, no. 504, pl. 19, 504, pl. 45, 504. Atalanti: Dakoronia 1996, 318, pl. 85, α.

[35] Triantafyllopoulou 1998, 414, pl. 157, α.

[36] Analytically for this type: Xanthopoulou 2010, 6-14, 100-162. For a

It has even been noted that, typologically, this form of metal lamp has similarities with metal lamps from the Hellenistic period.[37] The salient features of the metal counterparts of this category are a particularly oblong nozzle, which is tubular in form, a pear-shaped, slightly squat reservoir with concave walls which sometimes feature a ridge, a short neck and a flat or discoid base.[38] They have a metallic cover and present variations with regard to the form of the vertical handle and its decoration, the form of the cover, the base of the reservoir, the end of the nozzle and the formation of the fill hole. However, a close observation of some of these metal lamps and a comparison with clay lamps of the category under examination here, reveals obvious analogies: the oblong nozzle, the form of the main body, the cylindrical neck, the collar around the fill hole and the use of a vertical handle.[39] It is obviously not germane that these metal lamps were widespread in the eastern Mediterranean, where they formed -based on the related data- the majority among other types of metal lamps, and date from between the 5th and, probably even the 7th century AD.[40] In other words, their geographical and chronological context is comparable with those of the wheel-made lamps in question. It can therefore be reasonably suggested that while the adoption of this particular form of wheel-made lamp during this period could be derived from some clay prototypes of the past, it is more likely a result of a need to copy a well-diffused and more expensive metal lamp in a far more accessible material such as clay. This two-way relationship between clay and metal lamps is also suggested by the morphological peculiarities of clay lamps observed on metal lamps,[41] which is understandable in that both are everyday objects made to serve the same purpose. This bidirectional relationship is even noticeable among lamps made of the same material: clay. To use examples from the specific period, some mold-made lamps from Amorium[42] seem to have been influenced in the shape of their reservoir and disk by comparable wheel-made lamps from Anemurium.[43]

The wheel-made clay lamps of the first sub-type, which are extremely common in Crete, are quite distinct with their tall, narrow necks and multiple variations as to the shape of their rim. They have analogies with small, closed vessels in both glass and clay (Figures 2, 9) which date from the same period and are encountered not only in Gortyn, but also in Eleutherna or even Athens and Argos.[44] As other researchers have noted, this observation confirms the parallel production in the same workshops of pots of various shapes and of wheel-made lamps.[45] A particularly eloquent example is that of a lamp from Matala whose main body, tall wide neck and rim

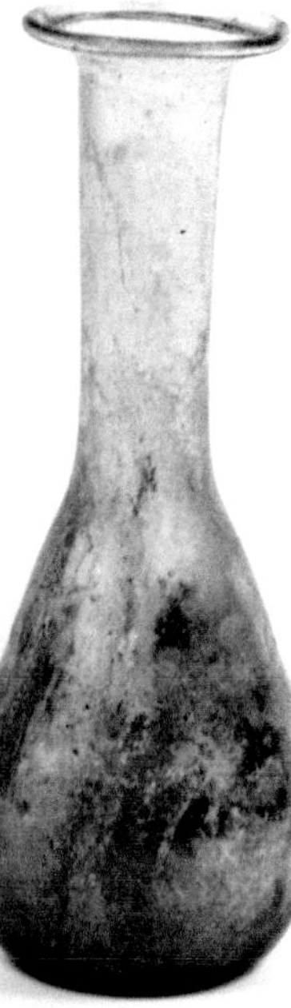

Figure 9. Glass vessel from Eleutherna, Crete [Inv. no. 501; © Eleutherna Excavations, Prof. Petros Themelis; published in: Stampolidis 2004, 302, no. 417 (P. Themelis)].

formation are more closely reminiscent of a jug than a lamp.[46] Since the manufacture of lamps of this category did not demand specialized equipment or labour, they were probably affordable for every level of society.

This category is found during the 6th and 7th centuries in parts of the eastern Mediterranean. It should be noted that in Algeria, for example, even if lamps from this category do exist during this period, these are very rare in comparison with wheel-made lamps of a different form.[47] Based on the maps presented above (Figures 4-8), it becomes obvious that the first three sub-types in particular were diffused in sites around the Aegean and the Near East, usually attributed to the local production of each site. The fifth and sixth sub-types seem to have had a more restricted diffusion, mainly in southern Palestine and Egypt, with rare examples occurring further to the north-west in other regions of the eastern Mediterranean. It thus seems that, during these two centuries (6th-7th), a specific form of wheel-made lamp was adopted in parallel with the more common use of mold-made lamps in many regions of the eastern Mediterranean. In terms of the second sub-type of this wheel-made lamp in particular, its diffusion in the Aegean and adjacent areas allows us to speak of a 'koinè' in this region during the 6th and 7th centuries, at least with regard to wheel-made lamps. As stated above, the earliest examples of wheel-made lamps with characteristics similar to those of this category seem to have appeared in

similar view, regarding the resemblance of these wheel-made lamps to metal ones: Baldini, Parello 2001, 163.

37 Xanthopoulou 2010, 6 and note 32 (with related bibliography).

38 Xanthopoulou 2010, 6.

39 Bernhard 1955, 381, no. 566, pl. CLXV, 566; Xanthopoulou 2010, 6; Motsianos, Bintsi 2011, 165-166, no. 53 (H. Zografou).

40 Xanthopoulou 2010, 6.

41 For a more detailed presentation of these: Xanthopoulou 2005, 303-307.

42 Gill, Tuğçe Şen 2003, 27, nos. 23-24, 42 fig. III/5, 23-24, 49, pl. III/10, 23.

43 Williams, Taylor 1975, 82, fig. 7, 83, fig. 8 drawing III, 84.

44 Gortyn: Albertocchi, Perna 2001, pl. CLXXXIV, X 1.1/1, II 1.1/1, II 1.1/4, pl. CLXXXII, II 2.2/1, II 2.4/1. Eleutherna: Stampolidis 2004, 196, no. 116 (P. Themelis). Athens: Tzavella 2010, 663, no. A 8413, 669, fig. 4β, A 8413. Argos: Aupert 1980, 431, nos. 252-259, 443-445, fig. 41, 253-254, 257-258, 445, fig. 48, 341-361.

45 Baldini, Parello 2001, 168; Motsianos 2010, 248.

46 Vallianou 1988, pl. 339, b.

47 Bussière 2000, 127, type V, 402 type V, nos. 7126-7129, pl. 137, 7126-7129 (However, the form of the handle of these lamps is different than that of lamps of the specific category). See also an example from central Tunisia in the Rivel collection, similar to these lamps as to its particular handle (Bussière, Rivel 2012, 256-257, no. 307).

central-eastern Europe in the 3rd to 5th centuries AD.[48] To date, the latest example of a lamp in this category is a glazed lamp from the 9th century found in Sabratha.[49] This category of lamp could have been an indirect source of inspiration for later Islamic clay wheel-made lamps common in the Near East, North Africa and the Iberian peninsula,[50] which are usually glazed, although metal Islamic lamps would have played a much more important role.

The oblong body, which is characterized by quite a large reservoir in the case of the first three sub-types, in combination with the oblong nozzle which is placed at an angle to the horizontal axis, were suitable for the storing of quite a large quantity of fuel in the interior. Furthermore, the cylindrical neck facilitated filling the lamp with fuel through the fill hole.[51] Ease of use must therefore have played a role in the more widespread use of these three sub-types in comparison with the more geographically restricted diffusion of the last two sub-categories, since the nozzle of these two types, which is placed at a very low point on the main body of the lamp and almost parallel to the horizontal axis, impedes the storing of any significant amount of fuel, despite their closed shape.[52] In particular, regarding the first sub-type, which was mainly prevalent in Crete, the tall, narrow, cylindrical neck kept rodents and insects away from the lamps, making them particularly efficacious.

To date, no reliable archaeological data exists to indicate that the production of sub-types 2 and 3 continued even later into the late 7th and 8th centuries. Only data regarding the first and the last two sub-types seems to indicate the continuation of their production into at least the 8th century.[53] However, it has been noted by D. Orssaud and J.-P. Sodini that it is mostly from this period onwards that wheel-made lamps with a round body and biconvex walls became prevalent in the Near East and Egypt. Their small, short nozzles, sometimes with an open reservoir, would influence lamp production further afield in parts of the Aegean and the Balkans, and the types would become quite common there during the Middle Byzantine period.[54] It can be suggested that the fourth sub-type[55] –which had quite a limited occurrence– constitutes an intermediary trial stage between this category and the former ones, such as those analyzed by D. Orssaud and J.-P. Sodini, which have round bodies.[56] O. Waagé has noted that lamps of this form (our sub-type 4) constitute a transitional stage for Middle Byzantine lamps and dates them to the 7th and

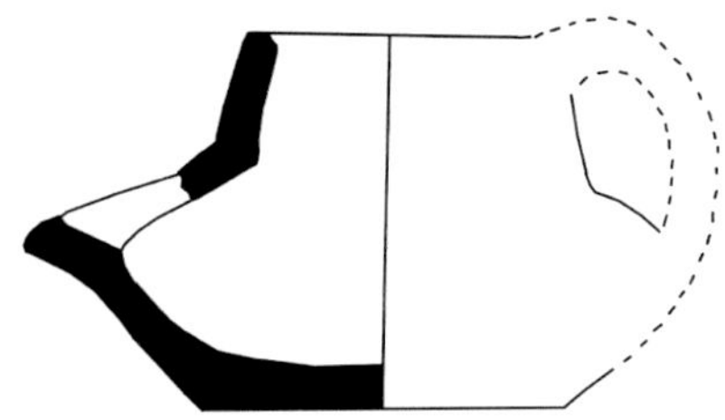

Figure 10. Lamp from Salamis, Cyprus (drawing by the author after Oziol 1977, pl. 58, 882).

8th centuries,[57] while D. Orssaud and J.-P. Sodini also consider examples of this type as 'prototypes' for their first group of lamps.[58] In terms of the transformation from an ovoid to a more round, slightly squat body, and of the small nozzle, which is incorporated in the walls of the reservoir, this sub-type does indeed reference later ones. However, the existence of a curtailed cylindrical neck around the central, fill hole and the absence of a clear distinction between the upper and the lower part of the main body of the lamp clearly call to mind earlier wheel-made lamps. The finds from Amathus, Cyprus, where lamps of sub-type 4, which date from after the mid 7th century, are found together with Orssaud's and Sodini's group 1 of lamps of the Middle Byzantine period, further enhances our view.[59] A few lamps from Salamis (Figure 10) and Amathus can be considered as other transitional forms between the category analyzed here and the Middle Byzantine forms; their high, broad, tubular neck reminds us of this category, while the circular body and its very small, attached closely to the body, nozzle recall later forms.[60] Lamps from El-Ashmunein in Egypt, though reminiscent of this category in having a large, quite high neck, could also be considered as transitional sub-types for later forms, given their more circular, compressed body and the small, triangular nozzle.[61]

Concluding remarks

The category of wheel-made lamp examined here, having been widespread a few centuries earlier in regions of central and eastern Europe primarily in the Balkans, became a basic category of wheel-made lamp during the 6th and especially in the 7th century in the Aegean region and adjacent areas in Asia Minor. This allows us to suggest the existence of a 'koinè' regarding the specific wheel-made lamps used in these areas. In contrast, in Cyprus, Syria, Palestine, Egypt and several parts of North Africa, various categories of wheel-made lamps with a round, slightly squat body and a very small nozzle were diffused during the same period, becoming prevalent there in the centuries following the Arabic conquest. It seems that under the influence of these, as presented by Orssaud and Sodini, the production of similar wheel-made lamps became more generalized later on, even in regions of present-day Greece and of Asia Minor.[62] This is how the wheel-made lamps

[48] For details, see above, note 32.
[49] Joly 1974, 58, no. 1329, 205, no. 1329, pl. LVIII, 1329.
[50] For characteristic forms of these lamps: Menzel 1954, 103, fig. 85, 10, 104, no. 661; Kubiak 1970, 10-14 types F-I, pl. 2, figs 9-12; Lassus 1958, 262 fig. 37; Vitelli 1981, 124, fig. 59, 1.876, 125; Motsianos, Bintsi 2011, 305-306, no. 220, 307, no. 223 (M. Sardi); Bussière, Rivel 2012, 316-319; Camacho Cruz 2015, 249-252, 250, fig. 5, 251, figs 7-9; Parent *et al.* 2018, 331, 335 fig. 31. For a similar opinion that lamps with similar characteristics constitute a predecessor to later forms, see also the opinion of D. Bailey (Bailey 1998, 148, X172-X173).
[51] The same remark is also made by R. H. Howland (Howland 1958, 115-116) in regard to Hellenistic lamps of a similar form.
[52] Colt 1962, 64.
[53] On the date of these forms see the information presented above.
[54] For these two basic categories: Orssaud, Sodini 1997, 63-72 and for this influence p. 71.
[55] For examples of this sub-type: Antioch: Waagé 1941, type 60, 78, no. 60a: 188. Libya, Sabratha: Joly 1974, 57, no. 1328, pl. LVIII, 1328.
[56] Orssaud, Sodini 1997, 63-64.
[57] Waagé 1941, type 60, 78, no. 69a: 188.
[58] Orssaud, Sodini 1997, 64, 66, figs 2, 6-7.
[59] For the Amathus finds: Prokopiou 1995, 262-263, figs 5-6.
[60] Salamis: Oziol 1977, pl. 58, 882; Amathus: Prokopiou 1995, 256, no. Π5κζ, 263, fig. 6, ζ.
[61] Bailey 1998, 148, X172-X173.
[62] Orssaud and Sodini 1997, 70-71. For a characteristic earlier exam-

with an elongated/oval body and oblong, tubular nozzle fell out of favor, leading to a renewed adoption of a 'koinè', though one that was much more widespread this time, including the regions of the Byzantine empire during the Middle Byzantine period and having direct analogies with lamps circulating during the same period in Arab-ruled parts of Near East and North Africa. Thus, the category of lamps presented here constitutes one of the fundamental missing links leading from the wheel-made lamps of the Roman period to those of the Middle Byzantine period. The more widespread use of wheel-made lamps with a round body and small nozzle from the 6th century AD should be linked to the conditions arising in Byzantium during the transitional period that followed. However, the evidence from Gortyn, where the first sub-type of wheel-made lamps with a high neck remained in use even during the 8th century AD raises the possibility that, in cases where the specific category of lamp had long been in use and formed part of the local pottery production, its production may well have continued for a longer period, during the years of transition to the Middle Byzantine period. Moreover, this argument is given further backing by the long use of the fifth sub-type, which remained in use into the 8th century AD in the Islamic world.

In conclusion, the above analysis can be added to I. Motsianos' argument that the mold-made clay lamps which have occupied researchers for so long were essentially a 'large and impressive parenthesis'[63] in the history of clay lamps during the Byzantine era. To gain a better understanding of the latter, it is perhaps time to turn to a more thorough study of the former, since 'In the beginning was... the wheel'.

Bibliography

Albertocchi, M. and Perna, R. 2001. Ceramica comune: vasi da mensa e da dispensa, in A. Di Vita (ed.) *Gortina V.3, lo scavo del Pretorio (1989-1995), vol. V.3* t. I, I Materiali*, 411-536. Padova: Bottega d'Erasmo.

Albertocchi, M. 2011. Lucerne, in A. Di Vita and M. Antonietta Rizzo (eds) *Gortina Agorà. Scavi 1996-1997* (Studi di archeologia cretese 9), 105-112. Padova: Bottega d'Erasmo.

Anselmino, L. and Pavolini, C. 1981. Terra Sigillata: Lampe, in A. Carandini (dir.) Anselmino, L., Pavolini, C., Saguì, L., Tortorella, S. and Tortorici, E. *Atlante delle forme ceramiche I. Ceramica fine romana nel Bacino mediterraneo (medio e tardo impero), Enciclopedia dell'arte antica, classica e orientale*, 184-207. Roma: Istituto della Enciclopedia Italiana.

Aupert, P. 1980. Objets de la vie quotidienne à Argos, in *Études Argiennes, Bulletin de Correspondance Hellénique, supplément* 6, 395-457.

Bailey, D. M. 1988. *A catalogue of the lamps in the British Museum, vol. III: Roman provincial lamps*. London: British Museum Publications.

Bailey, D. M. 1996. *A Catalogue of the lamps in the British Museum, vol. IV: Lamps of Metal and Stone and Lampstands*. London: British Museum Publications.

Bailey, D. M. 1998. *Excavations at El-Ashmunein V. Pottery, Lamps and Glass of the Late Roman and Early Arab Periods*. London: Published for the Trustees of the British Museum by British Museum Press.

Bairami, K. 2010. Ταφικό σύνολο παλαιοχριστιανικών λυχναριών από τη ροδιακή νεκρόπολη (Grave group of Early Christian lamps from the Rhodian necropolis), in D. Papanikola-Bakirtzis and D. Koussoulakou (eds) *Κεραμική της Ύστερης Αρχαιότητας από τον Ελλαδικό χώρο, Επιστημονική Συνάντηση, Θεσσαλονίκη 12-16 Νοεμβρίου 2006*, 249-269. Thessaloniki: Archaeological Institute of Macedonian and Thracian Studies (in Greek with English abstract).

Baldini, I. and Parello, G. 2001. Lucerne, in A. Di Vita (ed.) *Gortina V.3, lo scavo del Pretorio (1989-1995), vol. V.3* t. I, I Materiali*, 114-189. Padova: Bottega d'Erasmo.

Barrett, D. G. 2008. *The Ceramic Oil Lamps as an Indicator of Cultural Change within Nabataean Society in Petra and its Environs circa CE 106* (Gorgias Dissertations 32, Near Eastern Studies 8). Piscataway, N.J.: Gorgias; Oxford: Oxbow.

Benea, D. 1990. Lampes romaines de Tibiscum. *Dacia* 34, 139-168.

Bernhard, M. L. 1955. *Lampki Starożytne*. Warszawa: Sztuka.

Blondé, F., Gros, J.-S. and Pétridis, P. 2011-2012. La céramique au quotidien à Thasos de l'époque archaïque à l'époque protobyzantine, *Revue des Études Grecques* 124, 193-204.

Bonetto, J., Mazzochin, S. and Schindler-Kaudelka, E. 2010. Missione archeologica italiana presso il Teatro del Pythion a Gortyna (Creta): I dati preliminari della ceramica comune di età tardoantica, in S. Menchelli, S. Santoro, M. Pasquinucci and G. Guiducci (eds) LRCW3. *Late Roman Coarse Wares, Cooking Wares and Amphorae in the Mediterranean. Archaeology and Archaeometry. Comparison between Western and Eastern Mediterranean, vol. II* (British Archaeological Reports, Intern. Ser. 2185 [II]), 899-905. Oxford: Archaeopress, British Archaeological Reports, International Series 2185.

Bonifay, M. 2004. *Etudes sur la céramique romaine tardive d'Afrique*. Oxford: Archaeopress, British Archaeological Reports, International Series 1310.

Broneer, O. 1930. *Corinth IV.2: Terracotta lamps*. Cambridge Massachusetts.

Broneer, O. 1977. *Isthmia III: Terracotta Lamps*. Princeton, New Jersey: American School of Classical Studies at Athens.

Bussière, J. 2000. *Lampes antiques d'Algérie* (Monographie instrumentum 16). Montagnac: Ed. M. Mergoil.

Bussière, J. and Rivel, J.-C. 2012. *Lampes antiques de Méditerranée. La collection Rivel*. Oxford: Archaeopress, British Archaeological Reports, International Series 2428.

Camacho Cruz, C. 2015. Candiles de piquera. Uso y morfología en la Córdoba del siglo X, in M. J. Gonçalves and S. Gómez-Martínez (eds) *Actas do X Congresso Internacional a Cerâmica Medieval no Mediterrâneo, Silves 22 a 27. outubro'12, vol. I*, 248-252. Silves: Câmara Municipal de Silves & Campo Arqueológico de Mértola.

Chatzi-Vallianou, D. 1995. Λαξευτοί τάφοι στην επαρχία Πυργιωτίσσης. Τα παραδείγματα Φαιστού και Ματάλων, in N. E. Papadoyannakis (ed.) *Πεπραγμένα του Ζ' Διεθνούς Κρητολογικού Συνεδρίου, τόμ. Α2, Τμήμα Αρχαίο, Νέα Χριστιανική Κρήτη* 6-7 (1994-1995), 1007-1034. Rethymnon: Ιστορική και Λαογραφική Εταιρεία Ρεθύμνης.

Chrzanovski, L. 2013. La transition lychnologique byzantino-arabe en Egypte et au Proche-Orient: Lampes coptes et musulmanes de la collection Bouvier. *Peuce* 9, 177-228.

ple (end of 6th-early 7th century) of a wheel-made lamp from Thasos, that is related to their typology: Blondé, Gros, Pétridis 2011-2012, 200; Pétridis 2013, 50, 195 fig. 31. I would like to thank Associate Professor Platon Petridis, for the information on the date of this lamp.

[63] Motsianos 2010, 280.

Colt, H. D. 1962. Lamps, in H. D. Colt (ed.) *Excavations at Nessana (Auja Hafir, Palestine), volume 1*, 62-64. London: British School of Archaeology in Jerusalem.

Cvjetićanin, T. 2006. *Late Roman Glazed Pottery. Glazed Pottery from Moesia Prima, Dacia Ripensis, Dacia Mediterranea and Dardania* (Archaeological Monographies 19). Belgrade: National Museum.

Dakoronia, F. 1996. Οδός Εθνικής Αντιστάσεως και Σεραφείμ Κοκκόλα (οικόπεδο Β. Χουσιάδα). *Αρχαιολογικόν Δελτίον* 51, B 1, 318-319.

Deneauve, J. 1969. *Lampes de Carthage.* Paris: Editions du Centre National de la Recherche Scientifique

Des Courtils, J. *et al.* 2001, Xanthos, rapport sur la campagne de 2000. *Anatolia Antiqua* 9, 227-241.

Djuric, S. 1995. *The Anawati Collection Catalog I. Ancient Lamps from the Mediterranean,* Ontario.

Fiolitaki, A. 2011. Παρουσίαση της κεραμικής από το ναό των Αγίων Ασωμάτων στο Βαφέ Αποκορώνου, in M. Andrianakis (ed.) *Πεπραγμένα Ι΄ Διεθνούς Κρητολογικού Συνεδρίου (Χανιά, 1-8 Οκτωβρίου 2006), vol. B2, Βυζαντινή και Νεότερη Εποχή (Αρχαιολογία - Ιστορία Τέχνης)*, 87-107. Chania: Φιλολογικός Σύλλογος «Ο Χρυσόστομος».

Gerousi, E. 2010. Λυχνάρια από το νεκροταφείο της ΄Υστερης Αρχαιότητας και των πρωτοβυζαντινών χρόνων στην Περίσσα Θήρας, in D. Papanikola-Bakirtzis and D. Koussoulakou (eds) *Κεραμική της ΄Υστερης Αρχαιότητας από τον Ελλαδικό χώρο, Επιστημονική Συνάντηση, Θεσσαλονίκη 12-16 Νοεμβρίου 2006*, 217-231. Thessaloniki: Archaeological Institute of Macedonian and Thracian Studies (in Greek with English abstract).

Gerousi-Bendermacher, E. 2013. *Sepulkralkultur auf der Insel Thera (Santorin). Der spätantike Friedhof in Perissa und seine Ausgrabungsfunde unter besonderer Berücksichtigung der Tonlampen* (Marburger Beiträge zur Archäologie 1). Marburg: Marburger Winckelmann-Programm.

Giannikouri, A. and Zervaki, F. 2009. *Κάσος. Τα μνημεία - Οι αρχαιολογικές θέσεις - Η Συλλογή.* Athens: Archaeological Receipts Fund.

Gill, M. A. V. and Tuğçe Şen, N. 2003. Roman and Early Byzantine Terracotta Lamps, in C. S. Lightfoot (ed.) *Amorium Reports II. Research Papers and Technical Reports*, 25-63. Oxford: Archaeopress, British Archaeological Reports, International Series 1170.

Gregory, T. 1993. Additional Pottery, in J. Morganstern (ed.) *The Fort at Dereağzi and other Material Remains in its Vicinity: from Antiquity to the Middle Ages* (Istanbuler Forschungen 40), 135-139. Tübingen: Wasmuth.

Grünewald, M. 1984. Kleinfunde aus den Kenotaphgrabungen 1973 und 1974, in J. Ganzert (ed.) *Das Kenotaph für Gaius Caesar in Limyra. Architektur und Bauornamentik* (Istanbuler Forschungen 35), 23-64. Tübingen: Wasmuth.

Hadad, S. 2002. *The Oil Lamps from the Hebrew University Excavations at Beth Shean, Excavations at Beth Shean vol. 1* (Qedem Reports 4). Jerusalem: Institute of Archaeology; Hebrew University of Jerusalem.

Howland, R. H. 1958. *The Athenian Agora IV: Greek lamps and their survivals.* Princeton, New Jersey: American School of Classical Studies at Athens.

Joly, E. 1974. *Lucerne del Museo di Sabratha* (Monografie di Archeologia Libica 11). Roma: L'Erma di Breitschneider.

Katsioti, A. 2017. *The lamps of Late Antiquity from Rhodes: 3rd-7th centuries AD.* Oxford: Achaeopress Archaeology.

Kennedy, C. A., 1961-1963. The Development of the Lamp in Palestine. *Berytus* 14, 67-115.

Kubiak, W. 1970. Medieval Ceramic Oil Lamps from Fusṭāṭ. *Ars orientalis* 8, 1-18.

Kuzmanov, G. 1992. Die Lampen, in S. Uenze (ed.) *Die Spätantiken Befestigungen von Sadovec (Bulgarien), Ergebnisse der Deutch-Bulgarisch-Österreichischen Ausgrabungen 1934-1937*, 223-233. München: Beck.

Kuzmanov, G. 2002. Roman and Late Antique Lamps from Ratiaria. *Annuary of the Institute of Archaeology and Museum, BAS* 2, 242-261.

Lassus, J. 1958. L'archéologie algérienne en 1957. *Libyca* 6, 197-265.

Lazaridis, P. 1965. Ανασκαφαί Νέας Αγχιάλου. *Πρακτικά της εν Αθήναις Αρχαιολογικής Εταιρείας*, 10-23.

Lebessi, A. 1992. Το ιερό του Ερμή και της Αφροδίτης στη Σύμη Βιάνου. *Πρακτικά της εν Αθήναις Αρχαιολογικής Εταιρείας*, 211-230.

Lyon-Caen, C. 2005. Les lampes coptes du musée du Louvre, in L. Chrzanovski (ed.) *Lychnological Acts 1. Actes du 1er Congrès international d'études sur le luminaire antique (Nyon - Genève, 28.IX - 4.X.2003)* (Monographies instrumentum 31), 223-226. Montagnac: M. Mergoil.

Macalister, S. R. A. 1912. *The Excavation of Gezer, 1902-1905 and 1907-1909.* London: Murray.

Menzel, H. 1954. *Antike Lampen im Römisch Germanischen Zentralmuseum zu Mainz.* Mainz: Verl. des Römisch-germanischen Zentralmuseums.

Michelucci, M. 1975. *La collezione di lucerne del Museo Egizio di Firenze.* Firenze: Olschki.

Motsianos, I. 2005. Some Comments on Byzantine Wheel-made Lamps, in L. Chrzanovski (ed.) *Lychnological Acts 1. Actes du 1er Congrès international d'études sur le luminaire antique (Nyon - Genève, 29.IX - 4.X.2003)* (Monographies instrumentum 31), 247-251. Montagnac: M. Mergoil.

Motsianos, I. 2010. ΄Υστερη Αρχαιότητα: Το λυκόφως των κατασκευασμένων με μήτρα λυχναριών (Late Antiquity: the twilight of mouldmade lamps), in D. Papanikola-Bakirtzis and D. Koussoulakou (eds) *Κεραμική της ΄Υστερης Αρχαιότητας από τον Ελλαδικό χώρο, Επιστημονική Συνάντηση, Θεσσαλονίκη 12-16 Νοεμβρίου 2006*, 270-281. Thessaloniki: Archaeological Institute of Macedonian and Thracian Studies (in Greek with English abstract).

Motsianos, I. and E. Bintsi (eds) 2011. *Light on light, an illuminating story*, Exhibition Catalogue, Thessaloniki, Folklife & Ethnological Museum of Macedonia-Thrace, 31 October 2011-10 June 2012. Thessaloniki: Folklife and Ethnological museum of Macedonia and Thrace.

Olcay Uçkan, Y., Uğurlu, E., Gökalp, D. and Bursali, M. 2007. 2005 Yili Olympos yüzey araştimarsi. *Araştirma Sonuçlari toplantisi* 24.2, 123-134.

Orssaud, D. and Sodini, J.-P. 1997. Les lampes tournées de Qal'at Sem'an et leurs parallèles dans le bassin méditerranéen, in G. D. D' Archimbaud (ed.) *La céramique médiévale en Méditerranée, Actes du VIe congrès de l' Association Internationale pour l'Etude des Céramiques Médiévales en Méditerranée 2, Aix-en-Provence (13-18 novembre 1995)*, 63-72. Aix-en-Provence: Narration Editions.

Oziol, Th. 1977. *Salamine de Chypre VII, Les lampes du musée de Chypre.* Paris: De Boccard.

Oziol, Th. 1980. Quelques lampes de Constantia, in *Salamine de Chypre. Histoire et Archéologie. État des recherches, Lyon, 13-17 mars 1978*, 395-400. Paris: Éditions du Centre national de la recherche scientifique.

Oziol, Th.-J. and Pouilloux, J. 1969. *Les Lampes, Salamine de Chypre I.* Paris: De Boccard.

Öztaşkin, G. K. and Öztaşkin, M. 2012. Antalya-Olympos Mozaikli Yaplı, in K. Dörtlük, T. Kahya, R. Boyraz Seyhan and T. Ertekin (eds) *International Young Scholars Conference I. Mediterranean Anatolia, 04-07 November 2009, Antalya. Symposium Proceedings*, 329-346. Antalya: Suna & İnan Kıraç Research Institute on Mediterranean Civilizations.

Parent, F. with the collaboration of Djellid, A. and Chalah, F. 2018. Entre production et consummation: Place des Martyrs à Alger (résultats préliminaires), in Y. Hazırlayan and F. Yenişehirlioğlu (eds) *XIth Congress AIECM3 on Medieval and Modern Period Mediterranean Ceramics Proceedings, 19-24 October / 2015 Antalya, vol. II*, 311-337. Ankara: Koç Üniversitesi VEKAM.

Perlzweig, J. 1961. *The Athenian Agora VII: Lamps of the Roman Period*. Princeton, N.J.: American School of Classical Studies at Athens.

Pétridis, P. 2013. *Πρωτοβυζαντινή κεραμική του ελλαδικού χώρου*. Athens: Εκδόσεις Gutenberg.

Petrie, W. M. F. 1905. *Ehnasya 1904*. London: Egypt Exploration Fund.

Prokopiou, E. 1995. Αμαθούντα. Ανατολική νεκρόπολη. Τάφος οστεοφυλάκιο του 7ου μ.Χ. αιώνα. *Report of the Department of Antiquities in Cyprus*, 249-277.

Rahmani, L. Y. 1983. Finds from a Sixth to Seventh Centuries Site near Gaza II. Pottery and Stone Objects. *Israel Exploration Journal* 33, 219-230.

Reynolds, P. 1995. *Trade in the Western Mediterranean: A.D. 400-700: The Ceramic Evidence*. Oxford: Archaeopress, British Archaeological Reports, International Series 604.

Roman, C. A. 2000. Wheelmade lamps of Roman Dacia. *Acta Musei Napocensis* 37.1, 99-140.

Rosenthal, R. and Sivan, R. 1978. *Ancient Lamps in the Schlöessinger Collection*. Jerusalem: Institute of Archaeology.

Rubright, J. C. 1973. Lamps from Sirmium in the Museum of Sremska Mitrovica, in V. Popović and E. L. Ochsenschlager (eds) *Sirmium. Archaeological Investigations in Syrmian Pannonia, vol. III*. 45-80. Beograd: Archaeological Institute.

Stampolidis, N. C. (ed.) 2004. *Eleutherna: polis, acropolis, necropolis*. Athens: Ministry of Culture.

Sussman, V. 2009. *Greek and Hellenistic Wheel- and Mould-Made Closed Oil Lamps in the Holy Land. Collection of the Israel Antiquities Authority*. Oxford: Archaeopress, British Archaeological Reports, International Series 2015.

Sythiakaki, V. 2018. 231-236 Ομάδα υψίλαιμων λύχνων, in N. Ch. Stampolidis, E. Papadopoulou, I. G. Lourentzatou and I. D. Fappas (eds) *Κρήτη. Αναδυόμενες Πόλεις. Απτέρα - Ελεύθερνα - Κνωσός*, 327. Athens: Museum of Cycladic Art - Hellenic Ministry of Culture and Sports.

Triantafyllopoulou, P. 1998. Οδός Λεμεσού (οικόπεδο Π. Παπαχατζή). *Αρχαιολογικόν Δελτίον* 53, B 2, 414-415.

Tzavella, E. 2010. Κεραμική από αθηναϊκούς τάφους του τέλους της Αρχαιότητας και οι μαρτυρίες της για τον 7ο αι. στην Αττική, in D. Papanikola-Bakirtzis and D. Koussoulakou (eds) *Κεραμική της Ύστερης Αρχαιότητας από τον Ελλαδικό χώρο, Επιστημονική Συνάντηση, Θεσσαλονίκη 12-16 Νοεμβρίου 2006*, 649-670. Thessaloniki: Archaeological Institute of Macedonian and Thracian Studies (in Greek with English abstract).

Vallianou, D. 1988. Μάταλα, Βόρειο Νεκροταφείο. *Αρχαιολογικόν Δελτίον* 43, B 2, 538-542.

Vitelli, G. 1981. *Islamic Carthage. The Archaeological, Historical and Ceramic Evidence*. Carthage: Centre d'études et de documentation archéologique de Carthage.

Vroom, J. 2004. Late Antique Pottery, Settlement and Trade in the East Mediterranean: A Preliminary Comparison of Ceramics from Limyra (Lycia) and Boeotia, in W. Bowden, L. Lavan and C. Machado (eds) *Recent Research on the Late Antique Countryside* (Late Antique Archaeology 2), 281-331. Leiden: Brill.

Waagé, F. O. 1941. Lamps, in R. Stillwell (ed.) *Antioch on-the-Orontes, vol. III, The excavations 1937-1939*, 55-82. Princeton: Princeton University Press; London: Oxford University Press; The Hague: Nijhoff.

Williams, H. 1981. *Kenchreai, Eastern Port of Corinth V: The Lamps*. Leiden: Brill.

Williams, H. and Taylor, P. 1975. A Byzantine Lamp Hoard from Anamur (Cilicia). *Anatolian Studies* 25, 77-84.

Xanthopoulou, M. 2005. Lampes en metal, lampes en terre cuite: vies parallèles, in L. Chrzanovski (ed.) *Lychnological Acts 1. Actes du 1er Congrès international d'études sur le luminaire antique (Nyon - Genève, 29.IX - 4.X.2003)* (Monographies instrumentum 31), 303-307. Montagnac: M. Mergoil.

Xanthopoulou, M., 2010. *Les lampes en bronze à l'époque paléochrétienne* (Bibliothèque de l'Antiquité tardive 16). Turnhout: BE Brepols.

Xanthopoulou, M. 2015. Ένας αμφορέας της όψιμης πρωτοβυζαντινής περιόδου από την Ίτανο, in P. Karanastasi, A. Tzigounaki and C. Tsigonaki (eds) *Achaeological Work in Crete 3. Proceedings of the 3rd Meeting, Rethymnon, 5-8 December 2013, vol. B*, 585-593. Rethymnon: Εκδόσεις Φιλοσοφικής Σχολής Πανεπιστημίου Κρήτης / Εφορεία Αρχαιοτήτων Ρεθύμνου.

Yangaki, A. G. 2005. *La céramique des IVe-VIIIe siècles ap. J.-C. d'Eleutherna (Crète)*. Athènes: Εκδόσεις Πανεπιστημίου Κρήτης.

Yangaki, A. G. 2016. Pottery of the 4th-early 9th centuries AD on Crete: the current state of research and new directions, in J. E. Francis and A. Kouremenos (eds) *Roman Crete. New Perspectives*, 199-234. Oxford; Philadelphia: Oxbow Books.

Zavvou, E. 1999. Οδός Γκορτσολόγου 28 (Ο.Τ. 39, οικόπεδο Δ. Κρητικού). *Αρχαιολογικόν Δελτίον* 54, B 1, 164-165.

Zhuravlev, D. 2012. Bosporan Hellenistic 'jug-shaped' lamps, in L. Chrzanovski (ed.) *Le Luminaire antique. Lychnological Acts 3. Actes du 3e Congrès International d'études de l'ILA, Université d'Heidelberg, 21 - 26.IX.2009* (Monographies instrumentum 44), 365-371. Montagnac: M. Mergoil.

Lighting Devices found at Byzantine Amorium (Turkey)

Christopher S. Lightfoot
The Metropolitan Museum of Art, New York

Abstract: The lamp finds from the excavations at Amorium (modern Hisarköy, near Emirdağ, Afyonkarahisar) in central Anatolia have contributed important new evidence, especially with regard to the Byzantine period. They show that lighting devices of several different types continued to be used throughout the site's existence as a Byzantine city (late 5th–late 11th centuries). In addition to metal and glass lighting devices used in churches and other privileged contexts, small and relatively inexpensive terracotta lamps continued to be produced for use in daily life. The large assemblage of wheel-made middle Byzantine (10th–11th century) lamps sets the excavations at Amorium apart and shows how ubiquitous lamps remained in Byzantine society.

Keywords: Amorium, lamp, terracotta, bronze, glass.

Amorium existed as a Roman and Byzantine city for some twelve hundred years, stretching from the time that the kingdom of Pergamum was bequeathed to Rome by Attalus III (138–133 BC) to the last quarter of the 11th century AD, when Turkish settlers flooded into Anatolia after the battle of Manzikert in 1071.[1] During all that time the city and its inhabitants underwent numerous changes and transitions of a social, economic, political, military, and religious kind. Some of its checkered history is recorded in contemporary sources, but much new evidence has been revealed as a result of the excavations that have been conducted at the site since 1988. Among the many types of material that have been uncovered are numerous lamps and, because of the length and continuity of occupation at Amorium, it is possible to trace the development of and changes in the lighting equipment that was used at the city.[2]

As elsewhere throughout the Roman world, the most common form of lamp used at Amorium was the mold-made terracotta oil lamp. Plenty of lamp fragments have been found during the excavations, together with three intact examples, and a total of well over 100 examples have been published so far.[3] None of these mold-made lamps, all of which are slipped, can be dated later than the 7th century AD, although residual fragments do appear in later contexts.[4] It may be argued that the demise of the lamp industry in the 7th century was as much a consequence of the disruption to the supply of cheap olive oil (both its production and its transport) as it was the market collapse caused by a lack of demand for mass-produced goods such as lamps. Of course, this does not mean that during the late 7th to early 9th century –the so-called Byzantine 'Dark Age'– people did not use lighting devices. In churches in particular, hanging lamps and candles continued to be used throughout the period. In the home, too, some artificial light must have been available, and it is likely that candles became more popular. Metal lamps and candlesticks may also have had a place in well-to-do households.

Some important work has been done recently in cataloguing collections of Byzantine bronze lamps from Turkey and, although all there is to offer from Amorium is a single surface find (Figures 1-2, SF8402) from the 2009 season, the amount of material would seem to indicate that such lamps played an important role in Byzantine lighting.[5]

There are four bronze lamps in the Istanbul Archaeological Museum similar to the Amorium example but sadly none of them have provenances.[6] Amorium has provided some slight evidence for candles too; three metal objects can be identified as candlesticks or parts of candlesticks (Figure 3, SF2506, an iron candlestick; Figure 4, SF3919, a bronze dish; and Figures 5-6, SF4496, a bronze pricket), and significantly all three were found in the Lower City Church.[7] Also from the Church there is considerable evidence for the use of hanging lamps, not just in terms of fragments of the glass bowls themselves but also three attractive bronze suspension brackets (Figure 7, SF3924; Figure 8, SF8236; and Figure 9, SF8349).

Similar examples of hand-shaped lamp holders in the Istanbul Archaeological Museum have been published and attributed to a provenance in Turkish Thrace.[8] If correct, then the Amorium examples would be the only examples definitely known to have been found in Byzantine Anatolia. The last Amorium example (SF8349) however has no known parallel.

1 Lightfoot and Lightfoot 2007, 32-63.

2 An earlier version of this paper was presented in Turkish at the 12th Mediaeval and Turkish Period Excavations and Art History Symposium, held at Onsekiz Mart Üniversitesi Çanakkale on October 15-17, 2008; Lightfoot 2010. I am extremely grateful to Dr Ioannis Motsianos for inviting me to attend the Round Table in Thessaloniki and providing me with this opportunity to publish an updated English version.

3 Gill and Şen 2003; Schoolman 2012.

4 *Pace* Motsianos 2005, 247: 'mold-made lamps dating to as late as the 8th century have been found at Amorium.'

5 I wish to thank Ahmet Arıözsoy, one of the villagers at Hisarköy, who found and handed in the lamp.

6 Atasöy 2005, 32-4, nos. 61-64, all dated to the 6th century. Compare also Xanthopoulou 2010, 126-7, nos. LA 3.119-125, whose provenances include Beth Shean (Israel), Santa Chiara di Ballao (Sardinia), Ptuj (Croatia), and possibly Akhim (Egypt).

7 For iron candlesticks with folding legs, see Motsianos and Bintsi 2011, 168-9, nos. 56 and 58; and for parallels for the copper alloy candleholder (SF4496) but with squared prickets, see idem, 170-1, nos. 59 and 63.

8 Atasöy 2005, 87-91, nos. 146-154, with references to parallels in the Prähistorische Staatssammlung München, the Badisches Landesmuseum Karlsruhe, the Edirne Museum, and two collections in Istanbul.

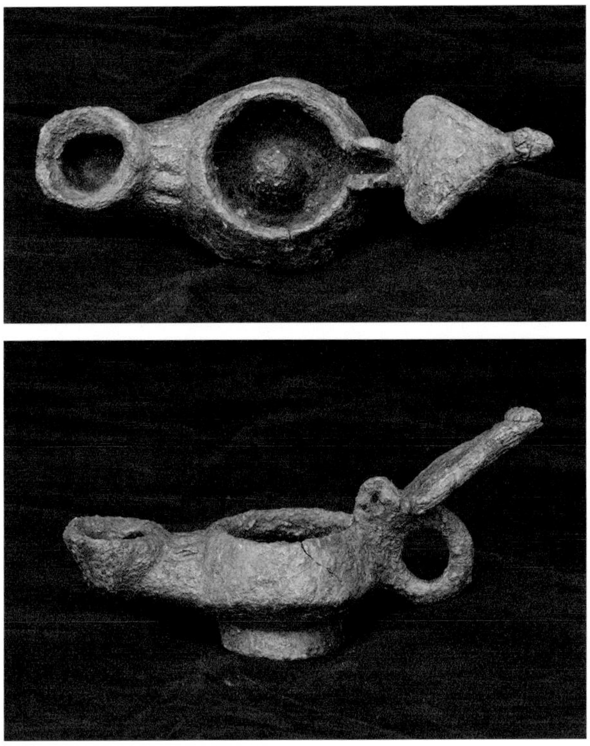

Figures 1-2. Bronze lamp, SF8402, L. 11.5 cm. Surface find, Lower City. Photo © Amorium Excavations Project.

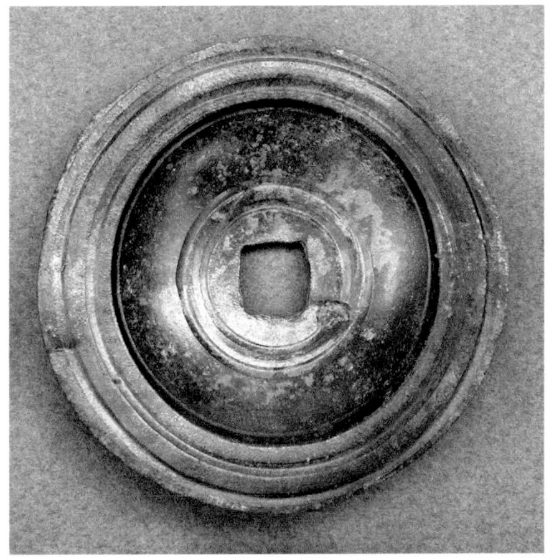

Figure 4. Bronze candlestick dish, SF3919, Diam. 6.8 cm. From the Lower City Church, surface find. Photo © Amorium Excavations Project.

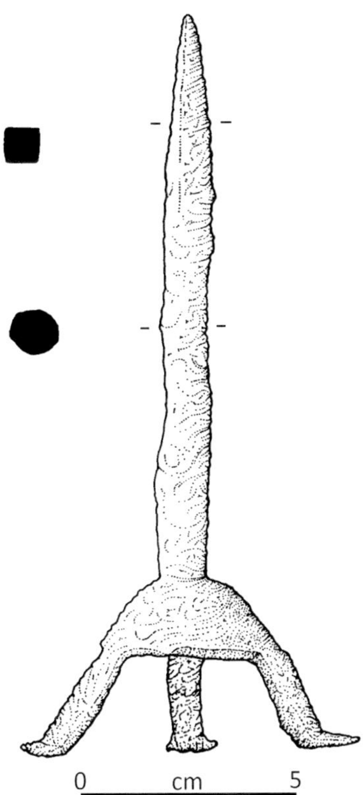

Figure 3. Iron candlestick, SF2506, H. 17.0 cm. From the Lower City Church, trench A4, context 58. Drawing © Amorium Excavations Project.

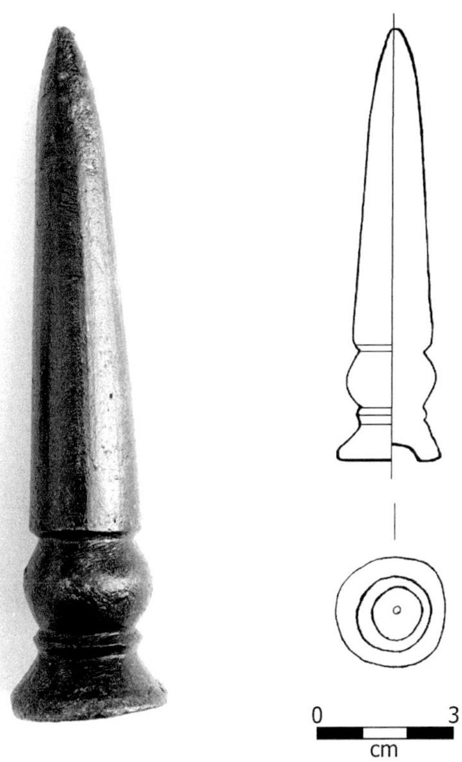

Figures 5-6. Bronze candlestick pricket, SF4496, L. 9.1 cm. From the Lower City Church, trench A8, context 3. Photo & Drawing © Amorium Excavations Project.

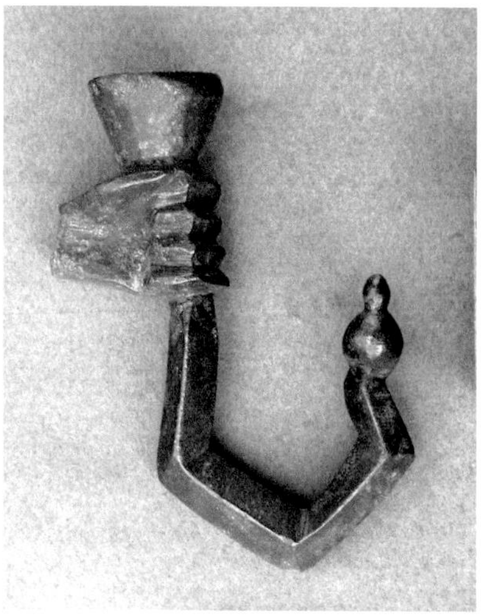

Figure 7. Bronze suspension bracket, SF3924, L. 6.25 cm. From the Lower City Church, trench A6, context 2. Photo © Amorium Excavations Project.

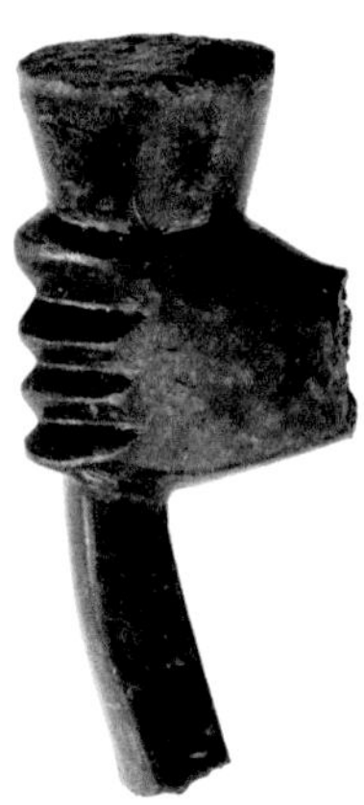

Figure 8. Bronze suspension bracket, SF8236, L. 4.95 cm. From the Lower City Church, trench A27, context 600. Photo © Amorium Excavations Project.

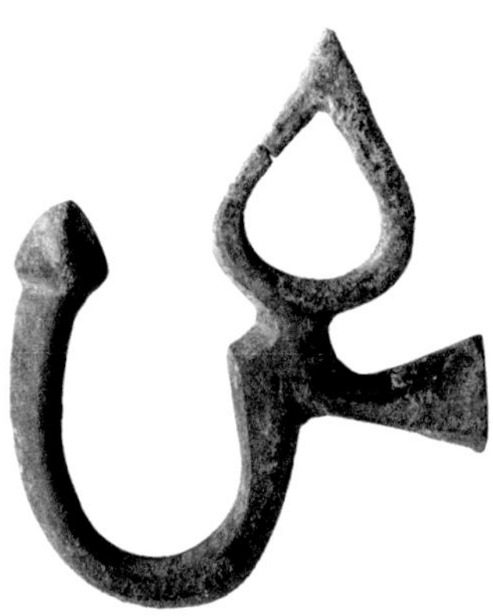

Figure 9. Bronze suspension bracket, SF8349, H. 6.0 cm. From the Lower City Church, trench A26, context 648. Photo © Amorium Excavations Project.

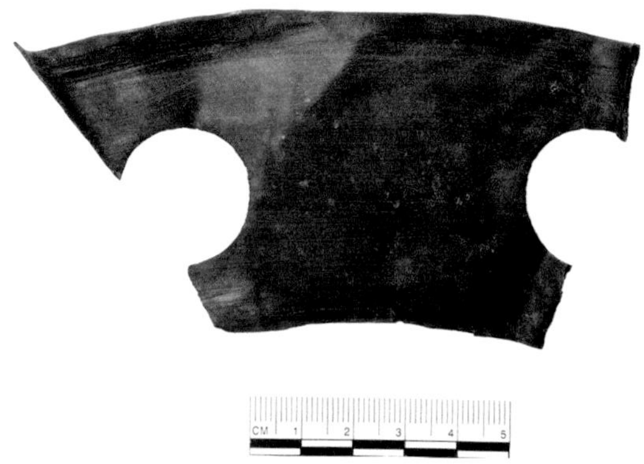

Figure 10. Copper alloy polykandelon fragment, SF8002, L. 11.68 cm. From the Lower City Church, trench A22, context 310. Photo © Amorium Excavations Project.

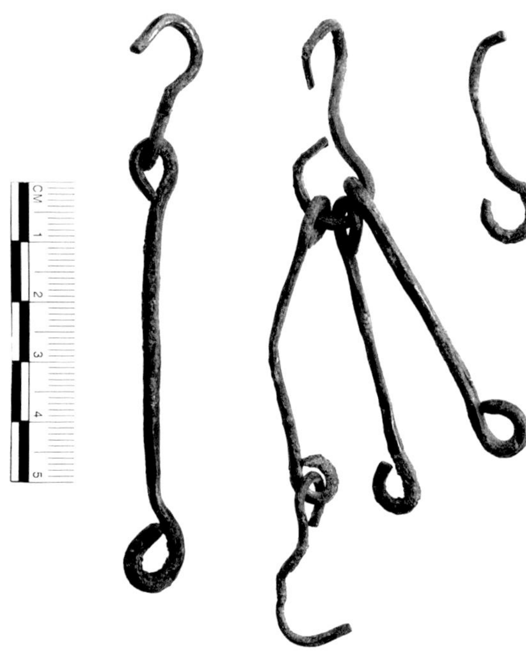

Figure 11. Copper alloy suspension hooks and links, SF8000, max. L. 11.12 cm. From the Lower City Church, trench A26, context 609. Photo © Amorium Excavations Project.

Amorium has produced a wealth of Byzantine glass finds, including large numbers of fragments that may be identified as belonging to glass lamps. They include both stemmed lamps such as would have been placed in metal polykandela and handled lamps that were suspended directly from chains. A fragment (Figure 10, SF8002), identified as part of a copper alloy polykandelon openwork disk, was also found at the Lower City Church in 2007. During excavations in 2007 and 2008 in a middle Byzantine annex, perhaps a tower, on the south side of the Church a large group of glass fragments, copper alloy links and hooks (Figure 11, SF8000), and lead

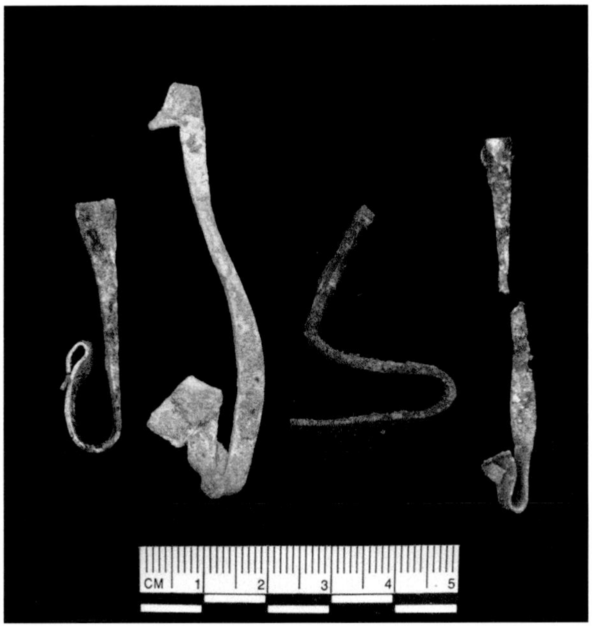

Figure 12. Lead wickholders, SF8358, max. L. 11.12 cm. From the Lower City Church, trench A26, context 648. Photo © Amorium Excavations Project.

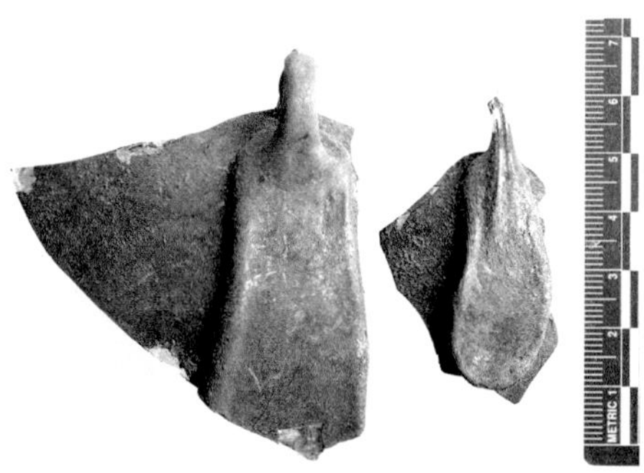

Figure 13. Glass hanging lamp fragments, max. L. 7.0 cm. From the Lower City Church, trench A26, context 620. Photo © Amorium Excavations Project.

wickholders (Figure 12, SF8358) belonging to hanging lamps were found, suggesting that the room may have been used to store such items of church equipment.[9] Although the glass lamps are very fragmentary, the handles are distinctive, comprising a ring or loop at the top and a long, flattened tail below (Figure 13). Some of the lamps also have decorative ring bases. They conform in shape and style, as well as in date, to examples found at the Church of St Nicholas at Myra and at sites elsewhere in Anatolia.[10] One may also compare a glass hanging lamp in the Musée du Verre Liege, said to be from Manisa and dated to *c.* the 9th century.[11] What is interesting,

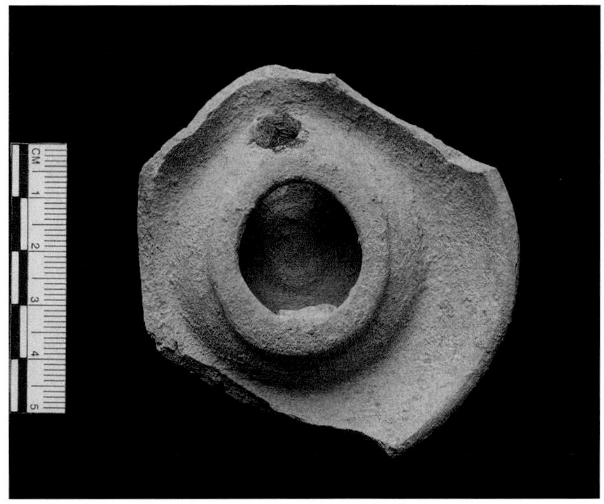

Figure 14. Terracotta wheel-made lamp, SF7500, Diam. 4.38 cm. From the Lower City Enclosure, trench XC-06, context 1009. Photo © Amorium Excavations Project.

however, is that this group appears to be closely related to contemporary Islamic glass lamps, exemplified by the finds of 'mosque lamps' found on the Serçe Limanı shipwreck of *c.* 1025.[12]

Likewise, the wheel-made terracotta lamps found at Amorium also bear comparison with some types of Islamic lamps. Although these issues are the subject for future research, the finds from Amorium provide significant new evidence, suggesting that the site may have played an important role in the transmission of goods, skills, and ideas between the Byzantine and Islamic worlds, especially in the 10th and 11th centuries.

A catalogue of some 56 wheel-made Middle Byzantine lamps from Amorium, found between 1988 and 1992, was published in 2003.[13] A second catalogue appeared in 2012, containing another 76 examples found during the seasons between 1993 and 2005, all of which belong to the three main, previously established types, although they also included a small number of green-glazed examples.[14] For these one may compare a green-glazed fragmentary lamp in Thessaloniki, also dated to the 10th–11th century.[15] During the four excavation seasons between 2006 and 2009 another 18 unglazed examples have been identified (Figure 14, SF7500), making a grand total of 150 middle Byzantine wheel-made lamps found at Amorium.[16] The overwhelming majority of these lamps are unglazed and,

[9] From the same deposit (context 648) as the bronze bracket and the wickholders, illustrated in figs 9 and 12 respectively, nine copper alloy coins were recovered. Eight of these are of the same type (SF8348, SF8359, SF8361, SF8362, SF8376, SF8378, SF8380, and SF8435): signed folles of Constantine X and Eudocia, class 2, dated 1059–1067.

[10] Olcay 2001, 86.

[11] Philippe 1990, 43 and fig. 54.

[12] Morden 2009, 372-3, esp. no. LP 235. The Serçe Limanı catalogue omits mention of the glass finds from Amorium; see Gill 2002, esp. 134, nos. 24-25; 160, nos. 306-308, and 261.

[13] Gill 2003.

[14] Lightfoot 2012, esp. 222-3, nos. 69-76. No fieldwork was conducted at Amorium in 2010–2012, but excavations resumed in 2013 under the direction of Assoc. Professor Zeliha Demirel Gökalp of the University of Anatolia, Eskişehir; see Demirel Gökalp, Erel, Tsivikis, and Yaşar 2016.

[15] Motsianos and Bintsi 2011, 156, no. 34.

[16] For the dating evidence for these lamps, see Lightfoot 2012, 225-6. At least four more fragmentary examples were found in 2014; pers. comm. Assoc. Professor Zeliha Demirel Gökalp.

to date, the few glazed examples do not include any with a brown glaze.[17] The size and uniformity of this assemblage from Amorium are very important for establishing that in Middle Byzantine times lamps were still an essential part of the secular scene, whether in the home or in the workplace, and were not just used in ecclesiastical settings. Moreover, although well over 100 Middle Byzantine tombs have been excavated in and around the Lower City Church at Amorium, not a single wheel-made lamp can be associated with a funerary context.[18] Amorium however is not the only site in Anatolia to have produced Middle Byzantine wheel-made lamps. Fourteen examples, including four green-glazed ones, have been published from Yumuktepe (Mersin) in Cilicia.[19] An unpublished example from the Yenikapı excavations in Istanbul, as befits a lamp from the capital, is much larger than the Amorium finds. Nevertheless, it is not surprising that Amorium has produced so many examples, given that it continued to be a large and flourishing Byzantine city during the 10th and 11th centuries.

How these lamps were used and what sort of fuel they contained are questions that were addressed in the discussion that accompanied the catalogue in *Amorium Reports* vol. 3.[20] It has also been argued that these lamps should not be regarded as oil lamps in the traditional sense since olive oil –at least at Amorium where it would have had to be imported– was a precious commodity, used only as fuel for lighting in special places and on special occasions.[21] However, it is still not possible to define more precisely when and where the wheel-made saucer lamp was first produced. At Amorium it would appear that such lamps were not yet in use in the early 9th century, although they were clearly ubiquitous there in Middle Byzantine times.[22] Likewise, the quantity of such finds at Amorium would seem to indicate local production, but there is as yet no proof of this. It is too early to state that they derive from a Byzantine rather than an Islamic prototype. Nevertheless, the wide distribution of saucer lamp variants from Italy to the Crimea, Syria, and Egypt[23] would suggest that Byzantium played a leading role in promoting its popularity and promulgating its use.[24]

Finally, another area in which it is hoped Amorium can play a major role is in the investigation of the transition from mold-made to wheel-made lamps. As yet there is no clear dating for the hand-made lamp fragments that have been found during the excavations,[25] nor sadly have any lamps at all been encountered in the sealed destruction levels associated with the sack of Amorium in AD 838.[26] However, it is certain that future work at the site will continue to shed new light on the subject of Byzantine lighting devices. In addition, new excavations that began in the summer of 2013 have focussed on the Upper City mound or tell, which probably served as the nucleus for the Middle Byzantine city and the subsequent Seljuk settlement that had been established at Amorium by the early 13th century AD. As yet the site has not produced any evidence for Seljuk lighting devices.

[17] For an example with brown glaze, found in the agora at Thessaloniki and dated to the 10th century, see Motsianos and Bintsi 2011, 209, no. 121, describing the saucer as 'the oil reservoir.'

[18] Motsianos 2005, 247: *Pace* '[wheel-made lamps] were still used in burial ceremonies.'

[19] Güngör, Köroğlu, and Laflı 2010, esp. 336-8, nos. 2-10.

[20] Lightfoot 2012, 225-6.

[21] Motsianos 2005, 247: *Pace* 'wheel-made lamps continued to use vegetable oils, mainly olive oil.'

[22] None have been found in any of the contexts that can be dated before the destruction of the city in AD 838.

[23] Kubiak 1970, 9-10, text fig. 6 a-b, and pl. 2, fig. 8.

[24] A fragmentary example in Thessaloniki has an olive green glaze that seems not dissimilar to that used on *Amorium Glazed Ware*; Motsianos and Bintsi 2011, 155, no. 32. For *AGW*, see Bohlendorf-Arslan 2004, 222-3; 425-8, nos. 400-413; pls 104-5.

[25] See Schoolman 2012, 199, no. 61.

[26] Ivison 2012, 60-5 for a full discussion of the archaeological contexts associated with this historical event.

Bibliography

Atasoy, S. 2005. *Bronze Lamps in the Istanbul Archaeological Museum.* Oxford: British Archaeological Reports, International Series 1436.

Böhlendorf-Arslan, B. 2004. *Die glasierte byzantinische Keramik aus der Turkei*, 3 Vols. Istanbul: Ege.

Demirel Gökalp, Z., Erel, A. C., Tsivikis, N., and Yaşar, H. Y. 2016. 2014 Yılı Amorium Kazısı, in 37. Kazı Sonuçları Toplantısı, Erzurum, 11-15 Mayıs 2015, vol. 3, 199-214. Ankara: T.C. Kültür ve Turizm Bakanlığı.

Gill, M. A. V. 2002. *Amorium Reports, Finds I: The Glass (1987–1997).* Oxford: British Archaeological Reports, International Series 1070.

Gill, M. A. V. 2003. Middle Byzantine Terracotta Lamps, in C. S. Lightfoot (ed.) *Amorium Reports, II: Research Papers and Technical Reports, 65-71.* Oxford: British Archaeological Reports, International Series 1170.

Gill, M. A. V. and Şen, N. T. 2003. Roman and Early Byzantine Terracotta Lamps, in C. S. Lightfoot (ed.) *Amorium Reports, II: Research Papers and Technical Reports*, 25-63. Oxford: British Archaeological Reports, International Series 1170.

Gungor, E., Koroğlu, G., and Laflı, E. 2010. Yumuktepe Pişmiş Toprak Kandilleri, in K. Pektaş *et al.* (eds) XIII. *Ortacağ Turk Donemi Kazıları ve Sanat Tarihi Sempozyumu Bildirileri, 14-16 Ekim 2009/* Proceedings of the XIIIth Symposium of Medieval and *Turkish Period Excavations and Art Historical Researches*, 333-41. Istanbul: Biltur.

Ivison, E. A. 2012. Excavations at the Lower City Enclosure, 1996–2008, in C. S. Lightfoot and E. A. Ivison (eds) *Amorium Reports 3: The Lower City Enclosure. Finds Reports and Technical Studies*, 5-151. Istanbul: Ege.

Kubiak, W. B. 1970. Medieval Ceramic Oil Lamps from Fusṭāṭ. *Ars Orientalis* 8, 1-18.

Lightfoot, C. 2010. Ortaçağ'da Aydınlatma Teknikleri ve Amorium'da Ele Geçen Buluntular, in A. O. Uysal, A. Yavas, M. Dündar, and O. Koçyiğit (eds) *XII. Ortaçağ-Türk Dönemi Kazıları ve Sanat Tarihi Sempozyumu, 15-17 Ekim 2008/ Proceedings of the XIIth Symposium of Medieval and Turkish Period Excavations and Art Historical Researches*, 41-9. İzmir: Çanakkale Onsekiz Mart Üniversitesi.

Lightfoot, C. S. 2012. Middle Byzantine Terracotta Lamps: 1993–2005, in C. S. Lightfoot and E. A. Ivison (eds) *Amorium Reports 3: The Lower City Enclosure. Finds Reports and Technical Studies*, 217-32. Istanbul: Ege.

Lightfoot, C. and Lightfoot, M. 2007. *A Byzantine City in Anatolia: Amorium, an Archaeological Guide*. Istanbul: Homer.

Morden, M. E. 2009. Lamps, in G. F. Bass *et al.* (eds) Serçe Limanı, vol. 2. *The Glass of an Eleventh-Century Shipwreck*, 361-74. College Station: Texas A&M University.

Motsianos, I. 2005. Some Comments on Byzantine Wheel-Made Lamps, in L. Chrzanovski (ed.) *Lychnological Acts 1. Actes du 1er Congres International d'etudes sur le luminaire antique (Nyon - Geneve, 29.IX - 4.X.2003)*, Monographies instrumentum 31, 247-51. Montagnac: Monique Mergoil.

Motsianos, I. and E. Bintsi (eds) 2011. *Light on light, an illuminating story*, Exhibition Catalogue. Thessaloniki: Folklife & Ethnological museum of Macedonia-Thrace.

Olcay, B. Y. 2001. Lighting Methods in the Byzantine Period and Findings of Glass Lamps in Anatolia. *Journal of Glass Studies* 43, 77-87.

Philippe, J. 1990. Reflections on Byzantine Glass, in *1st International Anatolian Glass Symposium, April 26th-27th, 1988*, 40-46. Istanbul: TŞCFAŞ.

Schoolman, E. 2012. Middle Imperial, Late Roman, and Early Byzantine Terracotta Oil Lamps: 2002-2005, in C. S. Lightfoot and E. A. Ivison (eds) *Amorium Reports 3: The Lower City Enclosure. Finds Reports and Technical Studies*, 193-215. Istanbul: Ege.

Xanthopoulou, M. 2010. *Les lampes en bronze à l'époque paléochrétienne*. Turnhout: Brepols.

Church Lighting in Byzantine Cherson

Stanislav Ryzhov and Tatyana Yashaeva
State Museum-Preserve 'Tauric Chersonesos'

Abstract: The article deals with the church lighting fixtures from the excavations of Cherson, located on the northern periphery of the Byzantine Empire. They are made mostly of bronze, but there are single instances of silver, iron, and glass. According to their design features, they are divided into six groups: polykandela, choroi, kandelai, lamnai, candelabra, and devices of mixed type. The accessories represented by fragments of massive castings, forged or thin wire chains, special brackets, connecting disks and candleholders are considered separately. It is noteworthy that in the late Byzantine churches of Cherson *polykandela* from an earlier time were used, the obsolete icon lamps were kept in special stores. Church lighting devices were used in Cherson not only during worship services or during burial rites, but also for home prayer in front of icons. Materials from the excavations of Cherson give new specimens of design for Byzantine church lamp options for positioning them in the church interior, and also indicate the existence of a tradition of the long-term use of these lighting devices as well as of many other church items.

Keywords: Byzantine Cherson, excavations, church lighting devices, *polykandela, choroi, kandelai, lamnai, candelabra.*

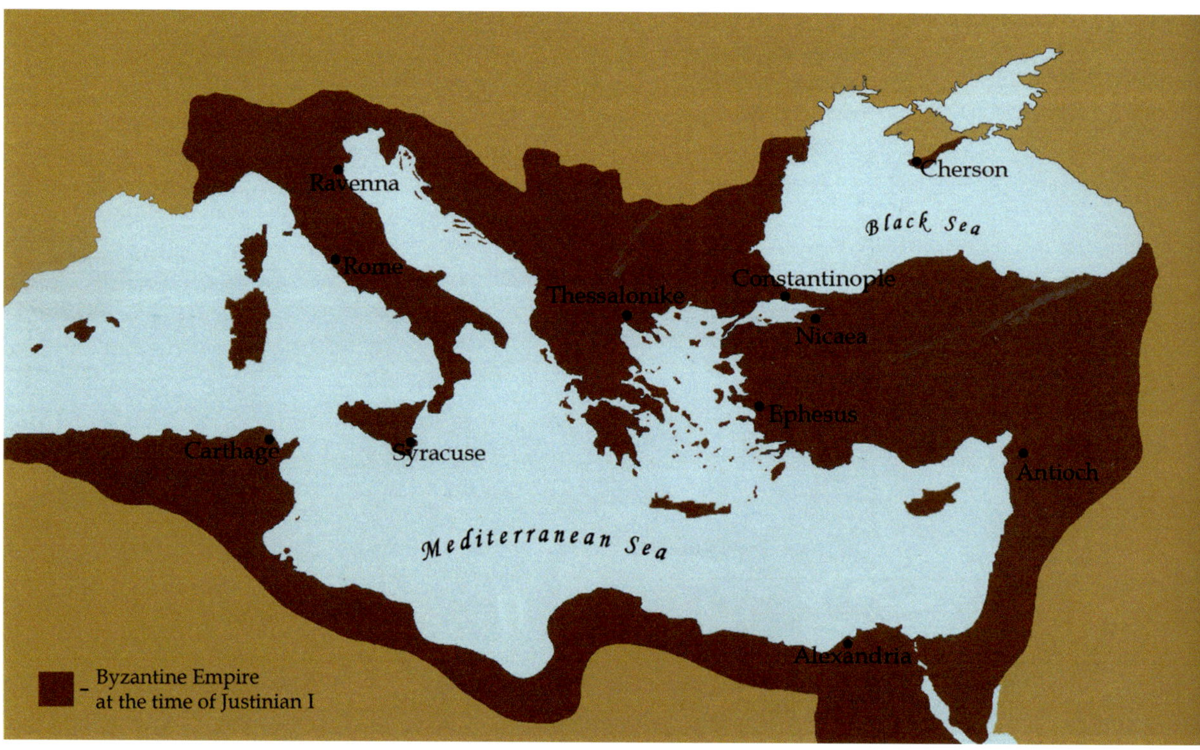

Figure 1. Map of the Byzantine Empire at the time of Justinian I (I. Gusakova, after a map in 'Byzanz. Pracht und Alltag', Bonn 2010, 20).

Located in Tauric on the northern shore of the Black Sea Chersonesos (Byzantine Cherson) was not only an outpost, but also a major Christian center on the northern periphery of the Byzantine Empire (Figure 1). During the over 190 years of excavations in the territory of Cherson, archaeologists have discovered about 70 Christian churches, as well as numerous associated architectural details, interior items and sacred implements. A distinct group among them consists of lighting devices. They not only played an important practical role in the church interior, but also had symbolic meaning. The Byzantine theologian of the 10th century AD Symeon the New Theologian wrote that 'a multitude of lighted candles symbolize the man's bright thoughts, enlightening him from inside; this is the light of the Holy Spirit'.[1]

Most of the church lighting devices from Byzantine Cherson are made of bronze, but there are single instances of silver, iron, and glass. Regardless of the material from which they

[1] Simeon Novyi Bogoslov 2011, 192.

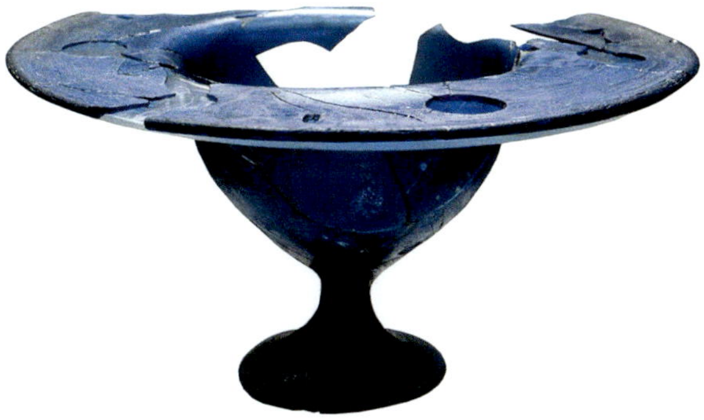

Figure 2. Early Byzantine glass *polykandelon* (K. Zykova).

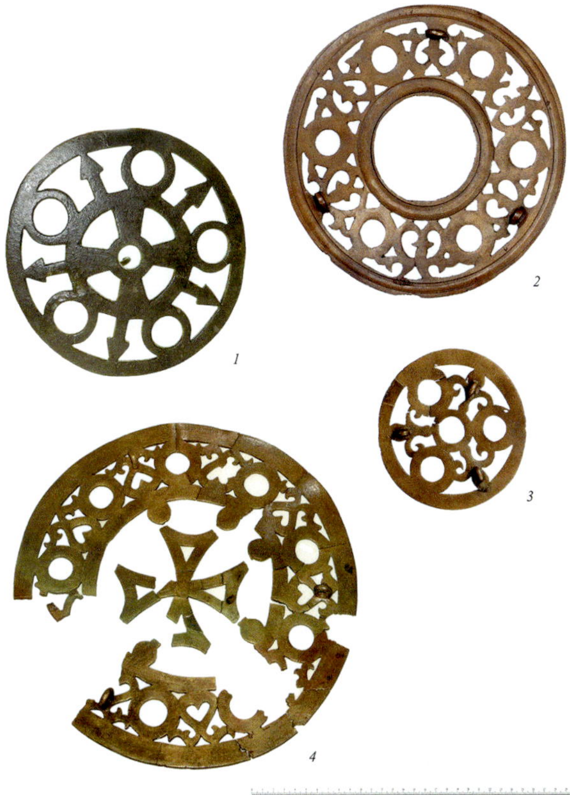

Figure 3. Bronze polykandela (K. Zykova).

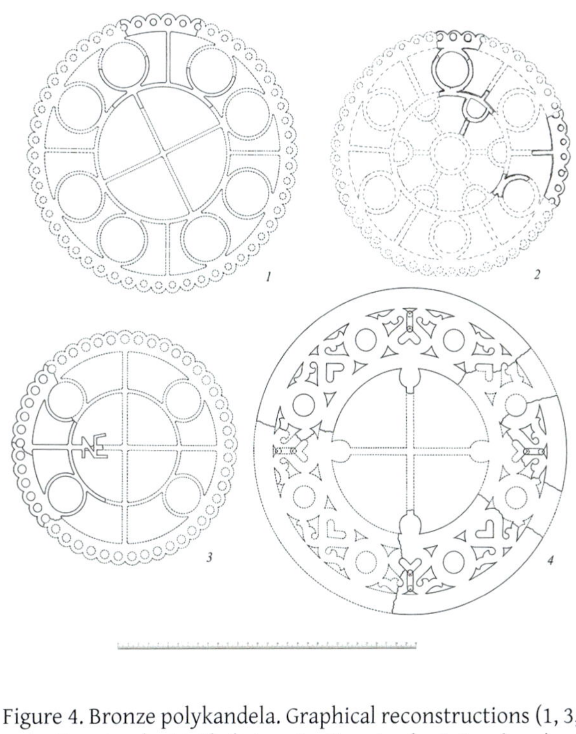

Figure 4. Bronze polykandela. Graphical reconstructions (1, 3, 4 – Drawing by M. Skubetov; 2 – Drawing by I. Gusakova).

were made, all the lamps can be divided into six groups based on their design elements.

Hanging Polykandela

The first group consists of hanging *polykandela* (πολυκάνδηλα),[2] which consist of round disks with round slotted cells for installing glass lamps. They are represented by one specimen made of blue glass and eight bronze disks.

The hanging glass polykandelon, dating from the third quarter of the 5th century AD, had six cells for the lamps. It was found near the Early Byzantine cruciform 'Temple with the Reliquary' in 1973 (Figure 2; cat. no. 1). Experts note its uniqueness, since even distant analogies have not yet been found for it.[3]

The Cherson bronze disks of hanging polykandela had a varied decor, different diameters and a different number of cells for the installation of glass icon lamps (Figure 3.1-4; 4.1-4; cat. nos. 2-9). As a rule, a directly proportional relationship is observed between the diameter of the disk and the number of cells: the larger the diameter, the greater the number of cells. The smallest disk with three cells had a diameter of 14.3 cm (Figure 3.3), the largest with nine cells, 3.10 cm (Figure 3.4). A part of the cast cross-shaped monogram of the 6th-7th centuries AD, made up of an anagram of Greek letters (perhaps NE) is preserved on the one of polykandela. It probably meant the name of the *κτήτορα* (founder) (Figure 4.3; cat. no. 8).[4] The monogram makes it possible to relate the manufacture of the polykandelon to the Early Byzantine period and to date to the same time the analogous Chersonesos' polykandela with scalloped edges which flank small punched circles The fragments of glass lamps with characteristic elongated narrow legs found together with them also point to an Early Byzantine date for the described specimens.[5]

[2] Describing church lighting devices, we mainly used the classification and terminology proposed in Bouras 1982, 479-481.

[3] Golofast 2001, 141.

[4] The authors are very grateful to Andrey Vinogradov and Michail Butyrskiy for the attribution of the monogram.

[5] Skubetov 1910, figs 2, 4, 8, 11; Golofast 2001, 138, figs 85.2, 85.12.

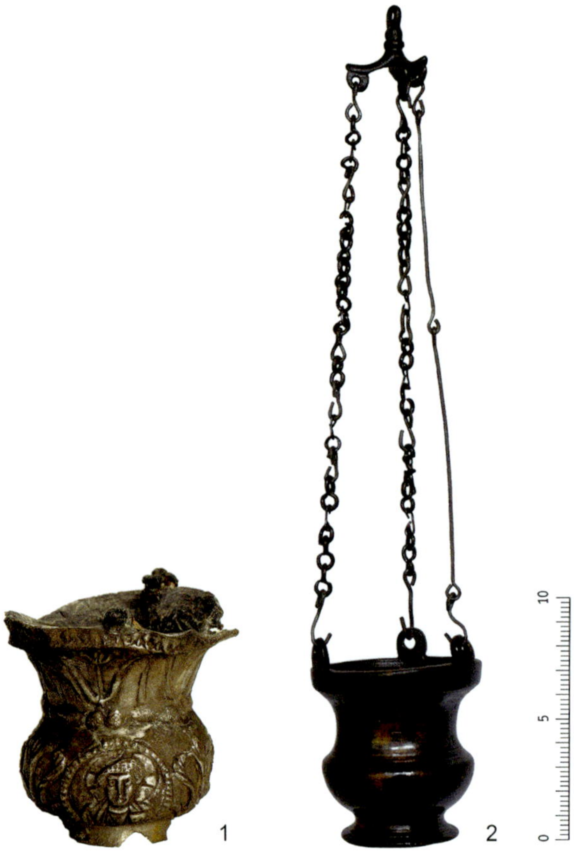

Figure 5. Early Byzantine openwork lamps of silver and bronze (K. Zykova).

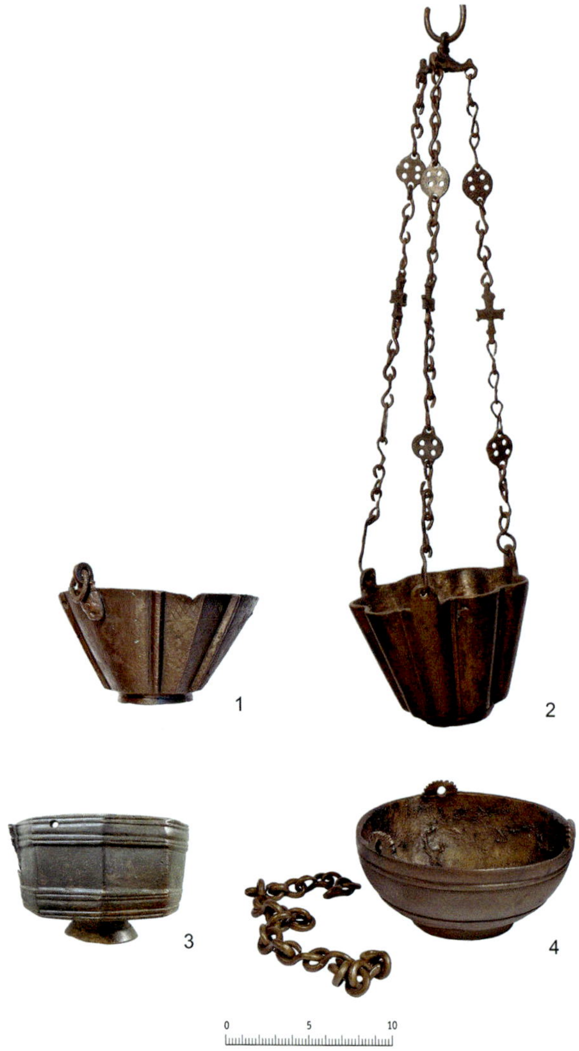

Figure 6. Late Byzantine bronze openwork lamps (K. Zykova).

Choroi

Hanging polykandela were used both independently and as a part of massive lighting devices called *choroi*. Such devices make up the second group. According to the interpretation of Symeon of Thessalonica, 'the circle of lamps represents the firmament and the circles of the planets'.[6] The surviving choroi from Cherson churches are only fragmentary, preserved only in the form of individual structural elements (Figure 9.2–3).

Kandelai

The third group includes single icon lamps -*κανδήλαι*/*kandelai*. According to Symeon of Thessalonica, they were 'indicating one of the Saints, before which they are hung'.[7] Six single *kandelai*, made of silver and bronze were found in Cherson. One of the silver icon lamps dates to the sixth century. It has preserved some round medallions with images of the young Christ and the apostles Peter and Paul (Figure 5.1; cat. no. 11). As was the blue polykandelon described above, it was found near the Early Byzantine 'Temple with the Reliquary', from which it probably originates. This light fixture is of particular interest also because it was likely a holy souvenir brought to Cherson from the holy places of Syria.[8] The second Early Byzantine icon lamp of the 6th-7th centuries is made of bronze and shaped like a cantharos. It was found in the crypt of the western necropolis of Cherson (Figure 5.1; cat. no. 12). The remainder of the lamps of this type are also made of bronze, but belong to the Late Byzantine period and come from residential areas (Figure 6.1-4; cat. nos 13-16).

Candelabra

The fourth group is represented by candelabra, made in the form of metal strips (sometimes with an arcuate central part), having spikes for fitting candles. Researchers call them *lamnai* or *kosmetaritzia*.[9] We will use the term *lamna*. The fragments of two *lamnai* were found in the central area of Cherson in one of the premises of a residential estate in a fire layer of the 13th century AD, where there was a store of obsolete church lighting devices (Figure 7.1).[10] A strip with an arc-shaped

[6] Simeon Fessalonikiiskii 1857, 195.
[7] Simeon Fessalonikiiskii 1857, 194.

[8] Zalesskaia 1997, 18, fig. 21.
[9] Bouras 1982, 480.
[10] Arkhiv Gosudarstvennogo muzeia-zapovednika 'Khersones Tavricheskii' 2004, 3677, 18–19.

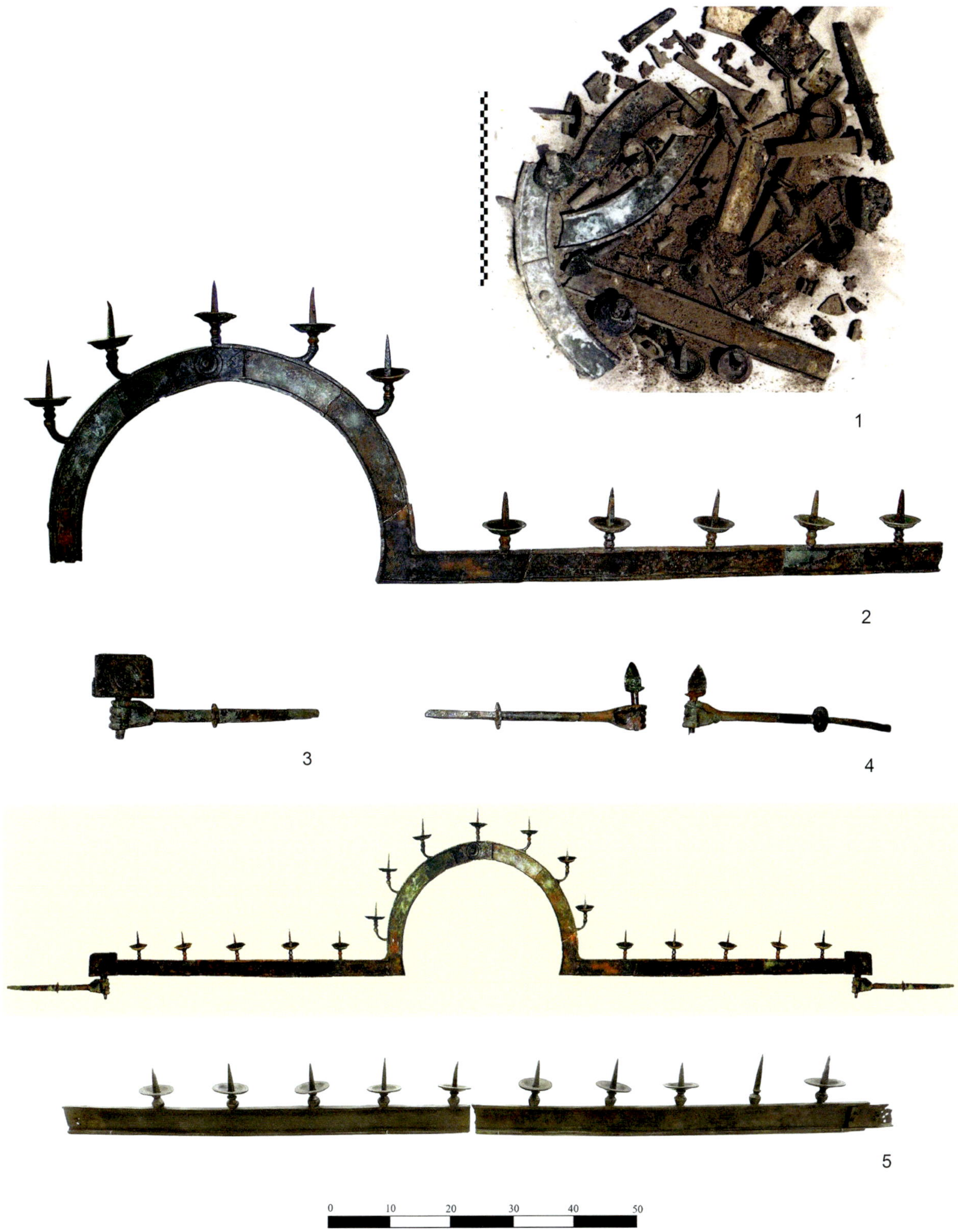

Figure 7. Late Byzantine bronze lamnai (K. Zykova).

central part (preserved length 144.3 cm, reconstruction 288.6 cm) remained from the one lamna (Figure 7.2; cat. no. 17). On the top surface there are cups with pointed rods for the installation of candles (ten preserved). The lamna was attached to the wall by means of two holders (the right one was preserved). The end of the holder adjacent to the luminaire was designed in the shape of a hand holding a rod with a rectangular bracket (Figure 7.3). At the other end of the holder, separated by a transverse disk, was a slightly tapered, rectangular in cross-section (the left wing of the lamna was lost). Only two holders were preserved from the second lamna (right and left), whose ends were also designed in the form of hands clasping a rod with a triangular brackets (Figure 7.4; cat. no. 18). Lighting devices with similar decor have been found on the territory of Bulgaria in the medieval cathedral church of Dorostolon, where they are dated to the end of the 10th-11th centuries AD.[11] Professor G. Atanasov notes their similarity with the Chersonesos' specimens and suggests that they most likely had one prototype and, possibly, were made in the same workshop or were contemporary and used same model. The fragments of three other bronze church lamp holders, decorated in the form of hands and supposedly of the 13th century, are in the collection of the Munich museum.[12]

Unfortunately, the Chersonesos block in which a store with fragments of bronze lamnai was found was not fully excavated, so we cannot yet determine from which church the lighting devices were taken. We also cannot determine if this storehouse was located at a parochial church or city monastery, residential manor or artisan workshop.

According to the researchers' observations, the lamnai could frame epistyles or icons in the Byzantine places of worship[13] or could be positioned above the altar barrier.[14] In 1891 during the excavations of a chapel on the east bank, fragments of a lamna were found directly on a wooden altar barrier. They consisted of fragments of a broad bronze strip with protrusions in the form of inverted truncated cones for placing cups with pointed rods for the attachment of candles and rotted wood from the object.[15]

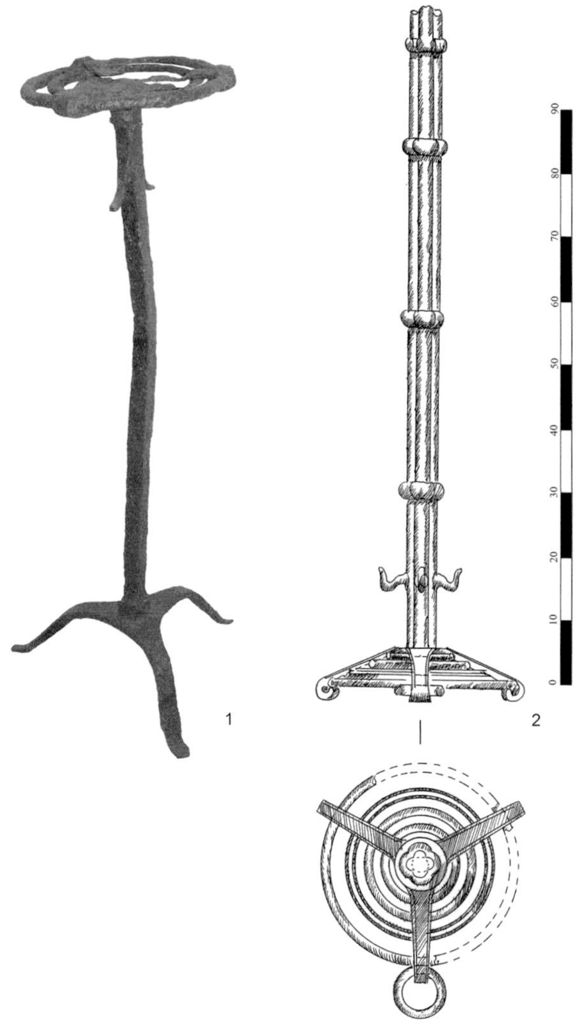

Figure 8. Iron candelabra (1 – K. Zykova; 2 – Drawing by M. Skubetov).

Footed Candelabra

The fifth group is represented by massive floor lamps – candelabra. In Cherson two iron-forged specimens were found, consisting of a tetrahedral cross-section of a rod with a double circle at the upper end, and three legs at the bottom. The first candelabrum was found in the agora and probably belonged to one of the churches there (Figure 8.1; cat. no. 20).[16] The second candelabrum was discovered in 1889 in a residential quarter of the western area of Cherson in a store of obsolete metal products (Figure 8.2; cat. no. 21).[17] Among the finds in this storehouse there were not only church utensils, but also broken household equipment, including fragments of anchors, plow tips and fishing hooks. Of particular interest are the 12 silver bars of the Novgorod type (13th century AD), which date the layer of fire and, accordingly, the candelabrum to the Late Byzantine era.

Lamps Designed for Use with Multiple Fuel Sources

The sixth group is the smallest. It is represented by one mixed-type lighting fixture, identified in the central area of the hill fort, which the Byzantines used in addition to their oil lamps and wax candles (Figure 9.1, cat. no. 22). Its upper part is made in the form of an equilateral cross, which was fastened on four chains to the dome or conch. There are two cups with pointed rods for candles on each branch of the cross. In the lower part there was a massive, round in cross-section hook for hanging a polykandelon. It is possible that such icon lamps were adapted to be held in the hand.[18] The symbolism of mixed-type lighting fixtures, which used both wax and oil, can be explained by the teachings of Symeon of Thessalonica, who wrote that wax, 'as produced by the bees after gathering nectar from a variety of flowers and trees, symbolically means an offering to God from the face of all creation', and the holy anointing oil 'is a sign of God's mercy to people, a sign of spiritual truths'. Thus, a symbolic meeting, a mystical dialogue between the Heavenly Father and his flock takes place as the light before the icon.[19]

Chains and Lighting Accesssories

The accessories for lighting fixtures are presented in a separate group. They are represented by fragments of massive

[11] Atanasov 2013, 132-145.
[12] Atanasov 2013, 136.
[13] Bouras 1982, 480.
[14] Atanasov 2013, 145, fig. 13.
[15] Kostsyushko-Valyuzhinich 1893, 20-21.
[16] Kostsyushko-Valyuzhinich 1892, 4.
[17] Kostsyushko-Valyuzhinich 1892, 14.
[18] Kostsyushko-Valyuzhinich 1905, 80, fig. 37.
[19] Simeon Fessalonikiiskii 1857, 15.

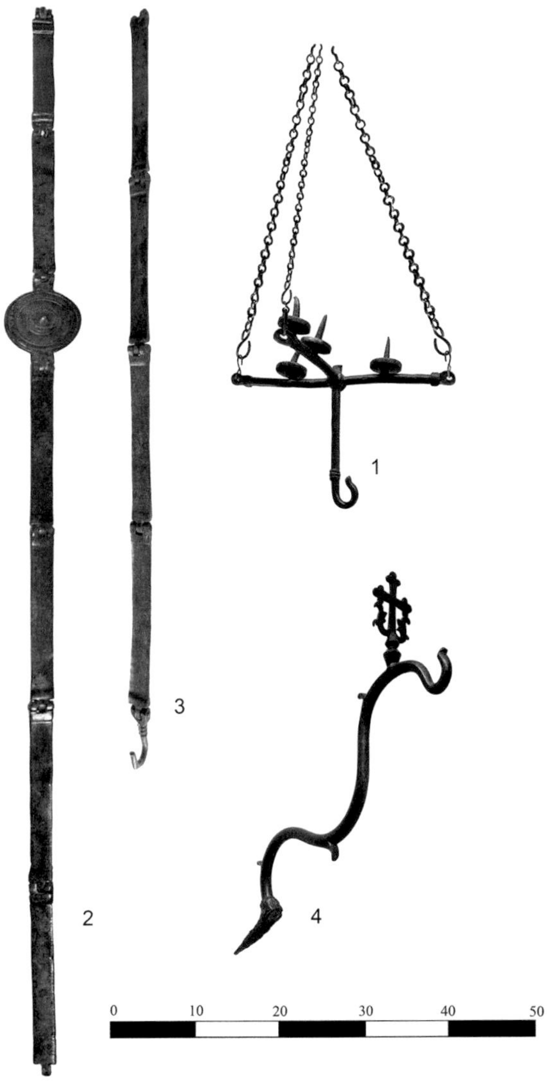

Figure 9. 1 – Bronze polykandelon of mixed type; 2, 3 – Bronze chains for the suspension of a choros; 4 – Bronze bracket (K. Zykova).

cast, forged or thin wire chains, special brackets with hooks, connecting disks, and holders.

Small polykandela and kandelai were sometimes hung on special curved brackets mounted on a wall (Figure 9.4; cat. no. 23). In terms of design features, the Chersonesos' bracket is close to the so-called dracondaria – dragon-shaped candlesticks, one of which is depicted on a twelfth-century fresco in the church of Evangelistria in the village Geraki.[20] But in contrast, the date of the Chersonesos' dracondaria can be somewhat expanded and attributed to the 10th-12th centuries AD, given the presence in its decor of a flourishing cross, whose popularity in Byzantine Cherson occurs at this time.

Chains, depending on the size and nature of the links, were used to hang different categories of lighting fixtures. Certainly, the choroi had massive cast bronze chains, and have analogies in Byzantine monuments of the 13th-14th centuries AD.[21] (Figure 9.2,3; cat. no. 24). A small bronze disk was a part of a choros or polykandelon (cat. no. 10). The disk is decorated with the slotted image of a cross with flared ends and large teardrops at the corners. On the one side there is an engraved cross (above) and a circular, poorly preserved Greek inscription which is not legible. On the other side there is a braided ornament. A similar connecting disk from the Athens Museum dates back to the 6th century AD.[22]

Researchers have observed that chains in the form of wide openwork bands became especially popular during the Palaeologan period[23] (Figure 10.6–8; cat. nos. 30–32). Chains with woven or straight wire, sometimes decorated with crosses or asterisks (Figure 10.1-5; cat. nos. 25-29), were intended for polykandela or single icon lamps. The lower ends of the chains were attached directly to the lamps, the upper ends to special holders.[24] The Cherson holders were predominantly six-armed (though there are single three-beam specimens) with a loop at the top and ears for attaching chains with hooks at the bottom of each extension. The ends of the rays and the faces between them were sometimes decorated with drop-shaped protrusions (Figure 11.1-5; cat. nos. 33–37). Such holders, as well as chains, could be used not only for lighting devices, but also for censers.

It is noteworthy that despite the large number of churches and chapels excavated on the hill, a relatively small number of lighting devices or their fragments were found. And all of them, as a rule, originated from Late Byzantine layers. However, it is well known that already in the Early Byzantine period there were basilicas and cruciform chapels in Cherson, which, of course, also needed illumination. Probably, in the churches of Late Byzantine period the lighting devices of earlier times were often used. Obviously, for this reason, the lighting devices which in other regions are dated Early Byzantine in Cherson we meet in the later layers. This allows us to verify the centuries-long use of lighting fixtures and other church utensils in Cherson's religious buildings. This is also evidenced by the traces of repairing of the chains, which were the most vulnerable part of the hanging icon lamps. The existence of such special stores for preserving out-of-use icon lamps speaks probably about the tradition of caring for church lighting devices, their restoration, long-term storage and transfer to subsequent generations that existed in Cherson.

With regret we have to admit that most of the lighting fixtures from Cherson do not have an archaeological context, so we cannot always establish from which church or chapel they originated or indicate their place in the church interior. Of the 37 luminaires examined, only seven were found in the sanctuaries or in their immediate vicinity. One specimen was found in a crypt, ten in city blocks, and three specimens were in special stores. But even the modest archaeological information that we have at our disposal suggests that church lighting devices were used in Cherson not only for church services or in funeral rites, but also for home prayer in front

[20] Bouras 1982, fig. 8.

[21] Frings 2010, 167, fig. 50.

[22] Papanikola-Bakirtzi 2002, 297, 284.

[23] Bouras 1982, 481.

[24] The article presents an incomplete sample of holders.

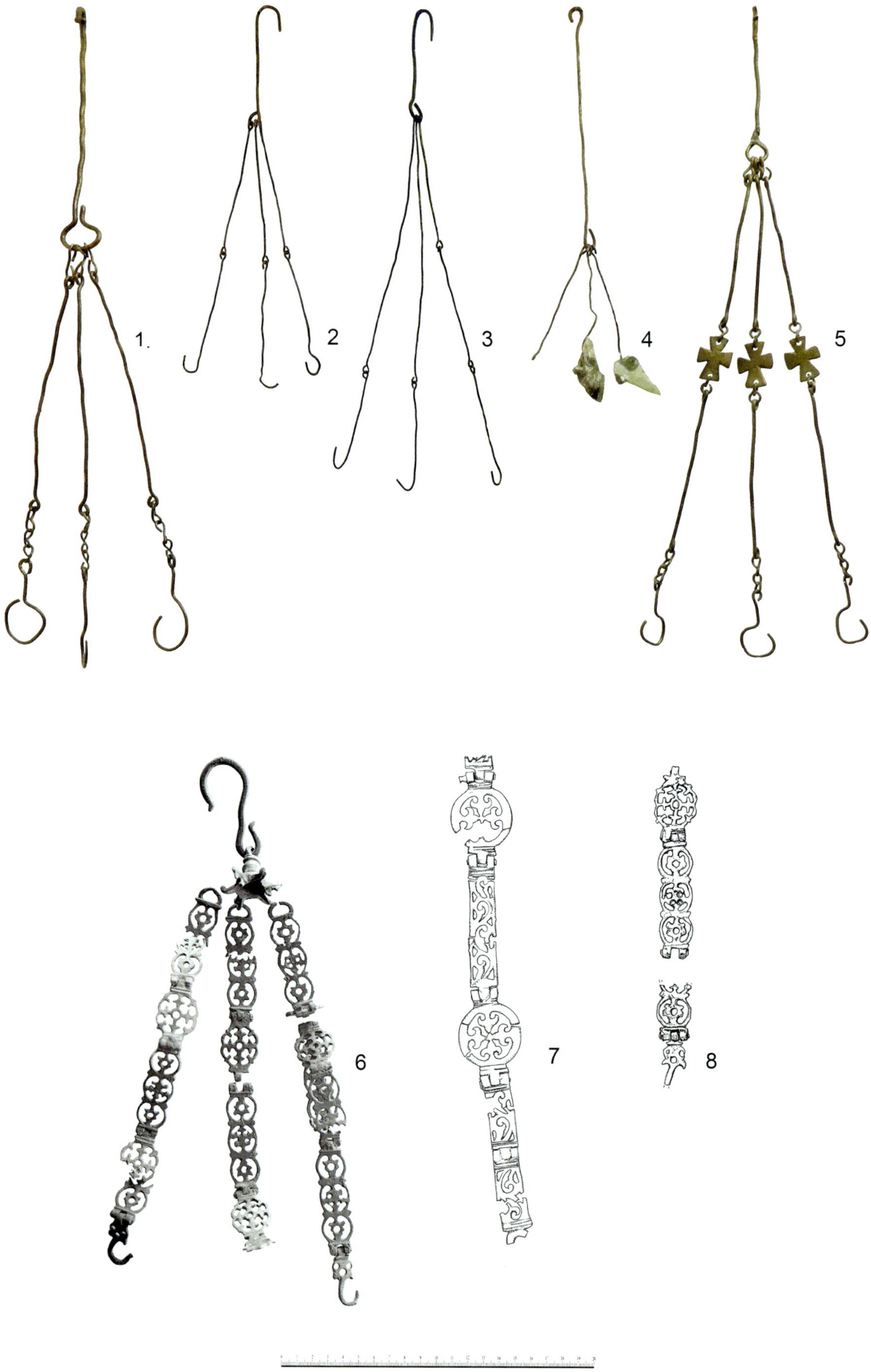

Figure 10. 1–5 – Bronze polyangistra; 6–8 – Bronze openworked chains for the suspension of a polykandelon (K. Zykova).

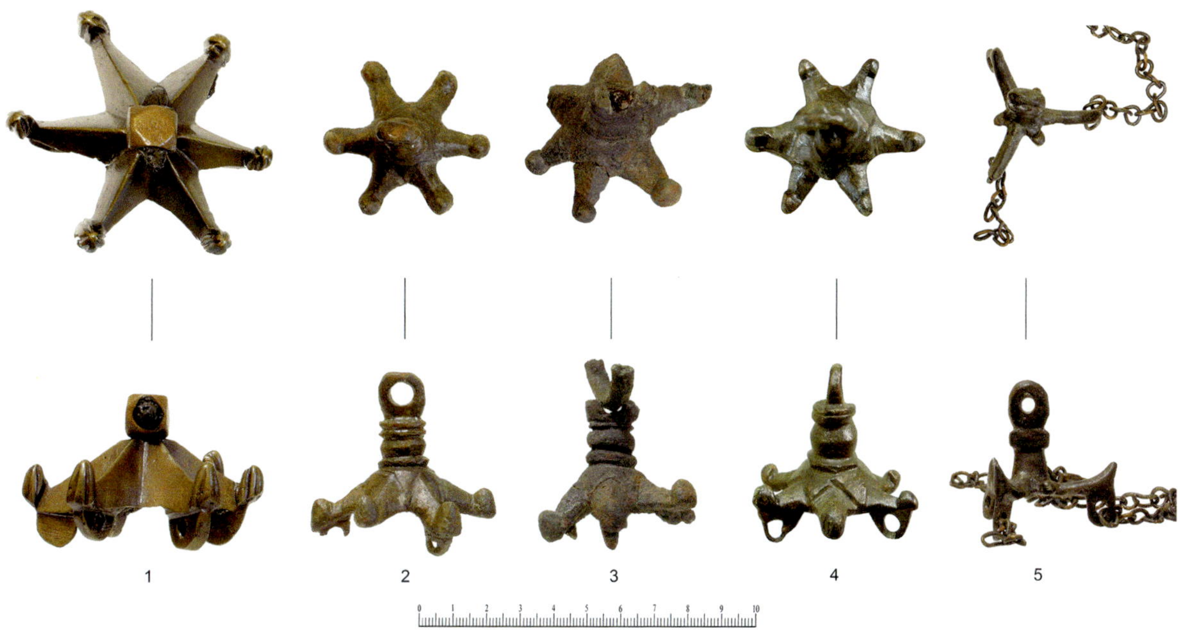

Figure 11. Bronze candleholders (K. Zykova).

of icons. It is probably for this reason that we find polykandela and icon lamps not only in churches and chapels, but also in crypts and in residential estates.

The artifacts from the excavations of Cherson give us not only rare specimens of design for Byzantine church lighting fixtures and insight into the options for placing them in the interior of the church, but also indicate the existence of a tradition of the long-term use of the lighting fixtures as well as many other church items.

Catalogue

1. Polykandelon[25] (Figure 2). Eastern Mediterranean. The third quarter of the 5th century AD. Glass. Height: 12.5 cm; diameter 24.5 cm. Central area, site of the ancient theater. Excavations of O. I. Dombrovskiy, 1973. Hanging polykandelon of blue glass on a tall foot with a hemispherical body and a broad, flat outturned rim with a thickened folded edge. Along the rim are six round holes for lamps of which four holes are preserved; between them are three small oval holes for suspension.

2. Polykandelon[26] (Figure 3.3). Eastern Mediterranean. The layer of fire of 13th century AD. Bronze. Diameter 14.3 cm. Northern area, block X-a. Excavations of S. G. Ryzhov, 1988. An openwork disk from a hanging polykandelon with three openings for icon lamps, between which are three fleurs-de-lys and three vertical loops for the attachment of chains.

3. Polykandelon[27] (Figure 3.1). Eastern Mediterranean. 13th century AD. Bronze, iron. Diameter 22 cm. Central area, block VII. Excavations of R. Ch. Loeper, 1913. An openwork disk from a hanging polykandelon. In the center is a cross with flared arms inside a circle, from which radiate five arrows with their points facing out; between the arrows are five circular openings for glass icon lamps. In the center of the cross is a smaller circular opening. Suspension loops for attachment to chains are missing.

4. Polykandelon[28] (Figure 3.2). Eastern Mediterranean. 13th century AD. Bronze. Diameter 26 cm. Northern area, block X. Excavations of S. G. Ryzhov, 1986. An openwork disk from a hanging polykandelon with six openings for icon lamps, between which are six stylized palmettes and three vertical loops for the attachment of chains.

5. Polykandelon[29] (Figure 3.4). Eastern Mediterranean. 13th century AD. Bronze. Diameter 31 cm. Excavations in the end of 19th century AD. An openwork disk from a hanging polykandelon decorated in the center with a cross with flared ends (some metal is lost), and around is a wide openwork band with nine openings for icon lamps.

6. Polykandelon[30] (Figure 4.2). (Reconstruction). Eastern Mediterranean. 6th-7th century AD. Bronze. Diameter 24 cm. The Main Street. Restoration work. S. G. Ryzhov. 2010. Fragments of an openwork disk from a hanging polykandelon with six openings for icon lamps. The edge is decorated with rounded scallops, in which small through holes have been made.

7. Polykandelon[31] (Figure 4.1). (Reconstruction). Eastern Mediterranean. 6th-7th century AD. Bronze. Diameter 27.3

[25] Golofast 2001, 141, fig. 8; Golofast 2009, 301-335; Yashaeva *et al.* 2011, 655, no. 476.
[26] Yashaeva *et al.* 2011, 507, no.166.
[27] Yashaeva *et al.* 2011, 507, no. 165.
[28] Yashaeva *et al.* 2011, 507, no. 167.
[29] Unpublished.
[30] Unpublished.
[31] Skubetov 1910, 42-47, fig. 6.

cm. Excavations inside a monastery manor, 1905. Fragments of an openwork disk from a hanging polykandelon with eight openings for icon lamps and four ears for hanging. The edge is decorated with rounded scallops, through which small holes have been made.

8. Polykandelon[32] (Figure 4.3). (Reconstruction). Eastern Mediterranean. 6th-7th century AD. Bronze. Diameter 22.3 cm. Excavations inside of monastery manor. 1905. Fragments of an openwork disk from a hanging polykandelon with six openings for icon lamps and four ears for hanging. The edge is decorated with rounded scallops, thorugh which small holes have been made. In the left half of the disk is preserved a part of a 6th-7th century AD cast cross-like monogram composed of Greek letters NE.

9. Polykandelon[33] (Figure 4.4). (Reconstruction). Eastern Mediterranean. 6th-7th century AD. Bronze. Diameter 31.5 cm. Excavations of Odessa Society of History and Antiquities, 1877–1878. Fragments of an openwork disk from a hanging polykandelon with eight openings for icon lamps and four ears for hanging; smooth edge.

10. Disk.[34] Eastern Mediterranean. 6th-7th century AD. Bronze. Diameter 9.8 cm. Small openwork disk in the center of which is a cross with flared arms with large finials at the corners. It was apparently part of a hanging choros with attached chains and supporting the disk of a polykandelon. On the one side of the disk is engraved a small cross at the top. A circular Greek inscription is poorly preserved and illegible. Along the central axis at the top and bottom of the disk are loops for the attachment of chains.

11. Eulogia icon lamp[35] (Figure 5.1). 6th century AD. Syria. Silver 960º. Height 8.8 cm; diameter 9.7 cm. Central area, near the 'Temple with Reliquary.' Excavations of K. K. Kostsyushko-Valyuzhinich, 1904. A hanging icon lamp on a ring foot, round body, wide flaring neck and a slightly outturned rim with three loops for the attachment of chains (several links are preserved). On the lower part of the icon lamp there are round medallions with relief images of a youthful Christ, the apostles Peter and Paul, and between them there are stylized lilies. The neck is decorated with a border of beads and vertical leaves. The rim is decorated with an egg-and-dart pattern.

12. Icon lamp[36] (Figure 5.2). 6th-7th century AD. Byzantium. Bronze, casting. Diameter 8 cm; height 7.3 cm; length of chains 24.7 cm. Western necropolis, script. Excavations of R. Ch. Loeper, 1914. Hanging icon lamp in the form of a handleless cantharos on a short stem, with a vertical rim and three vertical loops for the attachment of chains, two of which are composed of small links and the third consisting of long segments of wire (reconstructed).

13. Icon lamp[37] (Figure 6.1). 13th century AD. Eastern Mediterranean. Bronze. Height 6.3 cm. Diameter in a mouth: 11.6 cm. Northern area, block X. Excavations of S. G. Ryzhov, 1987. Hanging icon lamp in the form of a bowl on a low ring foot, with flaring walls divided into vertically-carinated segments separated by thick double vertical ridges. The rim is decorated with zones of reticulate ornament. There is one vertical loop for the attachment of a chain.

14. Icon lamp[38] (Figure 6.2). Eastern Mediterranean. 12th century AD. Bronze. Height 9.4 cm; diameter 12.2 cm. Length of chains 34.8 cm. Port area, block 1. Excavations of I. A. Antonova, 1963-4. Hanging icon lamp in the form of a relief bowl on a ring foot, widening toward the rim, from which rise three vertical loops for the attachment of three chains, each of which includes a cross with finials and two small flat disks decorated with four round openings arranged to suggest a cross.

15. Icon lamp[39] (Figure 6.4). Eastern Mediterranean. 13th century AD. Height 5.7 cm; diameter 13 cm. Length of chain 37 cm. Northern area, block VIII. Excavations of S. G. Ryzhov, 1985. Hanging icon lamp in the form of a low bowl on a wide ring foot with plain rounded walls and a wide mouth. From the rim rise three vertical loops for the attachment of chains. The upper edges of the semicircular loops are serrated. The external surface is decorated with a series of engraved horizontal bands.

16. Icon lamp[40] (Figure 6.3). Byzantine period. Cherson (?). Bronze. Diameter 10 cm. Height 7.9 cm. Excavations before 1888. Hanging icon lamp on a short foot, with nine vertical facets. At the top edge are three openings for hanging.

17. Lamna[41] (Figure 7.2,3). Eastern Mediterranean. 11th-13th centuries AD. Bronze. Length 144.3 cm. Central area, block XXXVI. Excavations of T. Yu. Yashaeva, S. G. Ryzhov, J. C. Carter, 2003. Fragments of a lamna in the form of a bronze band with a central arch. Along the upper surface are arranged concave horizontal disks in which a set prickets for candles (nine are preserved). The lamna was attached to a wall by means of two supports (only the right is preserved). One end of the support takes the form of a hand clasping a rod on which is set a rectangular bracket; the other end, separated by a transverse disk, consists of a rectangular-sectioned slightly pointed dowel (the left arm is lost).

18. Lamna supports[42] (Figure 7.4). 11th century AD. Bronze. Length 35.5 cm. Central area, block XXXVI. Excavations of T. Yu. Yashaeva, S. G. Ryzhov, J. C. Carter, 2003. One end takes the form of a fist clasping a rod topped with a triangular bracket (one side is lost); the other end, separated by a transverse disk, consists of a rectangular-sectioned dowel.

19. Lamna[43] (Figure 7.5). Eastern Mediterranean. 13th century AD. Bronze. Length 150 cm. Chapel on the steep east

[32] Skubetov 1910, 42-47, fig. 1.
[33] Skubetov 1910, 42-47, fig. 9.
[34] Yashaeva *et al.* 2011, 506, no. 164.
[35] Bank 1966, no. 86; Bank 1985, 284, pl. 81; Bank, Popova 1977, vol. I, 111, no. 152; Zaseckaja 1979, 162-163, fig. 83; Guarducci 1983, 454, pl. X, no. 18; Mango 1986, 102-103, 156-157; Zalesskaya 1997, 18, fig. 21; 2006/3, 66–67, no. 35; Piguet-Panayotova 1998, 44, no. 5.
[36] Golofast *et al.* 1991, 33, no. 18; Yashaeva *et al.* 2011, 504, no. 160.
[37] Golofast *et al.* 1991, 209, no. 224; Ryzhov 2001, 305, fig. 9.
[38] Golofast *et al.* 1991, 33, no. 18; Yashaeva *et al.* 2011, 505, no. 161.
[39] Yashaeva *et al.* 2011, 506, no. 163.
[40] Unpublished.
[41] Yashaeva *et al.* 2011, 504, no. 168.
[42] Yashaeva *et al.* 2011, 504, no. 169.
[43] Unpublished.

bank. Excavations by K.K. Kostsyushko-Valyuzhinich. 1891. Fragments of a lamna in the form of a bronze band. Along the upper surface are arranged concave horizontal disks in which a set prickets for candles (ten are preserved).

20. Candelabrum[44] (Figure 8.1). Cherson. Iron. Height 96 cm. The western part of the agora. Excavations by K.K. Kostsyushko-Valyuzhinich, 1891. It is a massive forged four-sided rod with a double circle at the top and three legs at the bottom.

21. Candelabrum[45] (Figure 8.2). Cherson. Iron. Height 107 cm. Western District. Store in a residential area. The excavation of Kostsyushko-Valyuzhinich, 1889. It is a massive forged rod, four-sided in cross-section, resting on a round stand. The upper part is lost.

22. An icon lamp of mixed type[46] (Figure 9.1). Eastern Mediterranean. 13th century AD. Bronze. Diameter 26x26 cm; diameter of the rod 1 cm. Central area. Excavations by K.K. Kostsyushko-Valyuzhinich, 1903. It is made of a massive rod, round in cross-section, in the form of an equilateral cross. At the ends of the arms there are loops for the attachment of chains, underneath there is a hook for the suspension of a polykandelon or a single icon lamp. On each arm of the cross are pairs of horizontal convex saucers in which are set prickets for candles. One arm and four saucers are lost.

23. Bracket[47] (Figure 9.4). Eastern Mediterranean. 10th-12th century AD. Bronze, iron. Length 41 cm. Diameter of the rod 1.53 cm. Central area. Agora. Excavations of K. K. Kostsyushko-Valyuzhinich, 1902. Bracket for the suspension of a polykandelon in a shape of a curving branch, round in cross-section, crowned with a flowering cross with barred arms. At the upper end there is a hook for the suspension, at the bottom end there is a soldered, pointed iron spike for the attachment to the wall.

24. Chain[48] (Figure 9.2,3). Eastern Mediterranean. 13th-14th centuries AD. Bronze. Total length 209 cm; width 2.5 cm; diameter of disk 8 cm. Excavations of K. K. Kostsyushko-Valyuzhinich, 19th century. Cast chain of the lamna composed of ten wide smooth links fastened with hinges; between the links is placed a round disk with umbo decorated with engraved concentric circles. At the lower end is the hook for the suspension.

25. Chains[49] (Figure 10.1). Cherson. Early Byzantine period. Copper. Length 40 cm. From the excavations of K.K. Kostsyushko-Valyuzhinich. Three chains on an extended hook for the suspension of a polykandelon. They are composed of two links: the top link is made of a long straight wire, the bottom one is made of a short braided chain with a hook at the bottom end.

26. Chains[50] (Figure 10.2). Cherson. Early Byzantine period. Copper. Length 25 cm. From the excavations of K.K. Kostsyushko-Valyuzhinich. Three chains for the suspension of an icon lamp on a short hook. They are composed of two links, made of thin straight wire with a hook for the suspension at the lower end.

27. Chains[51] (Figure 10.3). Cherson. Early Byzantine period. Copper. Length 31 cm. From the excavations of K.K. Kostsyushko-Valyuzhinich. Three chains for the suspension of an icon lamp on a short hook. They are composed of two links, made of straight wire with a hook for the suspension at the lower end.

28. Chains[52] (Figure 10.4). Cherson. Early Byzantine period. Copper. Length 25.5 cm. From the excavations of K. K. Kostsyushko-Valyuzhinich. Three chains for the suspension of an icon lamp on a long hook. They are composed of one link, made of straight wire with a hook at the lower end and fragments of a glass icon lamp.

29. Chains[53] (Figure 10.5). Cherson. Early Byzantine period. Copper. Length 38 cm. From the excavations of K. K. Kostsyushko-Valyuzhinich. Three chains on an extended hook for the suspension of a polykandelon. They are composed of three links: two upper links interconnected by four-armed crosses with slightly flared ends are made of long straight wire, the lower link is made of short braided wire with a hook.

30. Chains[54] (Figure 10.6). Eastern Mediterranean. 12th-13th centuries AD. Copper. The preserved length is 42 cm. Northern region of Cherson. Residential manor, room 4. Excavations of G. D. Belov, 1932. Three openwork chains composed of short wide links connected by hinges. The upper ends of the chains are attached to a six-blade holder; at the lower ends there is a hook for the suspension. The chains are decorated with four-armed crosses with barred arms and stylized six-pointed stars.

31. Chain[55] (Figure 10.7). (Reconstruction). Eastern Mediterranean. Early Byzantine period. Copper. Preserved length 7 cm. Main street. Restoration work of S.G. Ryzhov, 2010. A fragment of an openwork chain composed of short wide links connected by hinges. The chain is decorated with equal-armed crosses with curved arms within a circle.

32. Chain[56] (Figure 10.8). (Reconstruction). Eastern Mediterranean. Early Byzantine period. Copper. Preserved length 20 cm. Main street. Restoration work of S.G. Ryzhov, 2010. A fragment of an openwork chain composed of short wide links connected by hinges. The chain is decorated with four-armed crosses with barred arms and stylized six-pointed stars.

33. Candleholder[57] (Figure 11.1). Eastern Mediterranean. Found in the 13th century layer. Bronze. Diameter 71 mm.

[44] Kostsyushko-Valyuzhinich 1892, 4.
[45] Kostsyushko-Valyuzhinich 1892, 14.
[46] Kostsyushko-Valyuzhinich 1905, 80-81, fig. 3.
[47] Yashaeva, D. *et al.* 2011, 509, no. 171.
[48] Unpublished.
[49] Unpublished.
[50] Unpublished.
[51] Unpublished.
[52] Unpublished.
[53] Unpublished.
[54] Bank, Popova 1977, no. 572; Belov 1941, 245, fig. 77.
[55] Unpublished.
[56] Unpublished.
[57] Golofast, R. *et al.* 1991, 210, no. 225; Ryzhov 2001, 308, fig. 12 a.

Height: 4.8 cm. Northern area, block X-a, thirteenth-century fire layer. Excavations of S. G. Ryzhov, 1988 AD. A massive six-armed candleholder with a hole pierced at the top and suspension loops for chains at the bottom of each arm. The blade had a sharp rib with a protrusion at the end in the form of small cones. The upper suspension loop takes the form of a polyhedron.

34. Candleholder[58] (Figure 11.2). Eastern Mediterranean. Bronze. Diameter 5.6 cm, height 5.5 cm. Excavations by K.K. Kostsyushko-Valyuzhinich in late 19th century AD. Six-armed candleholder with suspension loop at the top and suspension loops for chains at the bottom of each arm (four are preserved). Each arm had a smooth edge with a rounded protrusion at the end.

35. Candleholder[59] (Figure 11.3). Eastern Mediterranean. Bronze. Diameter 5.5 cm, height 4.5 cm. Central area, block XXXVI. Storehouse of obsolete bronze lamps. Excavations T. Yu. Yashaeva, J. Carter, S. G. Ryzhov, 2003. Six-armed candleholder with suspension loop at the top and suspension loops for chains at the bottom of each arm.

36. Candleholder[60] (Figure 11.4). Eastern Mediterranean. Bronze. Diameter 4.5 cm; Height 4.5 cm. Northern district of Cherson. Residential manor, room 4. Excavations of G. D. Belov, 1932. Six-armed candleholder with suspension loop at the top and suspension loops for chains at the bottom of each arm.

37. Candleholder[61] (Figure 11.5). Eastern Mediterranean. Bronze. Height 4.5 cm; the greatest distance between the arms is 4 cm. Northeast district, near the 'Underground Temple'. Excavations of R. Ch. Loeper in 1909. Three-armed candleholder with suspension loop at the top and suspension loops for chains at the bottom of each arm (fragments of chains are preserved). The ends of the arms are pointed and bent upwards, between them are short protrusions.

Bibliography

Arkhiv Gosudarstvennogo muzeia-zapovednika 'Khersones Tavricheskii' 2004. 3677, 18-19.

Atanasov, G. G. 2013. Bronzova ruka – drzhatel na masiven sveshchnik (lamna, kozmetaritsiia?) ot srednovekovniia katedralen khram na Dristr (Silistra). *Preslav* 7, 132-145.

Bank, A. V. 1966. *Vizantiiskoe iskusstvo v sobraniiakh Sovetskogo Soiuza.* Moscow: Sovetskii khudozhnik.

Bank, A. 1985. *Byzantine Art in the Collection of Soviet Museums.* Leningrad: Aurora.

Bank, A. V., Popova, O. S. 1977. *Iskusstvo Vizantii v sobraniiakh SSSR: katalog vystavki.* Moscow: Sovetskii khudozhnik.

Belov, G. D. 1941. Raskopki v severnoy chasti Khersonesa v 1931-1933 godakh. *Materialy i issledovaniia po arkheologii* SSSR 4, 202-267.

Bouras, L. 1982. Byzantine Lighting Devices. *Jahrbuch der Österreichischen Byzantinistik* 32/3, 479-491.

Effenberger, A., Marsak, B., Zaseckaja, V. 1979. *Spätantike und Frühbyzantinische Silbergefässe aus der Staatlichen Ermitage.* Leningrad–Berlin: Staatliche Museen zu Berlin.

Frings, J. (ed.) 2010. *Byzanz. Pracht und Alltag. Kunst- und Ausstellungshalle der Bundesrepublik Deutschland, Bonn 26. Februar bis 13. Juni 2010: Katalog.* München: Hirmer.

Golofast, L. A. 2001. Steklo rannevizantiiskogo Khersonesa. *Materialy po arkheologii, istorii i etnografii Tavrii* 8, 97-260.

Golofast, L. 2009. Early Byzantine Glass from Tauric Chersonesos, in E. Lafli (ed.) *Late Antique / Early Byzantine Glass in the Eastern Mediterranean.* Izmir, 301-335.

Golofast, L. A., Romanchuk, A. I., Ryzhov, S. G., Antonova, I. A. 1991. *Vizantiiskii Kherson: katalog vystavki.* Moscow: Nauka.

Guarducci, M. 1983. *Scritti scelti sulla religione greca e romana e sul Christianesimo.* Leiden: Brill.

Kostsiushko-Valiuzhinich, K. K. 1892. Proizvodstvo arkheologicheskikh raskopok v Khersonese. *Otchet Imperatorskoi arkheologicheskoi komissii za 1889 g.*, 13-15.

Kostsiushko-Valiuzhinich, K. K. 1893. Proizvodstvo arkheologicheskikh raskopok v Khersonese. *Otchet Imperatorskoi arkheologicheskoi komissii za 1891 g.*, 2-20.

Kostsiushko-Valiuzhinich, K. K., 1905. Otchet o raskopkakh v Khersonese Tavricheskom v 1903 g. *Izvestiia Imperatorskoi arkheologicheskoi komissii* 16, 37-110.

Mundell Mango, M. 1986. *Silver from Early Byzantium. The Kaper Koraon and Related Treasures.* Baltimore: Trustees of the Walters Art Gallery.

Papanikola-Bakirtzi, D. 2002. *Everyday Life in Byzantium.* Athens: Kapon Editions.

Piguet-Panayotova, D. 1998. Three hexagonal decorated silver Censers and their artistic Environment. *Münchener Jahrbuch der Kunst* 3, 49, 7-34.

Ryzhov, S. G. 2001. Srednevekovye zhilye kvartaly X–XIII vv. v Severnom raione Khersonesa. *Materialy po arkheologii, istorii i etnografii Tavrii* 8, 290-311.

Novyi Bogoslov, S.. 2011. *Tvoreniia. Slova i gimny,* vol. 1. Moscow: Sibirskaia blagozvonnitsa.

Simeon Fessalonikiiskii 1857. *Pisaniia sviatykh ottsov i uchitelei Tserkvi, otnosiashchiesia k istolkovaniiu pravoslavnogo bogosluzheniia.* T. 3. Sochineniia blazhennogo Simeona, arkhiepiskopa Fessalonikiiskogo. St.-Petersburg.

Skubetov, M. I. 1910. Drevnekhristianskie bronzovye lampadnye podveski, naidennye pri raskopkakh v Khersonese. *Izvestiia Tavricheskoi uchenoi arkhivnoi komissii* 44, 42-44.

Yashaeva, T. Iu., Denisova, E. A., Gin'kut, N. V., Zalesskaia, V. N., Zhuravlev, D. V. 2011. *Nasledie vizantiiskogo Khersona / The Legacy of Byzantine Cherson.* Sevastopol', Ostin: Teleskop; Institute of Classical Archaeology.

Zalesskaia, V. N. 1997. *Prikladnoe iskusstvo Vizantii IV-XII vv.: opyt atributsii.* St.-Petersburg: Gosudarstvennyi Ermitazh.

Zalesskaia, V. N. 2006. *Pamiatniki vizantiiskogo prikladnogo iskusstva IV–VII vv. Gosudarstvennyi Ermitazh. Katalog kollektsii.* St.-Petersburg: Gosudarstvennyi Ermitazh.

[58] Unpublished.
[59] Unpublished.
[60] Bank, Popova 1977, 95; no. 572. Belov 1941, 245, fig. 77.
[61] Unpublished.

Problems of Methodology in the Dating of Byzantine Polykandela: The Benaki Museum Examples

Mara Verykokou
Assistant curator, Byzantine and Post Byzantine collection, Benaki Museum

Abstract: The purpose of this study is to address the issue of dating Byzantine *polykandela* (chandeliers) using the example of two objects from the Benaki Museum Byzantine collection. The question of dating will require considerable attention as the polykandela belong to a category of brass objects which comes into use in the 6th century and continue in use to the end of the 13th century. The particular objects were published in 2008 in the scientific journal of the Benaki Museum. At the end of the publication, the problem of chronology was quite evident. In this paper we will not go over the results of previous research but instead will try to establish the fact that the dating of polykandela with no inscriptions or stamps is a much more difficult process than was perceived until recently in the light of new publications.

Keywords: *polykandela*, lighting devices, Byzantine metalwork, Islamic metalwork, leaded bronze, chronology.

In the summer of 2008 Stylianos Liangis presented two Byzantine brass *polykandela* (chandeliers) to the Benaki Museum in memory of Iphigeneia Liangi. They were bought at auction from the art dealers Royal-Athena Galleries and Bonhams, who had dated them to the 6th–7th centuries AD. They entered the Museum's Byzantine collection with the inventory numbers 44724 and 44725 respectively and were transferred to the museum's Metal Conservation Laboratory, where they were conserved before being displayed in the Museum's permanent exhibition. A study was then undertaken with the aim of classifying them according to their construction and decoration in the relevant categories of lighting apparatus, using the wealth of comparative material available.[1] As the research progressed, some queries arose about the dating assigned to these objects by the two auction houses.

Why did the two auction houses date these particular objects to the 6th to 7th century? On what were they basing this relatively precise dating? How can an object with no stamps or inscriptions and of unknown provenance (as is the case with most of these types of object) be dated? What should be the basis for our research?

Broadly speaking it is not possible as yet to give a definitive answer on the issue of dating polykandela. The aim of this paper is not to throw doubt on any particular earlier studies and publications relating to the dating of lighting devices, but to set out the problems and questions that arose in the course of our research into the dating of the two polykandela belonging to the Byzantine Collection of the Benaki Museum.

In analysing the problems involved in dating these two objects I shall present other relevant examples, some comparisons between Byzantine and Islamic lighting devices, the chemical analysis of the two Benaki Museum polykandela and some remarks on trade in that period.

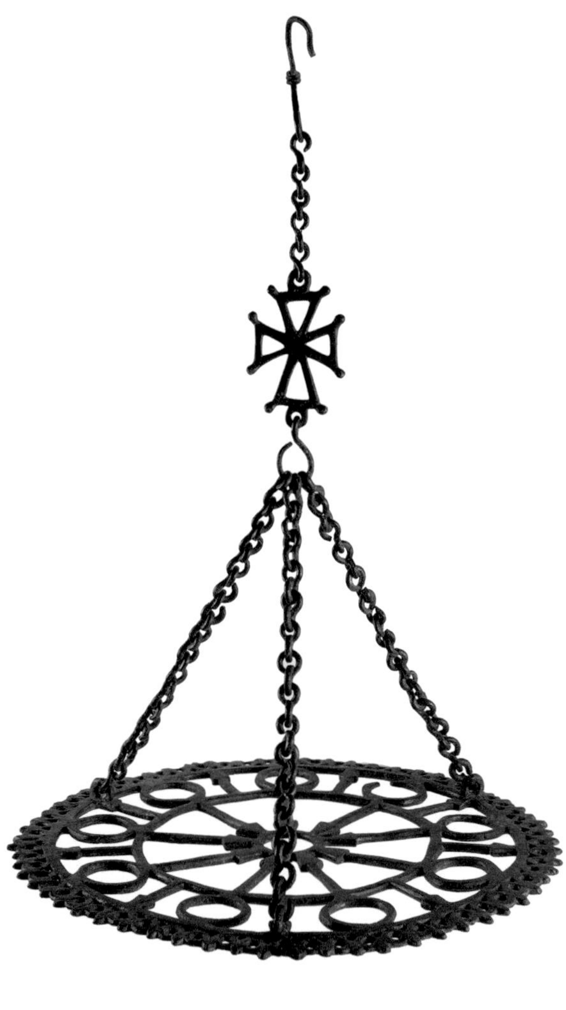

Figure 1. Polykandelon A, Benaki Museum, Inv. No. 44724. Photography by L. Kourgiantakis.

[1] Verykokou 2008.

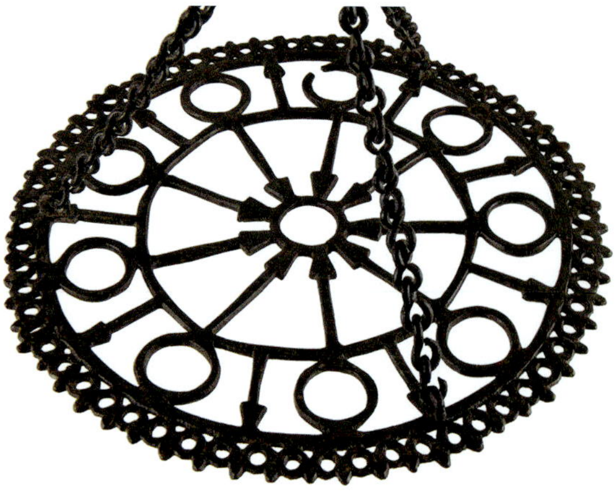

Figure 2. Detail of Polykandelon, Inv.No. 44724.

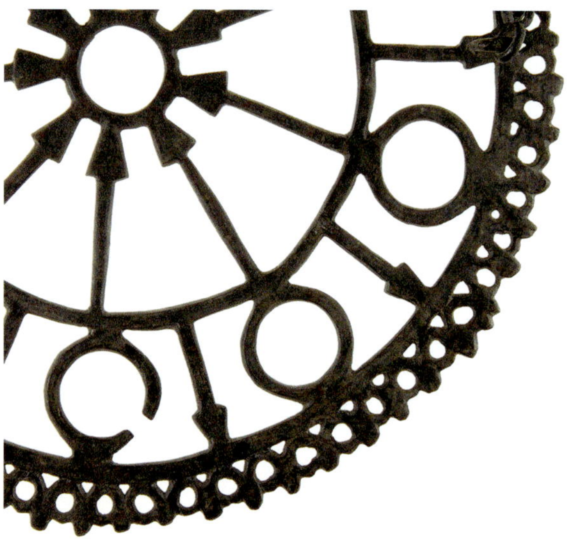

Figure 3. Detail of polykandelon A, Inv. No. 44724.

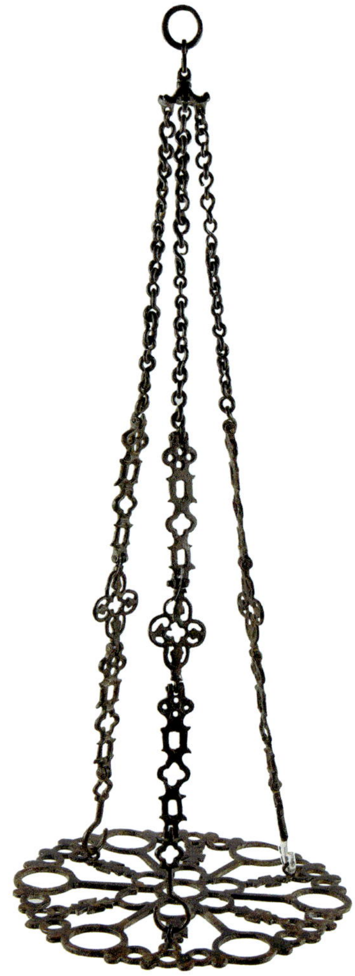

Figure 4. Polykandelon B, Benaki Museum, Inv. No. 44725. Photography by L. Kourgiantakis.

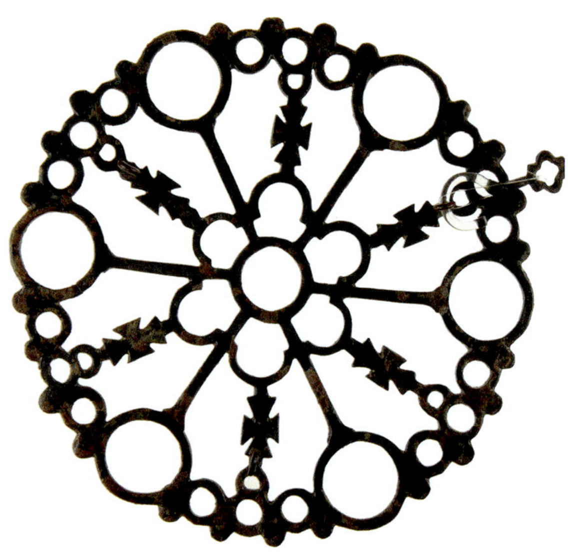

Figure 5. Polykandelon B, Benaki Museum, Inv. No. 44725. Photography by L. Kourgiantakis.

The first of the brass polykandela to be examined here (inv. no. 44724, hereafter polykandelon A) is cast and consists of a pierced disk, three chains and a suspension hook. The chain is 45 cm long and the disk has a diameter of 30.9 cm and has nine openings to hold glass lamps (Figures 1, 2, 3). On the upper surface of the disk are three welded rings which serve to attach the chains. The three chains end in a pierced Maltese cross. The chain which attaches the polykandelon to the suspension hook emerges from the top of the cross. The second brass lighting device (inv. no. 44725, hereafter polykandelon B) is also cast and consists of a pierced disk with three chains, ending in a suspension hook. It is 67.5 cm high and the disk is 27.5 cm in diameter. The disk has six holes arranged around the circumference to hold glass lamps. Three rings are soldered on to its upper surface and serve to attach the chain hooks (Figures 4, 5, 6, 7).

Extensive research in the Greek and foreign literature uncovered comparative material from well-known collections

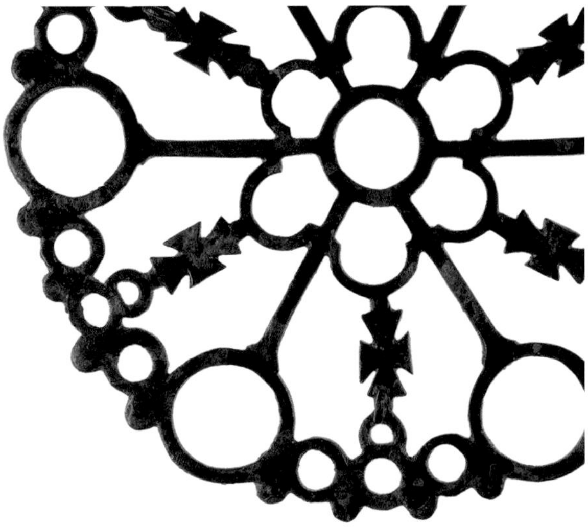

Figure 6. Detail of polykandelon B, Inv. No. 44725.

Figure 7. Detail of the chain of polykandelon B, Inv. No. 44725

and museums, which helped to classify our two objects. These polykandela had been classified on the basis of their decorative motifs, just as the two lighting devices in the Benaki Museum had originally been dated on the basis of their decoration. Firstly, polykandelon A (44724) can be grouped together with corresponding objects in the Dumbarton Oaks Collection, the British Museum, the Louvre, the Averoff Collection, from the Corinth excavations, the Archaeological Museum in Istanbul, the Malcove Collection, Kairouan Museum and finally the Archaeological Museum of Granada.[2] In its turn polykandelon B (44725) can be included in the same group as polykandela from the Dumbarton Oaks Collection, the Hermitage Museum, the Museum of Cairo, the Metropolitan Museum, the British Museum, the Walters Art Gallery, the Agios Titos Monastery in Gortyn, Crete and the Kairouan Museum.[3]

Its similarities with an example in the Louvre and another similar one in the Averoff Collection and with a polykandelon from Kairouan, all of which are precisely dated, would suggest a date in the 9th t0 11th centuries for polykandelon A.[4] The same applies to polykandelon B, which can be compared with one in the British Museum as well as with a large group of lighting devices from the Monastery of Agios Titos in Crete and the Kairouan polykandela.[5] To some extent the dating of the latter to the 9th-11th centuries is supported by comparative material from the excavations at Ravy and Beycesultan, and in particular by parts of chains used to suspend polykandela, which are dated to this period by archaeologists.[6]

The hitherto generally accepted literature and comparisons dated these lighting devices to the 6th to 7th centuries. Authors and scholars such as Ross, Buckton, Nesbitt, Metzger, Atasoy and Banck have published similar polykandela or parts of Byzantine lighting equipment (to which I have referred in an earlier article),[7] basing themselves in turn on the Lampsakos and the Kumluca Treasures for their dating.[8] The silver polykandela discovered in the excavations at Kumluca in Asia Minor have dedicatory inscriptions with the name of the patron who commissioned them, Bishop Eutychianos, and the control stamps of officials from the end of the reign of Justinian I (550-565 AD), make the dating of all the finds relatively easy. Similarly the Lampsakos Treasure has been dated to 565-578, because it bears stamps from the reign of the Emperor Justin II. Part of the Kumluca Treasure was made in a workshop in Constantinople out of extremely high quality silver, while the rest came from a local workshop, using the precious metal available in the area. The silver polykandela from Kumluca are examples of the amazing craftsmanship of the period. These treasures provide a dating reference point for similar objects and this is probably why most scholars have based their arguments on these objectts when dating the brass polykandela they published, assigning them in turn to the 6th-7th century. Lack of documentation as to the provenance of the brass polykandela or incomplete recording of the history of these objects only exacerbated this tendency.

[2] Ross 1962, 40-42; Lafontaine-Dosogne 1982, 161; Metzger 1974, 319-322, nos. 4-5; *Συλλογές Ευάγγελου Αβέρωφ: Ταξιδεύοντας στον χρόνο. 2000,* 90, fig.153 (V. Papadopoulou); Davidson 1952, 128, no. 860, pl. 63; Atasoy 2005, 103, no. 172; Campbell 1985, 58, no. 58; Marçais and Poinssot 1952, 448, fig. 103; Dodds 1992, 207 (M. C. Perez).

[3] Ross 1962, 40-42; Bank 1966, 335, no. 17; Strzygowski 1904, 297-98, nos. 9154-9157; Heilbrunn Timeline of Art History; Bouras and Parani 2008, 92, no. 28; Xanthopoulou 1998, 103-119; Xanthopoulou 1997, vol. I, 117, vol. II, 363, LU 5008, pls 262, 378.

[4] Metzger 1974, nos. 4-5; *Συλλογές Ευάγγελου Αβέρωφ* 2000, 90, fig. 153 (V. Papadopoulou); Marçais and Poinssot 1952, 449-50, figs 104, 105; Ward 1993, 68.

[5] Xanthopoulou 1997, vol I, 121-22, vol II, 380, LU 5.012; Xanthopoulou 1998, 103-119; Marçais and Poinssot 1952, 440-459, figs 104-105.

[6] Wright 2000; Rice 1955.

[7] Verykokou 2008.

[8] Ševčenko, Boyd and Mango 1993, 39-56; Ross 1962, 41.

The Benaki Museum's brass polykandela and the comparative material used from the collections of the British Museum, the Dumbarton Oaks Collection, the Louvre, the Hermitage, the Archaeological Museum of Istanbul, the Averoff Collection and the Corinth excavations have neither stamps nor inscriptions and in most cases no known provenance. It was deemed essential that Despina Kotzamani, conservator in the Metals Conservation Laboratory, should carry out the necessary analyses relating to the composition of the raw material on the Benaki Museum's polykandela to establish the manufacturing process and the metal used, and to date the objects on the basis of the alloys involved. The results produced some surprises in respect of the composition of the metal and an initial probable dating for these objects of the 9th-11th century.

First of all, the laboratory tests confirmed that the polykandela were made of leaded bronze. The octagonal parts of the suspension chains on both objects had a different metal composition, and consisted of a lead tetramer, a quaternary alloy, lead and copper. The presence of different ingredients in the alloys of the disks and the chains in both polykandela suggests that the craftsmen either chose them for their technical characteristics or simply used what was on hand. And the presence of arsenic in part of the finely wrought chain of polykandelon B would seem to confirm this. An initial assessment of the analyses showed that the use of leaded bronze is not consistent with dating the polykandela to the 9th-11th centuries, since at that time brass predominated. To produce the alloy found in this case a supply of tin is needed, which was very hard to find after the 4th and 5th centuries. However, though the practice of using quaternary alloys only emerged in the 3rd century, such products dominated the market for centuries to come. The presence of bronze rather than brass also suggests that the craftsman wanted to use a more precious metal to make these particular artefacts, regardless of whether it was original or recycled. The conclusions drawn from the analyses were published in the Benaki Museum Journal.[9]

Some scholars, wishing to study this particular category of objects, have turned to Islamic polykandela. Maria Xanthopoulou addressed the subject, when she made an initial examination of the lighting disks from the Monastery of Agios Titos in Crete and analysed the various types of polykandela known to date in her recently published doctoral thesis.[10] Basing herself on publications and studies such as those by Ross, Buckton, Nesbitt,[11] Metzger, Atasoy, Banck and Davidson, as well as the examples from Agios Titos and Kairouan, she proceeded to redate most of these objects from the Early Christian to the Middle Byzantine period. And having assembled a large array of such objects, she then grouped these lighting devices according to typology. She concluded that in some cases the polykandela could be found in the same form from the 6th to the 13th century. This could be put another way. Another reason why most scholars have hitherto dated most polykandela to the 6th-7th century is that these objects are durable and can remain in use for very long periods of time. Xanthopoulou, however, redated a large number of polykandela not just on the basis of the decoration on the disks from the Monastery of Agios Titos, but taking into account the Arab polykandela which have considerable similarities with the corresponding Byzantine ones, which were thought to be Early Christian and not Middle Byzantine.

It is interesting to note the similarities between 'Byzantine' and 'Islamic' polykandela. Let us take, for example the decorative motifs, both from the disks of the two polykandela belonging to the Benaki Museum and on the suspension chain of polyakandelon B. The decorative motifs on polykandelon A show similarities not only with the corresponding lighting devices from the Louvre, the Averoff Collection and the polykandelon from the Great Mosque of Medina Elvira, Granada, but also with a small section of a polykandelon disk from Kairouan.[12] Apart from the cross motif the decoration on the disk of polykandelon B is identical with that of two in particuar polykandela from Kairouan. The finely wrought chain of this same object has certain things in common with parts of the chains found at Ravy and Beycesultan.[13] It is clear that the casting technique which, according to the analyses carried out on the Benaki Museum objects, was used in the manufacture of both these examples, allowed the craftsman, whatever his religion, to create and adapt an object to the requirements of the patron, whether Christian or Moslem. This notion is endorsed by the fact that metalworking workshops were found in places where trade between the two religions was flourishing.

Between the 8th and the 11th centuries trade between the Byzantine Empire, Venice, the Fatimids and Egypt was thriving, with a break in AD 992, when the Emperors Basil II and Constantine VIII reduced the customs dues that the Venetians had hitherto paid the Byzantines.[14] And later, taking advantage of Alexios I's *chrysobull* (a pendent seal made of gold) issued in 1082, the Venetians criss-crossed Byzantine territorial waters without paying any taxes. Textiles, oil, wine, dairy products, perfumes and pharmaceutical goods, myrrh and precious stones were transported in Venetian ships all over the Mediterranean. Brass utensils, such as jugs, bowls and polykandela complete the list of products which traversed the Mediterranean and were traded on the coastline and inland.[15] And we cannot exclude the possibility that the latter were brought as gifts for the nobility and government officials. Indeed the example of the statue of St Theodore of Amaseia in St Mark's Square in Venice, which has a polykandelon disk as a halo like the polykandela from Kairouan confirms the exchanges made between the two cultures.[16] At this point it should be noted that metalworking workshops are not only found on the Mediterranean coast, but also inland as is apparent from the finds from the excavations at Amorium, Sardis and Beycesultan. In these areas lots of different kinds of metal objects have been found in places that can be identified as shops selling metalwork or metalworking workshops.

In trying to date the Benaki Museum polykandela we have been able to identify similar Byzantine and Islamic examples with which to compare them. It is also important to note

[9] Kotzamani and Aslani 2008, 107-114.
[10] Xanthopoulou 1997.
[11] Nesbitt *et al.* 1988, 53.

[12] Metzger 1974, vol 24, nos. 4-5; *Συλλογές Ευάγγελου Αβέρωφ* 2000, 90, fig. 153 (V. Papadopoulou); Marçais and Poinssot 1952, 448, fig. 103.
[13] Wright 2000; Rice 1955.
[14] Jacoby 2009.
[15] Mango 2009.
[16] Pincus 1992; Fortini-Brown 1997, 18-19.

that a large number of objects have been redated in the light of recent excavations and publications. Motifs and designs which had previously been described as 'Byzantine' have been shown to be closer to Islamic models, showing that the influence of these cultures on each other was quite strong, not just on the borders of the two empires but in their heartlands. Yet the nub of the problem remains. This study has shown up a lacuna in the current dating methodology and highlighted the fact that, despite efforts to redate the polykandela from the 6th or 7th century to the 9th-11th, some aspects still give cause for doubt. And this makes dating these objects a risky business – at least for the time being.

Bibliography

Atasoy, S. 2005. *Bronze Lamps in the Istanbul Archaeological Museum: An Illustrated Catalogue.* Oxford: Archaeopress, British Archaeological Reports International Series 1436.

Bank, A. V. 1966. *Byzantine Art in the Collections of the USSR. Leningrad.* Moscow: Sovietsky Khudozhnik.

Bouras, L. and Parani, M. G. 2008. *Lighting in Early Byzantium.* Washington, D.C: Dumbarton Oaks Research Library and Collection.

Campbell, S. D. 1985. *The Malcove Collection.* Toronto; Buffalo: University of Toronto Press.

Davidson, G. 1952. *Corinth XII: The Minor Objects.* Princeton: The American School of Classical Studies at Athens.

Dodds, J. D. (ed.) 1992. *Al-Andalus: The art of islamic Spain ; [in conjunction with the Exhibition Al-Andalus: the Art of Islamic Spain, held at the Alhambra, Granada (March 18 - June 7, 1992), and the Metropolitan Museum of Art.* New York, N.Y: Abrams.

Fortini-Brown, P. 1997. *Venice and Antiquity: The Venetian Sense of the Past.* New Haven (Conn.): Yale University Press.

Jacoby, D. 2009. Venetian Commercial Expansion in the Eastern Mediterranean, 8th-11th centuries, in M. M. Mango (ed.) *Byzantine Trade, 4th-12th centuries, Papers of the Thirty- eighth Spring Symposium of Byzantine Studies*, St. John's College, University of Oxford, March 2004, 371-391. Aldershot: Ashgate.

Heilbrunn Timeline of Art History. <https://www.metmuseum.org/toah/works-of-art/2002.483.7/>, viewed 15 November 2018.

Kotzamani, D. and Aslani, I. 2008. Τα χάλκινα πολυκάνδηλα του Μουσείου Μπενάκη, Epimetro. *Mouseio Benaki* 8, 107-115.

Lafontaine-Dosogne, J. 1982. *Splendeur de Byzance: 2 octobre-2 décembre 1982: Musées royaux d'art et d'histoire, Bruxelles.* Exhibition Catalogue. Bruxelles: Musées royaux d'art et d'histoire.

Mango, M. M. 2009, Tracking Byzantine silver and copper metalware, 4th-12th centuries, in M. M. Mango (ed.) *Byzantine Trade, 4th-12th centuries, Papers of the Thirty- eighth Spring Symposium of Byzantine Studies*, St. John's College, University of Oxford, March 2004, 221-236. Aldershot: Ashgate.

Marçais G. and Poinssot, L. 1952. *Objets Kairouanais, IXe au XIIIe siècle; reliures, verreries, cuivres et bronzes, bijoux. Notes et Documents XI-Facs. 2, Direction des Antiquités et Arts.* Tunis: Tournier.

Metzger, C. 1974. Trois lustres cultuels en métal. *Revue du Louvre* 24, 319-322.

Nesbitt, J. W., Spike, J. T. and Koortbojian M. 1988. *Byzantium: The Light in the Age of Darkness.* New York: Ariadne Galleries.

Pincus, D. 1992. Venice and the two Romes: Byzantium and Rome as a Double Heritage in Venetian Cultural Politics. *Artibus et Historiae* vol. 13, no. 26, 101-114.

Rice, D. S. 1955. Studies in Islamic Metal Work V, *Bulletin of the School of Oriental and African Studies, University of London*, 17.2, 206-231.

Ross, M. 1962. *Catalogue of the Byzantine and early medieval antiquities in the Dumbarton Oaks Collection, vol. 1: metalwork, ceramics, glass, glyptics, painting,* Washington, D.C.: Dumbarton Oaks center for Byzantine studies, Trustees for Harvard University.

Ševčenko, I. 1993. The Sion Treasure: The Evidence of the Inscriptions, in S. Boyd and M. M. Mungo (eds) *Ecclesiastical Silver Plate in Sixth-Century Byzantium*, 39-56. Washington, D.C.: Dumbarton Oaks Research Library and Collection.

Strzygowski, J. 1904. *Koptische Kunst. Catalogue général des antiquités égyptiennes du Musée du Caire.* Vienne: A. Holzhausen.

Συλλογές Ευάγγελου Αβέρωφ: Ταξιδεύοντας στον χρόνο. 2000, Exhibition Catalogue, Πινακοθήκη Αβέρωφ, Μέτσοβο, 9 Σεπτεμβρίου 2000-15 Ιανουαρίου 2001. Αθήνα: Ίδρυμα Ευαγγέλου Αβέρωφ-Τοσίτσα.

Verykokou, M. 2008, Καὶ ἔγεινε φῶς. Δύο μεσοβυζαντινά πολυκάνδηλα του Μουσείου Μπενάκη. *Mouseio Benaki* 8, 91-115.

Ward, R. 1993. *Islamic Metalwork.* London: British Museum Press.

Wright, G. R. H. 2000. Some Byzantine bronze objects from Beycesultan. *Anatolian Studies* 50, 159-170.

Xanthopoulou, M. 1997. Les luminaires en bronze et fer aux époques paléochrétienne et byzantine: typologie, technologie, utilisation. PhD dissertation, Université de Paris I Panthéon-Sorbonne.

Xanthopoulou, M. 1998. Le mobilier ecclésiastique métallique de la basilique de Saint-Tite à Gortyne (Crète centrale). *Cahiers Archeologiques* 46, 103-119.

Byzantine Metal Support Fittings for a Candlestick from the Monastery of Chilandar on Mount Athos

Paschalis Androudis
Aristotle University of Thessaloniki

Ioannis Motsianos
Museum of Byzantine Culture, Thessaloniki

Abstract: This paper aims to identify the initial use of two Byzantine metal support fittings found in the Monastery of Chilandar on Mount Athos. These metal supports were re-used as icon carriers. The presence of the two-headed eagle in their decoration is the only clue to enable us to date them in the Palaelogan period (14th century AD) taking into account some parallels in Palaeologan art.

Keywords: Monastery of Chilandar, Mount Athos, candlestick, metal supports, double-headed eagle.

Introduction

In the catholicon of the Monastery of Chilandar on Mount Athos, a few years ago it was reported that there was a pair of small metal objects in the lower frame of the icons of the Virgin and Child (Figures 1, 3, 5) and Saint George (Figures 2, 4, 6).[1] Made of copper alloy, the two objects are almost identical and bear the same decoration, a two-headed eagle surrounded by a circle. Both objects are in very good condition.

Description and decoration of the metal support fittings

The first support fitting of the catholicon of Chilandar (height: 13.4 cm, length: 11.7 cm, total thickness: 4.5 cm), (Figure 3), is a copper alloy blade folded in inverted Π (Figure 5). Below the blade is fixed a cylindrical shaft, which could itself be inserted in a horizontal support. The main face (the obverse) carries in the center a circular medallion, formed by a pair of concentric circles. This medallion, 11.6 cm in diameter, is perforated and decorated with a two-headed eagle (Figure 3). The four corners of the obverse face are also openwork. The stylized bird is represented in the so-called 'heraldic' position, with its wings open. In the center of its body is a hemispherical projection, identical to that seen on similar objects from the Basilica of Saint Titus of Gortyn.[2] The back plate is flat and without decoration. The second support fitting (Figures 2, 4 and 6) is exactly the same as the first one. Both objects appear to have been produced from the same mold.

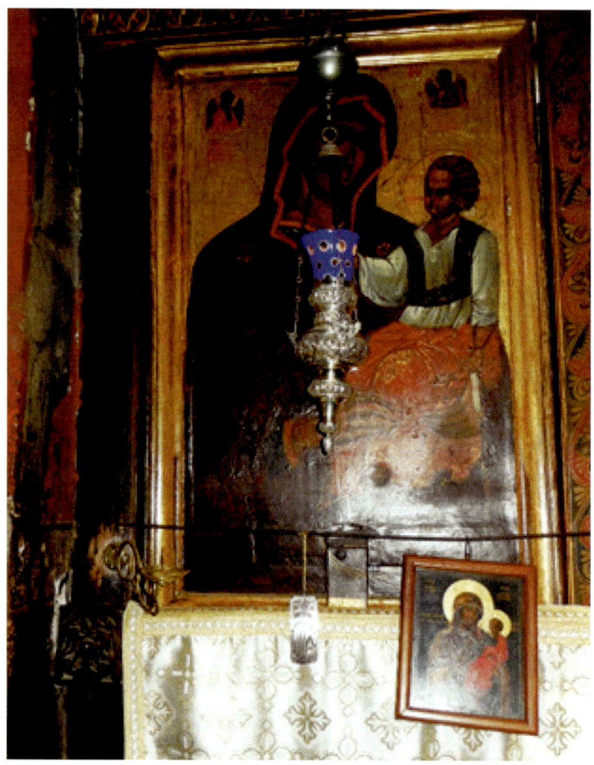

Figure 1. Icon of the Virgin and Child. Monastery of Chilandar on Mount Athos.

The provenance and the conditions of acquisition of these support fittings by the monastery of Chilandar are both unknown to us.

The eagle, which is the main decorative element on the objects, has its heads divergent but springing from a single neck. A curved feather separates the heads from the neck. On each head of the eagle the eye is large and almond-shaped, with the pupil placed in the center. The beaks are slightly curved. The wings, designated by two stylized hanging feathers, are not very wide; they are open, horizontal and perpendicular to the axis of the body. The tips ends curl slightly. The upper part of each wing is decorated with an incised volute, hollowed out either by a chisel or nitric acid. The feathers are well drawn and tend to deviate from each other to give the wings a slightly 'spread' look. The legs are not parallel to the axis of the body, but form with the trunk an angle of about 30°. The

[1] We would like to thank our colleague Petros Kapsoudas who provided recent photographs of the objects.
[2] Xanthopoulou, M. 1998, figs 8, 9.

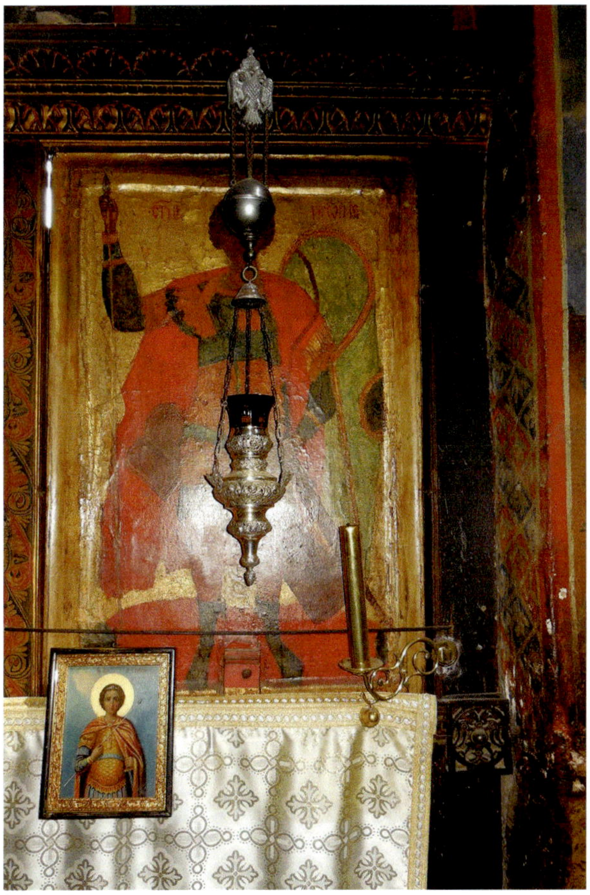

Figure 2. Icon of Saint George. Monastery of Chilandar on Mount Athos.

talons are well delineated and end in two claws. The tail is swollen and elongated, and the lower plume, has a small fan composed of four feathers which flare from an incised oval.

On the dating of the support fittings

The catholicon of the Serbian Monastery of Chilandar[3] is generally dated to the period 1300-1303 or between 1306 and 1311.[4]

We have no clear clues about the dating of the two objects, except what the presence of the two-headed eagle in their decoration might tell us. The double-headed eagle is a fabulous creation, whose oriental origin dates back to ancient times. As a symbol, it plays a preponderant role in pre-Islamic Turkish mythology. In the Middle Ages it is associated with oriental animal art, especially of the time of Seljuk of Rum and Turkoman (Artuqid and Zangid).[5] One example is the two-

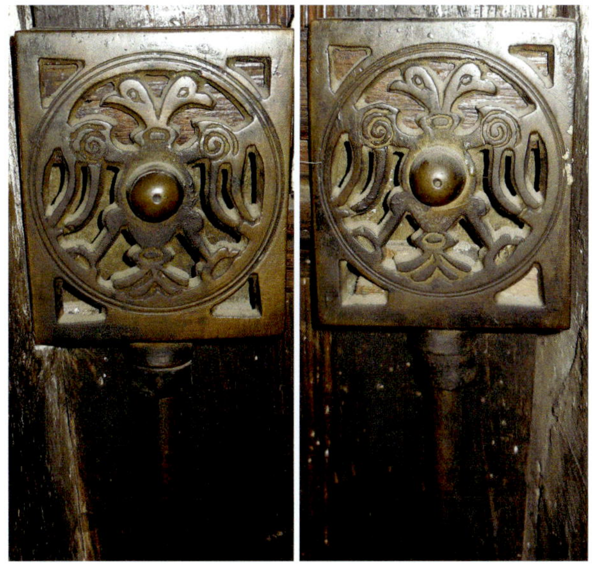

Figures 3, 4. Metal support fittings. Monastery of Chilandar on Mount Athos.

headed eagle of a former column (Eski Kapu) of the Konya Fortress, 1220 AD,[6] and another on the west side of the Çifte Minareli Medrese porch in Erzurum, *c.* 1250 AD.[7]

The appearances of two-headed eagles in the Byzantine world of the 11th century (slab of Beroe, the former Stara-Zagora in Bulgaria[8] and that of Miafarquin of Asia Minor of South-East),[9] are quite rare and undoubtedly related to oriental art. Also rare are the apparitions of the chimerical bird in the following centuries, e.g., in the detail of a thirteenth-century fresco in Prilep, North Macedonia.[10] However it is found in the 12th century in the Romanesque art of Western Europe.[11]

The meaning of the two-headed eagle in Byzantium is a subject that has been the cause of much discussion.[12] Some scholars have considered that the motif arrived in the Byzantine Empire either through the Seljuk Turks, or through the West via the Balkans. The double-headed eagle already appears in the reign of Andronicus II Palaeologus (1282-1328) in details of imperial portraits. Until now, the oldest example, dating from 1294, is found in the fresco that features the founders of the church of the Virgin Pantanassa in Philippiada (Epirus): the

[3] Cf. Nenadović 1997, 59-124; Korać 1998; Marković and Hosteter 1998; Ćurčić 2010, 597, 654-655, 664, 669, 674, 682.

[4] Marković and Hosteter 1998, 201-220.

[5] For the oldest known representations of two-headed eagles among the Seljuk Turks of Rûm and the Artukids, see Androudis 1999. The double-headed eagle of the Seljuks goes back to the beliefs and traditions of Central Asia where he was considered a sacred bird, a protective spirit, a guardian of the heavens, a symbol and power and fertility. Thus, eagle motifs in mosques, fortresses, palaces and Anatolian Turkish caravanserais can be seen as magical, protective presences, totems and symbols of strength. See Roux 1984, 44; Roux 1984b; Roux 1992, 82.

[6] Gierlichs 1996, 192-193, fig. 35.3.

[7] Sitting on top of a palm tree is seen a two-headed rapacious eagle. This eagle is therefore likely to correspond to Turkish-Mongolian mythology.

[8] *La Bulgarie médiévale* 1980, 104-105, fig. 151; Alchermès 1997.

[9] Strzygowski and Van Berchem 1910, 365-366, fig. 317.

[10] On the detail of a fresco depicting the garment of the patron of the ancient catholicon of the Monastery of St Michael the Archangel, see, Korunovski and Dimitrova 2006, 94, fig. 69.

[11] Androudis 2013.

[12] On the image of the two-headed eagle in Byzantium, see Lambros 1909; Bees 1912; Gerola 1934; Soloviev 1935; Spyridakis 1972-73; Fourlas 1980; Fourlas 1984; Chotzakoglou 1996. Also, see Androudis 2012, 33-34, notes 9-14; Androudis 2013b.

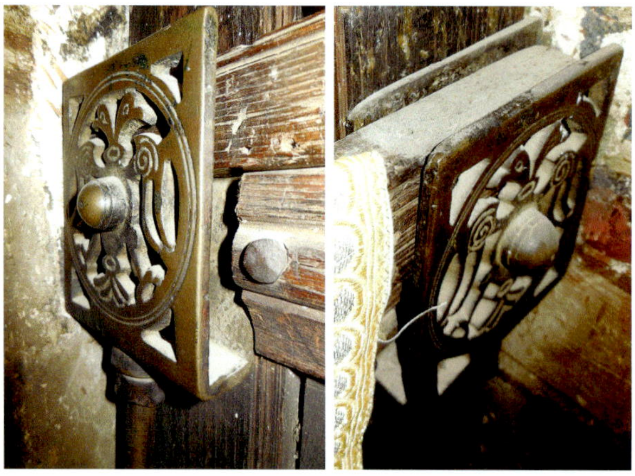

Figures 5, 6. Metal support fittings. Monastery of Chilandar on Mount Athos.

dress of Anna Palaiologina, niece of Michel VIII Palaeologus is dotted with medallions embroidered with pearls containing two-headed eagles.[13] As for the imperial portraits, the oldest certified copy dates back to 1301, in the miniature of the Chrysobulle of Monemvasia with the emperor Andronicus II Palaeologus.[14] Note however, that Balkan representations of the double-headed eagle are noticeably later than the researchers thought just a few decades ago.

As an imperial emblem of Byzantium, the two-headed eagle appears from the end of the 13th century in the clothes of the Greek emperors and their sons. According to the treatise of Pseudo-Kodinos (1347-1348),[15] eagles (with one or two heads) were very fashionable in the Byzantine court of the 14th century. Embroidered on the luxurious stuffs for the use of the imperial court, on the shoes of princes and despots, and on their saddles and tents.[16] They were also on the saddles and tents of the sebastocrators.[17] The author does not mention anywhere whether these eagles had one or two heads, but it may be supposed, since Eastern fabrics or Oriental-inspired items from the 11th-12th century featured especially two-headed eagles and not single-headed eagles.[18]

The iconography of these two-headed eagles is in keeping with the bird's flight. The wings are spread out, the legs and the claws are widely spread. The two necks have no common base; they are distant from each other and graciously turned aside. The heads of the eagle are quite raised. The hooked bills are accentuated, as well as the ellipsoidal eyes. The torso of the eagle is bulging. A node-shaped element marks the junction points with the heads and the tail.

It is notable that in the supports of Chilandar the heads of the eagle are less raised than earlier examples (late 13th-early 14th century), but remain round and a curved pen separates the neck. Also note that the eagle heads of our support fittings are not crowned, an iconographic particularity that appears in Byzantine art of a later date, during the second half of the 14th century.

In addition, the two-headed eagle of the two 'icon carriers' of Chilandar presents, with its presence in a medallion, several affinities with other double-headed eagles which appear enclosed in circles that adorn the luxury garments and ceremonial attire of the members (emperor, despot, Caesar) of the imperial Greek or Serbian families of the time.

As an example, eagles appear in medallions on a garment of Sebastocrator Constantine Palaeologus, brother of Emperor Michael VIII Palaeologus, in his portrait with his wife Irene Comnene Branaina Paleologina in a miniature in a manuscript in the Bodleian Library (Ms Lincoln College, 35, IV, dated between 1328 and 1344).[19] Constantine wears a long tunic in scarlet red (*ροῦχον*)[20] with a gilded decoration and a long brown mantle (*ταμπάριον*)[21] adorned with two-headed eagles, inscribed in circles (*χασδίαι*) and embroidered with gold.

In the Serbian world, the two-headed eagle appears in the art of the royal court at the time of King Milutin, just after his marriage to Simonis, daughter of the Greek emperor Andronicus II (1299).[22]

Additionally, the two-headed eagle of the support fittings of Chilandar reminds us of two-headed eagles which are depicted on the precious fabrics (*podéai*),[23] painted in frescoes of 14th century churches in Byzantine Macedonia and Serbia. Two-headed eagles in circles are also found in the decoration of some chandeliers discovered in Serbia (Decani, Psača).

At Mount Athos there are two-headed eagles, painted in gold, red and blue medallions, in details of the frescoes in the lateral Chapel of Saint Demetrios, the ancient catholicon of the Xenophon monastery, which illustrate *podéai*.[24] Two-headed eagles in medallions are also painted on folio 56r of the Greek Gospel of 1340/1 donated by the Emperor John VI Cantacuzenus (1292-1383) to the Monastery of Vatopedi[25] and also appear in the decoration of the bronze plates that cover the doors of the main entrance of the catholicon of the same monastery.[26]

[13] Vocotopoulos 2007, 61-66, figs 61-62; Vocotopoulos 2008, 73-80.

[14] A golden bull or chrysobull written on the occasion of the donation of the cathedral of Monemvasia in 1301. The eagle is illustrated in gold on the ends of the imperial suppedion. As on seals and coins the imperial portrait is considered as a guarantee of the authenticity and efficiency of the document. Cf. Binon 1938. Kalligas 2009, 365-366 (fig. 1), 367, 370, 373, 375-377.

[15] See Verpeaux 1966, 35-40.

[16] Verpeaux 1966, II, 143.5-145.13. On the despot's insignia, see Failler 1982.

[17] Verpeaux 1966, II, 147.14-148.12.

[18] Androudis 2015.

[19] It is the Typikon of Theodora Synadene for the Convent of the Mother of God Bebaia Elpis in Constantinople. For the portrait, see Spatharakis 1976, 192-193, fig. 134.

[20] This indoor garment, later called *ροῦχον*, was adorned with gold embroidery. See Verpeaux 1966, 143. 2-5, notes 2-3).

[21] The ancient comments of the Pseudo-Kodinos text agree to identify the mantle as *ταμπάριον*. See Verpeaux 1966, 143. 5-6: '... Τὸ ταμπάριον αὐτοῦ κόκκινον μετὰ μαργελλίων ...'.

[22] Babić 1996.

[23] Frolow 1938, 461-504.

[24] See Tsigaridas 1999, 81-102, figs 48, 49.

[25] Vatopédi, skevophylakion, no. 16. See Lamberz 1991, vol. II, 633-636; Lamberz 1996, 161-163; Lamberz 1996b, 567-569.

[26] Loverdou-Tsigarida 1996, vol. II, 497-499.

The presence of and features of the two-headed eagles on the obverse of the two objects of Chilandar allows us to comfortably locate their production in the first half of the 14th century.

Most probably the original objects which provided the two support fittings of Chilandar were created in the early 14th century, and were likely devices of the luminaire and would have constituted precious elements of the ecclesiastical equipment of the Palaeologan era. They are very rare objects of their kind. We do not even know how our support fittings arrived at Mount Athos and the Chilandar Monastery. They could have been the gift to the Athonite monastery from a Greek or Serbian ruler, or from another member of the high aristocracy of the time. As for the place of their production, the city of Thessaloniki could be attributed, subject to reservation.

On the use of the metal supports

In 1998 Maria Xanthopoulou published two similar medallion-like supports, which were discovered in 1901 during the excavation of the Basilica of Saint Titus in Gortyn, Crete. Each support consists of two parallel plates of copper alloy, placed vertically, connected at the bottom and carried by a truncated shaft. 'Banner supports' was the initial interpretation proposed by A. Orlandos.[27] M. Xanthopoulou also proposed their use as 'icon supports.' And the metal support fittings from the Monastery of Chilandar would have provided great evidence to confirm this interpretation. However, over the following years similar artifacts, which have been presented in exhibitions, published in exhibition catalogues and found in excavations, show to us that the metal support fittings from the Monastery of Chilandar were re-used as 'icon supports' or 'icon frame supports'.

The first time their initial use was 'interpreted,' at least as we know, occurred when similar supports were exhibited at the Archäologische Staatsammlung in Munich.[28] Photos of the exhibition presentation were included in the exhibition catalogue and denote their use as a part of the support of a candlestick which would have more than one pricket for candles.[29]

This interpretation was confirmed in the best way when fragments of a candlestick with many candle prickets was found in a thirteen-century destruction layer in Cherson. It was proposed that this candlestick was 'attached to a wall by means of two supports (the right is preserved). One end of the support takes the form of a hand clasping a rod on which is set a rectangular bracket'[30] similar to ones presented in this paper.

These elements offered to scholars the opportunity to interpret similar objects belonging to museum and private collections,[31] even if items are now isolated from their context. A whole series of metal brackets in the shape of a human arm ending in a fist were interpreted as bases for the rectangular brackets which supported the candlesticks. These constructions seem to have been widespread by the Middle Byzantine period. Also, the number of terms used for them in the Byzantine monastic typika[32] reflects their diversity, a fact which is confirmed by the artifacts themselves. The various lengths and thicknesses of the brackets in the shape of the human arm[33] indicate that they could support candlesticks and chandeliers of corresponding sizes and forms.[34]

We must consider that these constructions were sufficient to keep the luminaires at a safe distance from the side walls to which the arms were hinged so as to limit the fire risk and protect the walls from the soot of the lamps. Moreover, this arrangement could facilitate the movement of the candlesticks for cleaning. Also parts of these support systems imitating human arms and hands seems to continue into the Middle Ages, reflecting a practice which was also very common in Late Antiquity.[35] And it is a choice that still survives even today as a decorative and functional element.

To conclude, research on the lighting devices conducted during the last years has offered us the opportunity to interpret the initial use of some artifacts which were subsequently used in a different way. The metal support fittings from the Monastery of Chilandar on Mount Athos were originally associated with a candlestick. These items also indicate the widespread use of beeswax candles particularly from the Middle Byzantine period and confirm that objects used to support or suspend lighting devices developed in the Byzantine era are closely related to the evolution of the lighting means and, thus inevitably, are an immediate reflection of that evolution.

Bibliography

Alchermès, J. 1997. Two stone slabs, in E. Evans (ed.) *The Glory of Byzantium. Art and Culture of the Middle Byzantine Era. A.D. 843-1261*, 326-327, no. 220 B. New York: The Metropolitan Museum of Art.

Androudis, P. 1999. Origines et symbolique de l'aigle bicéphale des Turcs Seldjoukides et Artuqides de l'Asie Mineure. *Βυζαντιακά* 19, 309-345.

Androudis, P. 2012. Chapiteau de la crypte de la basilique de Saint-Démétrios à Thessalonique avec emblèmes de la famille des Paléologues. *Δελτίον Χριστιανικής Αρχαιολογικής Εταιρείας*, period Δ´ 33-34, notes 9-14.

Androudis, P. 2013. Les premières apparitions attestées de l'aigle bicéphale dans l'art roman d'Occident (XIe-XIIe siècles). Origines et symbolique. *Niš and Byzantium. Eleventh Symposium Niš, 3-5 June 2012*, 209-225. Niš: Univerzitet u Nišu.

[27] Orlandos 1926, 325, fig. 25.

[28] Reginek 2004, 102-103, no. 139-I.

[29] Wamser 2004, Vorwort (preface), XXIV-XXV.

[30] Yashaeva *et al.* 2011, 221, 508, no. 168. See, in this volume, Ryzhov and Yashaeva, fig. 7.

[31] Atasoy 2005, 87-91; Durand *et al.* 2010, 138, no. 43; Pitarakis 2015; Pitarakis 2016. Also, see Drauschke and Kirchhainer 2017. For the lighting devices of the same collection in general, see Kirchhainer 2017.

[32] L. Bouras was the first scholar who listed the various forms of Byzantine metal candlesticks and chandeliers taking into account mainly the Byzantine monastic typika and their depictions in the Byzantine Art. See Bouras 1981.

[33] Pitarakis 2016, 441.

[34] For the various types of metal support and suspension fittings and their Byzantine names, see in Motsianos 2011, 321-348 and Motsianos 2011b, 110-112.

[35] For this issue, see Motsianos 2011, 323-325.

Androudis, P. 2013b. Dalle avec aigle bicéphale, en provenance de l'enceinte byzantine de Trébizonde. *Δελτίον Χριστιανικής Αρχαιολογικής Εταιρείας*, period Δ΄ 34, 67-78.

Androudis, P. 2015. Double-Headed eagles on early (11th-12th c.) medieval textiles: Aspects of their iconography and symbolism, in M. RaKocija (ed.) *Proceedings of the International Congress: Nis and Byzantium XIV*, 315-341. Niš: Univerzitet u Nišu.

Atasoy, S. 2005. *Bronze lamps in the Istanbul Archaeological Museum: An illustrated catalogue*. Oxford: British Archaeological Reports, International Series 1436.

Babić, G. 1996. Les portraits des grands dignitaires du temps des tsars serbes. Hiérachie et idéologie, in *Byzantium and Serbia in the Fourteenth Century. International Symposium 3 (12-14 November 1993)*, 158-168. Athens: Institute for Byzantine Research, National Hellenic Research Foundation.

Bees, N. 1912. Zum Thema der Darstellung des zweiköpfige Adlers bei den Byzantinern. *Repertorium für Kunstwissenscheft* 35.1, 321-330.

Binon, S. 1938. L'histoire et la légende de deux chrysobulles d'Andronic II en faveur de Monembasie. Macaire ou Phrantzès? *Echos d'Orient* 37, 274-311.

Bouras, L. (1981). Byzantine Lighting Devices, in *Akten II/3 der XVI. internationaler Byzantinistenkongreß*, vol. 3, 479-491). Wien: Verlag der Akademie der Wissenschaften.

Chotzakoglou, C. 1996. Die Palaiologen und das früheste Auftreten des byzantinischen Doppeladlers. *Byzantinoslavica* LVII, 60-68, pls I-XII.

Ćurčić, S. 2010. *Architecture in the Balkans. From Diocletian to Suleyman the Magnificent (c. 300-1500)*. New Haven, Conn.; London: Yale University Press.

Durand, J., Giovannoni, D., Rapti, I. and Clavien, R. 2010. *Sainte-Russie: L'art russe des origines à Pierre le Grand*. Paris: Somogy Éditions d'art.

Drauschke, J. and Kirchhainer, K. 2017. Lampen und Leuchterzubehör aus Buntmetall, in F. Daim, B. Fourlas, K. Horst, and V. Tsamakda (eds) *Spätantike und Byzanz, Bestandskatalog Badisches Landesmuseum Karlsruhe Objekte aus Bein, Elfenbein, Glas, Keramik, Metall und Stein*, 73-79. Mainz: Verlag des Römisch-Germanischen Zentralmuseums.

Failler, A. 1982. Les insignes et la signature du despote. *Revue des Études Byzantines* 40, 171-186.

Fourlas, A. 1980. Adler und Doppeladler. Materialen zum 'Adler in Byzanz', in B. Kötting and A. Kallis (eds) *Philoxenia: Prof. Dr. Bernhard Kötting gewidmet von seinen griechischen Schülern*, 97-120. Münster: Aschendorff.

Fourlas, A. 1984. Adler und Doppeladler: Kunstgeschichtliche Zeugnisse zum palaiologischen Doppeladler, in D. Ahrens and J. Fink (eds) *Thiasos ton Mouson: Studien zu Antike und Christentum; Festschrift für Josef Fink zum 70. Geburtstag*, 179-190. Köln: Böhlau Verlag.

Frolow, A. 1938. La « Podéa », un tissu décoratif de l'Église Byzantine. *Byzantion* 13.2.

Gerola, G. 1934. L'aquila bizantina e l'aquila imperiale a due teste. *Felix Ravenna* 43, 7-36.

Gierlichs, J. 1996. *Mittelalterliche Tierreliefs in Anatolien und Nordmesopotamien: Untersuchungen zur figürlichen Baudekoration der Seldschuken, Artuqiden und ihrer Nachfolger bis ins 15. Jahrhundert*. Tübingen: Wasmuth.

Kalligas, H. 2009. The Miniatures in the Chrysobulls of Andronikos II for Monemvasia, in R. Shukurov (ed.) *Mare et litora. Essays presented to Sergei Karpov for his 60th Birthday*, 366-378. Moskva: Indrik.

Kirchhainer, K. 2017. Lampen und Leuchterzubehör aus Buntmetall, in F. Daim, B. Fourlas, K. Horst, and V. Tsamakda (eds) *Spätantike und Byzanz, Bestandskatalog Badisches Landesmuseum Karlsruhe Objekte aus Bein, Elfenbein, Glas, Keramik, Metall und Stein*, 90-92, nos. IV.50-IV.55, pls 55-59. Mainz: Verlag des Römisch-Germanischen Zentralmuseums.

Korać, V. 1998. King Milutin's church, in G. Subotić (éd.) *Hilandar Monastery*, 75-98. Belgrade: The Serbian Academy of Sciences and Arts Gallery.

Korunovski, S. and Dimitrova, E. 2006. *Macédoine Byzantine. Histoire de l'Art macédonien du IXe au XIVe siècle*. Paris: Thalia.

La Bulgarie médiévale. Art et civilisation, 1980. Exhibition Catalogue, Grand Palais, Paris, 13 June-18 August 1980. Paris: Association Française d'Action Artistique.

Lamberz, E. 1996. Νεώτερες έρευνες στις βιβλιοθήκες του Αγίου Όρους, in *Διεθνές Συμπόσιο Το Άγιον Όρος. Χθες-σήμερα-αύριο, Θεσσαλονίκη 29 Οκτωβρίου- 1 Νοεμβρίου 1993*, 161-163. Θεσσαλονίκη: Εταιρεία Μακεδονικών Σπουδών.

Lamberz, E. 1991. Die Schenkung des Kaisers Johannes VI. Kantakuzenos au das Kloster Vatopedi und die Schreibzentren Konstantinopels im 14. Jahrhundert, in *XVIIIe Congrès International d'Etudes Byzantines. Résumés des Communications*, Moscou, vol. II, 633-636.

Lamberz, E. 1996b. The Library of Vatopaidi and Its Manuscripts, in *Great Monastery of Vatopaidi. Tradition - History - Art.*, vol. II, 562-574.

Lambros, S. 1909. Ὁ δικέφαλος ἀετὸς τοῦ Βυζαντίου. *Νέος Ἑλληνομνήμων* 6, 433-473.

Loverdou-Tsigarida, K. 1996. Βυζαντινή Μικροτεχνία, in *Great Monastery of Vatopaidi. Tradition - History - Art*, vol. II, 497-499.

Marković, M. and Hosteter, W. 1998. Prilog hronologiji gradnje i oslikavanja hilandarskog katolikona. *Hilandarski Zbornik* 10, 201-220.

Motsianos, I. 2011. Φως ιλαρόν: ο τεχνητός φωτισμός στο Βυζάντιο (Joyful Light: The Artificial Lighting in Byzantium). Unpublished PhD dissertation, Πανεπιστήμιο Θεσσαλίας.<http://thesis.ekt.gr/thesisBookReader/id/31739#page/1/mode/2up>, viewed 15 December 2018.

Motsianos, I. 2011b. Artificial lighting during Byzantine and Post-Byzantine period, in I. Motsianos and E. Bintsi (eds) *Light on light: an illuminating story, exhibition catalogue, Thessaloniki, Folklife and Ethnological Museum of Macedonia-Thrace, 31. October 2011-11. June 2012*, 90-125. Thessaloniki: Folklife and Ethnological Museum of Macedonia&-Thrace.

Nenadović, S. 1997. *Osam vekova Hilandara. Građenje i građevine*. Beograd: Republički zavod za zaštitu spomenika kulture.

Orlandos, A. 1926. Νεώτεραι ἔρευναι ἐν Ἁγίῳ Τίτῳ τῆς Γορτύνης. *Επετηρὶς Εταιρείας Βυζαντινών Σπουδών* 3, 301-328.

Pitarakis, B. 2016, Bras de lumières sur le templon médiéval (XIe-XIIIe siècle): un dispositif en bronze inédit au Musée archéologique d'Edirne, in S. Brodbeck, A. Nicolaïdès, P. Pagès, B. Pitarakis, I. Rapti, E. Yota (eds) *Mélanges Catherine Jolivet-Lévy (=Travaux et Mémoires 20/2)*, 435-452. Paris.

Pitarakis, B. 2015. Dispositif de support de luminaire, in M. Martiniani-Reber (ed.) *Donation Janet Zakos. De Rome à Byzance*, 160-163, no. 55. Genève: Musée d'art et d'histoire; Milan: 5 continents.

Reginek, M. C. 2004. Teile einer Byzantinischen Kirchenausstattung, in L. Wamser(ed.) *Die Welt von Byzanz: Europas östliches Erbe : Glanz, Krisen und Fortleben einer tausendjährigen Kultur* Exhibition Catalogue, Archäologische Staatsammlung München-Museum für Frühgeschichte, München, 22.10.2004-3.4.2005, 102-103 (no 139-I). Stuttgart: K. Theiss.

Roux, J.-P. 1984. Le bestiaire de l'Islam. *Archéologia* 117, 38-47.

Roux, J.-P. 1984b. *La Religion des Turcs et des Mongols.* Paris: Payot.

Roux, J.-P. 1992. La sculpture figurative de l'Anatolie Musulmane. *Turcica* 24, 27-90.

Soloviev, A. V. 1935. Les emblèmes Héraldiques de Byzance et les Slaves. *Seminarium Kondakovianum* 7, 119-164.

Spatharakis, I. 1976. *The Portrait in the Byzantine Illuminated Manuscripts.* Leiden : E.J. Brill.

Spyridakis, G. 1972-73. Ο δικέφαλος αετός ιδία ως σύμβολον ή ως θέμα κοσμήσεως κατά την βυζαντινήν και μεταβυζαντινήν μέχρι των νεωτέρων χρόνων περίοδον. *Επετηρὶς Εταιρείας Βυζαντινών Σπουδών* 39-40, 162-174.

Strzygowski, J. and Van Berchem, M. 1910. *Amida.* Heidelberg: Carl Winter.

Tsigaridas, E. 1999. *Τοιχογραφίες της περιόδου των Παλαιολόγων σε ναούς της Μακεδονίας.* Θεσσαλονίκη: Πουρναράς.

Verpeaux, J. 1966. *Pseudo-Kodinos, Traite des Offices.* Paris: Editions du Centre national de la recherche scientifique.

Vocotopoulos, P. 2007. *Παντάνασσα Φιλιππιάδος.* Αθήνα.

Vocotopoulos, P. 2008. Η κτιτορική τοιχογραφία στο περίστωο της Παντανάσσης Φιλιππιάδος. *Δελτίον Χριστιανικής Αρχαιολογικής Εταιρείας*, period Δ΄ 29, 73-80.

Wamser, L. (ed.) *Die Welt von Byzanz: Europas östliches Erbe : Glanz, Krisen und Fortleben einer tausendjährigen Kultur.* Exhibition Catalogue, Archäologische Staatsammlung München-Museum für Frühgeschichte, München, 22.10.2004-3.4.2005. Stuttgart: K. Theiss.

Xanthopoulou, M. 1998. Le mobilier ecclésiastique métallique de la basilique de Saint-Tite à Gortyne (Crète centrale). *Cahiers Archeologiques* 46, 103-119.

Yashaeva, T., Denisova, E., Ginkut, N. and Zalesskaya V. 2011. *The legacy of Byzantine Cherson.* Austin, Texas: University of Texas at Austin: Institute of Classical Archaeology.

Once again on the 'Byzantine-Mediterranean Limoges'?

Mariela Inkova

National Museum of History, Sofia, Bulgaria

Abstract: This article examines a candlestick base newly acquired by the National Museum of History, Sofia, Bulgaria, which is said to originate from Mezdra, a region in Northwest Bulgaria. The obverse is decorated with four mythological creatures within medallions –two griffins and two dogs/lions. The base is cast in almost pure copper which is consistent with the subsequent operations –enameling and gilding. In the following stages the pattern was 'tremolo' engraved, then the enamels applied in *champlevé* technique and then the copper base was gilded. By its technological, stylistic and morphological characteristics, the base falls into a group of similar monuments –two candlesticks from Abegg Foundation, Switzerland, originating according to the archive data from Torcello, Italy; another candlestick from the Metamorphosis Monastery, Greece; and two censers, type katzion from Kamień, Poland and the Gallery of L. Blumka, New York, all most probably products of the same workshop. Despite the identical techniques, they differ from the Limoges champlevé enamels and their origin is associated with the eastern Mediterranean – Italy, the islands and even the Jerusalem Kingdom from the second half of the 12th century. The fact that the censers belong to the Orthodox 'liturgical instrumentarium' and that the iconography follows Byzantine patterns again suggest an attribution of their manufacture to the Byzantine East. The study comments on the function of the candlesticks of this type both in daily life and in sacred space –in church interior, monastery scriptoria and liturgical practices.

Keywords: candlestick, *champlevé* enamel, gilding, pure copper, griffin, dog/lion, National Museum of History, Sofia, Abegg Foundation, Riggisberg, Switzerland, Byzantium, Limoges, incense burners.

Introduction

The present study falls within the context of the large Byzantine East and Latin West problematics –traditions and innovations, contact zones, ateliers, craftsmen and recipients, etc. The literature on the subject is enormous and the international conferences and organized exhibitions related to Byzantine and West European art are numerous.[1] In connection with the subject of this study[2] we will mention the studies of Grabar[3] on the asymmetry of the relations between Byzantium and the West in the arts, of Buckton[4] on Byzantine enamels and western toreutics since the Carolingians, of Lafontaine-Dosogne[5] on Byzantine influences in the Rhine-Mosel area during the 8th and early 13th centuries, of Nelson on Byzantine 'borrowings' from the Western world,[6] etc. In contrast, an article by Durand, published several years ago, examines a wide range of influences in the field of metalworking during this era in the reverse direction: from the economically rising West in the 13th and 15th centuries[7] to the contemporaneous decline of the Eastern Empire. By specific examples the author focuses on Gothic innovations in the technology, shapes and stylistic decoration of the monuments, at the same time noting that the degree of their penetration depended on the recipient's taste. If fashion trends in shapes, technology and decoration are generic in the secular culture of the aristocracy, then in the fields of the sacred art innovations are more sporadic.

Well-known are the circumstances which determined the mutual influences in the field of art between the Latin West and Byzantine East in the so-called contact zones of the two cultures.[8] These were primarily the Byzantine possessions in Italy –Calabria, Apulia, Venetia and Sicily. After the 11th century and especially during the Crusades, the western presence in the East became tangible and even permanent. What we have in mind here are the territorial conquests, known in the French literature as *France de l'Outre mer* –the Frankish kingdoms in Syria and Palestine, the establishment of the Latin Empire, the Morea (Achaea) Despotate – Mystra, Koroni, Modon, Monemvasia, Clarentza, Patras, Corinth, the Venetian possessions in Crete, the islands of the Aegean and the Adriatic coast, the possessions of Genoa in the Levant and on the Black Sea coast, the islands of Chios, of the Angevins from Naples in Corfu, the possessions of the Knights Hospitallers in Rhodes and chiefly the Western presence in the capital Constantinople. To this, the diplomatic gifts from the West should be added, as well as the trade relations, etc. As a consequence, the examples typical of the Western world fashion accessories, traced by Durand, were either imported goods or semi-finished products, finalized in the East, according to the local taste.

Other products of thirteenth- and fourteenth-century Byzantine metalworking that used technological and decorative skills, so typical of the Gothic tradition, are rather the output of a western workshop, situated on one of the islands –Crete, Rhodes or Cyprus, or in some large

[1] Vokotopoulos 2007; Nelson 2004, 515-523; Wixom 1996, 435-449. A colloquium on the topic 'Relations et échanges entre Orient et Occident méditerranéens au XIIIe siècle. Le témoinage des progammes picturaux', held in Athens (I owe the information about the colloquium to Prof. E. Bakalova)

[2] In abbreviated form this study was reported at the XXII Byzantine Congress in Sofia in August 2011.

[3] Grabar 1984, 9-24.

[4] Bukton 1988, 13, 235-244.

[5] Lafontaine-Dosogne 1995, 181-192.

[6] Nelson 2007, 143-151.

[7] Durand 2004, 58, 333-354.

[8] Bozhilov 2002, 64-78.

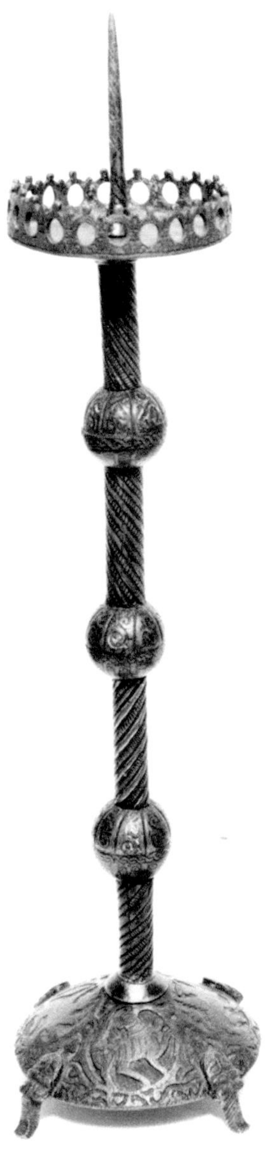

Figure 1. Candlesticks from the Abegg Foundation, Riggisberg, Switzerland: General appearance. (photo Abegg Foundation).

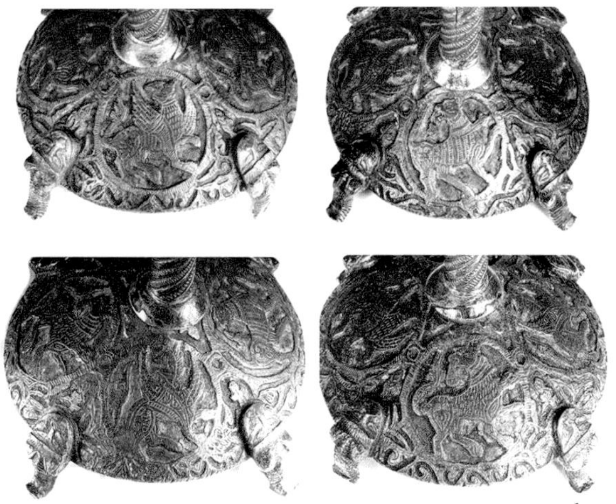

Figure 2. Candlesticks from the Abegg Foundation, Riggisberg, Switzerland: Ornamental decoration on the candlestick base, inv. n.8.186.72 (photo Abegg Foundation).

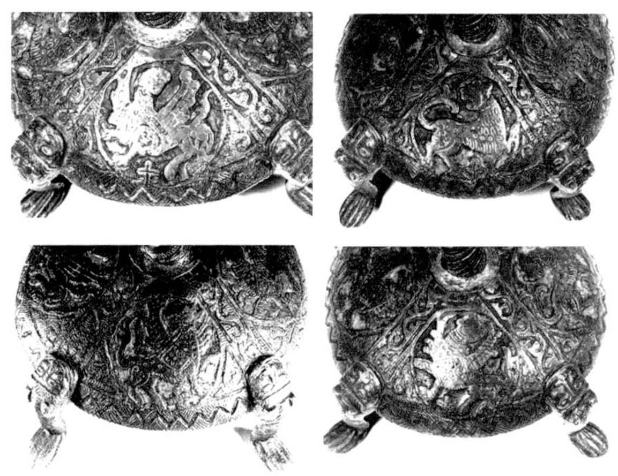

Figure 3. Candlesticks from the Abegg Foundation, Riggisberg, Switzerland: Ornamental decoration on the candlestick base, inv. no. 8.186.72; (photo Abegg Foundation).

cosmopolitan centers such as Constantinople, Thessaloniki and Mystra. Another similar center of the 14th and 15th century seems to have existed in Dalmatia (Zadar), strongly influenced by Venice.

A third group of objects, however, suggests the production of a Greek workshop, 'perfectly imitating Western products'.[9] According to Otavsky, Bouras and Swarzenski, whose opinion is based on stylistic and technological features, this group includes also two candlesticks from the Abegg Foundation in Riggisberg, Switzerland (Figures 1-3),[10] another candlestick almost identical to them from the Metamorphosis Monastery of the Meteora (Figure 4) and two standing incense burners with long handles, called in Greek κατσία or κατζία (katsia or katzia) and dated at the end of the 12th-13th century (Figure 5).[11] According to the research, they were used in the Byzantine funerary ritual or were placed in the church, in front of icons. Another argument concerning the mainly eastern liturgical practice to use this type of vessel is a sgraffito ceramic censer from the 12th century,[12] whose shape and twisted cord-like decorative motif imitates the above mentioned censers (Figure 6). In Otavsky's opinion, these liturgical objects differ from the numerous *champlevé* enamels, produced in the ateliers of Limoges and Castile, the Rhine-Mosel area and were widespread in Western and Central Europe during the 12th-13th centuries. Archival documents indicate that the candlesticks from the Abegg Foundation

[9] Durand 2004, 342.

[10] My sincere gratitude to Mr. Michael Peter from Abegg Foundation for the illustrations.

[11] Otavsky 1994, 239-240, 125-130; Otavsky 1994a, 137-143.

[12] Borboudaki 1999, 30, no. 8.

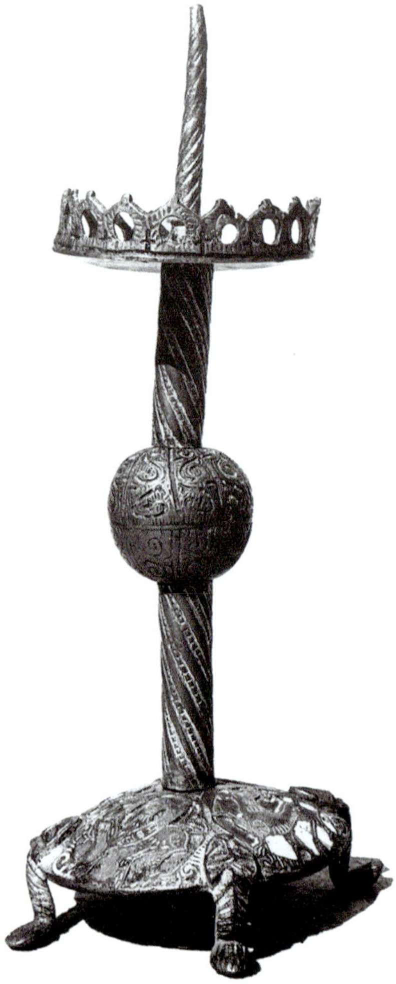

Figure 4. Candlestick from the Transfiguration monastery, Meteora, Greece (after K. Otavsky).

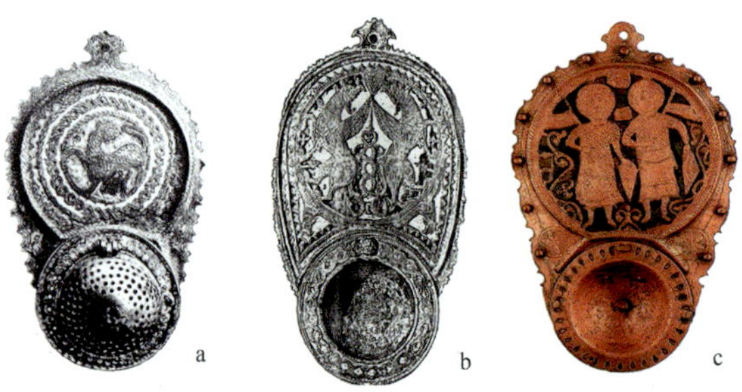

Figure 5. Incense burners, type *katzion* from:
a. The cathedral in Kamień, Poland (after W. Borchers);
b. The Gallery of L. Blumka, New York (a-b. after K. Otavsky);
c. The Benaki Museum (after A. Ballian).

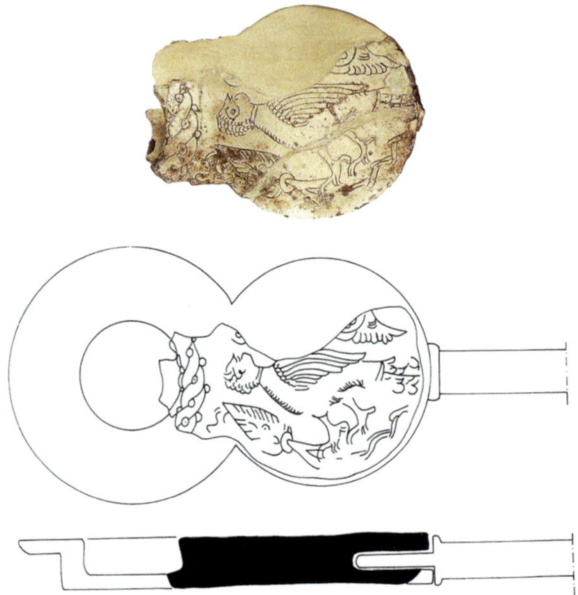

Figure 6. Ceramic incense burner, type *katzion* in sgraffito technique, 12th century (after M. Borboudaki).

belonged to Giovanelli's collection in Rome, claiming to have been found in a church in Torcello. Their origin is associated with the Mediterranean region (Spain, Italy) and even with the Jerusalem Kingdom according to Gautier who dates the candlesticks to the second half of the 12th century.[13] One of the two incense burners originates from Kamień, Poland, but it was lost after World War II (Figure 5a). In Borchers's view, based on some Byzantine and 'oriental' influences in the decoration, its place of origin should be sought in Southeast Europe.[14] The second censer belongs to the Leopold Blumka Gallery in New York (Figure 5b). A third similar censer is kept in the Benaki Museum, attributed by Bouras and Ballian to the second half of the 13th century (Figure 5c).[15] The stylistic, morphological and technological analogies of the incense burners and the three candlesticks, and particularly the fact that they belonged to the Eastern Orthodox 'liturgical instrumentarium',[16] distinguish these artifacts as an independent group whose origin should perhaps be sought in the area of the Byzantine East.

The aim of this study is to present a new artifacts which, in my opinion could also be included in this group. According to recent studies, this group 'illustrates' a mastery that is typical of western world technology and ornamentation, characteristic of both Romano-Gothic and Eastern art in its Byzantine version.

Description of the object

In 2007, the National Museum of History in Sofia purchased an object unusual for medieval Bulgarian toreutics, most probably part of church plate or an article of secular usage,

[13] Gautier 1987, 119-120, no. 124, pl. XCIII, 357.
[14] Borchers 1933, 38-39.
[15] Μπούρα 1981, 1, 67-68; Ballian 2004, 128-129, no. 65.
[16] Zalesskaya 1995, 672, note 20; Drandaki 2004, 128, no. 64.

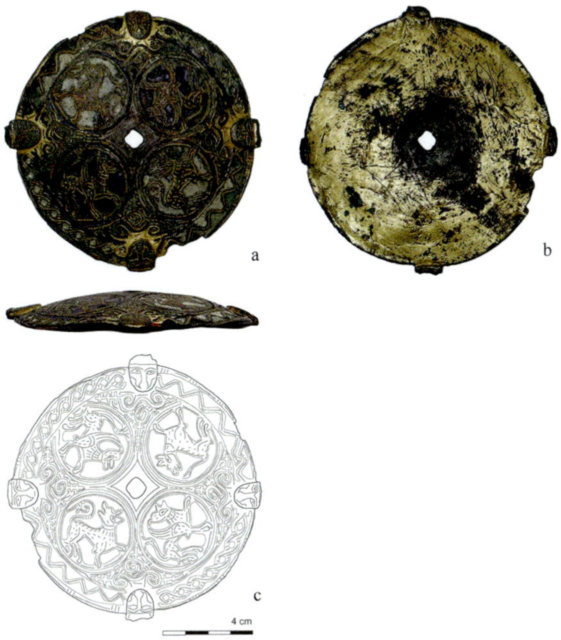

Figure 7. Candlestick base from the National Museum of History, Sofia, Bulgaria. Originating from the region of Mezdra, North-West Bulgaria; a. Obverse b. Reverse (photo I. Levicharov). c. Drawing (I. Nikolov).

richly decorated with polychrome enamels. The base has a round, slightly convex shape and a square opening in the middle (Figure 7a, c). Around the opening, two engraved lines form a square with concave edges. The obverse surface bears four medallions (two of them have truncated bases) in which mythological creatures are represented, all with heads turned backwards. In the two opposite medallions, eagle-headed griffins are depicted against a dark blue enamel background, in the other two -dogs/lions? on a white enamel field. The griffins are almost identical, placed in the middle of the roundels and represented by identical iconography -facing right, with lifted forelegs, heads turned backwards and tails curved over back. The sharp and raptorial beaks touch the feathers of the spread wing. Two incised lines depict a collar and below it cut in two levels, short and radial notches mark the mane. The right front leg is lifted high and the left stands firmly, organically connected with the wing, while the rear legs are represented one behind the other. The tail is twisted twice over the back, and a small knot shapes its tufted end reaching the contours of the medallion. With a few short incisions and three parallel tiny bows the artisan has skillfully marked the fur and the feathers of the wing. The differences are minor -one of the heads is more elongated and the ears are shown in a different and rather primitive way.

The animals in the other two medallions are depicted in the same pose - marching to the right with foreleg raised, rearward facing head and tail curved over the back. The short arched neck supports a relatively large head with a mouth gaping against the twisted tail. The dogs have similar collars and short 'zigzag' cuts represent the fur. The differences are again in the way the ears and the tails are depicted - in the one case the tail is S-twisted and terminating in a three-leaved small palmette, while in the other representation the tail ends in a simple hair tuft. The eyes are depicted in the same manner - almond-shaped and elongated, with open ends. An interesting detail is that the ears of the neighboring dog-griffin pairs are alike. Each medallion is outlined with engraved lines and one enamel strip. One of the medallions has a white enamel background and is surrounded by a blue frame, and the frame of the other is half blue, half white because of contamination with particles from the background. The medallions with a blue background are also framed differently -one in white and the other in the same blue color, possibly neglecting the originally conceived contrast. The area around the four roundels is entirely filled with leafless tendrils, tied as in a bunch and with spirally twisted ends, from which grow three- and five-leaved linear cross-like palmettes filled with light green enamel. In turn, the spirals are also twice outlined -by one engraved and one light green enamel line.

The entire composition is placed within a frame of alternating 'zigzag' and 'twisted' patterns, the latter consisting of two strings intertwined and 'tied' in a way to shape small squares in the middle. The cells of the 'zigzag'-like ornament are filled in with white (from inside) and green (outside) enamel, while that of the intertwined strings -with dark blue enamel from the inside and on the outside- as the enamel has altogether vanished -with maybe light green enamel judging by the particles preserved in one area. The small squares formed by the twisting strings are filled in with white enamel. In places, the outer frame of the border has double 'ties' represented by three incised lines. Four anthropo-zoomorphic faces or masks (?) are depicted en face and equidistant from each other on the periphery of the circle. The faces are U-shaped and turned toward the center of the base; the eyes follow the same manner of representation as the eyes of the animals -almond-like, elongated and with 'open' outlines. The nose is broad, rather beast-shaped, and sits above a high uplifted mustache. Over the forehead is presented a bicornuate embossed hat, a crown or maybe small ears. The original edges are lost but they probably continued into the four vertical feet of the base. We assume that after some damage, they were deliberately leveled out thus allowing the base to be reused.

The non-enameled parts of the supporting base have been gilded, which enhances the polychrome effect. The backside is also gilded (Figure 7b) and that circumstance suggests the function of the object as an article. Around the opening on the front and the back side of the base as well as on the head of one of the dogs there is spilled solder (Figure 8b), probably the result of an initial or later repair of a constructive nature. Dimensions: 11.09 x 11.21 cm; thickness of 0.21cm; opening 0.66x 0.67 cm.

According to the owner the object originates from the territory of the Tsakonitsa village, Mezdra region, North-western Bulgaria.

Condition of the candlestick base

The round metal base must have been in use for a long time; the feet and parts of the border piece are currently missing. The fact that the periphery of the four anthropo-zoomorphic faces was equally leveled suggests deliberate manipulation for a secondary use. All this had caused partial moldering of the enamel, particularly in some border sections and deletion

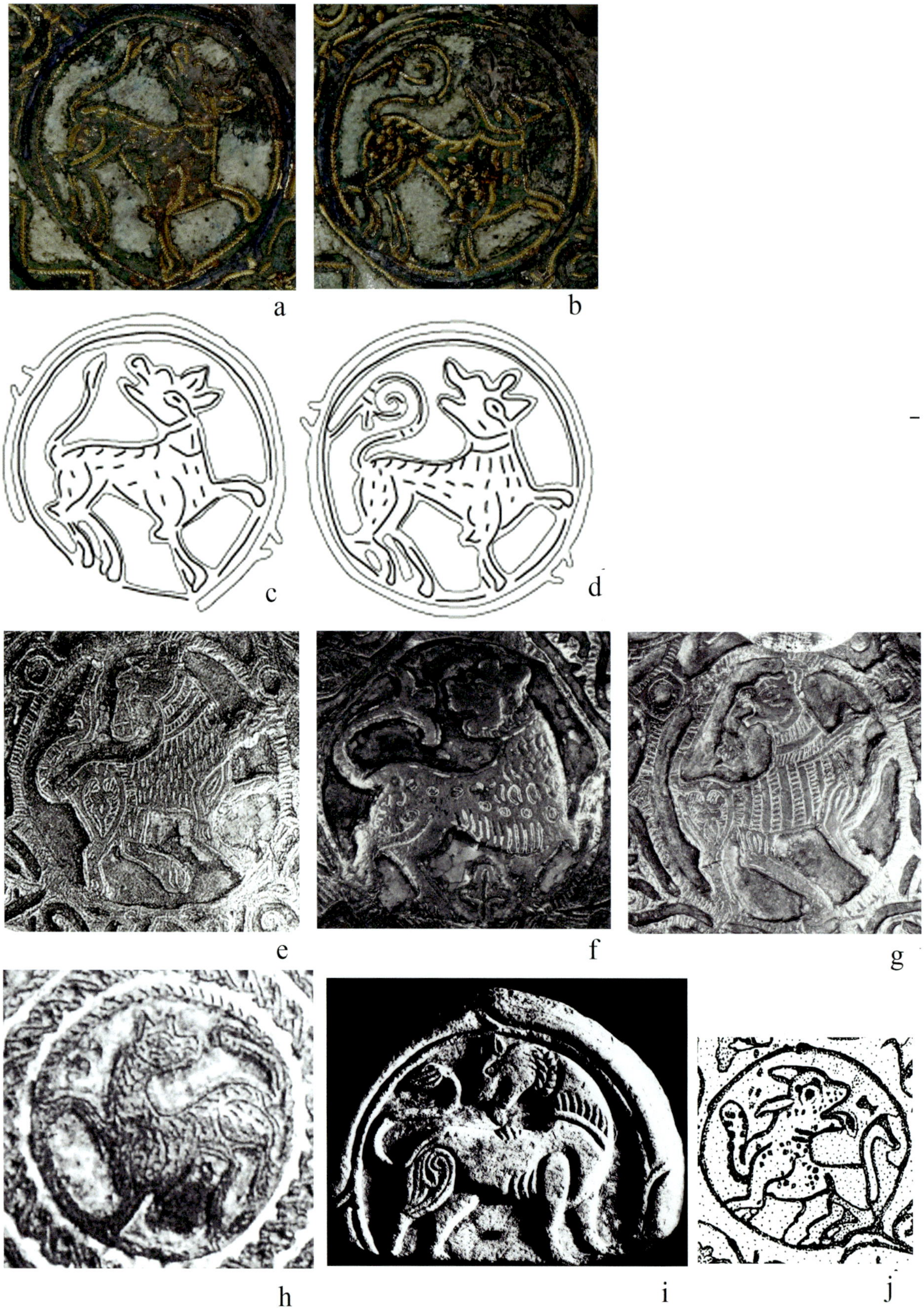

Figure 8. Images of a dog/lion on: a-d. The base from the National Museum of History, Sofia (a-b photo I. Levicharov, c-d I. Nikolov, detail); e-g. The candlesticks from the Abegg Foundation, Riggisberg (after K. Otavsky, detail); h. Incense burner from Kamień (after W. Borchers); i. Relief from Thessaloniki, 11th-12th century (after La Bulgarie); j. Medallion from the Museum of Alzey, Germany (after A. Frey).

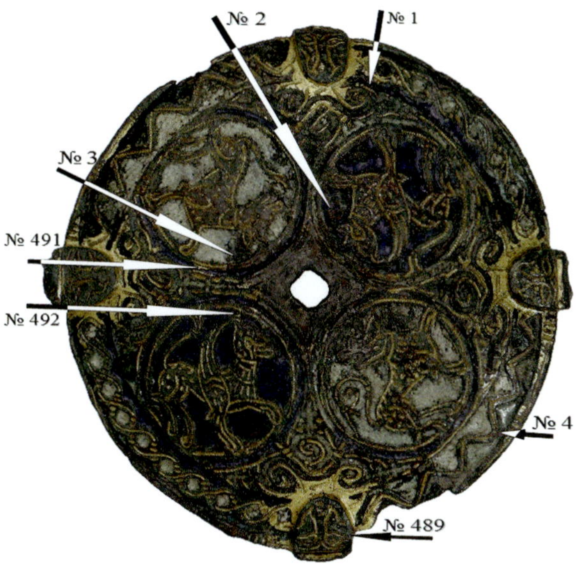

Figure 9. Topography of samples.

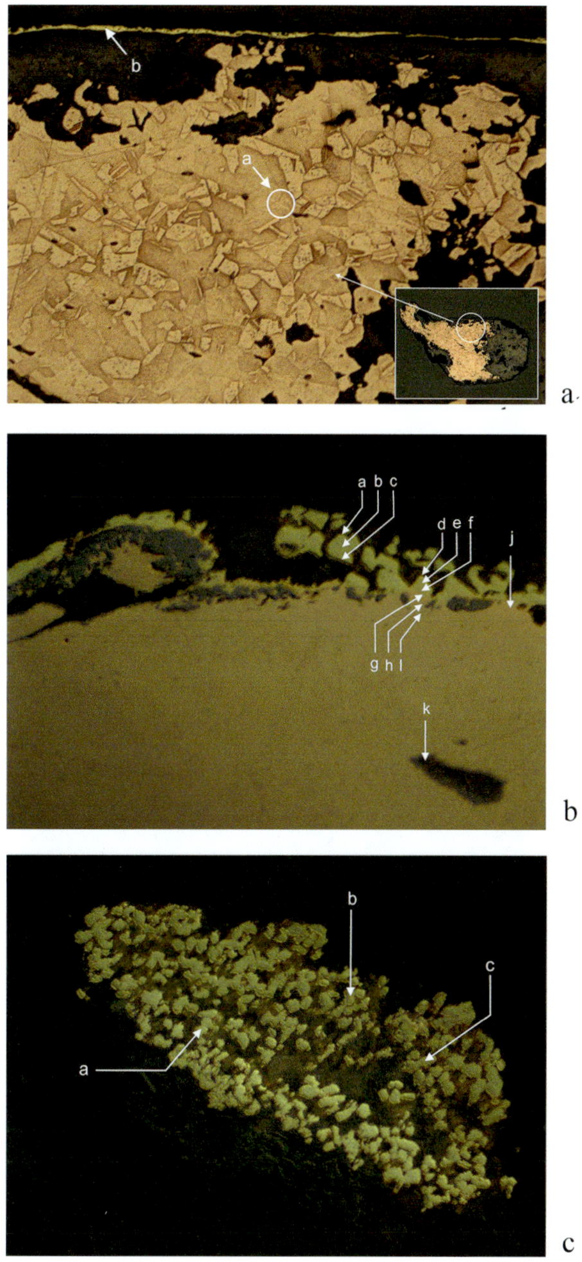

Figure 10. Cross-sections of samples from the base and the gilt of the product: a. no. 489; b. no. 491; c. no. 492 (V. Inkova).

of the gilt on the copper base. Traces of gilding are preserved on the reverse side of the base, while on the face the gilt survived only in the grooves of the engraving and around the 'masks'. The surface coating layer of the enamel is destroyed, and contains many pores and cracks due to devitrification. The most corroded is the green layer, whose whitish color is caused by loss of colorant. In some places there are blue or green inclusions in the white enamel as contaminants during the very process of enameling.

Technique of manufacture

Data from the underlying SEM-EDS analyses on cross-sections of the metal base shows that the base is cast of almost pure copper – 99.77%, (Figure 9, no. 489; Figure 10a; Figure 11 (Appendix 1), analysis 12, a).[17] This configuration gives rise to comparisons with materials from the Romanesque period kept in the British Museum and investigated by Oddy, La Niece and Strarford. The results of the analysis of over twenty-three enameled and gilded items indicates the typical use of alloys containing copper from 95.6% to 99.3%.[18] According to the authors, the purity of the material used (refined copper) is evidence of recognized and intended correlation with the subsequent decorative operations –enameling and gilding techniques. Consistency with the processes of enameling and gilding is also represented by two medallions with heraldic images of lions, which were found in the area of Tarnovgrad – the capital of the Second Bulgarian Kingdom. The medallions were made in some West-European workshop using the same technology of enamel *champlevé* on a copper base and were most probably an imported commodity. Average data from the spectral analysis in different parts of the candlestick base shows that the parts were molded virtually from almost pure copper Cu – 99.8 %, Pb – 0.1 %, Sb – 0.001-0.003%.[19]

The metallographic structure of the metal substrate (Figure 12) indicates that the object was cast and finished by means of short-lived plastic deformation.[20] Traces of hammering are seen on the back of the supporting base.

At the next stage, the pattern and all its details were applied with clearly defined lines on the front side. The lines contouring the animals and the specific details on them, the anthropo-zoomorphic images, the floral ornament and the border are *tremolo* engraved (Figure 13).[21] The fact that the

[17] I wish to thank Hr Stanchev, department head at *Eurotest Control* AD (Bulgaria) who conducted the analyses.
[18] Oddy *et al.* 1993, 102, 107-108; Raub 1993, 102, 107-108.
[19] Markov1995, 1, 41-46, bel. 2.

[20] Metallographic analysis and micrograph were made by Prof. V. Inkova, director of the Laboratory for Conservation and Restoration of the Trakia Foundation. I express my gratitude for the assistance.
[21] Ogden 1998, 15.

N analysis	Type of test	Analyzed material	Analyzed segment	Superfici	Fig. Обр.	Composition of elements (poid %)								
						Cu	Au	Ag	Hg	Sn	Pb	Zn	As	Sb
1	Thin-section 491	'grain' in the gilt 7/8 μм	Upper layer	Point	a	6.44	67.98	2.96	22.62	-	-	-	-	-
2			Center	Point	b	19.92	60.36	1.07	18.65	-	-	-	-	-
3			Lower layer	Point	c	23.24	57.74	1.21	17.82	-	-	-	-	-
4		Gold coating	Upper layer	Point	d	18.22	59.36	2.24	20.18	-	-	-	-	-
5			Center	Point	e	24.79	55.20	1.69	18.33	-	-	-	-	-
6			Edge of the base	Point	f	33.90	49.69	1.66	14.75	-	-	-	-	-
7		Copper base	1 μм at the base	Point	g	87.59	9.50	0.34	2.57	-	-	-	-	-
8			3 μм at the base	Point	h	97.67	1.54	-	0.79	-	-	-	-	-
9			5 μм at the base	Point	i	99.38	0.62	-	-	-	-	-	-	-
10			Under the gilding	12/12 μм	j	91.90	5.59	1.52	-	-	-	-	0.76	0.22
11		Inclusions in the base			к	78.94	-	-	-	-	8.59	-	0.68	11.79
12	Thin-section 489	Appliqué		300/300 μм	a	99.77	-	-	-	-	-	-	-	0.23
13		Gilding		Point	в	16.43	65.30	2.01	16.04	-	-	-	-	-
14	Thin-section 492	Gilding under the enamel	Grain	Point	a	11.92	69.55	2.25	16.27	-	-	-	-	-
15			Grain	Point	в	5.52	67.79	2.27	24.41	-	-	-	-	-
16			Grain	Point	c	14.16	64.86	1.98	19.00	-	-	-	-	-

Figure 11. Chemical composition detected by SEM-EDS.

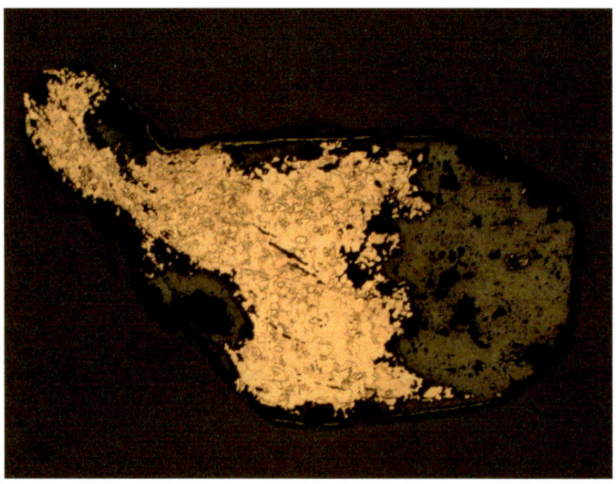

Figure 12. Cross-section of the copper base with known metallographic structure (x 50 dic), (V. Inkova).

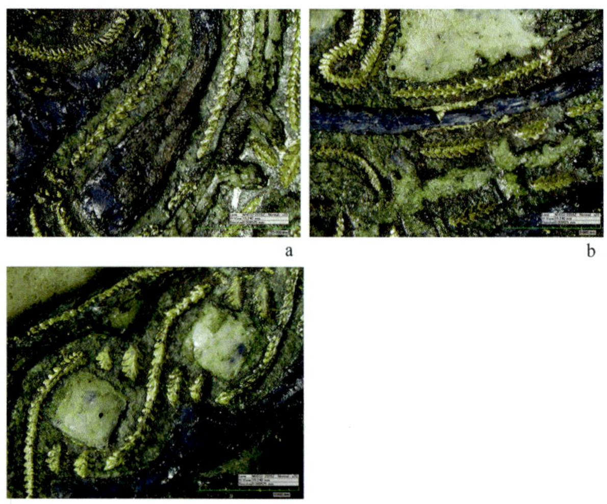

a b c

Figure 13. Engraved *tremolo* lines (x 20; x 20; x20), (photo. V. Inkova).

gilt survived only in the grooves, testifies that the engraving preceded the process of gilding. One of the above-mentioned medallions found in the area of Tarnovgrad was made in an identical way. With the second one, however, gilding precedes engraving.[22]

Only then did the master proceed to the application of the enamels. The areas of lost enamel indicate that the pits (cells or recesses) were shaped by cutting the metal away (Figure 14), a technique characteristic of *champlevé* enamels.[23] Three types of opaque enamels were used -white or milky white, dark blue and light green. To obtain preliminary information on the enamels and to determine the nature of the colorants, an elemental analysis of samples from the three color types was applied (Figure 15; Figure 9, nos. 1-4).[24] Their interpretation requires mandatory reporting of the effect of corrosion of metals on them, which determines the change of ratio between the components in the composition of the enamels. Evidence suggests that a frit compound of

[22] Ogden 1998, 43.

[23] Gautier 1972, 27-28; Tsaneva *et al.* 2008, 195.

[24] I take the opportunity to express my gratitude for the assistance of Sv. Tsaneva, Head of Central Laboratory for Conservation and Restoration in the NMH as well as of Hr Stanchev from Eurotest Control AD.

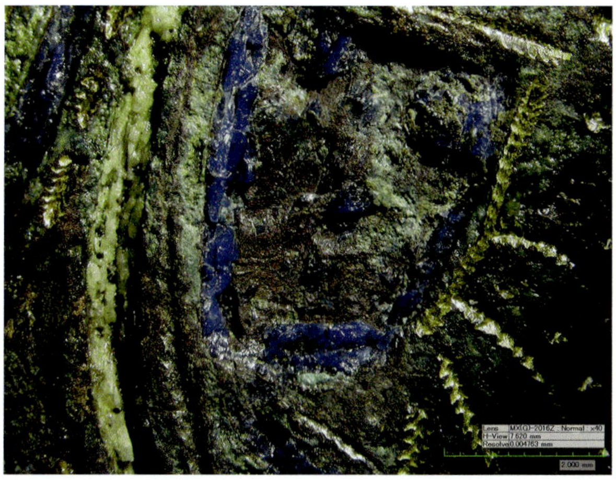

a

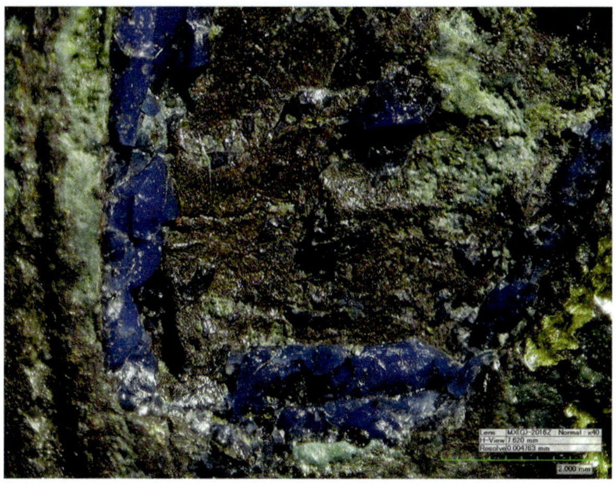

b

Figure 14. The 'pits' for the enamel, shaped by cutting the metal away (x 40; x 40), (photo. V. Inkova).

sodium-potassium-calcium-lead glass was used to make the enamels. It is noteworthy that all three configurations with different colored enamels contain CuO, SnO_2 and PbO, a sign suggesting the same recipe approach. Since the technique of enameling requires *in situ* melting of vitreous mass, probably the PbO was used to lower the melting point and reduce the coefficient of thermal expansion.[25] Moreover, Pb facilitates the leakage of enamel by 'wetting' the cells for it.[26] In all cases, SnO_2 is the matting element neutralizing the copper base. According to studies, the use of Sn for this purpose in the Mediterranean dates back to the 4th century BC.[27] The test results show that CuO is the colorant of the pale green enamel.[28]

Written sources on the manufacture of enamels are scarce. However, we find data on glass coloring in Marcus Vitruvius, in the Lucca Manuscript' from the second half of the 8th century AD and in the *Mappae Clavicula* from the 8th-9th century AD. Recipes for coloring in green by Cu are described in the Mappae –nos. 154, 155, 224,[29] also in the *Lucca Manuscript* -nos. A 4-7 A 7-8, A 25-30.[30] CoO was used for making the dark blue enamel on the item,[31] and for the white or milky white enamel -SnO was used, which in this case plays a double role– as a matting component and as a colorant,[32] The high percentage of CuO in the composition of the pigment is probably due to contamination of the sample with corrosion products from the copper substrate. We find information on the *lactei coloris* milky coloring in recipes nos. 156 and 160[33] from the *Mappae Clavicula*, while recipe no. 281 describes obtaining the color *albus*'.[34] Recipe A 9-10 from the *Lucca Manuscript*[35] provides information on 'other milky white colors'. Vitruvius also mentions 'milky glass'.

After the enamels were applied and fixed, the oxidized ambient surface was cleaned mechanically (by friction), which in some areas led to a 'slip' of metal on enamel (Figure 16). Given the high melting temperature of the enamels compared to the heating temperature used in gilding, the gilt of the copper substrate was the last technological step in manufacturing the candlestick base.

The conclusions made are based on microscopic observations and analyses of cross-section on the gilded metal base. The established data:

- high percentage of Hg (up to 24.41%- Figure 11, analysis 15, Figure 10c: a, b, c; Figure 9, no. 492);
- uneven thickness of the plating layer (from 0.5568 µm to 12.25 µm – Figure 17) and porous structure;
- grain surface gilding (Figure 18), both in flat and sunken areas; and
- evidence in places of gilt having penetrated beneath the enamel and into the metal base (Figure 19),

justify the conclusion that the method of amalgam gilding was used. The practice of applying gold amalgam in the gilding process (fire-gilding) is attested in ancient treatises such as the Leyden papyrus X from the 3rd century AD (Recipes nos. 54, 57),[36] the *Lucca Manuscript* from the 8th century AD (Recipes Z-18, A-7),[37] the *Mappae Clavicula* from the 8th-9th century AD (Recipes nos. 55, 219),[38] the 9th-10th-century AD treatise of Eraclius *De coloribus et artibus Romanorum* (Recipes nos. 14-16),[39] and recipes from Theophilus from the beginning of the 12th century AD (Recipes 35, 37, 38, 39);[40] as well as nos. 43-44 which describe the same technique of gilding but after applying niello.[41]

[25] Heck, Hoffmann 2000, 349, 350-355; Ogden 1982, 130.
[26] Brepol 1986, 18-20; Tsaneva *et al.* 2008, 203.
[27] Mason, Tite 1997, 41, 51-52, 55, 57-58.
[28] Ogden 1982, 129.

[29] Sharenkov 1988, 222, 234.
[30] Sharenkov 1988, 160, 154; Hedfors 1932, 3, 5.
[31] Ogden 1982, 129; Shchapova 1983, 35.
[32] Tsaneva *et al.* 2008, 195; Shchapova 1983, 39.
[33] Sharenkov 1988, 222.
[34] Sharenkov 1988, 244.
[35] Sharenkov 1988, 164; Hedfors 1932, 3.
[36] Sharenkov 1988, 148; Halleux 1981, 97.
[37] Sharenkov 1988, 180; Hedfors 1932, 52-53.
[38] Sharenkov 1988, 205-206, 233.
[39] Sharenkov 1988, 263.
[40] Sharenkov 1988, 327-328, 329-330; Theophilus 1933, ch. 35, 110; ch. 37, 112; ch. 38, 113; ch. 39, 114.
[41] Sharenkov 1988, 331; Theophilus 1933, ch. 43, 116-117; ch. 44, 117.

№ of analysis	Type of enamel	Test №	Chemical composition (w. %)										
			CoO	Na2O	Al2O3	SiO2	PbO	SnO	K2O	CaO	MnO	FeO	CuO
1	Dark blue	1	1.50	2.54	2.88	48.81	12.01	10.17	2.11	7.29	0.75	4.08	7.65
2	Dark blue	1	1.30	11.13	2.76	53.69	10.08	8.79	2.01	6.58	0.57	2.31	1.67
3	White (opaque)	2		2.98	3.12	38.44	15.79	5.34	0.51	2.92	<0.01	1.20	29.54
4	Light green	3		9.88	4.28	54.72	6.89	6.02	1.64	5.48	0.20	1.76	8.68

Figure 15. Chemical composition of the enamels, detected by SEM-EDS

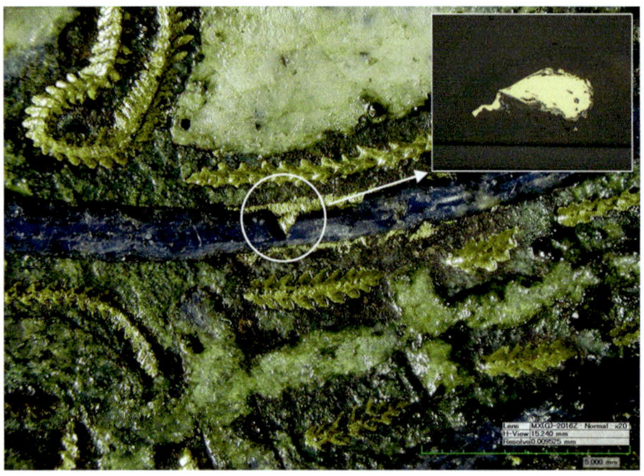

Figure 16. 'Slip' of metal on enamel as result of mechanical cleaning of the surface (x 100), (photo V. Inkova).

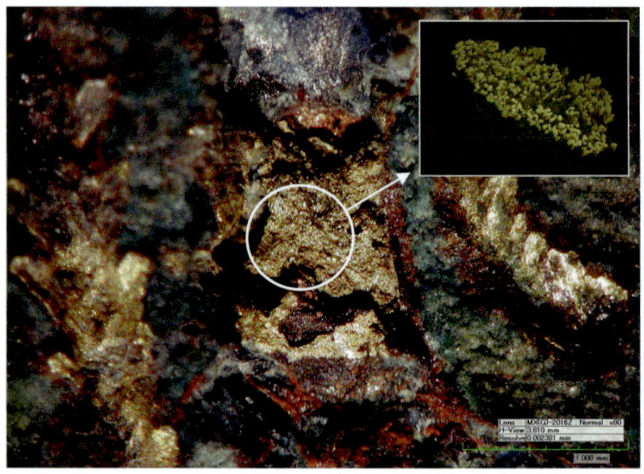

Figure 18. Grain structure of gilding (x 80; x 160), (V. Inkova).

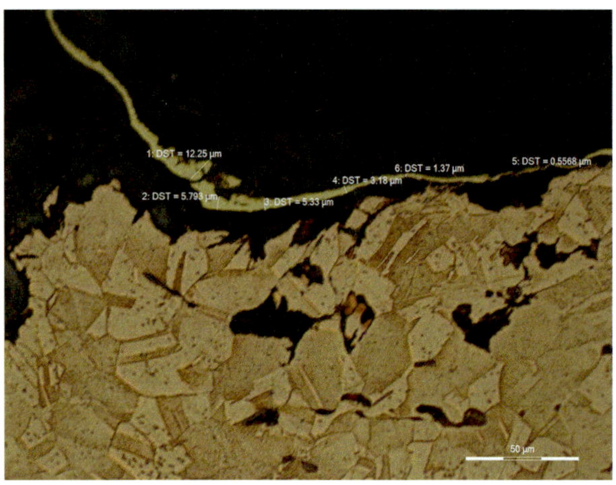

Figure 17. Thickness of the gild layer from 0.5568 to 12.25 μm (photo V. Inkova).

The amount of still preserved Hg, and the rate of diffusion of Hg (Figure 10b; Figure 9, no. 491; Figure 11, analyses 7-8) and Au (Figure 10b, Figure 9, no. 491; Figure 11, analysis 10) in the metal substrate suggest that a relatively short heating was applied at not too high a temperature (*c.* 400° C). The precision of the temperature-time regime of gilding requires further research to clarify the nature of diffusion and intermetallic compounds Au/Hg/Cu. Finally the gilded surface was carefully smoothed and polished.

The composition of the metal substrate –refined, alloyed copper– is not very suitable for casting a massive and solid product. However, it has several advantages that allow alteration in form by plastic deformation, especially for compliance with subsequent decorative techniques –clearing the cells for the *champlevé* enamels by cutting the metal away and fire (not cold adhesive sheet) gilding.[42] The correlation between the basic material and subsequently applied operations speaks to a good understanding of the properties of metals and therefore of a product manufactured in a center with established practices and traditions.

The technological scheme follows the cast-in-copper and gold-plated items, manufactured in *opus limovicense* and decorated with opaque *champlevé* enamel, which is typical of West European metalwork from the Romanesque and Gothic periods, and of the specified group including the candlesticks from Riggisberg, of the Meteora and the incense burners from Kamień, the Blumka Gallery and the Benaki Museum. This technology was rarely used in Byzantine metalworking during the 12th-14th century and was applied, according to Durand, not so much on copper as on silver wares.[43] According to Otavskys, the three censers are made of an alloy with as high a copper content as the item under discussion. It should

[42] Lead, tin and zinc do not allow quality gilding, see note 15. According to Ogden zinc hampers the strong link between the substrate and enamel (Ogden 1998, 134). Theophilus also draws attention to this in Chapter 62 and particularly in Chapters 66 and 68 of his treatise.

[43] Durand 2004, 348, figs 25-26.

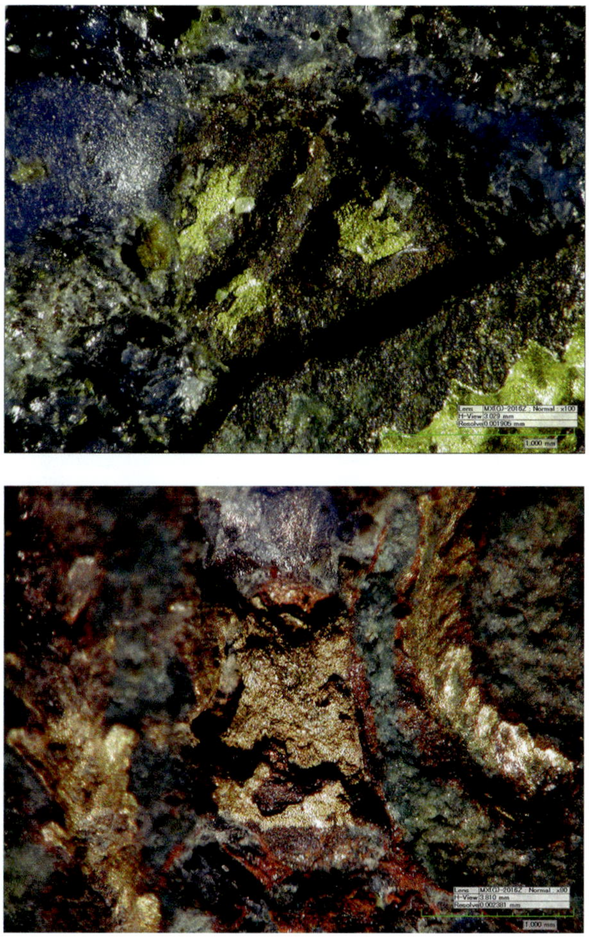

Figure 19. Penetration of gild into the metal base and under the enamel (x 100; x 80), (photo. V. Inkova).

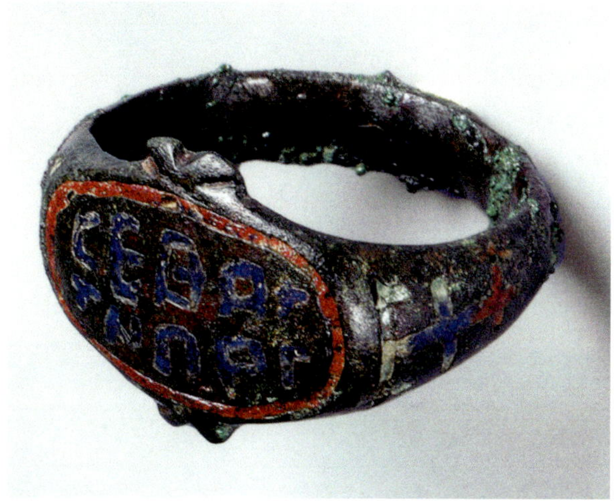

Figure 20. Inscribed finger ring from a necropolis in Aerino, Greece. Copper alloy, enamels, 11th-12th centuries (after A. Dina).

be noted, however, that the incense burner from the Benaki Museum imitates western technology since the enamel is laid not on the cast copper base but into the sunken grooves of the product obtained by matrix molding.[44] As we have already mentioned, the candlestick base from the National Museum of History is cast and additionally shaped by hammering.

Lack of technological research on the items of the specified group, however, impedes our detailed analysis. We can only indicate the identical color repertoire of the residual enamel -dark blue and green for the candlesticks from Riggisberg, and milky white and blue for the incense burner from Kamień. Apart from the Benaki censer, a finger ring may be added to the examples of *champlevé* (?) enamel on copper alloy substrate from the territory of the Empire. The ring has an inscription in Greek ΣΕΒΑΣ/ΤΝΠΡΓΓ -it is dated to the 11th-12th century AD and originates from a necropolis in the Aerino municipality, Magnesia, Greece (Figure 20).[45] Besides the identical blue and green colors, red is also present in the color range of the enamel. Another technological feature suggests connections of the discussed items with Romanesque and Gothic metalwork products. All representations have an additional engraving on the contours, which is characteristic of some products of Western European art from the 12th-13th century AD, made in the same technique.[46]

Morphology and construction

The comparison between the item from the National Museum of History and the bases of the candlesticks from Riggisberg and Meteora -their identical round shape with concave cross-section, decorative composition of four medallions with fabulous creatures on an enamel background, their strip-shaped linear band along the periphery broken by the so characteristic anthropo-zoomorphic faces, and the color range of the enamels- immediately suggest that the item under study had an identical function (Figure 21). Candlesticks with a similar construction consisting of separately cast elements (base, stem, spherical knops and a bobeche with a spike for the candle) subsequently assembled, were widespread in the West European practice early in the Romanesque period.[47] Some are equipped with a tripod base, others with a base of four feet. A hallmark in the morphology of the four candlestick bases is the constant presence of the anthropo-zoomorphic 'face' at the base of the feet. A direct parallel is found in several candlesticks made in *opus limovicense* with a strong Aquitaine-Spanish influence, relative to mid- to early second half of the 12th century AD and now housed in Florence, Saint Petersburg and Jerusalem (Figure 22a-b).[48] Their feet seem to grow out of the 'watching' frontal images/masks? of 'lions', with two palmette-like stylized eyes, a nose and especially a mane. The analogy points to a shared compositional principle. The genesis, however, is different. If, for the frontal animal imagery, the genesis may be seen in Late Roman and Sasanian furniture feet which are crowned with busts of lions and griffins, the 'faces' focused on the flame of the candle in the center from the candlesticks under study, reveal the impact rather of the stone sculpture of the Romanesque and Gothic periods of Western art.

[44] Ballian 2004, 128-129, no. 65; Iatridis 1981, 73-74; Μπούρα 1981, 68.

[45] Dina 2002, 442, no. 579.

[46] Gautier 1987, pl. CCXXXIX, no. 288; Kovács 1968, figs 4, 6, 9a, 10.

[47] Falke, Meyer 1935.

[48] Gautier 1972, 81-82, nos. 70-72; pl. XLII-180-181.

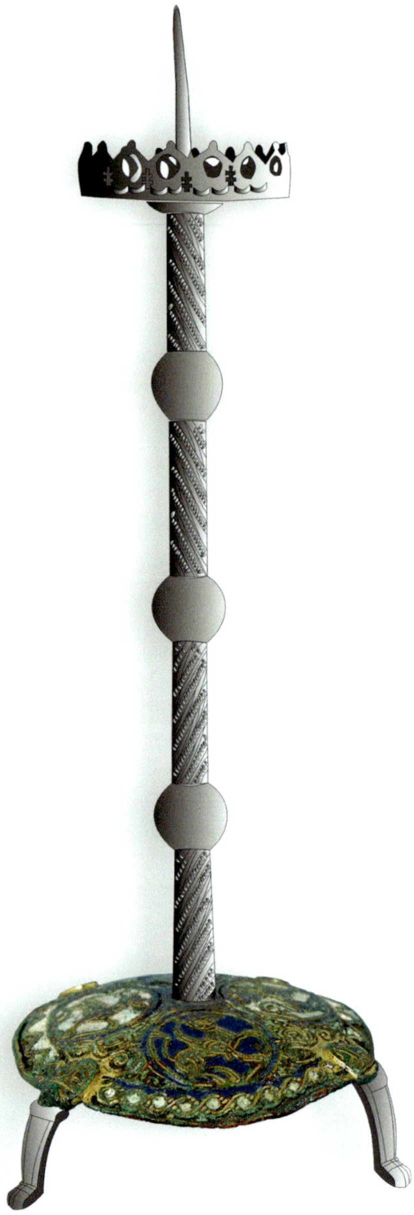

Figure 21. Reconstruction by the artist E. Boyadzhieva.

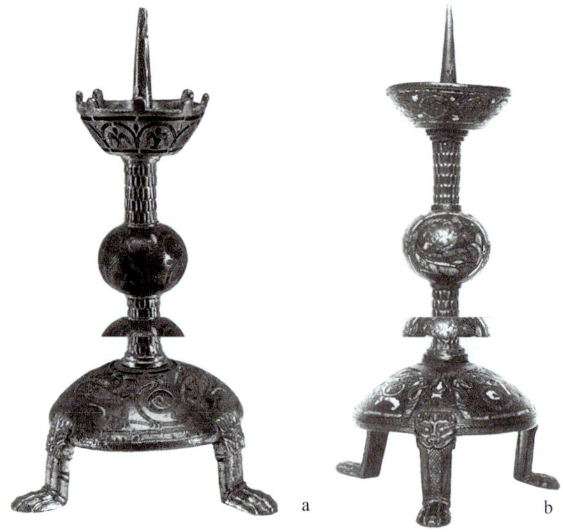

Figure 22. Candlesticks, 1150-1160: a. Florence, National Museum; b. Hermitage, Saint Petersburg, Russia (a-b after M.-M. Gautier).

Composition and decoration

The basic element in the decoration of the four candlestick bases is the representation of mythical creatures in medallions. The medallions on the bases from the National Museum of History in Sofia and from Meteora are adjacent roundels, while the medallions on one of the Riggisberg candlesticks are connected by smaller circles and those on another Riggisberg candlestick are separated by vertical bands. Such an organization of space we find on Sasanian and later on Byzantine silk fabrics from the 8th-11th century AD. The four-part composition of a circular space as in our case, either in sectors or medallions separated by vertical bands is, however, especially characteristic for Romanesque art too.[49]

On the white enamel background of two of the four medallions, the artisan has displayed a predatory animal, more resembling a dog than a lion (Figure 8a-d). A similar fabulous creature is also depicted within two of the medallions from the decoration of the Riggisberg candlesticks (Figure 8e-g), whose stylistic and iconographic features are comparable. Despite the common characteristics of the images, the details of the fur are rendered differently –in one of the images the fur zone is densely covered by short notches, while in the other image they are arranged in vertical rows, imitating a tiger's pelage. On the second candlestick, carved circlets render the fur coat of the animal, perhaps suggesting the representation of a leopard. So the overall silhouette, iconography and artistic approach probably indicate that the Riggisberg pair of candlesticks and the candlestick base from the National Museum of History in Sofia were made in the same workshop, while the differences suggest the competence of different masters. The most direct analogy is found on the incense burner from Kamień, where the image of the lion-dog is almost identical –the same posture, with arched neck and a head turned backwards and a front right paw raised high (Figure 8h). In the same way, short carved lines imply the pelage of the animal is like that of the dog represented on the candlestick from the National Museum of History, but only one of the ears is pointed. There are, however, some differences: the animal from the censer is marching right, and the neck is gracefully extended which allowed the artisan to depict the mane, left loose on one side. The tail, however, is shortened, without knots or palmettes, and barely thickened at the end, a characteristic feature of the lions. Comparative analysis identifies the hand of one skilled master, who, carving repeatedly one and the same image on different objects, freely changes the representation of certain details. In publications by western researchers the incense burner from Kamień is defined as a product of the Byzantine art, which draws attention to the eastern ornamentation. Indeed, the iconography of the image could be explained by

[49] Gautier 1987, 67-68.

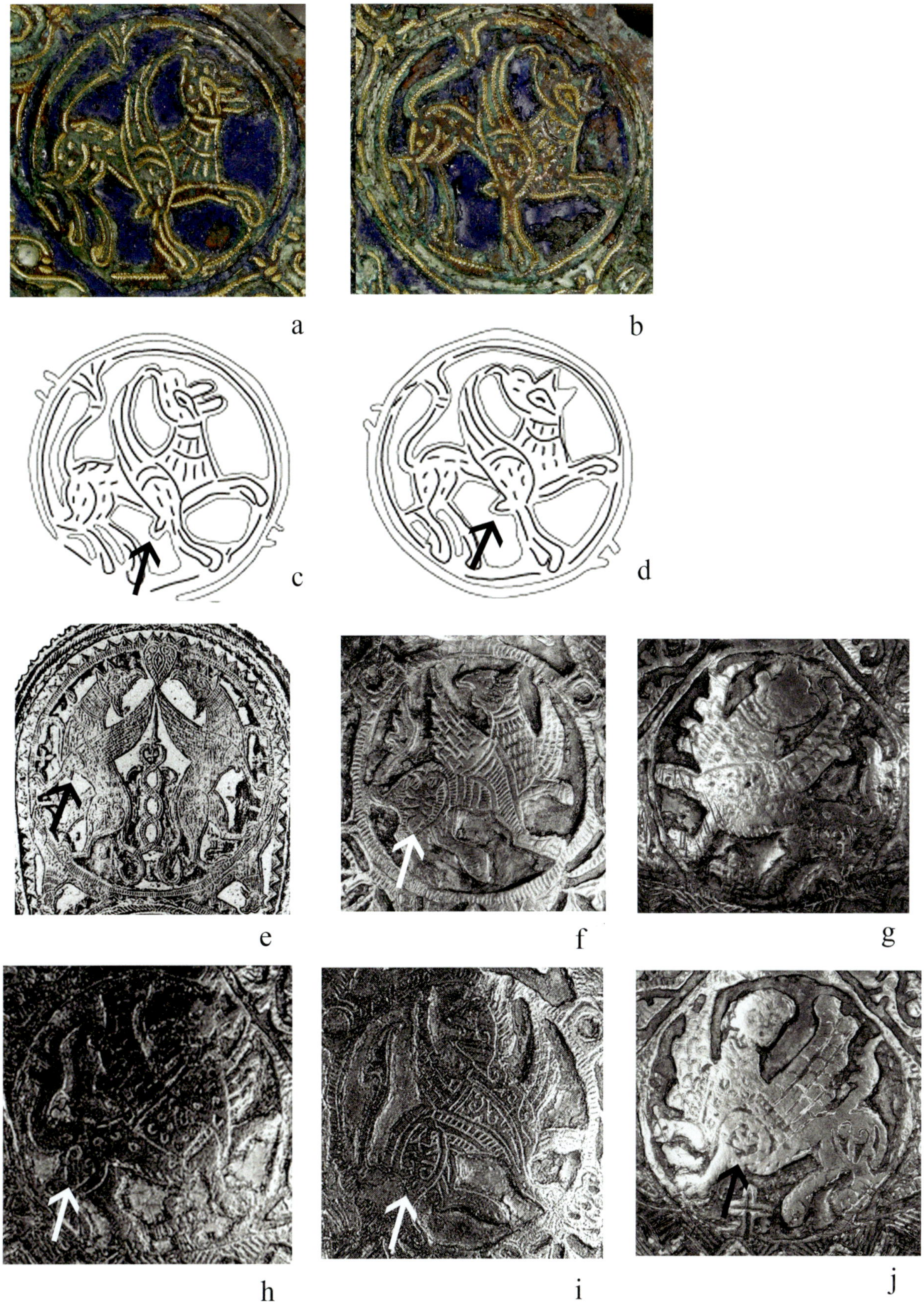

Figure 23. Images of griffins on: a-d. The candlestick base from the National Museum of History, Sofia (a-b photo I. Levicharov, c-d I. Nikolov, detail); e. Incense burner from the Gallery of L. Blumka; f-j. Candlesticks from the Abegg Foundation, Riggisberg (e-j (after K. Otavsky).

comparison with richly illuminated Byzantine manuscripts[50] and with silk fabrics decorated with similar representations in circular medallions, where the hunted game (most often a lion or panther) is running with a lifted front leg and a neck bent backwards. In this connection, the panther (?) on the silk fabric from the Museum of Vatican could be mentioned, which is in a similar posture and the hairs of its coat are suggested by short identical vertical lines.[51] Of the examples provided by metalworking suitable for comparison is one of the medallions with an image of a lion-dog with a head turned to its tail on a silver bracelet from the Louvre, dated to the 12th-3th century.[52] Closer, in terms of style, however, is the relief image of a lion in a medallion seen on a stone plate from Thessaloniki, also of Byzantine production but dated to the 11th-12th century AD (Figure 8i).[53] The different materials and technologies, however, change the characteristics of a certain representation, which impedes our analysis. Distant is the similarity with the lion-dog-in-roundel design of an open-work gold appliqué on a garment from Veliki Preslav, dated to the 10th century AD.[54]

So at this stage, direct parallels cannot be identified. Undoubtedly, after the 10th-11th centuries AD the Byzantine patterns that had penetrated to the Latin West began to be copied and to inspire greatly the figurative repertoire of Romanesque and Gothic art, including the enamel production of the Limoges ateliers. The comparative analysis, however, shows significant differences as well -the animal representations have a characteristic protruding chest, contrasting against the fine, helix-like curved feather tips of their wings. Nevertheless, a fairly close analogue from the 13th century AD is the image of a dog in the same posture on a decorative round piece from horse trappings, now in the Museum of Alzey, Germany (Figure 8j).[55]

The griffins in the other two medallions (Figure 23a-d) have direct stylistic and iconographic parallels with the images on the incense burner from the Blumka Gallery (Figure 23e) and the candlesticks from Riggisberg (Figure 23f-j). Although the griffins on the candlesticks are represented with outspread wings, the bodies of the mythical creatures are identical – they all are walking right with front left legs raised high, have short necks and heads bent to the feathers, while the front right leg is organically connected to the wing. The difference is again in the richer repertoire used in the decoration of the griffins' pelage from Riggisberg -short notches arranged in curved lines, tendrils and even a foliate medallion with inscribed five-leaf palmettes on the flanks of the beasts. In comparison, the griffins on the base from the National Museum of History are sparingly decorated. We cannot be clear whether these observations suggest the hand of different masters, but it is certain that this is the production of the same workshop. Apart from the different iconography and the rampant griffins on the incense burner, they are even closer to those from the discussed base. A very characteristic feature is the triangular bulge on the curve of the lifted front legs in the representation of the fabulous creatures from the two objects (Figure 23a-e) indicating an identical pictorial manner. Similar are the three-band collars, the double bows, separation of the thigh/flank and the wing of the animals, as well as the ears rendered as small triangles. We believe the identified similarities provide sufficient evidence that the candlestick base from the National Museum of History and the censer from the Blumka Gallery are most likely the work of the same master.

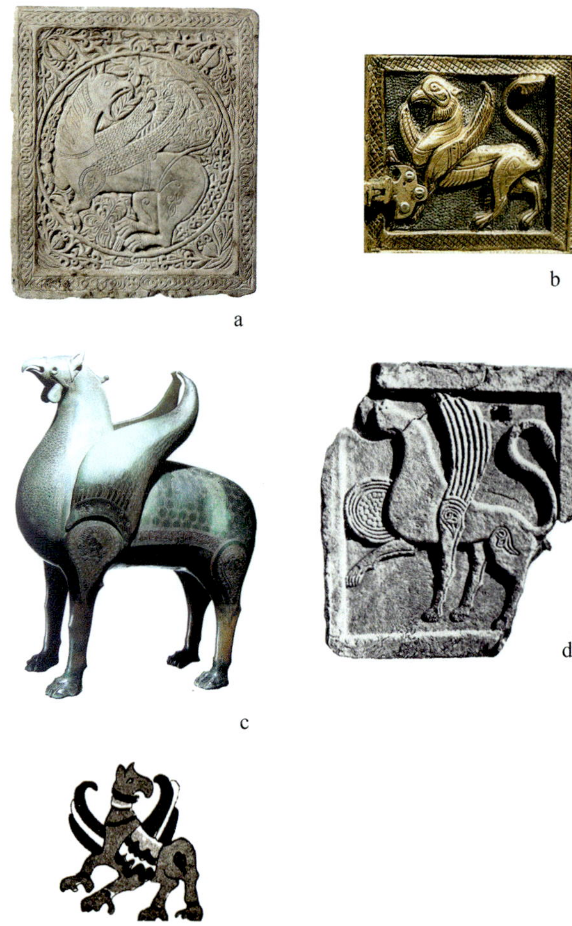

Figure 24. Images of a palmette within a medallion, on the flank of the animal from: a. Relief from Central Greece, 13th century (after S. Brooks). b. Incense burner from San Marco, 12th century (after D. Gaborit-Chopin); c. Griffin from Pisa, 11th century (after R. Ward); d. Relief from the island of Halki, 10th-11th centuries (after V. Darkevic); e. Cod. 213, 10th century, Sinai, 10th century (after K. Weitzmann).

The origin of the composition of griffins within circles again points to eastern (Sasanian) silk fabrics whose tradition continues well into the 13th century AD. A similar theme is represented in relief on a marble Byzantine slab from Central Greece or from the Balkans (Figure 24a).[56] The griffin is in the same posture: the head bent back, the front leg raised, and the hind legs static. The detailed rendering of the feathers and the wings with helicoid ends draws the image closer to the griffins on one of the Riggisberg candlesticks. Moreover, the

[50] Weitzmann 1935, pls XXVI-142; LXXI-430, 427.
[51] Muthesius 1997, 117, 19 A, M 40.
[52] Darkevic 1975, 272, fig. 387.
[53] Bulgarie 1980, 99, 104-105, no. 153.
[54] Stanilov 2010, 161-164, fig. 4.
[55] Frey 2007, 55, fig. 1.
[56] Brooks 2004, 112-113, no. 58.

both the griffin's flanks are decorated with palmettes either separate or inscribed in leaf-like medallions. Such decoration distinguishes also the mythological creatures on the candlesticks from the Abegg Foundation (Figure 23f-j), on the well-known censer from San Marco Cathedral (Figure 24b), supposedly produced in Constantinople, Venice or in some workshop in South Italy in the 12th century.[57] The genesis of the motif again shows the East, but this time it is found on an object with rich Arab *kufi* ornaments. We see it on the flank of the griffin from Pisa (Figure 24c).[58] The object is dated to the 11th century, and the assumptions about the place of its production vary between Spain, Sicily, Egypt and Iran. The same pattern decorates the flanks of the griffin on the plate from the island of Halki, dated also in the 11th century (Figure 24d)[59] and the lion's image on the plate of Thessaloniki (Figure 8i). Even earlier in date, from the 10th century, are parallels in Byzantine manuscripts and metalworking. In Cod. 213 from Sinai a griffin is depicted with a drop-like medallion on the hind flank (Figure 24e).[60] A five-leaf palmette in cloisonné enamel technique adorns the flanks of the griffins on the plates from the Preslav treasure.[61]

A closer parallel to the composition on the censer from the Gallery of Blumka (Figure 25a) we find on a stone medallion from the church of Ravanica, produced according to Sl. Ćurčić around 1377 and suggesting links with the art of the Middle Byzantine period (Figure 25b).[62] Here again the wing and the leg are organically linked, the feathers are 'separated' by the same double bows. The eyes are almond-shaped, the beak is curved and long, one front leg is raised high, and the decoration with the double collar is repeated. The same distinctive features are represented in the griffin's image on the stone relief from the island of Halki, dated to the 11th century,[63] which extends the Byzantine parallels. Somewhat similar features we find in the decoration of Pope Clement II's socks, undoubtedly a product of Constantinople from the first half of the 11th century (Figure 25c-d). The griffins are in mirror symmetry, with heads turned back and the same distinguishing sharp chins.[64] Between the heads of the griffins on the censer from the Blumka Gallery there is a leaf-like motif with an inscribed palmette. Under the wings, another palmette grows in a roundel, formed by the animals' intertwining tails. Such a compositional solution is offered on a spherical incense burner, a product of Romanesque art, where between the two symmetrically positioned birds the same foliate motif is depicted, surrounding the palmette.[65] It was made in the 12th century in Catalunya, in whose area the influence of both Limoges and the Orient were quite tangible.

Finally we draw attention to another common motif. Two of the fabulous creatures on one of the candlesticks from Riggisberg (Figure 26a-b) and the griffins on the silk of the reliquary of St Potencianus carry in their beaks twigs with tiny palmettes (?) (Figure 26c). The style, as indicated by the material (fabric and metal) is different, but the iconographic

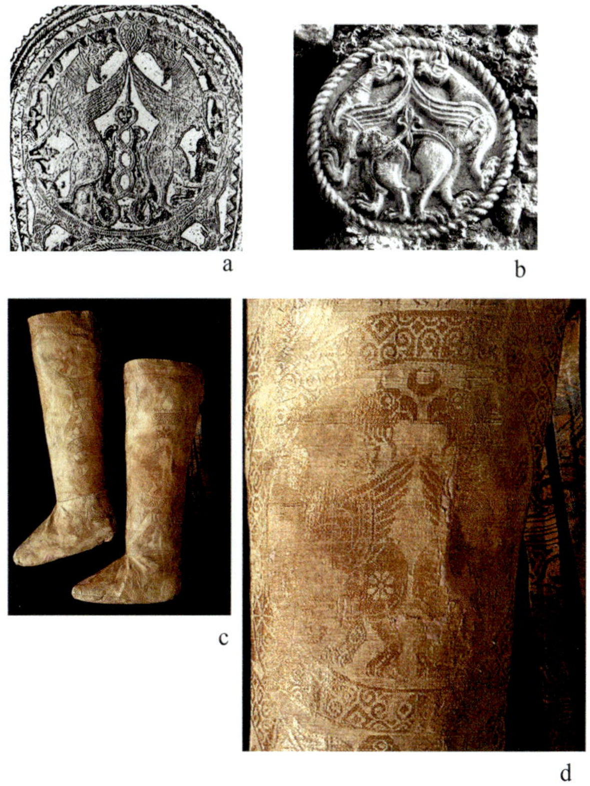

Figure 25. Images of griffins in mirror symmetry on: a. Incense burner from the Gallery of L. Blumka (after K. Otavsky); b. Relief from the church in Ravanica, Serbia, 1377 (after Sl. Ćurčić); c-d. Socks of Pope Clement II, first half of the 11th century (after A. Stauffer).

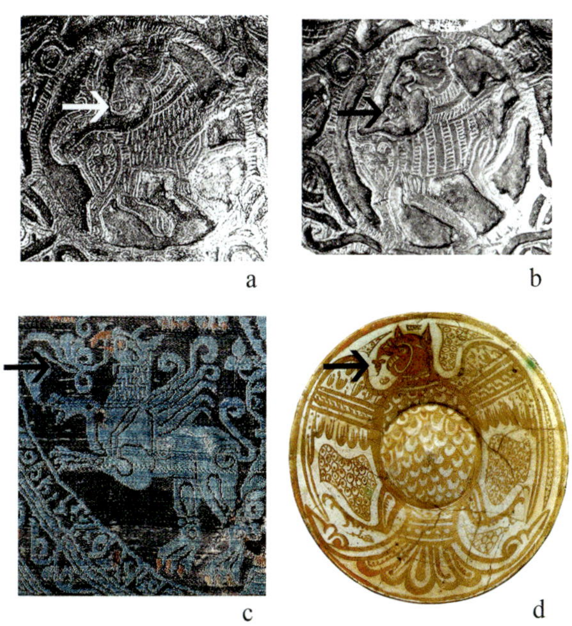

Figure 26. Images of mythical animals with a twig or a tiny palmette in the beak on: a-b. Candlesticks from the Abegg Foundation, Riggisberg (after K. Otavsky); c. Silk of the reliquary of St Potencianus, 12th century (after M. Martiniani-Reber). d. Glazed vessel, 10th century (after M. Jenkins-Madina).

57 Gaborit-Chopin 1984, 245-251; Urbani 2010, 349-350.
58 Ward 1993, 67, fig. 50.
59 Darkevic 1975, 192, fig. 297.
60 Weitzmann 1935, LXXX, 496.
61 Totev 1993, 33, fig. 1, 38-39, fig. 5.
62 Ćurčić 1995, 598, note 11, 603, fig. 4.
63 Darkevic 1975, 192, fig. 297.
64 Stauffer 2010, 172, no. 60.
65 Gautier, 1987, pl. XXX, 141-143.

source is found in Sasanian images of birds from the 6th-7th century,[66] carrying in their beaks a *pativ*, necklace or twig, as in the later versions in metalworking, e.g., the necklace of the Preslav treasure,[67] the Thessaloniki bracelets,[68] the tenth-century Byzantine manuscripts[69] and even Arab painted pottery from the Fatimid period (Figure 26d).[70] At this stage we cannot cite such parallels from Romanesque and Gothic art production. Having in mind the origin of the discussed characteristic motifs, the iconographic relationship of the griffins with Eastern art in its Byzantine and Arab versions, is more justified.

The comparative analysis of the images of carnivorous animals can be completed by comments on the ornamental motifs. The cord-like decoration of two twisted strings along the edge of the candlestick base is present also on three items of the examined group. On the censer from Kamień, such decoration frames the medallion of the lion-dog and the incense holder. On that from the Blumka Gallery it encircles the incense holder only, and on the Riggisberg candlesticks it appears on the spherical knop of the stem. The motive is widespread; it is familiar in Late Antiquity and later on in the Middle Ages. In the eastern artistic repertoire it is primarily used as a framing element on mosaics, in manuscripts or in the decoration of their bindings, less frequently in murals, sgraffito ceramics and mostly in stone sculpture in the wide range of the 8th-13th centuries. As an example we can cite the border mosaic decoration of the palace of Al-Walid II (743-744).[71] Its use is limited to textiles where it serves an identical function –to fill the frame of the medallion on the silk from the St Ambrogio church in Milan,[72] dated to the 8th-9th century. In post-iconoclastic manuscripts the cord-like ornament is used mostly for decorating initials[73] and less often in capital letters,[74] a practice, however, that can be traced in later monuments, for example, in a Menologion from the first half of the 14th-15th century and in a Tetra-gospel from the first half of the 13th century kept in the Library in Oxford.[75] The same ornament is stamped on the leather binding of a Constantinople gospel from the end of the 13th century, thus shaping its outer frame.[76] Worthy of note is also the use of the motif in the vertical bands of a glazed fourteenth-century pitcher from Morphou, Cyprus, where, as on one of the Riggisberg candlesticks, it is combined with a band of trailing foliage.[77] The most widely used instances of the motif are in twelfth-century stone sculpture, chiefly as vertical decoration and less frequently in the arches of the churches in Greece.[78] The same motif frames the tympanum over the small gate in the south façade of San Marco Cathedral in Venice,[79] which in Schneider's opinion, was made in the late 11th century AD by Greek masters. The most direct parallels are provided by the border of the gilt medallion on a silver vessel from the 12th century,[80] as well as by that of the round medallion on an encolpion from Chersonese, dated 10th-11th century.[81] On the wall-paintings the motif is quite schematized, resembling more the twisted strings of a rope. We see it in the Bachkovo monastery of the 12th century[82] and among Coptic monuments –in the church of the Red Monastery, the wall-paintings of which are from the 5th-14th century.[83] It should be noted that on both monuments the motif is combined with a zigzag-patterned band as is the case with the stand under discussion from the National Museum of History in Sofia.

Like the cord-like ornamention, the zigzag one is found on the four items in the group under discussion. On one of the Riggisberg candlesticks it runs along the periphery of the base, dividing the space into opposite triangles. It shapes the outer frame of the medallions on the censers from Kamień and the Gallery of Blumka. It should be noted that the zigzag motif is largely used in western metalworking of the Romanesque and Gothic periods, where it has an identical function –to outline most often round, rectangular or oval shapes. Examples are provided by the decorative plates appliquéd on the urn of St Domingo from Silos, dated in the 12th century,[84] by the plate with an image of Christ Blessing, a work of art of a Rhine atelier from the 13th century,[85] by a pair of thirteenth-century *gemellions* from France,[86] the chest of Valencia dated in the early 14th century,[87] another two *gemellions* from the National Museum in Budapest[88] etc. all made in the émail champlevé technique. We can also add here the above mentioned spherical censer from Catalonia with a zigzag band.

The zigzag pattern is still more widely used in the productions of Byzantine cultural circles. With manuscripts from the 10th to the 14th century it is usually in the header; where the triangular fields are mostly filled with palmettes, floral motifs or *kufi* ornaments.[89] Among these falls also a Greek tetra gospel from the 11th-12th century made in a South Italian atelier.[90] Chevrons surrounding a trefoil adorn the arch above the representation of the monk Liberius on a plate from San Ciriaco church in Ancona.[91] As noted by Schneider, the entire sculptural decoration of the church, dated from the third quarter of the 12th century, is executed in the Byzantine *champlevé* technique and follows eastern prototypes. The motif is also present in the mosaic decoration of 'Roger's room' in the palace of Palermo, dated *c.* 1170.[92] In Byzantine metal work it is used primarily as a frame as is the case with

[66] Trever and Lukonin 1987, 114, no. 24; 115, nos. 27, 79, 83.
[67] Totev 1993, 48, fig. 13.
[68] Haseloff 1990, 31-33.
[69] Weitzmann 1935, pls III-15, VI-27, XXVI-143, XXXVII-209, LXIV-385, LXX-419-420, LXXII-434, LXXV-458, LXXVII-476, XC-562-563, XCII-888, 592.
[70] Jenkins-Madina 1986, 416, no. 272.
[71] *L'art* 1979, 282.
[72] Muthesius 1997, 174, fig. 25 A, M 31.
[73] Weitzmann 1935, 87, fig. 94, XC-567, 566, XCIII-607.
[74] Weitzmann 1935, LXXXI-508.
[75] Hutter 1993, 202, fig. 587; 203, fig. 595; 79-80, 120, fig. 391.
[76] Hutter 1977, 96-98, 238, fig. 366.
[77] Zachariou 2002, 337, no. 379.
[78] Μπούρας, Μπούρα 2002, 73-55, 416-445, 453-493, 530–533; Voyatzis 1996, 40, no. 6.
[79] Schneider 1981, 169, fig. 52.
[80] Ballian-Drandaki 2003, 54, fig. 6 a, 57.
[81] Yashayeva 2010, 305, no. 381.
[82] Mavrodinova 1995, 278, 280, 1.
[83] Grossmann 2004, 58, 382, fig. 16.
[84] Gautier 1987, I, pl. XLVIII, no. 78, fig. 197, L, fig. 203.
[85] Barsali 1966, 41, fig. 14.
[86] Cherry 1991, 29, fig. 35.
[87] Cherry 1991, 31, fig. 49.
[88] Kovács 1968, 38, 41.
[89] Weitzmann 1935, XLV - 264, LXXXI - 506; Hutter 1977, 94-98, 237, fig. 364-58; Hutter 1982, Band 3.2, 34, fig. 123-32, 137, fig. 424-109, 140, fig. 445-118, 187, fig. 577-163; Hutter 1993, 202, fig. 589-45.
[90] Canart and Lucà 2000, 74.
[91] Schneider 1981, 137, 139, fig. 14.
[92] *L'art* 1979, 192.

a portable icon bearing the image of Panagia Hagiosoritissa[93] and a plate with the image of Theotokos with the Child, both dated from the 12th century AD.[94] We see the same motif in the sgraffito ceramics as an element of dress decoration[95] and even in glass jewelry it is the main ornamental motif.[96]

We also draw attention to a detail typical of zigzag decoration: the ribbed triangular fields on one of the candlesticks from Riggisberg. A similar manner of triangular surface decoration is particularly characteristic of Byzantine metalwork from the 12th century. We find it also in the bands at the base of the censer from Venice. Identically decorated are the lead myrrh flasks -*koutrouvia* with St Demetrios' image- undoubtedly a mass production of Thessaloniki in the 12th century.[97] Analogy can be found in stone sculpture from the period, including a border on a plate from the museum in Thebes,[98] and very probably the similar ribbed triangles imitate the wall-painting on the arch in the church of Theotokos on the Rock in Polyfengos, also dated to the 12th century.[99] This very characteristic pattern once again puts the group of monuments under discussion firmly in the context of the Eastern artistic tradition.

The 'leafless' scrolling tendrils of the third motif fit into the typical eleventh-thirteenth-century dry linear graphic repertoire of Western and Byzantine ornamentation. Spirals, with or without 'links', appear in the headers of Greek manuscripts in Calabria,[100] or fill the background fields in metalwork.[101] The nearest parallel to the composition of two symmetrical small helixes with a third one between them we find among the foliage trailers in the mosaic of 'Roger's room' in the palace of Palermo, where, incidentally, the same fierce animals are represented.[102] The style of decoration there is not so much influenced by Byzantine artistic tradition as by ancient and particularly by Arab cultural practices. The 'crosses' organically growing from the spirals, are actually stylized three- or five-leafed palmettes. The motif is repeated along the periphery of the candle pan of the Riggisberg candlesticks. In the first case it resembles the silhouette of a cross, while in the other it is more palmette-like. Its origin can be traced to earlier monuments of the Byzantine cultural circle. A similar and rather modest decoration is found on the shoulders of a Byzantine finger ring from the second half of the 9th century,[103] as well as on the aforementioned ring from Aerino, Greece. Very reminiscent of crosses are the twigs decorating a Hungarian crown from the middle of the 11th century, some of them growing in the same way out from a tendril.[104] At the same time, however, it should be noted that the composition with the anthropo-zoomorphic face or mask, from which tendrils seem to grow (Figure 27a), actually reproduces a motif, typical of the Romanesque sculpture decoration[105] (Figure 27b). The plastic image is perhaps the most telling sign of Romanesque influence in the production of the candlesticks, a practice which is rather unusual in the Byzantine tradition. The artistic repertoire of sculptural decoration in the abbey of Moissac, France, offers an almost identical 'face'.[106] The anthropo-zoomorphic representations on a capital in the gallery possess the same broad nostrils, curved moustaches and animal ears. The space beneath is occupied by tendrils 'tied' with double strips. The same face with the typical broad nose and stiff moustache is presented flat on the spherical knops of the candlesticks from Riggisberg.[107]

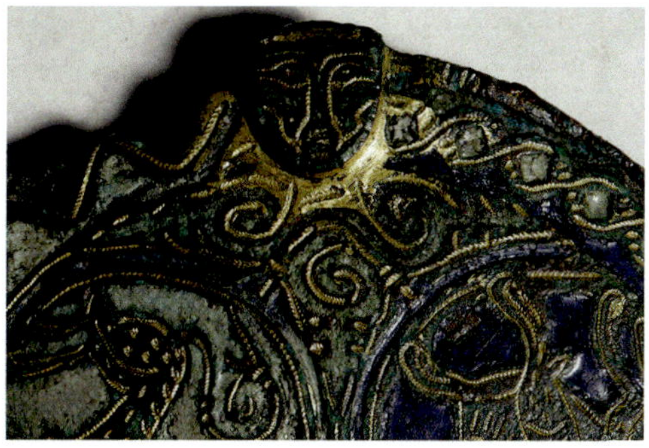

a

b

Figure 27. a. 'Faces' with scrolling patterns on the base from the National Museum of History (photo I. Levicharov); b. Romanesque frieze, Musée Cluny (after R. Hamann-M. Lean).

Three short notches on the outer border frame imitate 'double ties'. We see similar 'bundles' consisting of a few short incised lines on the periphery of a Byzantine sgraffito bowl in the Archaeological Museum of Kilkis, dated late 12th-early 13th century.[108] It is possible that the motif is part of the repertoire of the Byzantine ornamental system during that period. On the stand, however, the notches in the animals' tails, in the tendrils, the cord-like ornament and the frame are rationalized as single or double 'ties'.

[93] Vaklinova 1972, III, 13-18.
[94] Durand 1992, 332, no. 246.
[95] Koilakou 2002, 137-138, no. 149.
[96] Kephallonitou 2002, 582-583, no. 827.
[97] Gerasimova-Tomova 1987, 135-140; Gerasimova 1997, 140-155; Kanev 2001, 31-44; Totev 2010, 205-216, 411-416 and cit. there ref.; Bakirtzis 2002, 184, no. 203; Ghini-Tsofopoulou 2002, 185, no. 205; Papadopoulou 2002, 184-185, no. 204; Totev 2011, 135-148, nos. 1-2, 4, 6-12, 15-17, 20, 22-23, 28.
[98] Μπούρας, Μπούρα 2002, 537, fig. 538.
[99] Μπούρας, Μπούρα 2002, 527, fig. 530.
[100] Canart and Lucà 2000, 95 no. 35.
[101] Bank 1978, fig. 14, 5.
[102] L'art 1979, 192.
[103] Cheynet, Morrisson 1992, 309, no. 219.
[104] Maguire 1996, 211, 145.
[105] Hamann, Lean 1987, 393, 14.
[106] Hearn 1981, 125, fig. 93; Henry 1979, VI 9, fig. 4, a-b; VI 10, fig. 5, a-b, VI 12, fig. 6, e, pl. V.
[107] Otavsky 1994a, 140, fig. 9.
[108] Savopoulou 2002, 332-333, 368.

Finally we shall consider two more ornamental motifs on one of the Riggisberg candlesticks, which bind the items to the Eastern-Byzantine tradition. Immediately below the chin of each 'face' palmettes of different shapes develop. One of these patterns –a pointed central leaf with two side leaves symmetrically curved down– repeat the palmette above the door of the Venice censer. Identical pointed palmettes decorate the corners of a slab from the templon screen in the monastery of St Athanasius on Mount Athos, dated by Grabar to the 11th century.[109] The pointed leaves of the trailing foliage from the same candlestick, according to Gautier, are reminiscent of Arab bone-carved palmettes.[110]

We should also mention another shared feature in our stylistic and morphological analysis. As noted, the objects of the group consist of several cast, hammered and subsequently assembled elements, so that the height of the candlestick can be adjusted by adding or removing segments.[111] The handle of the incense burner from Kamień, the stems of the Riggisberg candlesticks and of another candlestick found during archaeological excavations in the Basilica of the Nativity in Bethlehem, have similar decoration imitating twisted ribbons.[112] This suggests the same approach in their manufacture, or perhaps association of the Bethlehem candlestick with the objects of this group and even with the same workshop (?) or rather reproduction of a model already established for such products. Moreover, the decorative style and the colors of the residual enamel –dark blue, azure and milky white– are similar. The applied decorative nostrum probably follows the 'twisted' stone columns, an interpretation supported by the direct analogy between the stem of the Riggisberg candlestick and the column in Saint Paul Trois Châteaux, where smooth and ribbed pseudo-twisted 'ribbons' alternate in the decoration.[113]

Comparative, morphological and stylistic analysis, the correlation between the different images, the equal approach in building a composition, the similar color repertoire –all show interconnections and identify the analyzed objects as products of the same workshop. The main distinctive feature of its production is the well mastered technique of *email champlevé* on copper substrate, typical for the West and the pictorial arsenal rather representative of the Byzantine ornamental system of the Komnenoi. Essential for their identification is the circumstance that this technique is applied in the incense burners of the type *katzion* used in Byzantine liturgical practice. In the East the introduction of the technology of filling troughs or cells in cheaper material such as copper with an opaque vitreous mass, was connected on the one hand with the economic weakness of the Empire during that period; and on the other hand, the Byzantine workshops undoubtedly enjoyed the experience of the Latin East (Frankish kingdoms in Syria and Palestine, the Latin Empire), whose imported goods and artistic output (the Limoges enamels as well) set new fashion trends in style and chronology. But if the art forms could have been copied, the use of the new technology required direct contact between producers in polycultural centers such as Constantinople, Thessaloniki, Mystras and/or the islands of the Aegean and Eastern Mediterranean.

In the specialized literature, the group of discussed liturgical implements is dated to the 13th century. This is consistent with the cultural and political situation in the Empire caused by the Crusades and explains the sequence of Gothic innovations in Byzantine metal work. Much of the adduced parallels, however, point to the 12th century and the Romanesque period style – therefore we cannot exclude the lower boundary extending to the second half of the 12th century, as suggested by Gauthier for the candlesticks from Riggisberg and by Borchers for the Kamień censer. That is what admits to another possible area, Venetia, Torcello, Ancona, South Italy and Norman Sicily, where the symbiosis between the Byzantine East, Latin West and the Arab influence is most notable. The candlestick base from the National Museum of History confirms Otavsky and Bouras's assumption of the increasing production of this workshop. We also do not exclude the fact that yet more of its works are to be found in monasteries.

Function and Symbolism

Candlesticks of this type were widely used during the Middle Ages, both in the profane sphere, in private homes, workshops, public buildings, and in the sacred space, in the church decoration, liturgical practices, monastic scriptoria, etc. If in the early period their use in the household and in the public sector was limited (in comparison with the widespread use of the cheaper oil lamps),[114] then later they were widely used, evidence for which are the pictorial, written and archaeological sources. A miniature from the 15th century illustrates their use in a metalwork shop (Figure 28),[115] and the images of people holding candlesticks represented on a chest from the British Museum[116] indicate their use in a domestic environment. The different heights of the candlesticks, ranging from 20 cm to 70 cm imply their customary location to be on the table or on the floor. The possibility of assembling the candlesticks from separate segments, i.e., of adjusting their height, points to a multifunctional use. The composition 'Return of Vasiliye, Argiro's infant son, to his parents' painted on the wall of the narthex in the Boyana church from 1259, represents a family table lit by a metal candlestick placed on it (Figure 29).[117]

The use of candlesticks in sacred spaces is inherited from ancient practice. Ammianus Marcellinus tells of burning candles in the temple of Apollo in Daphne,[118] and *Notitia dignitatum imperii* describes the candles carried before the dignitaries of the empire,[119] whence the use of candles during a service before the gospel or the bishops is derived. Burning of candles and incense in front of the statue of Constantine the Great was considered a pagan reminiscence and was fiercely criticized by the church historian Filostorgius.[120]

109 Grabar 1976, 68, pl. XXXIX, b.
110 Gautier 1987, pl. XXXIX, il. 173, no. 68.
111 Bouras and Parani 2008, 10. Compare the different heights of the Riggisberg and Meteora examples.
112 Bouras and Parani 2008, 120, pl. XCIII, 335, no. 125.
113 Hamann, Lean 1987, 397, 21.
114 Bouras and Parani 2008, 1.
115 Bonev and Doncheva 2011, 57, fig. 26.
116 Bouras and Parani 2008, 20, fig. 19.
117 Grabar 1978, 79, table LXXV; Popkonstantinov, 2009, 24-25, 31, fig. 12.
118 Cabrol, Leclercq 1914, 1613.
119 Cabrol, Leclercq 1914, 1614.
120 Bouras and Parani 2008, 21, note 104.

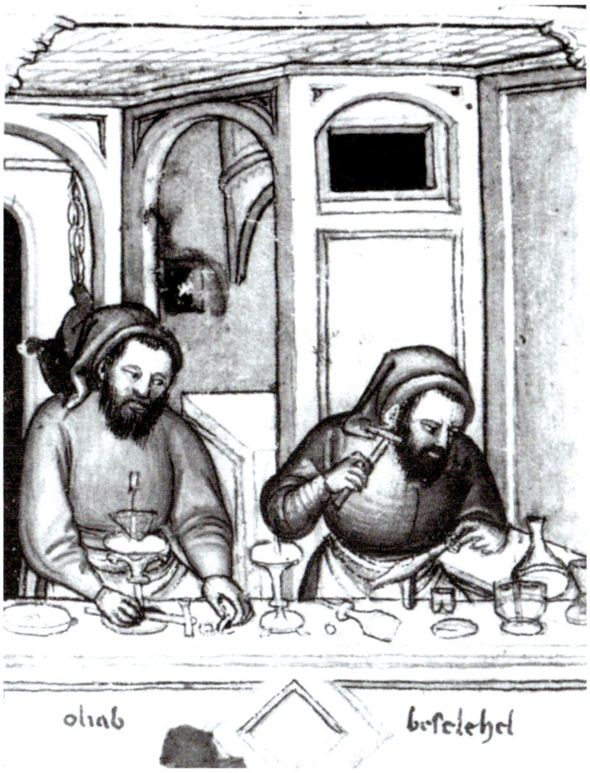

Figure 28. Metalworking shop, miniature, 15th century (after St. Bonev, St. Doncheva).

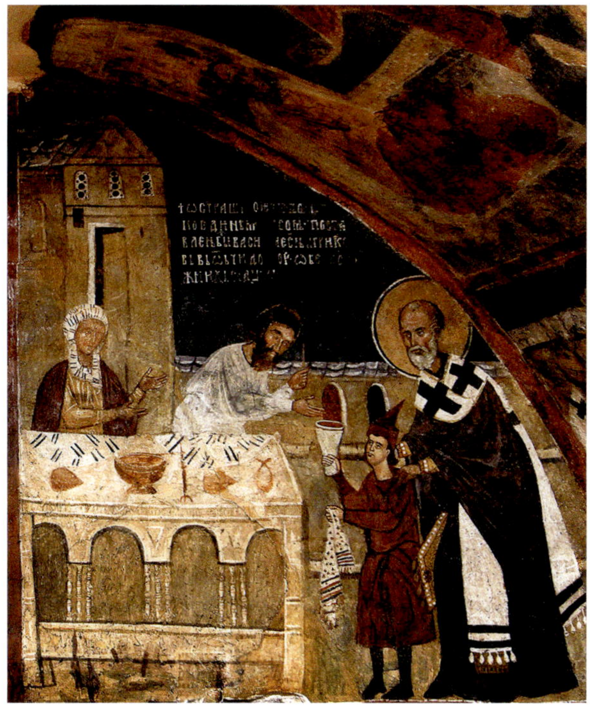

Figure 29. St Nikolas takes young Vasilios back to his parents from Arab captivity, a scene from the life of St Nikolas. Painting on the western wall of the narthex in the Boyana Church, Sofia, Bulgaria, 1259 (photo T. Dimitrov).

There is an opinion in the specialized literature that the use of light in the early Christian catacombs was prompted by purely pragmatic purposes. Subsequently, however, as a symbol of Christ (John I: 9, VIII: 12) and of the salvation of souls, light became an essential ritual component of the liturgy, which had its place in the planning of the church space. According to Christian theology, the church is seen as a 'paradise of light', which is why special attention was paid to its lighting not only by natural light, but also by various luminaries: candelabra, polykandela, candlesticks, chandeliers etc.[121]

Tall candlesticks of the *manualion* type were placed in various locations in the church, in front of the iconostasis, to the side, front or behind of the Holy Altar but always on the floor. An illustration of this practice is the scene 'Divine Service,' painted in the chapel of St George in the church of the Decani monastery, where two candlesticks are placed on the floor behind the Holy Altar (Figure 30a).[122] The written sources provide contradictory information about the presence of candlesticks on the Holy Altar. According to a Decree of Pope Leo IV, the Holy Table should be 'kept free of all items that were not necessary for the festive Mass'. Three centuries later, Pope Innocent III in '*De sacro altaris mysterio*' ordered two metal candlesticks to be placed 'on the Holy Table on both sides of the cross.[123] The practice of placing metal candlesticks on the Holy Altar is first attested in an eleventh-century miniature from the Visegrad gospel (Figure 30b),[124] but became an established practice only in the 15th-16th century.[125] However, an image of the Holy Altar on a silver icon casing from Ohrid, dated to the 12th century, illustrates the same custom in the Eastern Orthodox Church. In the center of the Holy Altar, a cross is placed, flanked by two metal candlesticks with lit candles (Figure 30c).[126]

According to miniatures in tenth-century Byzantine manuscripts large candlesticks flank the canopy, either placed on the ground (Figure 30d) or on its columns.[127]

Church records and typika of the monasteries describe various luminaries used in the churches. *Liber pontificalis* describes the gold and silver candlesticks donated by Constantine the Great to the Great Cathedral in Rome.[128] The Typikon of Gregory Pakourianos for the Monastery of the Mother of God Petritzonitissa in Bachkovo, mentions as among the implements donated to the Bachkovo monastery 'large copper hand candlesticks...other small ones – two items, a candlestick for the iconostasis – fully fixed up, with places for candles – one item'.[129] 'Two cast candelabra for six candles' also figure in Kali Bakuriani's legacy on behalf of the Holy monastery of Iviron on Mount Athos.[130] To this day various candlesticks are kept in the Armory of the monasteries. As previously stated, one of the candlesticks discussed here

[121] Skaltsis 2012, 269-274.
[122] Todich 2005, 434-435, no. 357.
[123] Cabrol, Leclercq 1913, 213; Springer 1981, 25-26.
[124] Springer 1981, A 13, A 17, A 78.
[125] Cabrol, Leclercq 1913, 212-214.
[126] Kondakov 1909, 271, fig. 185; Kalavrezou-Maxeiner 1985, no. 159.
[127] Weitzmann 1935, XV, 79; XVIII, 101.
[128] Bouras and Parani 2008, 2; Cabrol, Leclercq 1913, 211.
[129] Typicon 1968, 62.
[130] Symbattii 1968, 72.

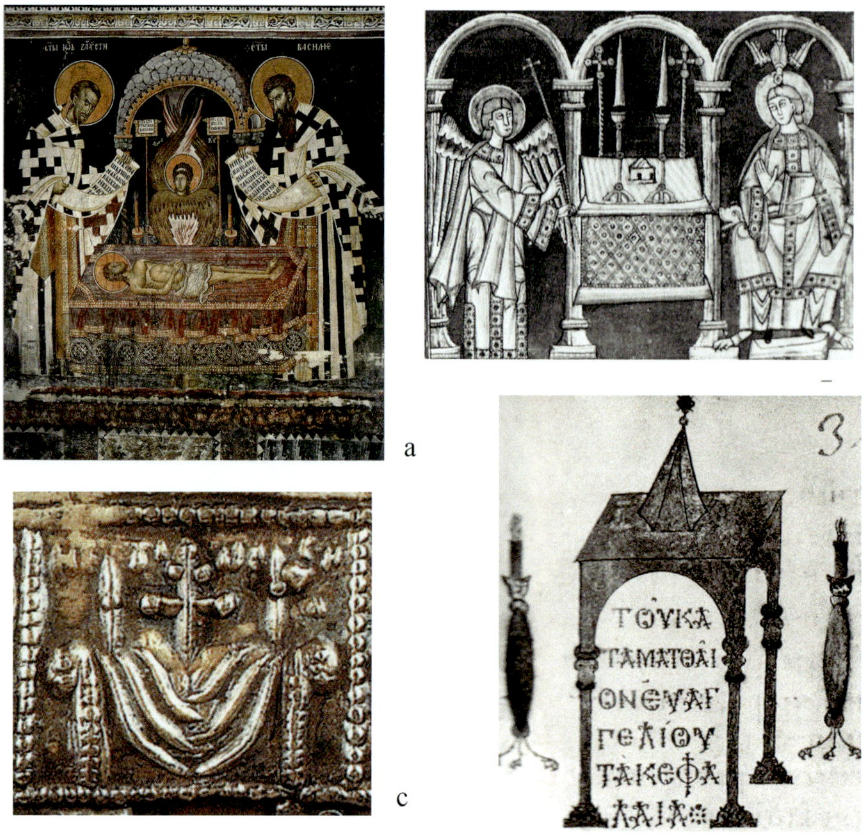

Figure 30. a. 'Service of Liturgy', mural painting from the chapel of St George in the Decani monastery (after Br. Todich); b. Image of the Holy Altar. Miniature from the Vishegrad gospel, 1085. (after P. Springer); c. Image of the Holy Altar from the icon casing with the 'Twelve feasts', 12th century (photo NMH); d. Miniature from Cod. Gr. 70, 10th century (after K. Weitzmann).

belongs to the Transfiguration Monastery on the Meteora, and another bronze candlestick from the 14th century to the Dechani monastery.[131] The painted wall decoration confirms that practice. Namely as a symbol of faith and salvation, the candlesticks are among the donations for churches and saints. In the 'Rescued *voivodes* carry gifts to St Nicholas' scene represented in the church of the Dechani monastery, two tall *manualia* candlesticks are depicted in the hands of the *voivodes'* saved from captivity,[132] while in the narthex of the Boyana church candlesticks flank the throne of the saint (Figure 31).[133]

Burning candles participate in all religious daily and devotional services, from baptism to funeral, thus symbolizing the transition from darkness to light.[134] Skaltsis indicates that the custom of holding a lit candle during Divine service was already established in the 4th century.[135] As a symbol of Christ, life and salvation, light and consequently lit candles are invariably present in the weekly liturgy and festive services, in various processions, such as *adventus*, ceremonies of welcoming emperors, ecclesiastical persons or relics, inauguration of churches, etc.[136] With purely pragmatic

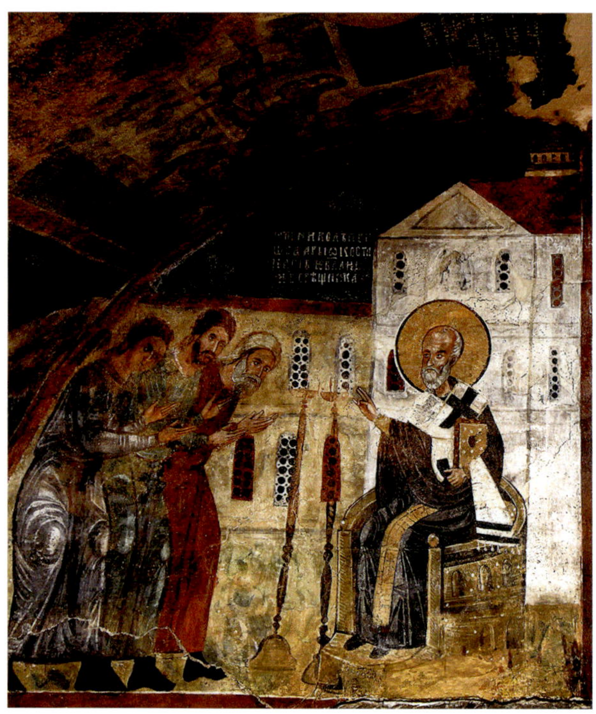

Figure 31. 'Rescued *voivodes* carry gifts to St Nicholas', a scene from the Life of St Nikolas, western wall of the narthex in the Church of Boyana, Sofia, Bulgaria, 1259 (photo. T. Dimitrov).

[131] Todich 2005, 49-50, no. 38; Shakota 1984, 185, fig. 9.
[132] Todich 2005, 475- 476, no. 392.
[133] Grabar 1978, 78, табл. LXXIV; Dimitrov 2008, 16, (25).
[134] Arhiepiskop Beniamin 1884, 351, 374, 416, 430; Psaltou 2012, 262.
[135] Skaltsis 2012, 271.
[136] Bouras and Parani 2008, 21; Galavaris 1978, 69-78.

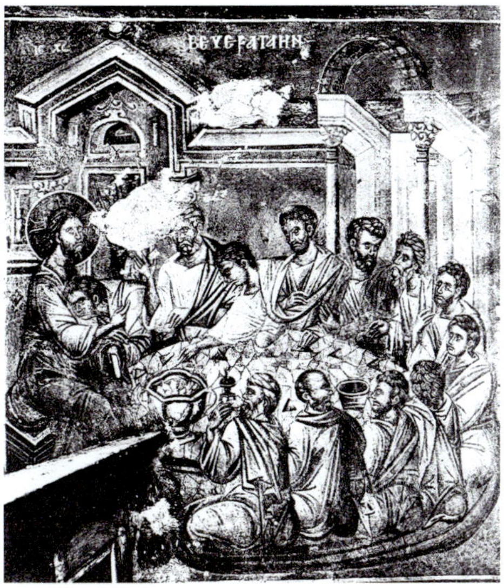

a

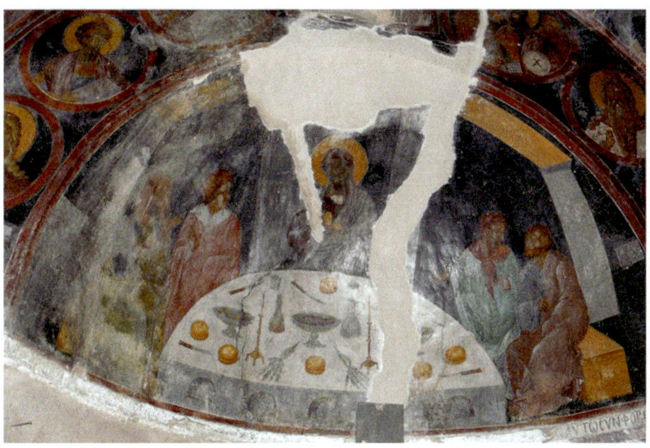

b

Figure 32. a. The 'Last Supper', wall-painting from the Church of St Nicetas near Skopje, beginning of 14th century (after G. Millet); b. Supper at Emmaus, wall-painting from the Church of St Antonios in Vrontisi Monastery, Crete (after I. Spatharakis).

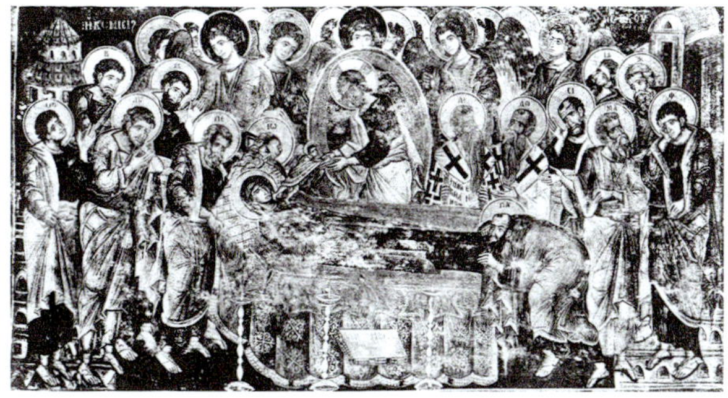

a

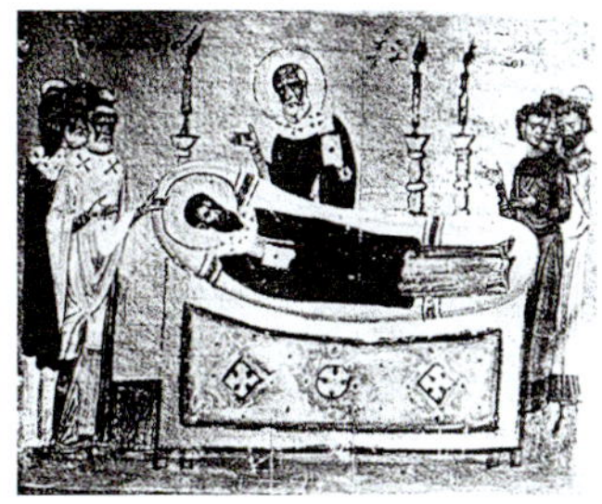

b

Figure 33. a. The Assumption of the Theotokos (Virgin), wall-painting from the Church of St Nikolas in Prilep (after G. Millet); b. The Assumption of St Vasilios, cod. I. 6 (University Library in Turin, after G. Galavaris).

functions tall candlesticks were used in monastic scriptoria as is seen in two tenth-century miniatures of cod. Add. 28815 with a representation of John the Evangelist,[137] and in the Dobreisho Gospel from the 13th century.[138]

In different compositions candlesticks are depicted in the murals in the churches. Depending on the context, they have specific meaning, but mostly, like other realities, refer to the actual performance of a ritual.

The soteriological semantics of the lit candle probably inspired the depiction of metal candlesticks in the mural paintings with 'Last Supper' compositions, where they most often illuminate the sacred table. As an example we point out the scene from the early 14th century in St Nicetas Church, near Skopje (Figure 32a),[139] from the St George Church in Staro Nagoričino[140] and from the Protaton Church, from the beginning of the 14th century, Philoteon (1540) and in the catholicon of the Lavra monastery (1535) on Mount Athos.[141] A tall *manualion* candlestick illuminates the Last Supper, as represented in the mural paintings of the Great Meteora Monastery.[142]

Two candlesticks also decorate the table in the composition 'Supper at Emmaus' on the half cupola of the apse of St Anthonios church in Vrontisi, Crete (Figure 32b).[143] The scene represents the recognition of Christ, expressed through the breaking of bread and the surprise of his disciples; thus the scene symbolizes the sacrament. According to Spatharakis, this interpretation is maintained not only by placing the scene in the half cupola, but is also reinforced by the entire iconographic program of the murals in the church. The Gates of Heaven with a procession of angels carrying candlesticks and heading for the 'Supper' are depicted there.[144] In this context, the candlesticks illuminating the table, again symbolizing Christ and His grace, salvation and the light in our souls, as indeed Simeon Metaphrastes mentions in his Prayers of Thanksgiving after Communion.

As a symbol of salvation and immortality of soul, candlesticks with lit candles also participate in funeral rituals. Their representation on early Christian sarcophagi is an expression of the same soteriological concept. The iconographic model was established in Late Antiquity and further developed through the introduction of different imagery in medieval

[137] Weitzmann 1935, XXV, 138.
[138] Musakova 2000, 195, 71.
[139] Millet 1962, pl. 41-2.
[140] Millet 1962, pl. 83-1.
[141] Millet 1927, pl. 26–2, pl. 95-1, pl. 124-2.
[142] Papanikola-Bakirtzi 2002, 226-227.
[143] Spatharakis 1996, 249-292, fig. 2-3.
[144] Spatharakis 1996, 261, 268.

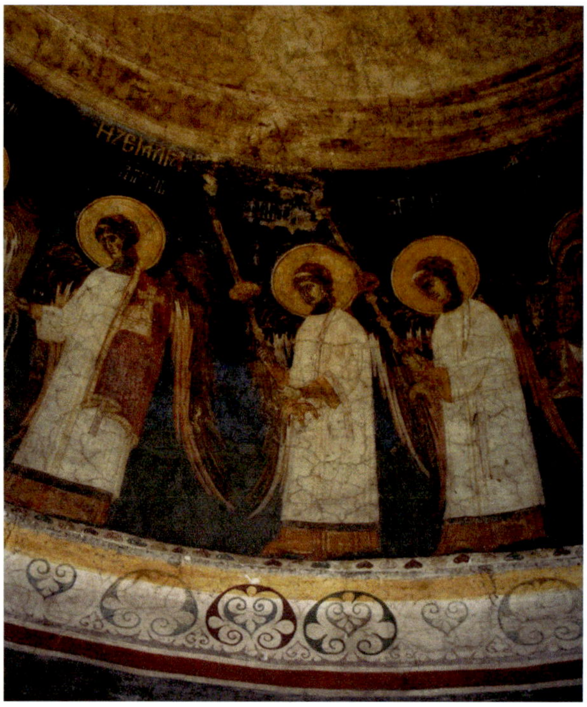

Figure 34. Divine Liturgy, wall-painting from the church in the Decani Monastery (after Br. Todich).

artwork. In the compositions of the 'Assumption', consistently two tall candlesticks with burning candles are depicted at the bed of the Theotokos. Similar tall *manualia* candlesticks with several spheres on the trunk can be seen in the hands of the angelic characters behind her bed. Examples here may be provided by the scenes with the Assumptions from the Sopocani Monastery from the end of the 13th and the beginning of the 14th century,[145] from the church of St Nikolas in Prilep from the end of the 13th or mid-14th century (Figure 33a),[146] from St Nikita church,[147] from the Churches St George in Staro Nagoričino,[148] Panagia Peribleptos in Mystra[149] and Saints Peter and Paul in Veliko Tarnovo.[150] In a depiction of the Assumption from Boyana Church two candlesticks are depicted in front of and another two behind her bed.[151] Thus arranged, they resemble the candlesticks in front of and behind the altar and confirm the semantic equivalence of Mother of Christ's bed with the Holy Altar. The origin of the iconography of the scene with the Assumption is derived from hymnological texts. The homilies of Simeon of Thessaloniki tell how the Theotokos herself lit a candle and a censer at her deathbed.[152] Subsequently, the scene became a model and was admitted in the life cycles of the saints. For example, we see the same tall *manualia* candlesticks depicted behind the bed in the scenes with St Basil's assumption in the miniatures from cod. I. 6 (Biblioteca Nazionale Universitaria di Torino. Figure 33b), cod. 146 (State Historical Museum, Moscow), cod. 107 (Vatopedi Monastery, Athos),[153] and in the Assumption of St Nikolas scene in the Markov Monastery.[154] According to Simeon of Thessaloniki, in the funeral rank burning candles symbolize the divine light with which every Christian is blessed at his/her baptism.[155]

The function of the candlesticks in the scene with the Adoration of the Victim (Melismos) is different. An early version of the scene (12th century) is represented in the altar apse of Bachkovo Ossuary, where in the windows are two large *manualia* (candlesticks).[156] Since the composition does not include the Lamb on the Holy Table, in Y. Boycheva's opinion namely the burning candles on the tall candlesticks and the *ripidia* (fans with which the deacon fans the oblation) in the angels' hands suggest simultaneity of worship in the Earthly and Heavenly Church.[157]

Most often candlesticks are found in the composition with 'Heavenly Liturgy'. In corresponding scenes from the 14th century in the churches of Panagia Peribleptos in Mystra[158] and in the Decani Monastery a procession is featured with angels half turned to the Holy Altar[159] and carrying tall candlesticks thus emphasizing the rituality of the procession and of the whole composition (Figure 34).

They have another meaning in the scene 'Ordination of Bishop Jacob (?)', painted in the apse of the Episcopal basilica St Nikolas in Melnik. Presented are Christ, Apostle Peter and an unknown bishop and above the window, together with two candlesticks, the Cup of Heavenly Manna and the Book of the Law. The representation of objects from the Old Testament tabernacle in this case implies the idea of continuity between the Old and the New Testament churches.[160]

The symbolism of candlesticks depends not only on the context in which they are used, but also it is largely determined by their decoration. The essence of the pictorial 'agenda' of the round base from the National Museum of History in Sofia is the four-shared composition of two eagle-headed griffins and two lions/dogs within roundels. In antiquity the griffin was a symbol of the Greek god of light, Apollo, which justifies its appearance on oil lamps. Besides, the griffin was known for its strong protective function.[161] This motivates its representation both on early Christian sarcophagi[162] and on antique lamps, where it probably acted as a guardian of light.[163] Similar apotropaic functions for the lion-dog explain its presence on a clay lamp from Malaga, Italy.[164] Perhaps, the cross-like detail surmounting the leafless tendrils on the

145 Millet 1957, Fasc. II, pl. 19; Djurić 1963, figs XXVII, XXIX-XXXIII.
146 Millet 1962, Fasc. III, 1962, pl. 27.
147 Millet 1962, pl. 45.
148 Millet 1962, pl. 99.
149 Millet 1910, pl. 101.
150 Grabar 1929, V. I, 276-277, pl. XLIII.
151 Grabar 1978, 52-53, fig. 4; Popkonstantinov 2009, 116, fig. 31.
152 Boycheva 2012, 149. I thank Y. Boycheva for the possibility to get acquainted with the manuscript of her work.

153 Galavaris 1969, VIII, 363; LXIII, 11; LXIII, 328.
154 Millet 1969, Fasc. IV, pl. 106-192.
155 Arhiepiskop Beniamin 1884, 416-417.
156 Bakalova 1977, 74 ff; Bakalova and Karageorghis 2003, 66-68.
157 Bakalova 1977, 74-78, 81; Boycheva, unpublished manuscript, 36.
158 Lazarev 1986, 562.
159 Todich 2005, 337, no. 259; 339, no. 261; 512 no. 429.
160 Boycheva 2012, manuscript, 48-49 and cit. lit. 62-63.
161 In Greek mythology the griffin guarded the gold of the Hyperboreans.
162 Cabrol, Leclercq 1925, 1814, fig. 5468.
163 Bouras and Parani 2008, 15-16, 19, 49-52, cat. 7-9.
164 Cabrol, Leclercq 1930, 1204, fig. 7127. Probably, such protective function has the mythical creature (possibly a leopard) displayed on a clay lamp from the Late Antiquity Ephesus (Ladstatter 2010, 268, no. 284).

base under study has the same function as that of the early Christian luminaries with similar images, which were then Christianized with the addition of a cross, Christograms, inscriptions or other religious symbols. As we have seen, the 'cross' is repeated also on the candle pan of the Riggisberg candlesticks. Indeed, in the 12th-13th century, as evidenced by the Chersonese candlestick,[165] the polykandela from the National Museum in Kiev[166] and the Museum in Metsovo,[167] the lion-dog and the griffin were stereotypically used in the decoration of various luminaries because of their apotropaic functions.

Bibliography

Archibishop Beniamin, 1884. *Novaia skrizhol.* Saint Petersburg.

Bakalova, E. and V. Karageorghis (eds) 2003. *The Ossuary of the Bachkovo Monastery.* Plovdiv: Pygmalion.

Bakalova, E. 1977. *Bachkovskata kostnica.* Sofia.

Bakirtzis, Ch. 2002. Lead myrrh flask (koutrouvion), in D. Papanikola-Bakirtzi (ed.) 2002. *Everyday Life in Byzantium,* exhibition catalogue Thessaloniki, White Tower, 2001, 184. Athens: Kapon Editions.

Ballian, A. 2004. Liturgical Implements, in H. C. Evans (ed.) *Byzantium: Faith and power, 1261-1557,* 117-124. New York: The Metropolitan Museum of Art.

Ballian, A. and Drandaki, A. 2003. A Middle Byzatine silver treasure. *Μουσείο Μπενάκη* 3, 47-80.

Ballian, A. 2004. Standing Censer with Military Saints, in H. C. Evans (ed.) *Byzantium: Faith and power, 1261-1557,* 128-129. New York: The Metropolitan Museum of Art.

Bank, A. 1978. *Prikladnoe iskusstvo Vizantii IX- XII vv.* Moskva: Izd. 'Nauka'.

Barsali, I. 1966. *European enamels.* London: Cassell.

Bezkorovaina, J. 2010. Polikandelon, in *Byzanz: Pracht und Alltag; Kunst- und Ausstellungshalle der Bundesrepublik Deutschland, Bonn ; 26.2. - 13.6.2010,* 326-327. München: Hirmer.

Bonev, St. and Doncheva, St. 2011. *Starobalgarski proizvodstven tsentar pri s. Novosel, Shumensko.* Veliko Tarnovo.

Borboudaki, M. 1999. Incense burner (katsio), in D. Papanikola-Bakirtzi (ed.) *Byzantine Glazed Ceramics. The Art of Sgraffito.* 30, Athens: Archaeological Receipts Fund.

Borchers, W. 1933. *Der camminer Domschatz.* Stettin: Leon Sauniers Buchhandlung.

Bouras, L. and Parani, M. G. 2008. *Lighting in Early Byzantium.* Washington: Dumbarton Oaks Research Library and Collection.

Boycheva, Y. 2012. Liturgichni predmeti vav vizantiyskata ikonografia prez XI-XV v., (rakopis). Sofia.

Bozhilov, V. 2002. Iztok i Zapad v Evropeiskoto srednovekovie. *Srednovekovna hristianska Evroppa: Iztok i Zapad. Tsennosti, tradicii, obshtuvane,* 64-78. Sofia.

Brepol. E. 1986. *Hudozhestvennoe emaylirovanie.* Leningrad.

Brooks, S. 2004. Relief Depicting a Griffin, in H. C. Evans (ed.) *Byzantium: Faith and power, 1261-1557,* 112-113. New York: The Metropolitan Museum of Art.

Bukton, D. 1988. Byzantine Enamel and the West. *Byzantinische Forschungen* 13, 235-244.

Bulgarie, 1980. *La Bulgarie médiévale. Art et civilisation.* Paris: Association Française d' Action Artistique.

Cabrol, F. and Leclercq, H. 1913. *Dictionnaire d'archéologie chrétienne et de liturgie,* vol. 3.1. Paris: Letouzey et Ané.

Cabrol, F. and Leclercq, H. 1914. *Dictionnaire d'archéologie chrétienne et de liturgie,* vol. 3.2. Paris: Librairie Letouzey et Ane.

Cabrol, F. and Leclercq, H. 1925. *Dictionnaire d'archéologie chrétienne et de liturgie,* vol. 6.2. Paris: Librairie Letouzey et Ane.

Cabrol, F. and Leclercq, H. 1930. *Dictionnaire d'archéologie chrétienne et de liturgie,* vol. 9.1-2. Paris : Librairie Letouzey et Ane.

Canart, P. and Lucà, S. 2000. *Codici greci dell'Italia meridionale.* Roma: Retablo.

Cherry, J. 1991. *Medieval Decorative Art.* London: British Museum Press.

Cheynet, J.-C. and Morrisson, C. 1992. Bague de Basil le Parakoimômène, in *Byzance: L'art byzantin dans les collections publiques françaises; Musée du Louvre, 3 novembre 1992-1er février 1993,* 309. Paris: Bibliothèque Nationale.

Ćurčić, S. 1995. Some Uses (and Reuses) of Griffins in K. Weitzmann, C. Moss and K. Kiefer (eds) *Byzantine East, Latin West: Art-historical studies in honor of Kurt Weitzmann,* 597-605. Princeton (N.J.: Department of art and archaeology, Princeton University.

Darkevic, V. P. 1975. *Svetskoe iskusstvo Vizantii: Proizvedenija vizantijskogo chudožestvennogo remesla v Vostočnoj Evrope X-XIII veka = Byzantine Secular Art in the 12th and 13th Centuries. Moskva: Iskusstvo.*

Dimitrov, B. 2008. *Boianskata tsarkva 'Sv. Nikola I sv. Panteleymon'* Sofia.

Dina, A. 2002. Copper alloy ring in D. Papanikola-Bakirtzi (ed.) 2002. *Everyday Life in Byzantium,* exhibition catalogue Thessaloniki, White Tower, 2001, 442. Athens: Kapon Editions.

Djurić, V. J. 1963. *Sopoćani.* Beograd: Srpska književna zadruga i Izd. preduzeće 'Prosveta'.

Drandaki, A. 2004. Handle of Standing Censer (Katzion), in H. C. Evans (ed.) *Byzantium: Faith and power, 1261-1557,* 128. New York: The Metropolitan Museum of Art.

Durand, J. 1992. Plaque: Vierge à l'Enfant, in *Byzance: L'art byzantin dans les collections publiques françaises; Musée du Louvre, 3 novembre 1992-1er février 1993,* 332. Paris: Bibliothèque Nationale.

Durand, J. 2004. Innovations gothiques dans l'orfèvrerie byzantine sous les Paléologues. *Dumbarton Oaks Papers,* 58, 333-354.

Falke, O. and Meyer, E. 1935. *Romanisce Leuchter und Gefässe Giessgefässe der Gotik.* Berlin, 1935.

Frey, A. 2007. Ein mittelalterlicher Anhänger aus dem Museum Alzey. *Sonderdrück aus Alzeyer Geschichtsblätter,* 36, 55-63.

Gaborit-Chopin, D. 1984. Lampe oder Rauchergefäss in Form eines Kupelbaues, in H. Hellenkemper (ed.) *Der Schatz von San Marco in Venedig,* 245-251. Köln: Röm.-German. Museum.

Galavaris, G. 1969. *The illustrations of the liturgical homilies of Gregory Nazianzenus.* Princeton (N.J.): Princeton University Press.

Galavaris, G. 1978. Some Aspects of Symbolic Use of Lights in the Eastern Church. Candles, Lamps and Ostrich Eggs. *Byzantine and Modern Greek Studies* 4, 69-78.

Gautier, M. 1972. *Émaux du moyen âge occidental.* Paris: Office du livre.

[165] Yashayeva 2010, 304, no. 379.

[166] Bezkorovaina 2010, 326-327, no. 441.

[167] Ballian 2004, 123-124, figs 5.7-5.8.

Gautier, M.-M. 1987. *Emaux méridionaux. Catalogue international de l'Oeuvre de Limoges. Tome I, L'Epoque romane.* Paris: Éditions du Centre National de la Recherche Scientifique

Gerasimova-Tomova, V. 1997. Überblick über Ampulen. *Miscelanea Bulgarika* II, 140-155.

Gerasimova-Tomova, V. 1987. Dva arheologicheski pametnika ot Iztochnite Rodopi. *Rodopski sbornik* VI, 135-140.

Ghini-Tsofopoulou, E. 2002. Lead myrrh flask (koutrouvion), in D. Papanikola-Bakirtzi (ed.) 2002. *Everyday Life in Byzantium*, exhibition catalogue Thessaloniki, White Tower, 2001, 245-251. Athens: Kapon Editions.

Grabar, A. 1929. *La Peinture religieuse en Bulgarie.* Mâcon: Impr. Protat frères.

Grabar, A. 1976. *Sculpture byzantines du Moyen Age.* Paris: Editions A. et J. Picard.

Grabar, A. 1978. *Boianskata tsurkva: L'eglise de Boana.* Sofia: Naouka i izkoustvo.

Grabar, A. 1984. L'Asymétrie des relations de Byzance et de l'Occident dans le domaine des arts au Moyen Age, in I. Hutter and H. Hunger (eds). *Byzanz und der Westen: Studien zur Kunst des europäischen Mittelalters,* 9-24. Wien: Verlag der Österreichischen Akademie der Wissenschaften.

Grossmann, P., Brooks-Hedstrom, D. and Abdal-Rassul M., 2004. The Excavation in the Monastery of Apa Shenute (Dayr Anba Shinuda) at Suhag. *Dumbarton Oaks Papers* 58, 371-384.

Halleux, R. 1981. *Les alchimistes grecs: Papyrus de Leyde, Papyrus de Stockholm, fragments de recettes.* Paris: Belles Lettres.

Hamann, R. and Lean, M. 1987. Künstleraunen im Mittelalter. in *Skulpture des Mittelalters. Funktion und Gestalt,* 385-452. Weimar.

Haseloff, G. 1990. *Email im frühen Mittelalter: Frühchristliche Kunst von der Spätantike bis zu den Karolingern.* Marburg: Hitzeroth.

Hearn, M. 1981. *Romanesque sculpture: The revival of monumental stone sculpture in the eleventh and twelfth centuries.* Ithaca, N.Y: Cornell University Press.

Heck, M., P. Hoffmann, 2000. Coloured opaque glass beads of the Merovingians. *Archaeometry* 42, 2, 341-357.

Hedfors, H. 1932. *Compositiones ad tingenda musiva.* Uppsala: Almqvist & Wiksell.

Henry, F. 1979. Romanesque Arches decorated with Human and Animal Heads, in G. Zarnecki. *Studies in Romanesque Sculpture,* VI 1-35. London: Dorian Press.

Hutter, I. 1977. *Corpus der byzantinischen Miniaturenhandschriften,* Band 1. Stuttgart: Hiersemann.

Hutter, I. 1982. *Corpus der byzantinischen Miniaturenhandschriften,* Band 3.2. Stuttgart: Hiersemann.

Hutter, I. 1993. *Corpus der byzantinischen Miniaturenhandschriften,* Stuttgart, Band 4.2. Stuttgart: Hiersemann.

Iatridis, B. 1981. Μέτρηση των συστατικών των κραμάτων τριών θυμιατηρίων. *Αρχαιολογία* 1, 73-74.

Jenkins-Madina, M. 1986. Bowl with Eagle, in E. Evans and D. Wixon (eds) *The glory of Byzantium: Art and culture of the Middle Byzantine era: A.D. 843-1261,* 416-417. New York: The Metropolitan Museum of Art.

Kalavrezou-Maxeiner, I. 1985. *Byzantine icons in steatite: Text.* Wien: Verlag der Österreichischen Akademie der Wissenschaften.

Kanev, Iv. Shest novootkriti solunski ampuli-kutruvii ot XII-XIII v. *Arheologicheski vesti,* 2, 31-44.

Kephallonitou, F. 2002. Glass bracelets, in D. Papanikola-Bakirtzi (ed.) 2002. *Everyday Life in Byzantium,* exhibition catalogue Thessaloniki, White Tower, 2001, 582-583. Athens: Kapon Editions.

Koilakou, Ch. 2002. Glazed bowl with a hunting scene, in D. Papanikola-Bakirtzi (ed.) 2002. *Everyday Life in Byzantium,* exhibition catalogue Thessaloniki, White Tower, 2001, 137-138. Athens: Kapon Editions.

Kondakov, N. 1909. *Makedonia. Arheologicheskoe puteshestvie* (Macedonia. archaeological journey). Saint Petersburg: Izdanie otdeleniya russkago yazyka i slovesnosti Imperatorskoy Akademii Nauk.

Kovács, É. 1968. *Limoges champlevé enamels in Hungary.* Budapest: Corvina Press.

L'art de Byzance et de l'Islam, 1979. Paris; Bruxelles: Elsevier Sequoia.

Ladstatter, S. 2010. Lampe, in J. Frings (ed.) *Byzanz: Pracht und Alltag; Kunst- und Ausstellungshalle der Bundesrepublik Deutschland, Bonn; 26.2.-13.6.2010,* 268. München: Hirmer.

Lafontaine-Dosogne, J. 1995. L'influence artistique byzantine dans la région Meuse-Rhin du VIIIe au début du XIIIe siècle, in Weitzmann, K., Moss, C. F., Kiefer, K. *Byzantine East, Latin West: Art-historical studies in honor of Kurt Weitzmann,* 181-192. Princeton (N.J.: Department of art and archaeology, Princeton University.

Lazarev, V. 1986. *Istoria vizantyskoy zhivopisi.* Moscow.

Maguire, H. 1996. Enamel Plaques Medallions 'The Crown of Constantine IX Monomachos,' in E. Evans and D. Wixon (eds) *The glory of Byzantium: Art and culture of the Middle Byzantine era: A.D. 843-1261,* 210-112. New York: The Metropolitan Museum of Art.

Markov. N. 1995. Edin nov medallion s izobrazhenie na lav ot Tarnovsko. *Arheologia* 1, 41-46.

Mason, R., and Tite, M. 1997. The beginnings of tin-opacification of pottery glazes. *Archaeometry* 39.1, 41-58.

Mavrodinova, L. 1995. L'ornamentation dans la peinture murale en Bulgarie médieéval: Principes décoratifs, motifs, parallèles, in Weitzmann, K., Moss, C. F., Kiefer, K. *Byzantine East, Latin West: Art-historical studies in honor of Kurt Weitzmann,* 277-283. Princeton (N.J.: Department of Art and Archaeology, Princeton University.

Millet, G. 1910. *Monuments Byzantins de Mistra.* Paris: Leroux.

Millet, G. 1927. *Monuments de l'Athos.* Paris: E. Leroux.

Millet, G. 1957. *La peinture du Moyen Age en Yougoslavie. Serbie, Macédoine et Monténégro.* Paris: Boccard.

Μπούρα, Λ. 1981. Παλαιοχριστιανικά και βυζαντινά Θυμιατήρια του Μουσείου Μπενάκη. *Αρχαιολογία* 1, 65-70.

Μπούρας, Χ. and Μπούρα, Λ. 2002. Η Ελλαδική Ναοδομία κατά τον *12ο αιώνα.* Αθήνα: Εμπορική Τράπεζα της Ελλάδος.

Musakova, E. 2000. Tetravangelo di Dobrejšo, in V. Pace (ed.) *Tesori dell'arte cristiana in Bulgaria,* 195-197. Sofia: Borina.

Muthesius, A. 1997. *Byzantine silk weaving AD 400 to AD 1200.* Vienna: Fassbaender.

Nelson, R. 2004. Byzantium and the Rebirth of Art and Learning in Italy and France, in E. Evans and D. Wixon (eds). *The glory of Byzantium: Art and culture of the Middle Byzantine era: A.D. 843-1261,* 515-523. New York: The Metropolitan Museum of Art.

Nelson, R. 2007. High Justice: Venice, San Marco and the Spoils of 1204, in P. Vokotopoulos (ed.) *Byzantine Art in the Aftermath of the Fourth Crusade: The Forth Crusade and its Consequences,* Proceedings of an International Conference, Athens 2004, 143-151. Athens: Academy of Athens.

Oddy, W., La Niece, S. and Stratford, N. 1986. *Romanesque Metalwork. Copper Alloys and their Decoration.* London: British Museum Publications.

Ogden, J. 1982. *Jewellery of The Ancient World*. New York: Rizzoli.

Ogden, J. 1998. The Jewellery of Dark Age Greece: Construction and Cultural Connections, in D. Williams (ed.) *The Art of the Greek Goldsmith*, 14-17. London : British Museum Press.

Otavsky, K. 1994. Two Greek Candlesticks in the Abegg-Foundation, in *Θυμίαμα στη μνήμη της Λασκαρίνας Μπούρα* vol. 2, 239-240. Athens: Benaki Museum.

Otavsky, K. 1994a. Zwei gräko-byzantinische Kerzenleuchter in der Abegg-Stiftung. *Riggisberger Berichte* 2, 137-143.

Papadopoulou, V. 2002. Lead myrrh flask (koutrouvion), in D. Papanikola-Bakirtzi (ed.) 2002. *Everyday Life in Byzantium*, exhibition catalogue Thessaloniki, White Tower, 2001, 184-185. Athens: Kapon Editions.

Papanikola-Bakirtzi, D. (ed.) 2002. *Everyday Life in Byzantium*, exhibition catalogue Thessaloniki, White Tower, 2001, 582-583. Athens: Kapon Editions.

Popkonstantinov, K. 2009. *Zograf Vasilie i Boianskata tsarkva sled 750 godini*. Sofia.

Psaltou, S. 2012. Understanding the self through the religious signifiers of light, in I. Motsianos and E. Bintsi (eds) *Light on light, an illuminating story, Exhibition Catalogue*, 261-265. Thessaloniki: Folklife & Ethnological Museum of Macedonia-Thrace.

Raub, Ch. 1993. How to coat objects with gold – Pliny, Leyden Papyrus X, Mappae Clavicula and Theophilus seen with a modern chemist's eyes, in C. Eluère (ed.) *Outils et ateliers d'orfèvres des temps anciens*, 101-116. Saint-Germain-en-Laye: Société des Amis du Musée des Antiquités Nationales et du château.

Savopoulou, Th. 2002. Th. Savopoulou. Glazed bowl, in D. Papanikola-Bakirtzi (ed.) 2002. *Everyday Life in Byzantium*, exhibition catalogue Thessaloniki, White Tower, 2001, 332-333. Athens: Kapon Editions.

Schneider, U. 1981. Zwei mittelalterliche Chorschranken im Dom San Ciriaco von Ancona. Beiträge zur Geschichte der Inkrustationkunst in Byzanz und Italien. *Acta Historiae Artium* XXVII, 9-186.

Shakota, M. 1984. *Dechanska riznitsa*. Beograd.

Sharenkov, A. 1988. *Starinni tractate po tehnologia i tehnika na zhivopista* vol. I, Sofia.

Shchapova, Y. 1983. *Ocherki istorii drevnego steklodeliia*. Moscow.

Skaltsis, P. 2012. The illumination of the Church in the Christian liturgical tradition, in I. Motsianos and E. Bintsi (eds) *Light on light, an illuminating story, Exhibition Catalogue*, 266-276. Thessaloniki: Folklife & Ethnological Museum of Macedonia-Thrace.

Spatharakis, I. 1966. An exceptional representation of the Supper at Emmaus in the church of St. Antonios at Vrontisi, Crete, in I. Spatarakis. *Studies in Byzantine Manuscript Illumination and Iconograph*, 249-292. London: Pindar Press.

Springer, P. 1981. *Kreuzfüsse: Ikonographie und Typologie eines hochmittelalterlichen Gerätes*. Berlin: Deutscher Verlag für Kunstwissenschaft.

Stanilov, S. 2010. Der Phau und der Hund. Zwei goldene Zierscheiben aus Veliki Preslav, in F. Daim and J. Drauschke (eds) *Byzanz: das Römerreich im Mittelalter, Teil 3, Peripherie und Nachbarschaft*, 159-172. Mainz: Verlag des Römisch-Germanischen Zentralmuseums.

Stauffer, A. 2010. Caligae Clemens'II, in J. Frings (ed.) *Byzanz: Pracht und Alltag; Kunst- und Ausstellungshalle der Bundesrepublik Deutschland, Bonn; 26.2.-13.6.2010*, 172. München: Hirmer.

Symbattii 1968. Symbattii Calisque testsmenta. *Fontes Graeci historiae Bulgaricae* VII, 67-78.

Theophilus 1933. *On Divers arts*. New York, 1933.

Todich, Br. 2005. Freske dechanske tskve, in Br. Todich and M. Chanak-Medich (eds) *Manastir Dechani, 325-516*. Beograd.

Totev, K. 2010. Ampuli kutruvii s izobrazhenia na sv. Dimitar i sv. Bogoroditsa ot Balgaria, in *Godishnik na Sofiyskia universitet, Tsentar za slaviano-vizantiyski prouchvania 'Ivan Duychev,'* vol. 95 (14), 205-216. Sofia.

Totev, K. 2011. *Thessalonican Eulogia found in Bulgaria (Lead Ampules, Encolpia and Icons from the 12th-15th centuries)*. Veliko Tarnovo.

Totev, T. 1993. *Preslavskoto sakrovishte*. Sofia.

Trever, K. and Lukonin, V. 1987. *Sasanidskoe serebro*. Moscow.

Tsaneva, S., Kashieva, E. and Dimitrov, Y. 2008. Arheometrichni prouchvania na ukraseni s email predmeti ot rimsko vreme ot fonda na NIM: nachalo. *Izvestia na Nationalnia Istoricheski musei*, XVIII, 192-206.

Typicon 1968. Typicon Bacuriani. *Fontes Graeci historiae Bulgaricae* VII, 39-66. Sofia.

Urbani, M. 2010. Räuchergefäss in Form eines Fünf-Kuppelbaus, in J. Frings (ed.) *Byzanz: Pracht und Alltag; Kunst- und Ausstellungshalle der Bundesrepublik Deutschland, Bonn; 26.2. - 13.6.2010*, 349-350. München: Hirmer.

Vaklinova, M. 1972. Vizantiyska zlatna ikona – relikviar. *Arheologia* 3, 13-18.

Vokotopoulos P. (ed.) *Byzantine Art in the Aftermath of the Fourth Crusade: The Forth Crusade and its Consequences*, Proceedings of an International Conference, Athens 2004. Athens: Academy of Athens.

Voyatzis, S. 1996. Fragments from Two Templon Screens, in E. Evans and D. Wixon (eds). *The glory of Byzantium: Art and culture of the Middle Byzantine era: A.D. 843-1261*, 40. New York: The Metropolitan Museum of Art.

Ward, R. 1993. *Islamic Metalwork*. London: British Museum.

Weitzmann, K. 1935. *Die byzantinische Buchmalerei des 9. und 10. Jahrhunderts*. Berlin: Mann.

Wixom, W. 1996. Byzantine Art and the Latin West, in E. Evans and D. Wixon (eds) *The glory of Byzantium: Art and culture of the Middle Byzantine era: A.D. 843-1261*, 435- 449. New York: The Metropolitan Museum of Art.

Yashayeva, T. 2010. Enkolpion mit der Darstellung der Gottesmutter, in J. Frings (ed.) *Byzanz: Pracht und Alltag; Kunst- und Ausstellungshalle der Bundesrepublik Deutschland, Bonn; 26.2.-13.6.2010*, 305. München: Hirmer.

Yashayeva, T. 2010a. Kerzenhalter, in J. Frings (ed.) *Byzanz: Pracht und Alltag; Kunst- und Ausstellungshalle der Bundesrepublik Deutschland, Bonn; 26.2.-13.6.2010*, 304-305. München: Hirmer.

Zachariou, E. 2002. Glazed jug, in D. Papanikola-Bakirtzi (ed.) 2002. *Everyday Life in Byzantium*, exhibition catalogue Thessaloniki, White Tower, 2001, 337. Athens: Kapon Editions.

Zalesskaya, V. 1995. Quelques bronze byzantins à Chersonese (les liens avec l'Asie Mineure et les Balkans aux XIIe-XIVe siecles), in K. Weitzmann, C. Moss and K. Kiefer (eds) *Byzantine East, Latin West: Art-historical studies in honor of Kurt Weitzmann*, 669-677. Princeton (N.J.: Department of Art and Archaeology, Princeton University.

Considerations on the Function and Usage of Pottery Lamps, Inspired by Finds from the Forum of Thessaloniki

Georgios Velenis and Stavros Zachariadis
Aristotle University of Thessaloniki

Abstract: Experiments, analytical techniques, tradition and literal sources, supplemented by the archaeological evidence, provide us with adequate information concerning the use and function of closed lamps. The essential elements of a lamp can be summed up in four words: vessel, fuel, wick, oxygen. Olive oil is the most important lamp fuel while several other vegetal oils, butter, animal fat and even crude mineral oil were used for the same purpose. Wicks were mainly vegetal produced from a variety of plants either with or without any processing while a mineral wick is also mentioned. Technological advances are critical for the evolution of the form of the vessel while the lamp type adopted in every case is indicative of consumption and illumination habits. A series of roughly made lamps from the Forum of Thessaloniki, imitations of Asia Minor and Samian lamps, are indicative that technological features could overcome stylistic and decorative criteria.

Keywords: lamps, lamp fuel, lamp wick, lampmakers, illumination, Thessaloniki.

During the study of the lamps from the Forum of Thessaloniki[1] a series of questions rose, considering the function and usage of these vessels. It came as a surprise that information on these matters is limited, despite the extensive literature concerning typology and chronology. The most extensive and up to date study on this subject seems to be the unpublished dissertation of D. Moullou[2] while the work of D. Bailey[3] has been widely used by many scholars.

In the first paragraph of his book '*Greek and Roman pottery lamps*', D. Bailey defines the physical requirements of a lamp as follow: container, fuel, wick, air supply.[4] Everything else is simply refinements and technological improvements.[5]

The last of the necessities described above is relevant to the function of a lamp only if it concerns the presence of oxygen, and it has nothing to do with the so called air holes. These holes, often found on the disk, the shoulder or the rear part of the nozzle, are not related to the proper function of a lamp, which would burn in the same way even if they were absent, as has been demonstrated on experimental basis with exact copies of African, Attic and Asia Minor lamps.

If one wishes to simplify even more the essential elements of a lamp, these would be summed up in four words: vessel, fuel, wick, oxygen. Every vessel containing suitable fuel and a wick rising through it could function as a lamp, even if it were an open pot or a closed bottle (Figure 1).

In the case of a bottle the fuel would be safer from accidental spillage but its level would fall rapidly, causing the flame to weaken and eventually go out. The same stands for every tall and narrow vessel. The ability of a wick to transport fuel vertically by capillary action is limited, thus the higher above the fuel level the flame stands the less intense it gets. On the contrary, in broad vessels the level drops much slower, allowing the wick to burn longer without the need of a refill, however the fuel is much more likely to get spilled and it is also exposed to animals, such as rats and insects.

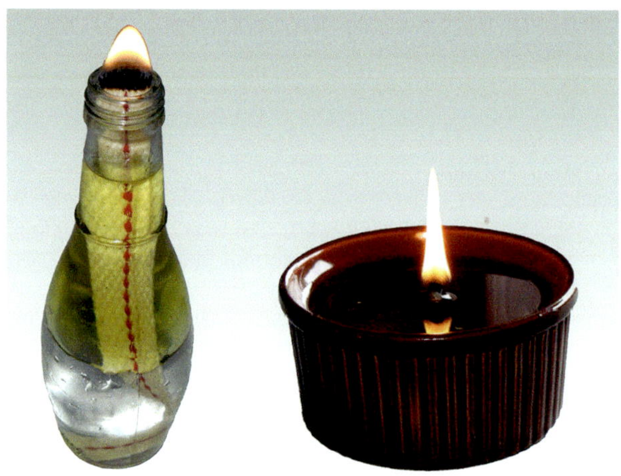

Figure 1. Bottle and open clay pot functioning as lamps.

A closed vessel, leaning to its side, would be the best option, combining a broad surface and a protected container. This seems to be the case in the shape of a typical lamp. In fact, a bottle, with an upturned neck placed horizontally, functions perfectly as a lamp (Figure 2).

Literal sources and the results of studies employing various analytical techniques provide us with information about the fuels used during antiquity. Among the fluid ones, except for olive oil[6] which was the most common, especially in

[1] A small part of this assemblage is published in the catalogue of the exhibition '*Light on light*', see Velenis, Zachariadis 2011.
[2] Moullou 2010.
[3] Bailey 1972.
[4] Bailey 1972, 9.
[5] Motsianos 2011a, 122.
[6] Olivieri 1935, i.100.1-14.

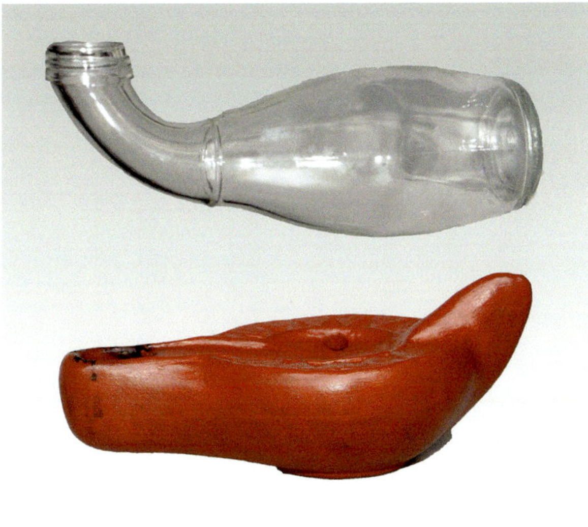

Figure 2. Glass bottle with an upturned neck and a reproductionj lamp of North African type.

the Mediterranean,[7] other oily substances are mentioned. Vegetable oils such as castor oil,[8] sesame oil,[9] linen oil[10] and oil from the plant *Lawsonia Intermis*, commonly known as henna[11] are encountered. The use of these fuels, as well as of oil coming from safflower[12] and radish seed or other plants of the *Brassicaceae* family,[13] has been verified by analytical techniques. Even an oily substance produced from calabash (*Lagenaria vulgaris)* is considered suitable for lamp fuel.[14]

Apart from vegetal oil, crude mineral oil was used when available[15] while Jewish sources mention fish oil.[16] Additionally it is possible that essential oils or other augmentations were utilized, though they were probably intended for scenting and not for illumination.[17]

Solid fuels included butter,[18] animal fat,[19] and resin.[20] However, only butter seems to have been used as an illuminant. Animal fat, according to Cyranides was burned during specific rituals while resin is designated by Aëtius for medical purposes. Nevertheless, those references imply that these substances could also function as illuminants. Besides, fat was the principal fuel in Eskimo lamps,[21] whose basic structure is not very different than that of the byzantine open lamps. Apart from the above, the use of beeswax has been attested in open lamps from Minoan Crete.[22] Most of these substances were probably less suitable than olive oil, either because of the bad smell, as in the case of linen oil,[23] or, in the case of resin, butter and animal fat, the smoke they would emit.[24] According to our experiments the visual effect of olive oil seems brighter[25] and produces less smoke and odor than butter while the more fluid the fuel is the better it seems to burn. Moreover some fuels, such as mineral oil, were extremely flammable.[26]

Along with oil, salt was sometimes added in lamps. Forbes provides a series of reasons for this practice.[27] In this way all moisture was removed from the fuel in order to prevent spluttering when the oil was almost finished.[28] Additionally it gives a bright yellow color to the flame and prevents oil from overheating and self igniting. The use of salt in illuminating devices is mentioned by Herodotus[29] while the validity of this thesis is enhanced by the results of analytic techniques.[30]

While liquids could be used in both closed and open lamps, when it comes to solid fuels open ones are more appropriate. This functional difference could explain the scarcity of lamps in regions such as western Macedonia or central Thessaly. In cities like Veroia and Larisa, despite their importance during the Late Roman era, pottery lamps are rather rare finds. In these areas the production of olive oil is virtually impossible. On the other hand these regions traditionally depended on livestock. In this context butter and animal fat would be abundant and would be preferred as lamp fuel over the imported, thus more expensive, olive oil. The same principal stands for other oil deprived territories.[31] Analysis in lamps from Olbia verifies this assumption.[32] In this case various open vessels would be used as lamps. This practice was not unknown in Greece, wherever oil was scarce, until a few decades ago.[33] However their identification of these pots in Late Roman contexts did not preoccupy archaeologists until now.

The use of solid fuels seems to be the reason why some lamps have their entire disk removed.[34] A few examples have been found in the Forum of Thessaloniki. In one of them burn marks can be noticed around the broken disk both on the upper and lower surface (Figure 3).[35]

Different kinds of oil were used in the same area. Analysis on lamps from Sagalassos (southwestern Turkey) reveals the presence of different types of fuel in the same lamps, either as a result of mixing oils or because they were used

[7] Bailey 1972, 11.
[8] Wellmann 1907, 4.161.1.7-8; Legrand 1932, 2.94.5-10; Vogel and Fischer 1888, I.34; Meineke 1877, 17.2.5.
[9] Olivieri 1935, vii.80.16-17.
[10] Nau 1907-1913, 151.5
[11] Olivieri 1935, vii.69.27-28.
[12] Kimpe *et al.* 2001, 95.
[13] Copley *et al.* 2005; Romanus *et al.* 2008, 793; Colombini *et al.* 2005, 897.
[14] Wellmann 1907, 4.150.5.4-6.
[15] Wellmann 1907, 1.73.6-7; Giannini 1965a 139.2.1-3.
[16] Forbes 1958, 145; Bailey 1972, 11.
[17] Moullou 2010, 155-157.
[18] Marrou *et al.* 1960, 1.6.51.3.3-4.
[19] Kaimakis 1976, 2.40.10-11.
[20] Olivieri 1935, i.322.12-13.
[21] De Laguna 1940, 63-68.
[22] Evershed *et al.* 1997.
[23] Nau 1907-1913, 151.5.
[24] Wellmann 1907, 1.72.4.1-4; 2.72.2-3.
[25] Our study did not involve physical measurements of light properties. The superiority of olive oil over other fuels tested refers to the impression and not the intensity of the light.
[26] Wellmann 1907, 1.73.2.3-4.
[27] Forbes 1958, 153.
[28] Rabe 1896, 217. 3-4.
[29] Legrand 1932, 2.62.1-5.
[30] Bussière 1973, 257.
[31] Crnobrnja 2008, 409, 412.
[32] Garnier *et al.* 2009, 55.
[33] Moullou 2010, 143, note 702.
[34] Motsianos 2010, 278, fig. 3; Bailey 1988, pl. 158, no. Q1436; Bonifay 2004, fig. 234, no. 1.
[35] Velenis, Zachariadis 2011, 207, no. 118.

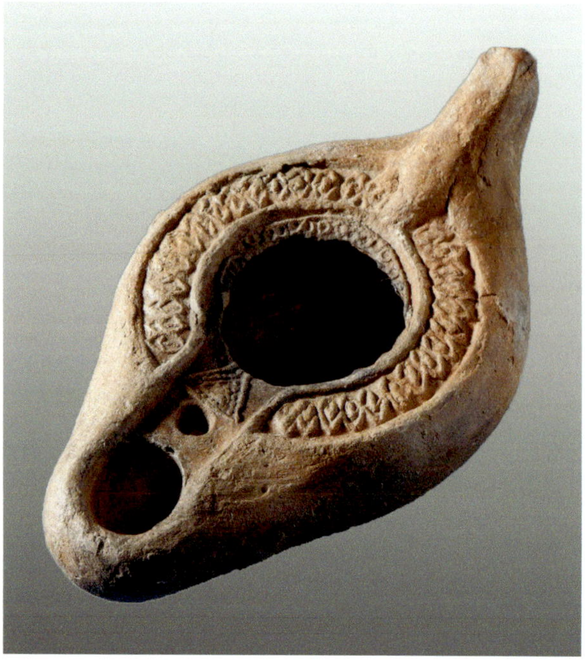

Figure 3. Intact oval lamp with chipping over the greater part of the disk.

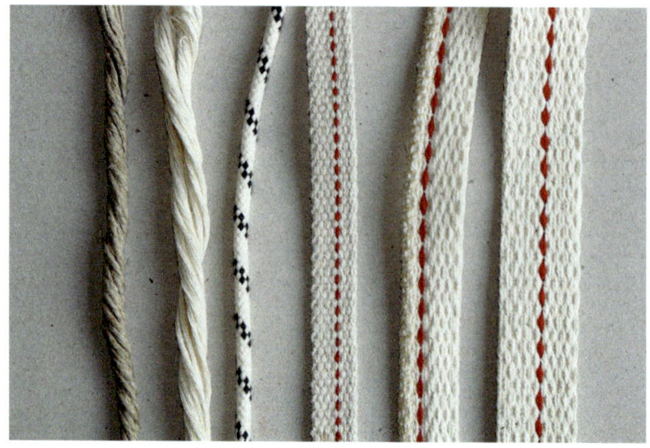

Figure 4. Various types of wicks, starting from the left, twisted wick made of cannabis and cotton threads, modern cylindrical and flat, woven lamp wicks.

alternatively. Even animal fat was traced along with vegetal oils. Nevertheless, olive oil remains the predominant fuel.[36]

Households could gather the rejected amount of oil from everyday cuisine in order to use it for illumination. Our experiment has shown that cooked oil functioned as well as the raw one. If this was common practice it could explain the co-existence of vegetable and animal fat in Sagalassos lamps, indicating that meats were cooked in vegetal oil and the remnants were reemployed as lamp fuel. The same principal would stand for products rendered useless for any reason. Aged olive oil, which was no longer edible,[37] oil from wild olive trees[38] and even the oily substance left in storage containers[39] could be used in lamps, although this last case should be placed under the solid or viscous fuel category.

Interpreting all the above information, one should reconsider the thesis, presented by many scholars, that using a lamp means burning food. Of course this was true in some cases, but it seems that using the same oil for food and illumination was not very common, and doing so was considered evidence of extreme poverty.[40]

The wick is the second perishable element of a lamp, thus the information available is limited, mostly originating from literary sources, experiments and tradition. Remnants of wicks are found along with pottery lamps in some rare cases.[41] In the Forum of Thessaloniki a Hellenistic lamp was found with remnants of a wick, consisting of threads with clear evidence of twisting.

Wicks could be either cylindrical or flattened, made mostly of vegetal fabrics (Figure 4). The commonest type was the one produced by threads twisted together.[42] During Roman times wicks were produced by the *stuppatores* with threads from the nuclei of various plants.[43] The most popular was probably flax, widely known as linen. In most cases only the worst quality of linen would be used for wicks.[44] Softer linen originating from the area of Tarsos is also mentioned.[45] However in these cases the use of the term ἐλλύχνιον (lamp wick) is probably not referring to a wick but to a piece of soft cloth for medical purposes, used for applying various substances on the skin. The castor plant[46] is also mentioned while esparto grass (*Macrochloa tenacissima* and *Stipa tenacissima),* known for its use in ropes,[47] could be used in the same way[48] for lamp wicks. Hemp (cannabis) is considered essential,[49] judging by the fact that in many cases it is mentioned as analogous of linen.[50] Nevertheless, though hemp was utilized for the fabrication of ropes,[51] its use for ἐλλύχνια (lamp wicks) is only mentioned in the context of medical procedures,[52] just like the soft linen fabric.

Variations of *schoenus*, are mentioned by Pliny as suitable for wicks.[53] However it seems that it was mainly used for torches

[36] Kimpe *et al.* 2001.
[37] Olivieri 1935, i.100.7-11.
[38] Migne 1857, vol. 26, Athanasius Theologus: Orationes tres contra Arianos, 26.32.29.
[39] Moullou 2010, 146.
[40] Migne 1857, vol. 47, Joannes Crysostomus scriptor ecclesiasticus: Ad Demetrium de compunctione (lib. 1), 403, 32-33.
[41] Perltzweig 1961, 76, no. 48; Bailey 1988, 274, pl. 56, no. Q2272; Bussière 1973, 255-256.
[42] Kühn 1827, 13.858.1-2; Giannini 1965 36.1.3-4.
[43] Mayhoff 1892, 19.16-18; Hermansen 1982, 125-126.
[44] Mayhoff 1892, 19.17.4-8; Hermansen 1982, 122, 124-125; Moullou 2010, 136.
[45] Olivieri 1935, viii.1.16; Kühn 1825, 10.954.9.
[46] Olivieri 1935, vi, 3, 40.
[47] Meineke 1877, 3.4.9.20-21.
[48] Hermansen 1982, 122.
[49] Moullou 2010, 136; Hermansen 1982, 122.
[50] Bethe 1900, 7.72.5-6; Wellmann 1907. 3.148.1.1-6; Wellmann 1907. 3. 148. 1-2.
[51] McCown 1922, 20.9-10; Stallbaum 1825, 1.107.29; Schneider 1908, 159. 5; Wellmann 1907, 3.148.1.1-6; Wellmann 1907, 3.148.1-2.
[52] Oder and Hoppe 1924, 96.26.5-6.
[53] Mayhoff 1892, 21.112.1-21.114.3.

or candles and not in closed lamps.[54] He is probably referring to the plant *Schoenus nigricans,* commonly known as black bulrush,[55] which seems to be the same as the plant *θρύον* (reed, rush).[56]

For our study, experiments were conducted with wicks made of hemp and cotton. The use of cotton in antiquity seems possible, since it is mentioned in several sources along with linen for producing cloth.[57] Although none of these sources refer to wicks or lamps, some scholars suggest the use of cotton wicks,[58] a thesis that seems plausible, although not likely.[59]

All the above refer to vegetal threads. According to Moullou the use of wool is not possible since it would burn rapidly.[60] Nevertheless the examination of a wick sample from northern Africa seems to favor the use of animal fabrics.[61]

Apart from the twisted wicks woven ones could also be employed, pieces of old cloth, even of old sails,[62] or stripes of papyrus[63] could function for the same purpose. This is verified by the archaeological finds from the Ancient Agora of Athens, since the wick that was found preserved in a 1st century AD lamp is actually made of a small piece of woven cloth, though its material is not described.[64]

In the same way would be used the only mineral wick mentioned, probably referring to asbestos, whose use for fabrics is known[65] and stripes of these cloths could be used as wicks imperishable by fire.[66]

Additionally a series of plants are mentioned whose leaves, or other parts, could be used as wicks. The most common one is the plant *φλόμος* (mullein, Latin verbascum), also mentioned as *φλομίς* or *θρυαλλίς* (a plant which, like our rush, was used for making wicks, perhaps mullein).[67] Its use was so extensive that it became synonymous with lamp wick.[68] In fact the exact plant described cannot be easily identified. Most probably the term *φλομίς* refers to one of the many varieties of the plant *phlomis* genus of the *Lamiaceae* family, commonly known as Jerusalem Sage or Lampwick Plant.[69] Alternatively it could be interpreted as the plant *Ballota acetabulosa* of the same family.[70]

However, it should not be confused with the plant *Verbascus*[71] which was probably also used for the same purpose,[72] and seems to be the same as the plant *θάψιον* known for its use in lamps.[73]

In close connection to lamps is mentioned the plant *antirrhinum*,[74] commonly known as dragon flower,[75] more often described as *λυχνίς*.[76] The word *θρυαλλίς* is also associated with *αρνόγλωσσον*, identified as *plantains* of the *plantago* family.[77] However it seems more possible that it is just a synonym for *φλομίς*.[78]

All the above plants bear hairy leaves which could be used in a lamp. In fact every material able to provide capillary action, in order to bring the fuel to the flame, can be included amongst possible wicks, since even small branches from olive trees are mentioned for this purpose.[79]

Bailey claims that the thickness of the wick does not cause much difference to the strength of the flame, although the consumption of fuel seems to be analoguous to that.[80] Experiments, both in the case of our study and other projects,[81] have proven that the strength of the flame depends highly on the size of the wick.

The exposed length is vital, but bears specific limitations, since a very long wick will produce smoke. On the other hand the thickness of the wick can be decisive for the flame size[82] without affecting the smokeless function of the lamp. According to our experiments it seems that the critical dimension is not diameter itself but the exposed surface, thus, the thinner but wider woven wicks produced a wider and higher flame (Figure 5). One should always bear in mind that a larger and stronger flame means faster fuel consumption. The fact that thicker wicks would consume more is often mentioned in literal sources.[83]

Other important factors for the proper function of the lamp are the diameter of the wick hole and the oil temperature. It has been proven that the wick should not take over the entire surface of the wick hole but it should float in it, providing better burning conditions. The temperature of the nozzle is crucial, and most importantly, the temperature of the oil reaching the nozzle through the wick. It seems that the higher it is the better the fuel burns. All these parameters defined the evolution of the clay lamp, as well as some of its functions. Lampmakers were aware of these properties and utilized the features of each lamp type, depending on the circumstances in every era and geographic region.

[54] Beckby 1965, 6.249.1-2; Forbes 1958, 153.
[55] Hesychius states that the plant *οξύσχοινος* mentioned in Plini (Mayhoff 1892, 21.112.5) is the same as the plant *μελαγκρανίς* (Latte 1953, mu.620.1) which is identified as *Schoenus nigricans* or commonly known as black bulrush.
[56] Liddell and Scott 1996, entry *θρύον*.
[57] Bethe 1900, 7.75.1-7.76.3; Hort 1916, 4.7.7.1-14.
[58] Bouras and Parani 2008, 3.
[59] Moullou 2010, 139-140.
[60] Moullou 2010, 139, note 683.
[61] Bussière 1973, 255-256.
[62] Olivieri 1935, vii.69.27.
[63] Olivieri 1935, vii.80.16.
[64] Perltzweig 1961, 76, no. 48.
[65] Giannini 1965 36.1.1-36.4. 2.
[66] Spiro 1903, 1.26.7.5-6.
[67] Olivieri 1935 i.405.1-5; Wellmann 1907. 4.103.2.7-10; Kaimakis 1976, 1. 13. 3-8; Wellmann 1907. 4.103.3-5; Latte 1953, phi.642.1-2; Porson 1822, Phi. 651.14-15; Kühn 1826, 12.150. 11; Heiberg 1921, 7.3.21.19
[68] Bethe 1900, 6.103.6; 10.115.9.
[69] Gennadios 1914, 999 entry *φλομίς*; Liddell and Scott 1996, entry *φλομίς, -ίδος*.
[70] Gennadios 1914, 999 entry *φλομίς*; Gennadios 1914, 169 entry *βαλλωτή*; Moullou 2010, 138.

[71] Wellmann 1907. 4.103.1. Gennadios 1914, 999-1000 entry *φλόμος*.
[72] Bailey 1972, 10.
[73] Delatte 1927, 389.9.
[74] Kühn 1830, 19.82.12; Wellmann 1907, 4.130.1.1-2; Wellmann 1907, 4.129.1-2.
[75] Gennadios 1914, 127, entry *αντίρρινον*.
[76] Wellmann 1907, 3. 101. 1.
[77] Liddell and Scott 1996, entry *θρυαλλίς, -ίδος*.
[78] Gennadios 1914, 423 entry *θρυαλλίς*.
[79] Migne 1857, vol. 68, Cyrillus Theologus: De adoratione et cultu in spiritu et veritate, 68.612.28-29.
[80] Bailey 1972, 10.
[81] Sunderland *et al.* 2011; Moullou, Topalis 2011a.
[82] Sunderland *et al.* 2011, 2492, 2496.
[83] Kenneth 1968, 56-59; Von Leutsch 1851, 1.37.7-10; Hayduck 1907, 253.9-10.

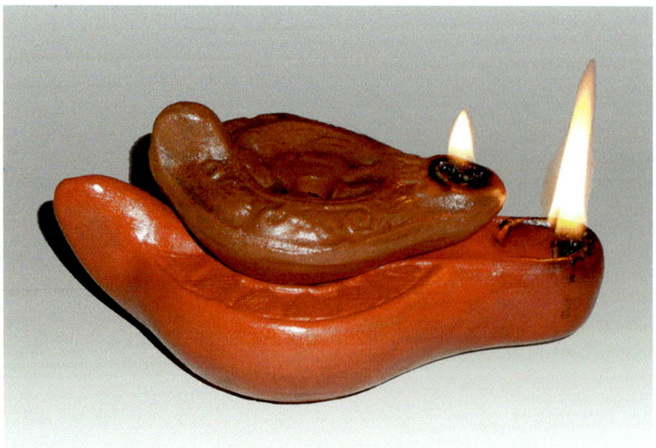

Figure 5. Comparison of the flame size produced by a cylindrical cotton wick (top) and a wider woven one (bottom).

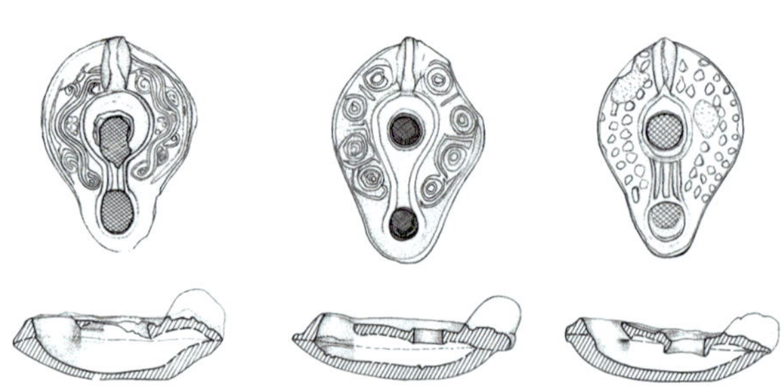

Figure 6. Imitations of Samian and Asia Minor type lamps from the Ancient Agora of Thessaloniki (drawings by Ch. Koundouras).

According to our experiments an Attic lamp of average size requires almost 30 gr of olive oil every three and a half or four hours, if using a cotton wick with a diameter of 0.5 cm. Using the same type of wick, African lamps with almost twice the capacity of the previous example can last for 7-8 hours without needing refill. On the contrary a smaller *planta pedis* type lamp, whose capacity does not exceed 20 gr, can last only for a little more than two hours. This means that the average consumption is almost 8-9 gr per hour.

These results are more or less consistent with the ones presented by Bailey,[84] Hermanns[85] and Wunderlich.[86] However it seems that consumption with the use of a cotton wick is almost double what it would be if a linen wick had been employed.[87] Apparently African lamps with the wide ovoid wick holes could bear thicker wicks and seem more suitable for wider, woven ones, which provide almost twice a stronger flame but also consume double the amount of fuel for the same duration of operation.

It is not clear if larger containers are indicative of longer duration or stronger flame. This could be the object of a separate study. Nevertheless different sizes of lamps could be interpreted as reflecting either different illumination habits, thus longer use during the day, or different consumption habits, depending on the availability of fuel.

Special attention should be paid to the functional purpose of the funnel connecting the disk to the nozzle. The thinner the clay is at this point the warmer it gets, creating a heating tunnel for the fuel before it reaches the flame. In the case of African lamps this funnel is especially thin and longer than in other lamp types, thus functioning more effectively in heating the liquid. Additionally, the height of African lamps is smaller than that of Attic types. As a result North African lamps can consume the entire amount of oil enclosed in the vessel more easily, even with a low quality wick, without being obstructed by the limitations of height mentioned in the beginning of this paper. These advanced features suggest that the transition from other lamp types to the African ones, either by importing, or by imitating them, was not based solely on aesthetic criteria but also on the technological supremacy of one shape over the other.

In the assemblage from the Forum of Thessaloniki a group of seventh-century lamps is identified as locally made (Figure 6). They are imitating Asia Minor and Samian shapes and decorative patterns.[88] The internal height of their reservoir is remarkably small, almost 1-1.5 cm, not very different than that of African lamps. This seems to be, more or less, the optimum height needed in order for the entire amount of oil to reach the nozzle, without any further intervention and without causing any significant weakening of the flame.[89] Taking into account the results of our experiments we could assume that these vessels, though roughly made and decorated, are not the products of inexperienced craftsmen but objects orientated to the maximum exploitation of the fuel.

In the case of lamps with deep reservoirs the exhaustion of the fuel could only be accomplished by the addition of water as Heron describes.[90] Evidence of this technique is provided by a copper lamp from a private collection.[91] Its lower part is highly oxidized because of the water used in order to raise the fuel level[92] while the rest of its surface is preserved intact.

Along with lamps came a series of minor tools, useful for their operation. A copy of an early twentieth-century copper lamp could be rather intuitive (Figure 7).[93] It comes together

[84] Bailey suggests that a lamp would require almost 31 gr of oil for three and a half hours using a linen wick of small caliber, without mentioning a specific thickness of the wick (Bailey 1972, 10).

[85] Hermmans mentions that 50 ml (45 gr) of olive oil would burn for six hours and 37 minutes (Hermanns 2004, 8).

[86] Wunderlich defines the consumption of a lamp to 8 gr of oil per hour (Wunderlich 2003, 256).

[87] Moullou 2010, 144.

[88] Velenis, Zachariadis 2011, nos. 105, 113, 117, 120.

[89] The use of an Attic lamp resulted as well in the exhaustion of the fuel but as oil level decreased so did the intensity of the flame.

[90] Schmidt 1899, 2.24.

[91] Motsianos, Bintsi 2011, 166-167, no. 54 (entry author Karagianni F.).

[92] Since oil is lighter than water the fuel floats over the water level. Adding water does not affect the flame but raises the fuel level efficiently, thus allowing the entire amount of oil to be used, even in tall, narrow vessels.

[93] An identical copper lamp with a detailed description is published in Motsianos and Bintsi 2011, 367-368, no. 279 (entry authors Koutsoudaki, Bintsi).

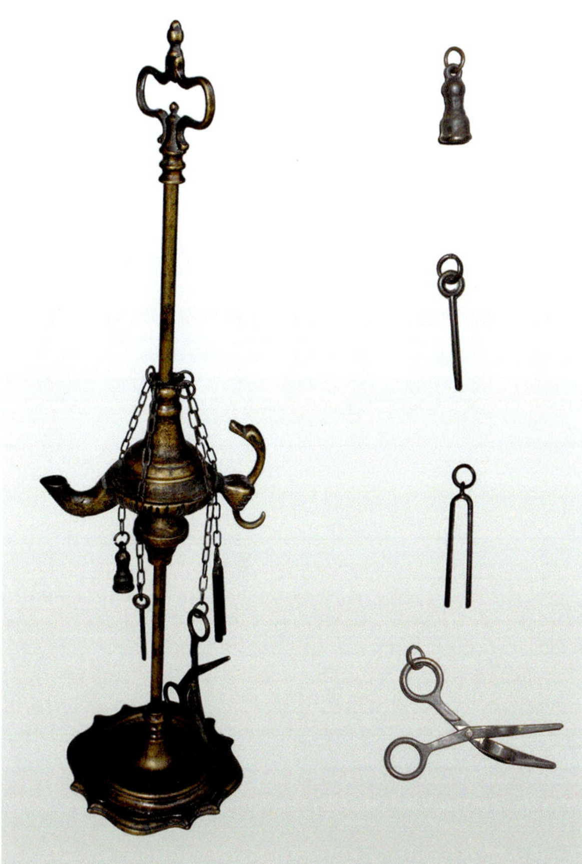

Figure 7. Copper cast made lamp with its accessories.

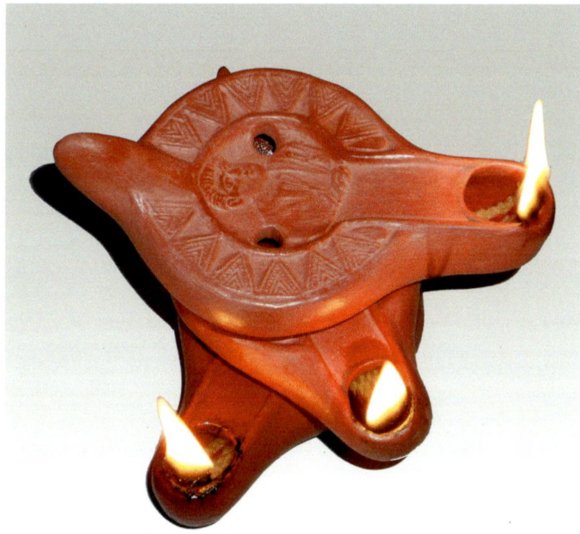

Figure 8. Stacking African lamps while they are burning.

with a small pair of scissors, for snuffing the wick, a pair of tweezers[94] and a pin for manipulating the wick as well as a small cup shaped object, like a small thimble for extinguishing the flame.

The most important one is the small pin with pointy edge, probably used since antiquity. It could be made of any material and it was used in order to poke the wick, to pull it forward, in order to increase the intensity of the flame, or push it back into the lamp, in order to obtain a smokeless result or to save fuel. Even a small branch could serve this purpose[95] while some would have arranged the wick simply with their fingers.[96] The small hole that many lamps bear on the rear of the nozzle, described by most as an 'air hole'[97] seems to be related to this pin. Nevertheless opposing opinions on its use had been presented rather early on in the literature.[98] The function of this small opening is not securely identified. The position, shape and size of these holes make them, in many cases, impractical to use in order to handle the wick, as it has been proposed. However they could be rather useful as a resting point for the pins previously described. The fact than in many cases these pins are found inside the fill holes,[99] instead of the openings on the nozzle, is not controversial since not all lamps bear these holes and neither is it uncommon for individuals to use things in a different manner than they were intended to be used. These holes were not actually necessary. This is obvious not only by their omission in many lamp types but also by the fact that in many cases of Attic lamps they are left unpierced, though they were clearly present in the archetype used.[100]

Filling vessels known as ἐπαρυστρίδες (name for vessels to pour oil into a lamp)[101] were supplementary to the use of lamps. They were made of various materials, adopted different shapes, and probably most had a small spout. They could be either guttus type vessels[102] or small open cups[103] known as υποχυτῆρες (name for vessels to pour oil into a lamp)[104] or ἀρύταινες.[105] Even small trefoil jugs could be employed for this purpose, as well as other similarly shaped vessels.

In order to enhance the lighting effect lamps with more than one nozzle or more than one lamp in various arrangements would be used. The form of multi-nozzle lamps could be indicative of this arrangement.[106] Attic lamps seem better placed one next to the other, while Late Roman African lamps are fit for stacking one over the other in arrangements with

[94] Migne 1857, vol. 68, Cyrillus Theologus: De adoratione et cultu in spiritu et veritate, 68.852.6; Wolska-Conus 1968, 5.32.15t.
[95] MacDowell 1971, 249.
[96] MacDowell 1971, 250-251.
[97] Perltzweig 1961, 5, 71; Karivieri 1996, 21.
[98] For the discussion on the use of these holes, see Broneer 1930, 9-17.

[99] Delattre 1889, 10, fig. 4; Broneer 1930, 11, note 2.
[100] For example: Perlzweig 1961, nos. 41, 46, 102, 105, 130, etc. Velenis, Zachariadis 2011, no. 81.
[101] Laga and Steel 1980, 63.238; Latte 1953, epsilon, entry 4239.1; Pusey 1868, 2.330.22; Migne 1857, vol. 68, Cyrillus Theologus: De adoratione et cultu in spiritu et veritate, 68.852.6.
[102] Dauterman-Maguire *et al.* 1989, 62, no. 11; Themelis 2004, 720-721; Motsianos 2011a, 121; Moullou 2010, 213. (Hellenistic guttus type vessels have been found in the excavations of Thessaloniki's Forum bearing burn marks in the front, where the spout is placed, indicative of their use while the lamp was burning, in order to replenish the oil).
[103] Motsianos, Bintsi 2011, 179-180, no. 73 with numerous parallels and discussion over the use of these vessels (entry author: Antonaras).
[104] Tittmann 1808, upsilon.1770.1; Porson 1822, Upsilon.633.1-2; Bachmann 1828, upsilon 400.20; Rahlfs 1935,52.18.3-4.
[105] Adler 1928, Alpha.4065; Jones and Wilson 1969, 1091a. 2-3.
[106] Perlzweig 1961, 159, no. 2031, pl.33; Bonifay 2004, figs 235-236.

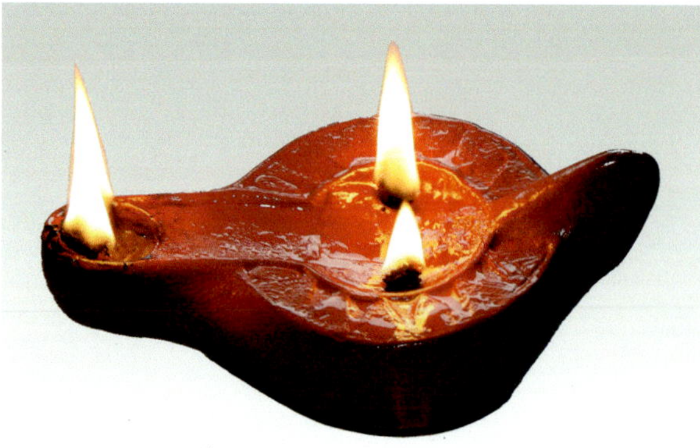

Figure 9. Using the filling holes as nozzles.

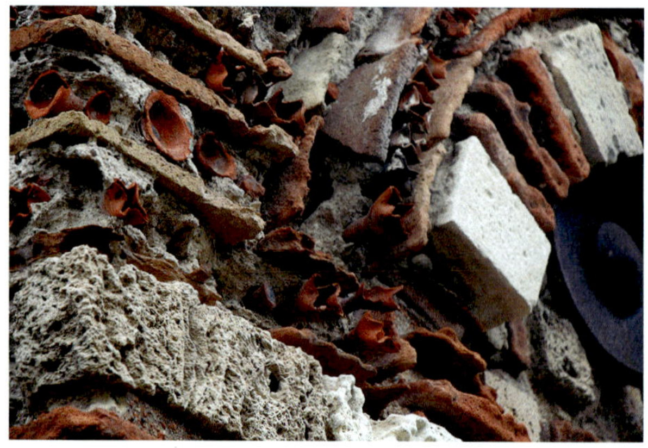

Figure 11. Use of fyalostomia as decorative elements in the Church of Archangels at Mesembria (photo by Fl. Karagianni).

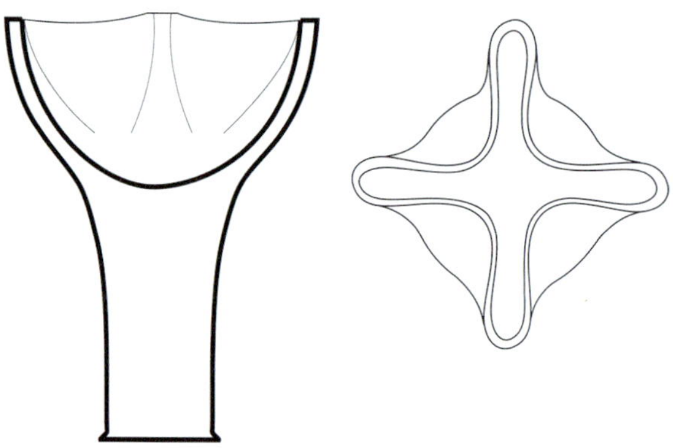

Figure 10. Fyalostomion (section/plan).

a slightly different axis (Figure 8). Additionally more wicks could be employed in the same lamp, simply by inserting a small wick in the filling holes of the disk (Figure 9).The habit of stacking lamps might be the inspiration for Byzantine and Post Byzantine glazed lampstands with more than one oil reservoir and pinched nozzles on different axes.

Amongst these open lamps we should probably consider the quatrefoil clay architectural elements of some Byzantine churches, known as *φυαλοστόμια* (fyalostomia) (Figures 10, 11). There is no doubt that the use of these elements on the exterior of buildings dated to the 13th and 14th centuries AD is purely decorative. Nevertheless, when these are found in small numbers, either in excavation contexts, or incorporated in earlier buildings, for example, in the church of St Anargyroi in Kastoria, it seems possible that these vessels were originally a kind of open lamp, able to have 1-4 wicks, used solitarily or in *polykandela* (chandeliers). In fact some of them bear traces of burn residue and their shape closely resembles that of some glass *kandelai*.[107]

Lampmakers are sometimes described as potters.[108] Probably the profession was not very esteemed in classical Athens since it seems that it was intended only for foreigners[109] and slaves.[110] During Roman times this seems to have changed. The owners of the workshops were dignified members of Roman society. The administrators, thus the craftsmen in charge of a workshop, could have been slaves, freed, or free men. Nevertheless, they were considered rather privileged, since they were trusted by the owners, acting as their representatives and in some cases they could even become the owners of a workshop.[111]

The social status of lampmakers in Roman and Late Roman epochs must have risen significantly. This could explain the elaborate markings indicating each workshop on most of the Athenian lamps. Some miniature lamps, found in a third-century burial in Thessaloniki,[112] could belong to the cloth of a lampmaker, probably used as buttons. The need of the deceased to state his profession in this way is indicative of a certain status, probably originating from the close relationship between pottery production and state authority.[113]

Lamps were essential in everyday life.[114] They are included amongst the necessities provided to every visitor.[115] Lamps and wicks were offered during feasts to the attendants[116] while the lack of lamp light was considered indicative of extreme poverty.[117]

It has been suggested that even a single lamp could provide enough lighting for a room.[118] Our own experiments confirmed that one lamp provides adequate light even for demanding

[107] See Motsianos 2011a, 109, fig. 28.iii.

[108] Jones and Wilson 1969, 1304b.1-2; Kaibel and François 1887, 11.48.32 - 33; Harris 1980, 142.

[109] Dalmeyda 1930, 3.5.4.

[110] Von Arnim 1893, 15.22.6-8.

[111] Karivieri 1996, 272-273 with further literature.

[112] Papanikola-Bakirtzi 2002, 554, no. 762 (entry author Pelekanidou).

[113] Velenis *et al.* 2009.

[114] Migne 1857, vol. 46, Gregorius Nyssenus theologus: Dialogus de anima et resurrectione, 46.104.13-18.

[115] Rose 1886, 8.49.631.5-8.

[116] Stallbaum 1825, 2.183.42-3; Kaibel and François 1887, 10.25.12-13. Porson 1822, Omicron. 321.20-21; Latte 1953, Omicron.332.1- 2; Bethe 1900, 6.21.6-8.

[117] Nauck 1889, 421.1; Long 1964, 1.81.8; Migne 1857, vol. 51, Joannes Crysostomus scriptor ecclesiasticus: De futurae vitae deliciis, 351, 55-56; Motsianos 2011a, 116.

[118] Moullou, Topalis 2011a.

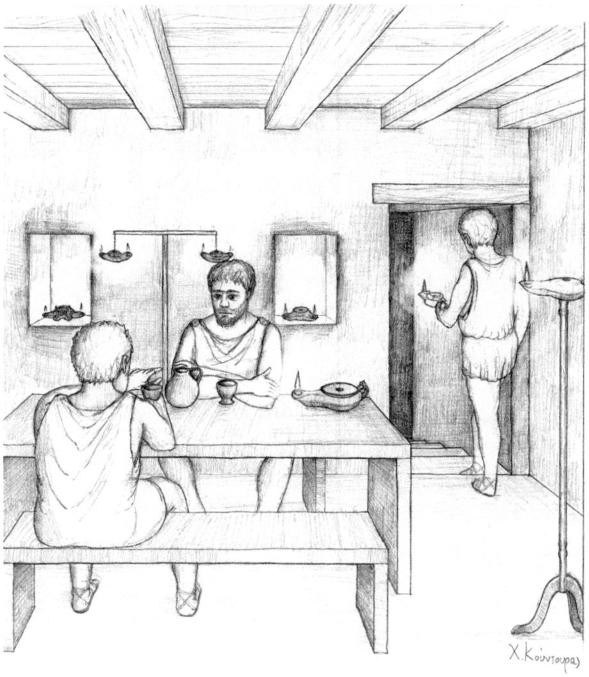

Figure 12. Representation of a late Hellenistic tavern (drawing by Ch. Koundouras).

tasks such as writing and reading, as long as the lamp is close to the person using it. Nevertheless it seems that it was customary to use more than one lamp for proper illumination of a room or a building (Figure 12).[119] Ten lamps would easily provide adequate illumination for an entire house.[120] They could be placed on lampstands,[121] hung by small chains[122] or simply placed on furniture, shelves or niches[123] in the room.[124]

In addditon to lighting the interior of buildings, lamps were used to mark tombs, graves,[125] churches,[126] and probably placed into lanterns in order to protect them from the wind.[127]

Additionally they were used along city streets,[128] although for this purpose other means of illumination, beyond the scope of this study, seem more suitable.

Lamps were used as portable lighting devices as well, usually placed inside lanterns[129] or hung from a chain or simply carried by hand for short distances.[130] Soldiers used a special device called a *ὀβελισκολύχνιον* (oveliskolychnion)[131] consisting of a lantern hung on the end of a stick.[132] For short distances a lamp could also be used by simply placing one's finger over the filling hole. By this means the fuel cannot be spilled while the flame is protected from strong currents of air.[133]

Finally lamps were necessary equipment for those working underground. Mine workers used lamps[134] placed in niches cut into the rock. The *fossores* in the catacombs are depicted carrying a lamp[135] while in the catacomb of Comodilla many lamps are fixed by mortar along the corridors[136]. Additionally the use of lamps, fixed onto the heads of mine workers[137], probably by the use of leather straps, is mentioned. In fact, the connection between lamps and the mining industry was so intense that many of its products are characterized by the adjective *λυχνίτης* (lychnitis).[138]

Lamps played an important role not only in people's lives but also for another species of the animal kingdom. Mice have been a natural enemy for lamps, since they liked to feed on the oil and fibrous wicks.[139] The invention of closed lamps made it harder for the rodents to steal the precious oil. Nevertheless this innovation did not seem to stop them. The smarter ones would use their tail in order to get the oil out of the narrow openings.[140] The scene of a mouse trying to eat the wick is depicted even on lamps[141] as a reminder to the owner of the forthcoming danger or simply because it was funny to view. Even recent Greek oral tradition presents us with a warning concerning these tiny wick thefts.[142]

'Πάησε ο ποντικός / άρπαξε το φιτίλι /από μέσα από το καντήλι / που 'φεγγε και κένταγε / η κόρη το μαντίλι'.

There came the mouse /and grabbed the wick / from inside the lamp / that was shedding its light / as the daughter embroidered her handkerchief.

Bibliography

Adler, A. (ed.) 1928. *Suidae lexicon*, 4 Vols. *Lexicographi Graeci* 1.11.4. Leipzig: Teubner.

Allen, T. W. (ed.) 1912. *Homeri opera*. vol. 5. Oxford: Clarendon Press (repr. 1969).

Bachmann, L. (ed.) 1828. *Anecdota Graeca*, vol. 1. Leipzig: Hinrichs.

[119] Richard and Bonnet 1903, 26.17; Bailey 1972, 11.

[120] Migne 1857, vol. 49, Joannes Crysostomus scriptor ecclesiasticus: Ad populum Antiochenum (homiliae 1-21) 91, 14-16.

[121] Bethe 1900, 10.115.7; Migne 1857, vol. 72, Cyrillus Theologus: Commentarii in Lucam (in catenis), 72.709.22-24; Doutreleau 1962, 1.288.3-4

[122] Kaibel and François 1887, 11. 48.28-30; Preger 1907, 4.32, 25-26.

[123] Husselaman 1979, 47.

[124] Dauterman-Maguire *et al.* 1989, 58, 79, no. 25; Xanthopoulou 2005, figs 1, 6.

[125] Boissevain *et al.* 1895, 67. 9.2.2; Bailey 1972, 11-12.

[126] Maas 1929-1930, 282.24-26.

[127] Boissevain *et al.* 1895, 67.9.2.2.

[128] Herdt. Hist.2.62.1-5; Bailey 1972, 11.

[129] Kaibel and François 1887, 15.58.19, 21, 37; Stallbaum 1825, 1.264.3-4; Woelfflin and Melber 1887,1.12.1.5-6; Wendland 1901 23.11-12.

[130] Iorio 2005, fig. 3-8.

[131] Bethe 1900, 6.103.7-8, 10.118.1; Thurn 1973, Leo 6.21. 9-10.

[132] Tsitoyridoy 1986, pl. 35.

[133] Migne 1857, vol. 51, Joannes Crysostomus scriptor ecclesiasticus: Peccata fratrum non evulganda, 358, 33-36; Dauterman-Maguire *et al.* 1989, 58, 79, no. 25.

[134] Von Leutsch 1851, 10.79.4-6.

[135] Testini 1966, fig. 100-102; Bouras and Parani 2008, fig. 9.

[136] See Marconi Cosentino *et al.* 1993.

[137] Müller 1855, 25.16-17; Vogel and Fischer 1888, 3.12. 6.3-4.

[138] Meineke 1877, 17.3.11.5 - 6; Liddell and Scott 1996, entry *λυχνίτης*. Darrouzès 1960, 7.21; Berthelot and Ruelle 1888, 2.406.10-11; Van Dieten 1975, John 2.30.25; Kayser 1870, 2.14.50; Miller 1855, 4.2.1t; Halleux and Schamp 1985, 7.1, 28.5, 29.3.

[139] Allen 1912, 178-180; Migne 1864, 316.37-42.

[140] Lambros 1885, 2.368.1-4.

[141] Motsianos 2011b, 122; Dauterman-Maguire *et al.* 1989, 58, 78, no. 24.

[142] Photioy and Lytis 1995.

Bailey, D. 1988. *A catalogue of the lamps in the British Museum, vol. 3, Roman provincial lamps.* London: British Museum Publications.

Bailey, D. 1972. *Greek and Roman pottery lamps.* London: The trustees of the British Museum.

Beckby, H. (ed.) 1965. *Anthologia Graeca: griechisch-deutsch.* 4 Vols. Munich: Heimeran.

Berthelot, M. and C. É. Ruelle (eds) 1888. *Collection des anciens alchimistes grecs,* vol. 2. Paris: Steinheil.

Bethe, E. (ed.) 1900. *Pollucis onomasticon,* 2 Vols. *Lexicographi Graeci.* Leipzig: Teubner.

Boissevain, U. P., H. Smilda and W. Nawijn (eds) 1895. *Cassii Dionis Cocceiani Historiarum romanarum quae supersunt.* 3 Vols. Berlin: Weidmann.

Bonifay, M. 2004. *Etudes sur la céramique romaine tardive d' Afrique.* London: Archaeopress, International series 1301.

Bouras, L. and Parani, M. G. 2008. *Lighting in early Byzantium.* Washington, D.C: Dumbarton Oaks research library and collection.

Broneer, O. 1930. *Corinth IV.2: Terracotta lamps, Results of excavations conducted by the American school of classical studies at Athens.* Cambridge, Massachusetts: Harvard University Press.

Bussière, J. 1973. Sur une mèche de lampe. *Antiquités Africaines* 7, 255-257.

Colombini, P. M., Modugno, F. and Ribechini, E. 2005. Organic mass spectrometry in archaeology: evidence for Brassicaceae seed oil in Egyptian ceramic lamps. *Journal of Mass Spectrometry* 40, 890-898.

Copley, M. S., Bland, H. A., Rose, P., Horton, M., and Evershed, R.P., 2005. Gas chromatographic, mass spectrometric and stable isotopic investigations of organic residues of plant oils and animal fats employed as illuminants in archaeological lamps from Egypt. *Analyst* 130, 860-871.

Crnobrnja, N. A. 2008. Economic aspects of the use of roman oil lamps in Moesia Superior, in M. Chichikova and D. Gergova (eds) *Phosphorion: studia in honorem Mariae Čičikova,* 408-412. Sofia: Prof. Marin Drinov Academic Publishing House.

Dalmeyda, G. (ed.) 1930. *Andocide: Discours.* Paris: Les Belles Lettres.

Darrouzès, J. (ed.) 1960. *Épistoliers byzantins du xe siècle. Archives de l'orient chrétien* 6. Paris, Institut Français d'Études Byzantines.

Dauterman-Maguire, E., Maguire, P. H. and Duncan-Flowers, J. M. 1989. *Art and holy powers in the early Christian house. Exhibition catalogue.* Urbana-Chicago: University of Illinois Press.

De Laguna, F. 1940. Eskimo Lamps and Pots. *The journal of the Royal Anthropological Institute of Great Britain and Ireland* 70, 53-76.

Delatte, A. (ed.) 1927. *Anecdota Atheniensia et alia,* vol. 2. Paris, Droz.

Delattre, P. R. 1889. *Les Lampes Antiques du Musee de Saint Louzs de Carthage.* Lille: Société de Saint Augustin.

Doutreleau, L. (ed.) 1962. *Didyme l'Aveugle sur Zacharie,* 3 Vols. *Sources chrétiennes* 83, 84, 85. Paris: Éditions du Cerf.

Evershed, P. R., Vaughan, J. S., Dudd, N. S. and Soles, S. J. 1997. Fuel for thought? Beeswax in lamps and conical cups from late Minoan Crete. *Antiquity* 71, 979-985.

Forbes, J. R. 1958. *Studies in ancient technology,* vol. 6, Leiden, E. J. Brill.

Garnier, N., Rolando, C., Munk-Høtje, J. and Tokarskia, C. 2009. Analysis of archaeological triacylglycerols by high resolution nanoESI, FT-ICR MS and IRMPD MS/ MS: Application to 5th century BC–4th century AD oil lamps from Olbia (Ukraine). *International Journal of Mass Spectrometry* 284, 47-56.

Gennadios, G. P. 1914, *Lexicon Phytologicon,* Athens.

Giannini, A. (ed.) 1965. *Paradoxographorum Graecorum reliquiae.* Milano: Istituto Ed. Ital.

Halleux, R. and J. Schamp (eds) 1985. *Les lapidaries grecs.* Paris, Les Belles Lettres.

Harris, V. W. 1980. Roman Terracotta Lamps: The Organization of an Industry. *Journal of Roman Studies* 70, 126-145.

Hayduck, M. (ed.) 1907. *Eustratii in analyticorum posteriorum librum secundum commentarium. Commentaria in Aristotelem Graeca* 21.1. Berlin: Reimer.

Heiberg, J. L. (ed.) 1921. *Paulus Aegineta,* 2 Vols. *Corpus medicorum Graecorum.* Leipzig: Teubner.

Hermanns, H. M. 2004. *Licht und Lampen im westgriechischen Alltag: Beleuchtungsgerät des 6.-3. Jhs. v. Chr. in Selinunt,* Rahden / Westf.: Leidorf.

Hermansen, G. 1982. The Stuppatores and their guild in Ostia. *American Journal of Archaeology* 86, 121-126.

Hort, A. (ed.) 1916. *Theophrastus. Enquiry into plants,* 2 Vols. Cambridge, Mass.: Harvard University Press.

Husselaman, M. E. 1979. *Karanis Excavations of the University of Michigan in Egypt, 1928-1935: Topography and Architecture a Summary of the Reports of the Director Enoch.* Michigan.

Iorio, V. 2005. I rinvenimenti degli scavi del '700 e quelli recenti di candelabri e lucerne nel Tempio di Iside a Pompei. In Chrzanovski L. (ed.) *Lychnological Acts 1, Actes du 1er Congrès international d'étude sur le luminaire antique (Nyon - Genève 29/9 - 4/10/2003),* 167-172, Monographies instrumentum 31. Montagnac: M. Mergoil.

Jones, D. M. and N. G. Wilson (eds) 1969. *Prolegomena de comoedia. Scholia in Acharnenses, Equites, Nubes. Scholia in Aristophanem* 1.2. Groningen, Wolters Noordhoff.

Kaibel, G. and L. François (eds) 1887. *Athenaei Naucratitae Dipnosophistarum: libri XV.* Vol. I. Leipzig, Teubner.

Kaimakis, D. (ed.) 1976. *Die Kyraniden.* Meisenheim am Glan: Hain.

Karivieri, A. 1996. *The Athenian lamp industry in late antiquity,* Papers and monographs of the Finish Institute at Athens vol. V. Helsinki.

Kayser, C. L. (ed.) 1870. *Flavii Philostrati opera,* vol. 1. Leipzig: Teubner.

Kenneth, J. D. (ed.) 1968. *Clouds.* Oxford: Clarendon press.

Kimpe, K., Jacobs, A. P. and Waelkens, M. 2001. Analysis of oil used in late Roman oil lamps with different mass spectrometric techniques revealed the presence of predominantly olive oil together with traces of animal fat, *Journal of Chromatography A,* 937, 87-95.

Kühn, C. G. (ed.) 1825. *Claudii Galeni opera omnia,* vol. 10. Leipzig: Knobloch.

Kühn, C. G. (ed.) 1826. *Claudii Galeni opera omnia,* vol. 11-12. Leipzig: Knobloch.

Kühn, C. G. (ed.) 1827. *Claudii Galeni opera omnia,* vol. 13. Leipzig: Knobloch.

Kühn, C. G. (ed.) 1830. *Claudii Galeni opera omnia,* vol. 19. Leipzig: Knobloch.

Laga, C. and C. Steel (eds) 1980. *Maximi confessoris quaestiones ad Thalassium,* 2 Vols. *Corpus Christianorum. Series Graeca* 7 & 22. Turnhout: Brepols.

Lambros, S. P. (ed.) 1885. *Excerptorum Constantini De natura animalium libri duo. Aristophanis Historiae animalium epitome, subiunctis Aeliani, Timothei aliorumque eclogis. Commentaria in Aristotelem Graeca,* suppl. 1.1. Berlin: Reimer.

Latte, K. (ed.) 1953. *Hesychii Alexandrini lexicon*, vols. 1-2. Copenhagen: Munksgaard.

Legrand, P. E. (ed.) 1932. *Hérodote Histoires.* 9 Vols. Paris: Les Belles Lettres.

Liddell, H. G. and Scott, R. 1996. *A Greek-English Lexicon, revised and augmented throughout H. S. Jones.* Oxford,: Oxford University Press (electronic edition).

Long, H. S. (ed.) 1964. *Diogenis Laertii vitae philosophorum*, 2 Vols. Oxford: Clarendon Press.

Maas, P. 1929-1930. Die ikonoklastische Episode in dem Brief des Epiphanios an Johannes. *Byzantinische Zeitschrift* 30.

MacDowell, D. M. (ed.) 1971. *Wasps.* Oxford: Clarendon Press.

Marconi-Cosentino, R. and Riacciardi L. 1993. *Catacomba di commodilla: lucerne ed altri materiali dalle galerie 1, 8, 13.* Roma: L' erma di Bretschneider.

Marrou, H.-I. and M. Harl (eds) 1960. *Clément d'Alexandrie. Le pédagogue*, 3 Vols. *Sources chrétiennes* 70, 108, 158. Paris: Éditions du Cerf.

Mayhoff, C. (ed.) 1892. *C. Plini Secundi Naturalis Historiae Libri XXXVII.* Vols. 1-5. Leipzig: Teubner.

McCown, C. C. 1922. *The testament of Solomon.* Leipzig: Hinrichs.

Meineke, A. (ed.) 1877. *Strabonis geographica*, 3 Vols. Leipzig: Teubner.

Migne, J. P. (ed.) 1864. *Apophthegmata patrum.* Paris: Lutet.

Miller, E. (ed.) 1855. *Manuelis Philae Carmina*, 12 Vols. Paris.

Motsianos, I. 2010. Ύστερη Αρχαιότητα: Το λυκόφως των κατασκευασμένων με μήτρα λυχναριών (Late Antiquity: the twilight of mouldmade lamps), in D. Papanikola-Bakirtzis and D. Koussoulakou (eds) *Κεραμική της Ύστερης Αρχαιότητας από τον Ελλαδικό χώρο*, Επιστημονική Συνάντηση, Θεσσαλονίκη 12-16 Νοεμβρίου 2006, 270-281. Thessaloniki: Archaeological Institute of Macedonian and Thracian Studies (in Greek with English abstract).

Motsianos, I. and E. Bintsi (eds) 2011. *Light on light, an illuminating story, Exhibition Catalogue.* Thessaloniki, Folklife & Ethnological Museum of Macedonia-Thrace.

Motsianos, I. 2011a. Artificial lighting during Byzantine and Post-Byzantine period, in I. Motsianos and E. Bintsi (eds) *Light on light, an illuminating story*, Exhibition Catalogue, 90-125. Thessaloniki: Folklife & Ethnological Museum of Macedonia-Thrace.

Motsianos, I. 2011b. Φως Ιλαρόν: ο τεχνητός φωτισμός στο Βυζάντιο (Joyful light: the artificial lighting in Byzantium), Unpublished PhD dissertation, University of Thessaly (in Greek).

Moullou, D. 2010. Τεχνητός φωτισμός στην Αρχαία Ελλάδα (Artificial lighting in Greece), Unpublished PhD dissertation, University of Crete (in Greek).

Moullou, D. and Topalis, F. 2011b. An example of room lighting from Classical Greece, in I. Motsianos and E. Bintsi (eds) *Light on light, an illuminating story*, Exhibition Catalogue, 64-66. Thessaloniki: Folklife & Ethnological Museum of Macedonia-Thrace.

Moullou, D. and Topalis, F. 2011a. Illuminating the past: measuring the efficiency of means of lighting, in I. Motsianos and E. Bintsi (eds) *Light on light, an illuminating story*, Exhibition Catalogue, 58-63. Thessaloniki: Folklife & Ethnological Museum of Macedonia-Thrace.

Müller, K. (ed.) 1855. *Geographi graeci minores. Volumen primum.* Geographi Graeci Minores. Paris: A. Firmin Didot.

Nau, F. (ed.) 1907-1913. Histoires des solitaires égyptiens, Apophthegmata patrum (Varia): Apophthegmata (collectio anonyma) (e cod. Coislin. 126). *Révue de l'Orient Chrétien* 12-14, 17-18.

Nauck, A. (ed.) 1889. *Tragicorum Graecorum fragmenta.* Leipzig: Teubner, 1889.

Oder, E. and K. Hoppe (eds) 1924. *Corpus hippiatricorum Graecorum*, vol. 1. Leipzig: Teubner.

Olivieri, A. (ed) 1935. *Aetii Amideni Libri medicinales i-iv.* Leipzig, Berlin: Teubner.

Olivieri, A. (ed), 1950. *Aëtii Amideni libri medicinales v-viii, Corpus medicorum Graecorum*, vol. 8.2. Berlin: AkademieVerlag.

Papanikola-Bakirtzi, D. (ed.) 2002. *Everyday Life in Byzantium*, Exhibition Catalogue, Thessaloniki, White Tower, October 2001 - January 2002. Athens: Kapon Editions.

Perlzweig, J. 1961. *The Athenian Agora VII: Lamps of the Roman period, first to seventh century after Christ.* Princeton: American School of Classical Studies at Athens.

Photioy, S. and Lytis, N. 1995. *Δημοτικά τραγούδια από τη βόρειο Ήπειρο.* Αθήνα: Νεφέλη.

Porson, R. (ed.) 1822. *Φωτίου τοῦ πατριάρχου λέξεων συναγωγή.* Cambridge: Cambridge University Press.

Preger, T. (ed.) 1907. *Scriptores originum Constantinopolitanarum. 2: Ps.-Codini Origines continens.* Leipzig: Teubner.

Pusey, P.E. (ed.) 1868. *Sancti patris nostri Cyrilli archiepiscopi Alexandrini in XII prophetas.* Oxford: Clarendon Press.

Rabe, H. (ed.) 1896. *Anonymi et Stephani In Artem rhetoricam commentaria: consilio et auctoritate Academiae Litterarum Regiae Borussicae. Commentaria in Aristotelem Graeca 21.2.* Berlin: Reimer.

Rahlfs, A. (ed.) 1935. *Septuaginta*, vol. 2, Stuttgart: Württembergische Bibelanstalt.

Richard, A. and Bonnet, M. 1903. *Acta Apostolorum Apocrypha*, vol. 2.2. Leipzig: Mendelssohn.

Romanus, K., Van Neer, W., Marinova, E., Verbeke, K., Luypaerts, A., Accardo, S., Hermans, I., Jacobs, P., De Vos, D. and Waelkens M. 2008. Brassicaceae seed oil identified as illuminant in Nilotic shells from a first millennium AD Coptic church in Bawit, Egypt. *Anal Bioanal Chem* 390, 783-793.

Rose, V. (ed.) 1886. *Aristotelis qui ferebantur librorum fragmenta.* Leipzig: Teubner.

Schmidt, M. 1861. *Hesychii Alexandrini lexicon*, vols. 3-4. Jena: Mauk-Dufft.

Schmidt, W. (ed.) 1899. *Heronis Alexandrini opera quae supersunt omnia*, vol. 1. Leipzig: Teubner.

Schneider, R. (ed.) 1908. *Griechische Poliorketiker.* Berlin: Weidmannsche buchhandlung.

Stallbaum, G. (ed.) 1825. *Eustathii archiepiscopi Thessalonicensis commentarii ad Homeri Odysseam*, 2 Vols. Leipzig: Weigel.

Sunderland, B. P., Quintiere, G. J., Tabaka, A. G., Lian, D. and Chiu, C. 2011. Analysis and measurement of candle flame shapes, *Proceedings of the Combustion Institute* 33, 2489-2496.

Testini, P. 1966, *Le catacombe e gli antichi cimiteri cristiani in Roma.* Bologna: Capelli.

Themelis, P. 2004. Ονοματολόγιο σκευών και αγγείων, in Δρούγου, Σ., Ν. Ζαφειροπούλου and Ε. Κυπραίου (eds) *ΣΤ' επιστημονική συνάντηση για την ελληνιστική κεραμική (Βόλος 17-23 Απριλίου 2000)*, 715-724. Athens: Archaeological Receipts Fund.

Theodoridis, C. (ed.) 1982, *Photii patriarchae lexicon*, vol. 1 (Α-Δ). Berlin: De Gruyter.

Thurn, J. (ed.) 1973. *Ioannis Scylitzae synopsis historiarum. Corpus fontium historiae Byzantinae 5. Series Berolinensis.* Berlin: De Gruyter.

Tittmann, J. A. H. (ed.) 1808. *Iohannis Zonarae lexicon ex tribus codicibus manuscriptis*, 2 Vols. Leipzig: Crusius.

Tsitoyridoy, A. 1986. *Ο ζωγραφικός διάκοσμος του Αγίου Νικολάου Ορφανού στη Θεσσαλονίκη*. Θεσσαλονίκη: Κέντρο Βυζαντινών Ερευνών.

Van Dieten, J. (ed.) 1975. *Nicetae Choniatae historia*, pars prior [*Corpus fontium historiae Byzantinae* 11.1. *Series Berolinensis*. Berlin: De Gruyter.

Velenis, G. and Zachariadis, S. 2011. Lamps from Thessaloniki's Agora, in I. Motsianos and E. Bintsi (eds) *Light on light, an illuminating story*, Exhibition Catalogue, 183-210. Thessaloniki: Folklife & Ethnological Museum of Macedonia-Thrace.

Velenis, G., Poyloy-Papadimitrioy, N. and Zachariadis, S. 2009. Λυχνάρια της ύστερης αρχαιότητας και των πρώιμων βυζαντινών χρόνων από την αρχαία Αγορά Θεσσαλονίκης. *Το Αρχαιολογικό Έργο στη Μακεδονία και τη Θράκη 23*, 273-283.

Vogel, F., and K. T. Fischer (eds) 1888. *Diodori bibliotheca historica*. 5 Vols. Leipzig: Teubner.

Von Arnim, H. F. A. 1893. *Dionis Prusaensis, quem vocant Chrysostomum, quae exstant omnia*. 12 Vols. Berlin: Weidmann.

Von Leutsch, E. L. (ed.) 1851. *Corpus paroemiographorum Graecorum*, vol. 2. Göttingen: Vandenhoeck & Ruprecht.

Wellmann, M. (ed.) 1907. *Pedanii Dioscuridis Anazarbei de materia medica libri quinque*, 3 Vols. Berlin: Weidmann.

Wendland, P. (ed.) 1901. *Alexandri in librum De sensu commentarium, Commentaria in Aristotelem Graeca* 3.1. Berlin: Reimer.

Woelfflin, E. and J. Melber (eds) 1887. *Polyaeni strategematon libri viii*. Leipzig: Teubner.

Wolska-Conus, W. (ed.) 1968. *Topographie chrétienne*. 3 Vols. *Sources chrétiennes* 141, 159, 197. Paris: Éditions du Cerf.

Wunderlich, H. C. 2003. Light and economy - An Essay about the economy of prehistoric and ancient lamps, in L. Chrzanovski (ed.) *Nouveautés Lychnologiques-Lychnological News 1*, 251-263. Hauterive NE: LychnoServices

Xanthopoulou, M. 2005. Lampes en métal, lampes en terre cuite: vies parallèles, in L. Chrzanovski (ed.) *Lychnological Acts 1, Actes du 1er Congrès international d'étude sur le luminaire antique*, Nyon - Genève 29/9 - 4/10/2003, 136-139. Montagnac: M. Mergoil.

Difficulties and Preconditions for the Utilitization of Glass in Lighting in Byzantium

Ioannis Motsianos

Museum of Byzantine Cutlure, Thessaloniki

Abstract: The beginning of the use of glass *kandelai* in Late Antiquity was an innovation in artificial lighting because a major problem was overcome that had prevented glass lamps, similar in shape to Roman clay and metallic lamps, from operating successfully. The positioning of the burning end of the wick at the nozzle of the lamp would cause the glass at the nozzle to break. The invention of metal and clay wickholders not only allowed the use of glass vessels as lighting devices but also influenced the shape of the glass *kandelai* and the evolution of their typology, both of the individual lamps and the ones that were placed in polykandela, and led to a huge variety of devices made especially for the support and suspension of wicks.

Keywords: Glass lamp, *kandela*, glass *kandela*, *Ballota acetabulosa*, wick, metal wickholder, clay wickholder.

Introduction

The free blowing technique of shaping glass was invented in the 1st century BC Step by step free-blown glass replaced earlier techniques and enabled glassware to be utilized as a medium to support artificial lighting. Earlier, glass vessels could have been used for lighting a fire or they could contain oil to feed the lamps and the torches, so that their light would burn longer. It seems, however, that some time passed until the moment when the glass vessels themselves were used as lighting devices. The use of glass vessels as functional lighting devices emerges in the 4th century AD and then causes a technological revolution in artificial lighting that affected not only the form of the light devices but also led to a variety of means of wick supports designed specifically for them.

This article will illustrate the difficulties encountered when exploiting glass for artificial lighting. Because of the special features of glass as well as the decisions made towards overcoming the obstacles in making glass lamps, these new glass lamp devices, together with *polykandela* (chandeliers) were the most widespread types of lamps in the Byzantine world from the 4th to the 7th century AD and in general during the whole Byzantine period.

Operating mode of the glass lamps and the terms used for them

Glass lamps of the Roman period, similar in shape to the clay and metal lamps of the same era, seem to have been difficult to operate. The juxtaposition of the burning end of the wick and the nozzle could cause the latter to break. Glass lamps of Roman times seem to have been unsuccessful experimentations but at the same time furthered steps towards the use of lighting devices made of glass. This is why lamps seem to be among the first attempts to exploit the free-blown glass techniques to create glass lighting devices.

The term glass lamp can be used to describe both lamps with a nozzle and nozzleless ones. But in using it we usually mean a type of nozzleless open lamp, that is, without a horizontal projection to support the wick. In this type of lamp the wick, where the fuel actually burns, is contained within the rim of the vessel and not above or outside it as in standard oil lamps (*λύχνοι/lychnoi*). The wick in the nozzleless open lamp was in a vertical position, whereas in ordinary lamps the wick was horizontal. In order to distinguish them we can call them using their Byzantine name *κανδήλαι/kandelai* (singular *κανδήλα/kandela*). This terminology choice has been perpetuated, since even today the term in Modern Greek describes these lighting articles. The word comes from the transcription of the Latin *candela*, a general term that denoted everything that used to shine, something similar to the English word lamp and obviously echoes the fact that in the Early Byzantine period no particular terms were developed to describe these objects but instead general terms were used for this purpose.

Preconditions for utilizing glass in lighting in Byzantium

The transparent walls of glass lamps which could diffuse the light in all directions were at the same time fragile, and the lamp would shatter if the heat was too high or the flame came too close to its walls. This was a handicap to the use of glass vessels as lamps, and explains why they were used as lamps only in the Late Roman period. Moreover, the aforementioned requirements largely determined how glass lamps worked.

As Paulinus of Nola mentions[1] and Leontius Presbyter of Constantinople (4th century AD) implies,[2] the glass *kandelai* were filled with olive oil and water. A depiction of a kandela in a miniature of an Armenian manuscript (Vienna, Mekhitarian Library Ms. 697, gospel fragment, 6r., late 10th century)[3] is a very good example in which three different colors were probably chosen to declare the different materials: one denotes the transparent glass, and two others the oil and the water. Adding water, which has an extremely high heat capacity, to the bowl of a glass lamp was one way of protecting it from overheating and breaking. At the same time adding water to a lamp vessel reduces the volume of oil needed to feed the wick efficiently.[4]

[1] Walsh 1978, 213-214.

[2] Datema 1987, 347-365 (vv. 458-464).

[3] Brentjes 1982, fig. 212.

[4] I thank Karen Garnett for this suggestion.

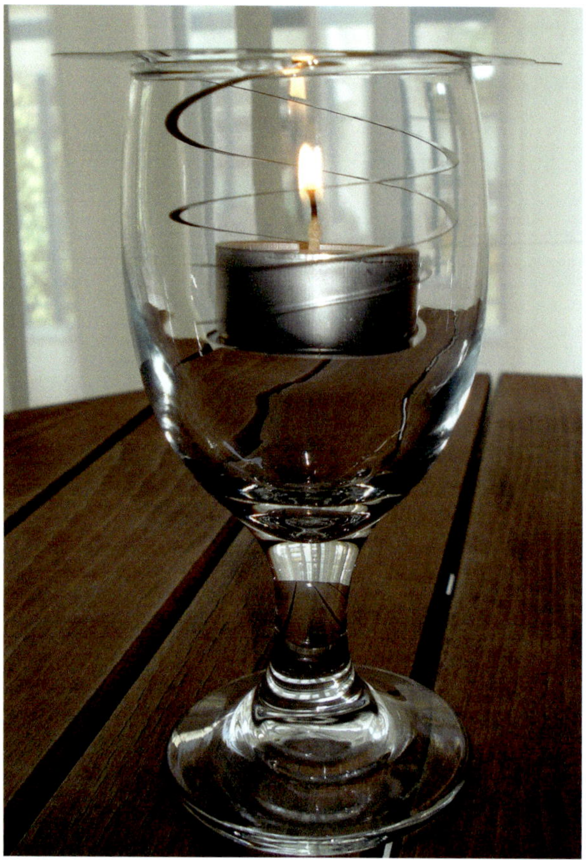

Figure 1. Modern candleholder.

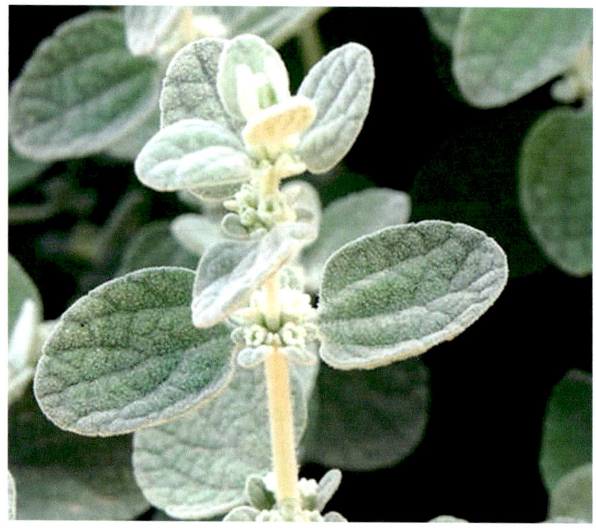

Figure 2. *Ballota acetabulosa.*

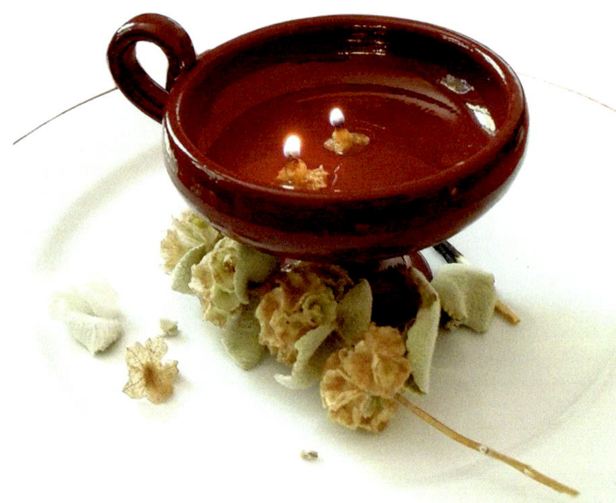

Figure 3. An open clay vessel with vegetal wicks operating as a lamp (*Ballota acetabulosa*) (Photo by Karen Garnett).

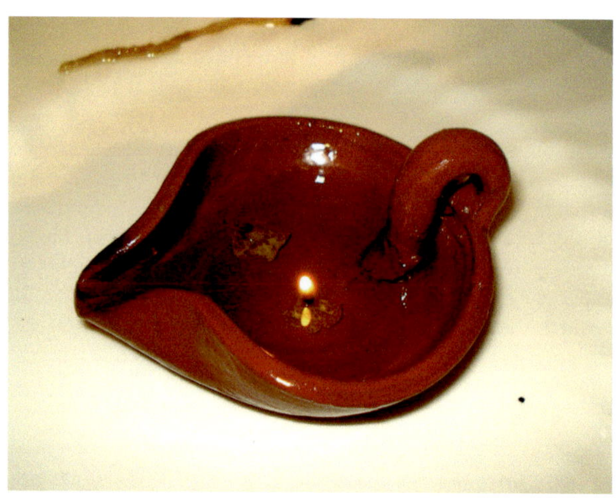

Figure 4. A copy of Byzantine open clay lamp operating with a vegetal wick (*Ballota acetabulosa*) (Photo by the author).

Another difficulty was the need to prevent the wicks of glass lamps from touching the walls of the vessel. It is worth noting that even today's fixtures for fireplace lamps must keep the flame away from the glass so that it does not break (Figure 1).

Various writers record the use of the calyces of garden horehound (*Ballota acetabulosa*) blossoms[5] (Figures 2-4) as lamp wicks, while the plant Dioscorides called *Phlomis tertia*, also referred to as *λυχνίτης* or *θρυαλλίς* (little flame), which abounds in Greece, was also used for the same purpose, as is reflected in its common names signifying 'lamp-plant', 'lamp-lighter', 'candlewick', etc. Moreover, it is confirmed by experiments that Ancient or Byzantine open clay vessels (Figures 3, 4) could operate as lamps with this type of wick. However, even in this case as far as the glass kandelai are concerned if the flame was too close to the glass the glass would inevitably break.

There was, therefore, the need to stabilize the wick inside the kandelai in such a way that it would not come into contact with the glass walls. So it was necessary to add a wickholder[6] to ensure this vital condition for the operation of glass kandelai and their protection against breakage, as well as most likely provide their optimal performance, as the wickholders at the same time served to keep the wicks in a

[5] Gennadiou 1914, 999-1000. For references to *phlomos*, see Schmidt 1867, 1527; Porson and Dobree 1822, 651.
[6] Foy 2007, 178.

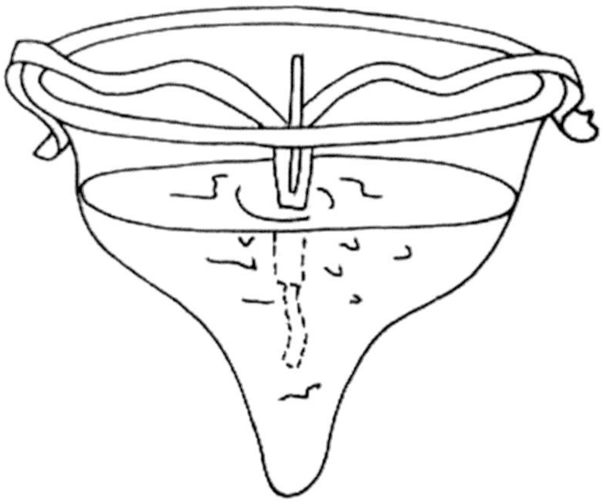

Figure 5. Metal wickholders in the shape of Y (drawing by Ch. Mallia).

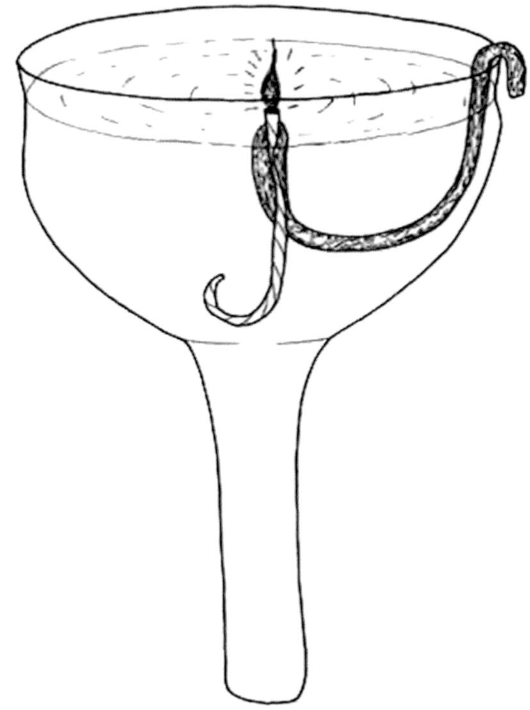

Figure 6. S-shaped metal wickholder (drawing by Ch. Mallia).

vertical position. All these facts seem to have imposed the invention of wickholders.

The use and variety of wickholders are confirmed by archaeological finds and their representations. According to their form and methods of function we can distinguish at least five different types of wickholders.

Y-shaped metal wickholders (Figure 5)

The Y-shaped metal plate is characterized as the most common type of wickholder.[7] The wick was attached to the cylinder formed in the middle of the holder, and the two ends were supported by the lips of the lamp thereby stabilizing the wick vertically in the center of the kandela[8] and preventing its immersion in the oil.[9]

The majority of these wickholders are made of copper alloy but there are some examples made of lead as well.[10] This type was extremely widespread in the Middle East, Greece and Turkey, and many examples have been found in excavations dating from the 6th to the early 8th century AD.[11]

A depiction of a wickholder of this type on a sixteenth-century wall painting in the Church of the Dormition of Mother of God at Kurdali in Cyprus[12] offers us an idea how it was used.

S-shaped metal wickholders (Figure 6)

The S-shaped wickholder consists of a metal plate, one end of which is bent to be fixed to the rim of the glass kandela while the other end is folded to create a ring through which the wick passes.

For this type of wickholder, as with the previous one,[13] Danièle Foy points out that it was in use on all the Mediterranean coasts but seems to be more widespread in the eastern part of the Mediterranean.[14] Many examples were found in excavations dating from the 6th to the early 8th century AD.[15]

I consider that the bronze wickholder found in Elaiussa Sebaste with a lamp in the form of a plate dating to the 7th century AD[16] belongs to this type. As suggested by the scholars, one end of the wickholder would rest on the rim of the glass plate, but in my opinion, unlike the restoration proposed in the drawing (Gençler 2009b, fig. 5), the wickholder was bent almost in the middle to rest on the bottom of the vessel, and then the edge was upturned to hold the tip of the wick over the surface of the olive oil to feed the flame of the glass kandela. I do not exclude the case in which the wickholder would be placed in a nearly horizontal position with a slight inclination so that the cylindrical hole where the wick was placed would be in a vertical position. Also, the size of the vessel and the length of

[7] Bouras and Parani 2008, 4; Kondoleon, 2000, 83, fig. 5.
[8] Sodini and Kolokotsas 1984, 187.
[9] Kondoleon 2000, 83, fig. 5.
[10] Foy 2011, 216-217.
[11] Foy 2011, 216-217; Sodini and Kolokotsas 1984, 187; Russell 1982, 149; Stern 1984, fig. 11; Ballance *et al.* 1967, 129, fig. 51, no. 53; Kollias 1974, fig. 149β; Olcay 2001, 80; Dunn 2007, 531, fig. 17.1, nos. 25, 27; Bouras and Parani 2008, 102, cat. no. 34.
[12] Stylianou 1985, 146, fig. 75. Also see the painting Georges de la Tour *c.* 1635-37 'The Magdalen with the Smoking Flame,' today in the Los Angeles County Museum of Art.

[13] Danièle Foy includes in the same type the metal wickholders that form a cylinder through which the wick passed and from this cylinder one or two metal plates extended whose ends were folded to the edges of the glass kandela mouth to stabilize the support. See Foy 2011, 215-216.
[14] Foy 2011, 232; Crowfoot and Harden 1931, 207; Alliata 1990, 247-261, fig. 8; Foy 2003, 82, note 83; Piccirillo and Alliata 1994, fig. XXXII; Sodini and Kolokotsas 1984, 187, fig. 70b, note 13; Dunn 2007, 531, nos 25, 27.
[15] Foy 2011, 216-217.
[16] Gençler 2009b, 297, fig. 5.

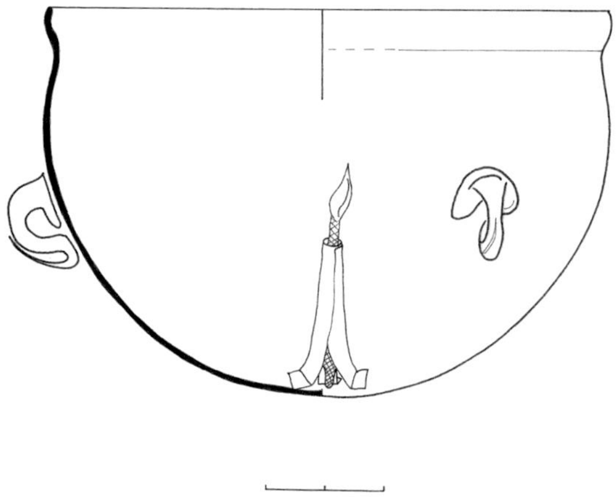

Figure 7. Tripod-shaped metal wickholder (drawing by Ch. Mallia according to Foy 2011, 237, fig. 12).

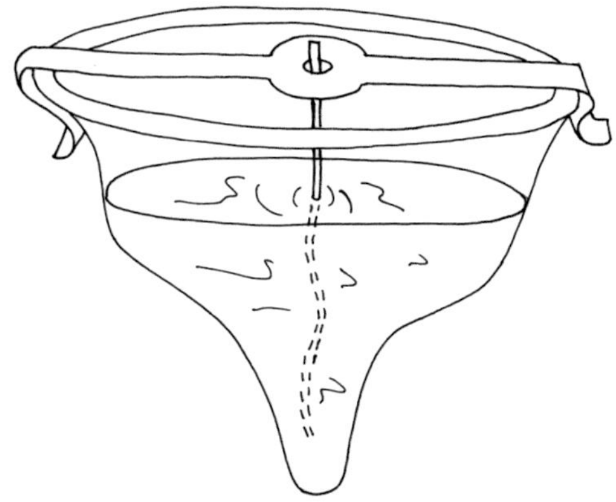

Figure 8. Strap-shaped metal wickholder placed on the rim of the glass candela (drawing by Ch. Mallia).

the support would justify the simultaneous use of more than one wickholder to turn this glass kandela into a chandelier.

S-shaped wickholders appear to have been common in the Middle Byzantine period, as the archaeological finds suggest[17] and are depicted in manuscripts of the Middle and Late Byzantine period.[18] Depictions of S-shaped wickholders on icons dating to the Post Byzantine period[19] attest their use even during that period.

Tripod-shaped metal wickholders (Figure 7)

These supports consist of a central cylindrical section at the bottom of which three or four sheets are formed to create a base, similar to a tripod, on the bottom of the kandelai. Wickholders of this type are also made of copper alloy and lead. The cylinder through which the wick was passed was in many cases perforated to let the wick be impregnated with oil more easily;[20] if we take into account that water was added to the kandlela, the tip at the bottom of the wick would be immersed in water and not in olive oil. So this cylinder configuration was necessary to make the kandela work better. Objects like this have been found in Rome, Tunisia,[21] Sardinia[22] and Sicily.[23]

In this type may belong the wickholder described by Paulinus of Nola: 'The triple base of lead is joined at the centre, from which the hollow top rises, filled with fuel for the oily wick. The dripping tow ignites to form a slender torch'.[24]

A wickholder depicted on an icon of 1770 from Albania[25] is reminiscent of the description of Paulinus of Nola. However, you must consider that, since the position of this type of wickholder was under the surface of the oil, the depiction of it was too difficult to render either in manuscripts or on icons.

Strap-shaped metal wickholders (Figure 8)

These metal plates were usually made of lead but rarely also of copper alloy or iron. They usually are found in the western Mediterranean in sixth-century excavations.[26] The two ends were adapted to the lips of the glass kandela, and there was a hole in the middle which supported the wick which passed through the hole. The middle part of the plate was usually oval or round. These wickholders were intended for deep kandelai and ensured that the flame would burn to the top of them.[27]

A disadvantage of this support was that the flame burned above the rim of the glass kandela and thus was more vulnerable to wind currents. Moreover, the use of the glass vessel whose walls could function as a lens for better lighting was not fully exploited. The other types of wickholders let the flame burn at or below the level of the rim. But if there was sufficient slack in the arms, then the flame could potentially be lower than the glass rim, but there would be a compensatory loss of fuel capacity.

Two examples of 1800 AD glass kandela from France,[28] in which the wick rests with a special metal construction at the center of the kandela mouth gives us a clear picture of how this type of wickholder works. Another example dating to the 20th century AD comes from Russia.[29]

[17] Bouras and Parani 2008, 4.

[18] Bouras and Parani 2008, fig. 5; Lowden 1999, fig. 168; Oretskaia 2008, 349; *Bizancio en Espana* 2003, 182; Spatharakis, Greek manuscripts, vol. 1. Djurić *et al.* 1990, fig. 81.

[19] *Treasures of Mount Athos 1997*, 145; Pirovano, Arapi 2002, 124, no. 54.

[20] Foy 2011, 219.

[21] Foy 2007, 172, fig. 109.

[22] Martorelli, Amante-Simoni 1986, 161-170, 180, figs CVII, XCVII, nos. 38, 39; Martorelli 2000, 41, nos. 74-76, fig. IX.

[23] Branciforti 2006, 163, no. 11; Orsi 1895, 498; Orsi 1896, 343.

[24] Walsh 1978, 214.

[25] Tourta 2006, no. 50.

[26] Foy 2011, 235; Foy 2003, 80-82, fig. 89; Foy 2007, 172, 178, fig. 110. Zitrides 2005.

[27] Foy 2011, 219.

[28] Kaiser 2003, 94-95.

[29] Chrzanovski 2003, 102, no. 148.

The survival of the use of similar wickholders, usually improvised, mainly in rural areas and especially in cemeteries is confirmed by oral testimonies both in Greece and in Cyprus until the 20th century AD.

Clay wickholders (Figures 9, 10)

Clay disks with a hole in their center found in the Bet Shean synagogue in association with glass kandelai stems led to the conclusion that they were intended to be used as wickholders in kandelai with a hollow stem[30] (Figure 9). Some finds from Elaiussa Sebaste (Mersin-Ayaş), in Turkey seem to verify this aspect.[31] Certainly the clay disks were of a sufficient diameter[32] to keep the lighted wick at a safe distance from the fragile glass walls. These clay wickholders seem to have been in use during the 7th century in the eastern Byzantine provinces.[33]

Danièle Foy suggested that clay wickholders could also be used in cup-shaped kandelai and that in this case the larger surface would touch the bottom (obviously to ensure better stability) in contrast to the way the clay wickholders were positioned in kandelai with a hollow stem.[34] However, as these objects could not float in the oil, it would be difficult for them to work in this manner because they would hold the flame at a very low height and any wick folded under the clay disk would create an imbalance of the support.[35]

It seems that the glass cylinder added at the bottom of the glass kandelai played a role in fixating the base for the clay wickholders (Figure 10). In other words by adding this glass cylinder they configured a tube similar to ones that were in glass candelai with hollow stems where they could support the clay wickholder turning the glass vessel into a glass kandela.

Based perhaps on G. M. Crowfoot and D. B. Harden,[36] it appears that some scholars have considered the glass cylinder appearing on the bottom of some glass kandela as a wickholder, as in the case of the 4th-7th century kandela from the Byzantine Collection of Dumbarton Oaks.[37]

Glass tubes, fused to the center of the bottom of glass lamps are only present in the Middle East, from the 5th to 7th century AD.[38] The earliest known examples come from Gezer and Ophel (Israel) and date to the 5th and 6th centuries AD.[39] It is noted that, although this configuration appears in the East, it features a group of glass kandelai that were almost certainly produced in Gortyn between the 5th and 7th centuries AD.[40]

However, how could the wick be fixed inside it? Even if we suppose that a piece of wood or reed could manage its fixation the danger of breaking introduced by the heat of the flame would remain.

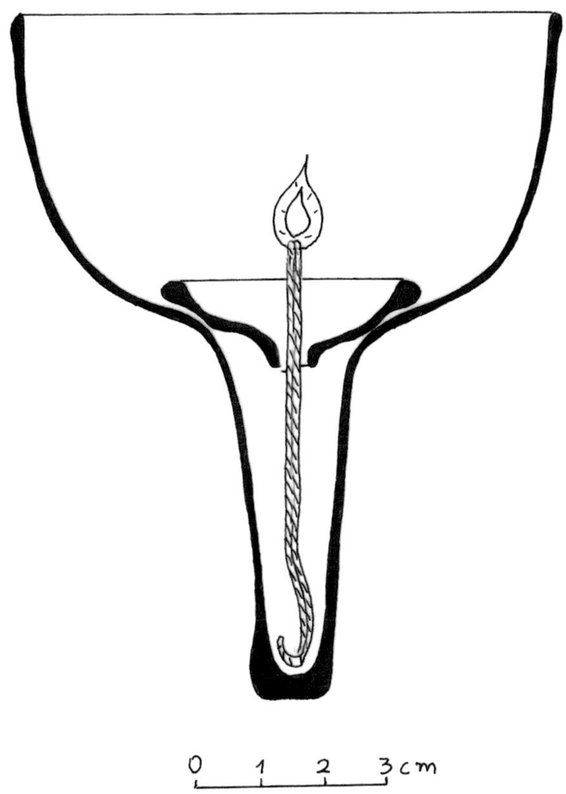

Figure 9. Clay wickholder in kandela with a hollow stem (drawing by Ch. Mallia according to Gençler 2009b, 296, fig. 4.7).

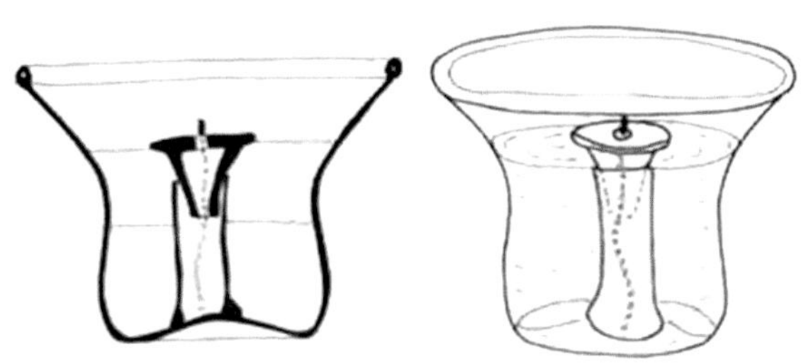

Figure 10. Clay wickholders based on a glass tube (drawing by Ch. Mallia).

[30] Peleg and Reich 1992, 160.
[31] Gençler 2009b, 297, fig. 4.
[32] Their average diameter was around 2.7 cm while the maximum was 3.4 cm.
[33] Foy 2011, 232.
[34] Foy 2011, 215.
[35] Peleg and Reich 1992, 160, note 2 where a combination of a nut with a clay wickholder is proposed. In this case the nut could prevent the wick from sinking in the oil and the clay wickholder could protect the nut from the flame.
[36] Crowfoot and Harden 1931, 207-208 and 201: 'This was the provision of a vertical glass tube, fixed on to the centre of the base, inside, to hold the wick'.
[37] See Bouras and Parani 2008, 3-4, no. 15; Dauterman-Maguire1989, no. 9. As far as the comment in Foy 2011, 211-212 I would like to note that I don't consider it possible that these kandelai could be used as candlestands. The height of the flame would vary depending on how much the candle had been burned. Certainly residual waxes would have been found in these containers. The molten candle will stain the glass walls.
[38] Foy 2011, 232.
[39] Forbes 1966, 147.
[40] Uboldi 1995, 93, note 2 where she refers to publications for glass kandelai of this type.

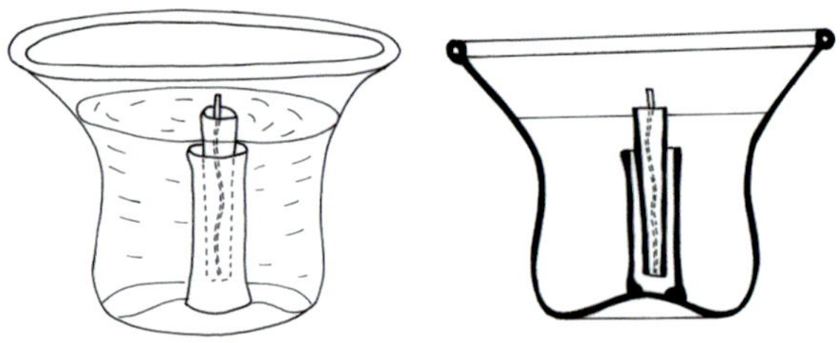

Figure 11. Suggested reconstruction of metal wickholders put in a glass tube (drawing by Ch. Mallia).

Moreover, a wick made of vegetal fibers would be dipped in the oil and water and its end could possibly be stabilized in the cylinder. But, if there was no other accessory, even if the wick did not sink, its flaming end would wander freely at the surface of the oil and could at any time dangerously approach the fragile walls of the glass kandela. It seems therefore that this construction was intended to support a clay wickholder for sure, without excluding the use of a metal one (Figure 11).

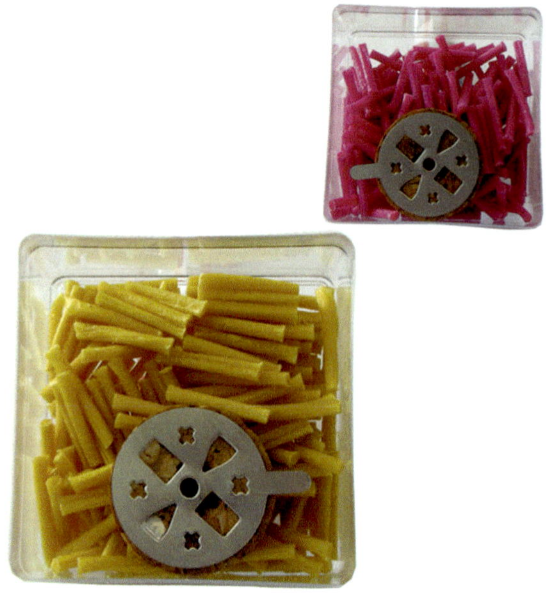

Figure 12. Modern wickholders and wicks.

Byzantine words used for the wickholders

The term *ἁπτότουβον* or *ἁπτρότουβον*, *ἀνδρότουβον*[41] and *ἁφτρότουβον*, which is interpreted as a tube holding a wick, could be related to the metal wickholders,[42] in particular, the Y and S-shaped metal wickholders and possibly the tripod-shaped metal wickholders. It is a compound word whose first part 'ἁπτρίν' is related to the wick, and the second 'τουβίν'[43] describes the cylindrical form of the support. When Theodore the Studite mentions: ''Εὰν μὴ καθαίρῃ τὰ πολυκάνδηλα ἢ τὰ κρεμαστάρια ἢ τὰ τουβία, ὥστε μὴ ἐξ ἰώσεως ἀχρειωθείη, ἀνὰ μετανοίας ρ'',[44] he makes clear that the monks had to depose and clean the polykandela, their hanging accessories and the metal wickholders to prevent them from rusting. In this case it appears that the word *τουβίον* could also refer to the wickholder.[45]

Moreover, the term *μολυβδίς* (*μολιβδίς*, *μολυβίς*), etymologically related to lead, refers not only to lead weights for nets but also to a lead wickholder:[46] 'Πλύνεται ἡ κανδῆλα. καὶ γεμίζεται ἔλαιον κανδῆλα μεγάλη χωροῦσα λίτρας ἑπτά, κι' ἕκαστος καταστένει τὸ ἴδιον μολυβίδιν μετὰ ἁπτρίου. κι' ἀλλάσουσιν οἱ ἐπάνω ἱερεῖς, κι' ἄρχεται.'[47]

One method for supporting a wick could potentially be the forerunner of the current wickholders[48] made of cork and a metal plate and commonly found in industrialized packs of wicks (Figure 12) available in Greek grocery stores. Also, it has been suggested that the wick could be stabilized on a piece of cork (*Quercus suber*) or on a piece of tree bark,[49] but we should probably rule out the possibility of similar wickholders made of wax pieces.[50] So far, I did not manage to find any indications regarding the use of cork as a wick support during the Byzantine era.

Finally, a drawing on a brick of the Late Antique period, from Bargala in North Macedonia, has been interpreted as a glass kandela hanging by three chains of metal, under an arch.[51] In the center there are two vertical elements which have been considered wicks. However, it is not entirely clear whether they are vegetal wicks or a combination of wick and support.

Conclusions

In conclusion, it is apparent that the great advantage of wickholders was that they allowed the flame to continue to burn over or near the mouth of the glass kandela even when the fuel level was lowered and they prevented the flame from getting too near the walls of a glass vessel.

The first glass objects used as kandelai were drinking vessels. The presence of a glass cylinder (Figure 10, 11) on the bottom of a glass vessel is an indication of its use exclusively as kandela. Also, the glass kandelai with a hollow stem were mainly if not exclusively to be used in polykandela. Moreover, the glass cylinder attached to the flat bottom of the kandelai was possibly an addition designed to create a socket for wickholders similar to the arrangement that existed in lamps with a hollow stem.

41 Du Cange 1688, vol. 2, 1589.
42 Trapp E. *et al.* 1994-2005, 190.
43 Cf. the Latin word *tubus*.
44 Talbot 2005, 142.Y
45 Du Cange 1688, vol. 2, 1588.
46 Lampe, Liddell 1961, 876.
47 Goar 1730, 346.
48 Crowfoot and Harden 1931, 207. According to the authors similar constructions were clearly in use in 1931 (the time when the article was written).

49 Uboldi 1995, 93.
50 Running an experiment I realized that beeswax can float in water especially if its shape is flat, but it appears to be submerged in the oil. If it was nevertheless used as a wickholder, both pieces of wood and beeswax would have to be protected by metal foil or plates to prevent them from burning.
51 Chevalier 1999, 194, fig. 7c. Also see Chrzanovski 2006, fig. 1.

The addition of more than one glass cylinder to the bottom of a glass kandela, as in the case of a glass kandela found in Samarra,[52] enabled an increase in light intensity. Such a configuration for an oil lamp with the addition of clay wickholders could explain the several flamed wicks illustrated in a miniature of the 11th century AD (Manuscript of the homilies of Iakovos Kokkinobaphos, Vatican City, Biblioteca Apostolica Vaticana, Vat. gr. 1162 fol. 92),[53] if such glass kandelai were also manufactured during the Middle Byzantine period. Otherwise, the wicks could rely on tripod metal wickholders seated on the bottom of the kandela, which seems to be the most likely.

Clay wickholders found with corresponding bronze and lead ones in the Early Christian city of Elaiussa Sebaste in Turkey, as well as the fact that six different types of glass kandelai came from the same site, allows us to assume the possibility that the residents would choose different types of wickholders for different types of glass kandelai.[54] The use of clay supports on certain types of kandelai does not exclude the parallel use of metallic wickholders in other kandelai. However, it is not yet possible to associate each type of wickholder exclusively with a particular type of glass kandela.[55]

Finally, the invention of wickholders not only influenced the shape of the glass kandelai and the evolution of their typology, both of the individual kandela and the ones that were placed in polykandela, but also led to a huge variety of devices made especially for the support and suspension of wicks.

Bibliography

Alliata, E. 1990. Ceramica bizantina e omayyade di 'Uyum Musa. *Liber Annicus* 40, 247-261.

Ballance, M., Boardman, J., Hood, S., Bialor, P. G. and Clutton-Brock, J. 1967. *Excavations in Chios, 1952-1955*. London: Thames and Hudson.

Beretta, M. and G. Di Pasquale (eds) 2004. *Vitrum. Il vetro fra arte e scienza nel mondo romano*. Firenze: Giunti.

Bizancio en España: de la Antigüedad tardía a El Greco 2003. April-June 2003, Museo Arqueológico Nacional. Madrid: Secretaría de Estado de Cultura, Subdirección General de Información y Publicaciones.

Bouras, L. and Parani, M. G. 2008. *Lighting in Early Byzantium*. Washington: Dumbarton Oaks Research Library and Collection.

Branciforti, M. G. (ed.) 2006. *L'area archeologica di Santa Venera al Pozzo-Acium: antiquarium*. Palermo: Regione siciliana, Assessorato regionale dei beni culturali, ambientali e della pubblica istruzione, Dipartimento dei beni culturali, ambientali e dell'educazione permanente.

Brentjes, B., Mnazakanjan, S. X., Stepanjan, N. S. and Daniltschenko, S. 1982. *Kunst des Mittelalters in Armenien: Kulturgeschichte, Architektur, Plastik, Wandmalerei, Buchmalerei, angewandte Kunst*. Wien: Schroll.

Chevalier, P. 1999. Les luminaires paléochrétiens de Dalmatie romaine. *Vjesnik za Arheologiju i Historiju Dalmatinsku*, 90/91, 161-199.

Chrzanovski, L. 2003. *Lumière! : l'éclairage dans l'antiquité : une exposition du Musée romain de Nyon*, Suisse, 15 mai 2003 - 30 avril 2004. Nyon: Musée romain; Milano: Edizioni Et.

Chrzanovski, L. 2006. L'illuminazione artificiale nel mondo tardoantico, *Dacia* 50, 361-388.

Ciarallo, A. and E. de Carolis (eds) 2001. *Pompéi, Nature, sciences et techniques*. Milan: Electa.

Crowfoot, G. M. and Harden, D. B. 1931. Early Byzantine and Later Glass Lamps. *The Journal of Egyptian Archaeology* 17(3/4), 196-208.

Datema, C., Allen, P. 1987. *Leontii presbyteri Constantinopolitani Homiliae*. Turnhout: Brepols; Leuven: University Press.

Dauterman-Maguire, E., Maguire, H. P. and Duncan-Flowers, M. J. 1989. *Art and holy powers in the early Christian house*. Urbana: University of Illinois Press.

Djurić, V., Ćirković, S. and Korać V. 1990. *Pećka Patrijaršija*. Beograd: Jugoslovenska revija.

Du Cange, Ch. 1688, *Glossarium ad Scriptores Mediae et Infimae Graecitatis*, I-II. Lyon: apud Anissonios, Joan. Posuel, & Claud. Rigaud.

Dunn, A. 2007. Small Finds, in A. H. S. Megaw (ed.) *Kourion: excavations in the episcopal precinct*, 527-554. Washington D.C.: Dumbarton Oaks Research Library and Collection.

Forbes, R. J. 1966. *Studies In Ancient Technology, vol. VI, Heat and Heating, Refrigaration, Light*. Leiden: Brill.

Foy, D. 2003. Le verre en Tunisie : l'apport des fouilles récentes tuniso-françaises. *Journal of Glass Studies* 45, 59-89.

Foy, D. 2007. Lampes de verre: Antiquité tardive et moyen âge, in L. Chrzanovski, P. Kaiser (eds) *Dark ages? Licht im Mittelalter/L'éclairage au moyen âge*, 170-180. Milano: Edizioni Et.

Foy, D. 2011. Les porte-mèche des lampes en verre de l'Antiquité tardive, *Provence historique, Hommages à Jean Guyon, LXI*, fasc. 243-244, janvier-juin, 207-239.

Gennadiou, P. G. 1914. *Λεξικόν Φυτολογικόν*, Αθήνα.

Treasures of Mount Athos 1997. Thessaloniki: Holy Community of Mount Athos; Museum of Byzantine Culture.

Gençler, G., Ç. 2009a. Early Byzantine glass finds from Elaiussa Sebaste (Ayash/Mersin - Turkey), in A. Antonaras and D. Ignatiadou (eds) *18th Congress of the International Association for the History of Glass 21-15 September 2009, Program and abstracts*, 65. Thessaloniki.

Gençler G., Ç. 2009b. Early Byzantine glass finds from Elaiussa Sebaste (Ayash/Mersin – Turkey), in A. Antonaras and D. Ignatiadou (eds) *Annales du 18e congrès de l'Association Internationale pour l'Histoire du Verre*, 292-299. Thessaloniki.

Goar, J. 1730. *Euchologion sive Rituale Graecorum*. Venice: Ex typographia Bartholomaei Javarina.

Goethert-Polaschek, K. 1977. *Katalog der römischen Gläser des Rheinischen Landesmuseums Trier*. Mainz: Philipp von Zabern.

Kaiser, P. 2003, *Öllampen aus der Sammlung von Hermann Müller im Historischen Museum Olten*. Olten: Historisches Museum Olten.

Kollias, H. 1974. Η ανασκαφή εις 'Άφωτην' Καρπάθου. *Πρακτικά της εν Αθήναις Αρχαιολογικής Εταιρείας* 129, 201-206.

Kondoleon, Ch. 2000. *Antioch: the lost ancient city*. Princeton: Princeton University Press.

Lampe, G. W. H., Liddell, H. G. 1961. *A Patristic Greek lexicon*. Oxford, New York: Clarendon Press.

Lowden, J. 1999. *Πρώϊμη Χριστιανική & Βυζαντινή Τέχνη*. Αθήνα: Καστανιώτης.

[52] Crowfoot and Harden 1931, pl. XXX, no. 53.

[53] Stornajolo 1910, 37.

[54] Gençler 2009a.

[55] Foy 2011, 215.

Martorelli, R. and Amante-Simoni, C. 1986. I corredi funerari e la suppellettile metallica. Cultura, Materiali e Fasi Storiche del Complesso Archeologico di Cornus. Primi risultati di una ricerca, in *L'Archeologia Romana e Altomedievale nell'Oristanese. Atti del Convegno di Cuglieri (22-23 giugno 1984)*, 161-170; 178-186. Taranto: Editrice Scorpione.

Martorelli, R. 1986. I corredi funerari e la suppellettile metallica, in *L'archeologia romana e altomedievale nell'oristanese: atti del Convegno di Cagliari (22-23 giugno 1984)*, 161-189. Taranto: Editrice Scorpione.

Martorelli, R. 2000. I materiali metallici e gli oggetti di corredo, in A. M. Giuntella (ed.) *Cornus I,2. L'area cimiteriale orientale. I materiali - Mediterraneo tardoantico e medievale. Scavi e ricerche* 13,2, 23-50. Oristano: S'Alvure.

Motsianos, I. 2011a. Lighting the fire in Byzantium, in I. Motsianos and E. Bintsi (eds) *Light on light, an illuminating story*, Exhibition Catalogue, 79-89. Thessaloniki: Folklife & Ethnological Museum of Macedonia-Thrace.

Motsianos, I. 2011b. Artificial Lighting during Byzantine and Post-Byzantine Period, in I. Motsianos and E. Bintsi (eds) *Light on light, an illuminating story*, Exhibition Catalogue, 90-125. Thessaloniki: Folklife & Ethnological Museum of Macedonia-Thrace.

Motsianos, I. 2012. Torches as lighting devices in Byzantium, in L. Chrzanovski (ed.) *Le Luminaire antique, Lychnological Acts 3, Actes du 3e Congrès International d'études de l'ILA*, Université d'Heidelberg, 21-26.IX.2009, 263-280. Montagnac: Monique Mergoil.

Motsianos, I. 2015. On Byzantine artificial lighting: particularities and survivals in today's culture, in *Proceedings of the Balkan Light 2015*, 16-19 September 2015, 277-288, Athens.

Olcay, Y. 2001. Lighting Methods in the Byzantine Period and Findings of Glass Lamps in Anatolia. *Journal of Glass Studies* 43, 77-87.

Oretskaia, I. 2008. On the Miniatures of the Four Gospels NLR, Gr. 801, in A. V. Zakharova and O. g. S. Popova (eds) *Obraz Vizantii: sbornik statei v chest' O.S. Popovoi*, 335-356. Moskva: Severnyi palomnik.

Orsi, P. 1895. Siracusa. Nuove esplorazioni nelle catacombe di S. Giovanni nel 1894, in *Notizie degli Scavi di Antichità*, 477-521.

Orsi, P. 1896. Siracusa. Di una necropoli dei bassi tempi riconosciuta nella contrada 'Groticelli', in *Notizie degli Scavi di Antichità* 1896, 334-356.

Peleg, M. and Reich, R. 1992, Excavations of a Segment of Byzantine City Wall of Caesarea Maritima, *Atiqot* 21, 137-170.

Piccirillo, M. and Alliata, E. 1994. *Umm al-Rasas Mayfaàh. I, Gli scavi del complesso di Santo Stefano*. Jerusalem: Studium Biblicum Franciscanum.

Pirovano, C., Arapi, M. 2002. *Percorsi del sacro: icone dai musei albanesi*, exhibition catalogue, Vicenza, Gallerie di Palazzo Leoni Montanari, 15 settembre - 1 dicembre 2002. Milano: Electa.

Porson, R. and Dobree, P. P. 1822. *Photiou Tou Patriarchou Lexeon Synagoge*, Volume 2: E Codice Galeano Descripsit.

Russell, J. 1982. Byzantine instrumenta domestica from Anemurium. The significance of context, in R. L. Hohlfelder (ed.) *City, town and countryside in the early Byzantine era*, 133-154. New York: East European Monographs Boulder-Columbia University Press.

Schmidt, G. M. 1867. *Hesychii Alexandrini Lexicon: Ailiou Diogeneianou Periergopenētes*. Jenae: Dufft.

Schmidt, M. 1867. *Hesychii Alexandrini Lexicon: Ailiou Diogeneianou periergopenētes* (Editionem minorem-Editio altera ed.). Jenae: Sumptibus Hermanni Dufftii.

Sodini, J.-P. 1993. La contribution de l'archéologie à la connaissance du monde byzantin (IVe-VIIe siècles). *Dumbarton Oaks Papers* 47, 139-184.

Sodini, J.-P. and Kolokotsas, K. 1984. *Aliki II, La basilique double*. Athènes: École Française d'Athènes.

Stern, E. M. 1984. Antikes Glas in der Südtürkei. *Glastechnische Berichte* 57, 132-139.

Stornajolo, C. 1910. Miniature delle Omilie di Giacomo Monaco (Cod. Vatic. gr. 1162) e dell'Evangeliario greco Urbinate (Cod. Vatic. Urbin. gr. 2). Rome: Danesi.

Stylianou, A., Stylianou, J. A. 1985. *The painted churches of Cyprus: treasures of Byzantine art*. London: Trigraph for the A.G. Leventis Foundation.

Talbot, A. M. 2005. Evidence about Byzantine Glass in Medieval Greek Texts from the Eighth to the Fifteenth Century, *Dumbarton Oaks Papers* 59, 141-145.

Thomas, J. P., A. Constantinides-Hero and C. Giles (eds) 2000. *Byzantine monastic foundation documents: a complete translation of the surviving founders' typika and testaments*. Washington, D.C.: Dumbarton Oaks Research Library and Collection.

Tourta, A. (ed.) 2006. *Icons from the Orthodox communities of Albania*. Thessaloniki: European Centre for Byzantine and Post-Byzantine Monuments.

Trapp, E. *et al.* 1994-2005. Lexikon zur byzantinischen Gräzität besonders des 9.-12. Jahrhunderts, vols I-IV (α -κώφευσις), vol. V (λ -παλιάνθρωπος). Wien: Verlag der Österreichischen Akademie der Wissenschaften.

Uboldi, M. 1995. Diffusione delle lampade vitree in età tardoantica e altomedievale e spunti per una tipologia. *Archeologia medievale. Cultura materiale, insediamenti, territorio* 22, 93-145.

Walsh, P. G. 1978. *The poems of St. Paulinus of Nola*. New York: Paulist Press.

Zitrides, C. 2005. Metal objects, in S. T. Stevens, A. V Kalinowski and H. van der Leest (eds) *Bir Ftouha: a Pilgrimage Church Complex at Carthage*, 463-473. Portsmouth: Journal of Roman archaeology.

A Linen Wick from the Northern Church at Shivta, Israel

Naama Sukenik
Israel Antiquities Authority

Yotam Tepper
Zinman Institute of Archaeology, University of Haifa, Mount Carmel,
Haifa 3498838, Israel and the Israel Antiquities Authority.

Abstract: The use of wicks in ancient times enabled the lighting of private and public houses and was vital for ancient life. However, only a few wicks have been identified in their archaeological context. In a reexamination of the Shivta Collection of the Colt Expedition (1933-1936, 1938) at the site a small tube was found, made of copper alloy, which has been dated to the Byzantine period and still contains the remains of a wick. The wickholder would have been used in a floating-wick oil lamp and it probably belonged to a type of glass lamp that was typical of the Byzantine and Early Islamic periods. The fiber of the wick was preserved because of the arid climate in the Negev and also because of the association of the wick with the metal tube, which prevented the destruction by micro-organisms. The fibers of the wick were identified by microscopic analysis as linen (*Linum usitatissimum* L.), based on the morphology of the internal layers of the fibers. The choice of flax as a raw material for making wicks seems very reasonable. Flax was a very common agricultural crop in the Land of Israel and was used as one of the important raw materials in the weaving industry until the Islamic period.

Keywords: Wick, wickholder, Shivta, Byzantine period, Early Islamic period, linen, *Linum usitatissimum*.

Introduction

An archaeological expedition was conducted by D. H. Colt between 1933-1936 and 1938 on the site of Shivta, a Byzantine village located in the Negev Desert about 100 kilometers southwest of Jerusalem (Figure 1). The village was probably founded at the beginning of the 2nd century AD and abandoned in the Early Islamic Period.[1] The findings from the Colt excavations, which were never published, were reexamined recently[2] in a comprehensive study which was carried out by Haifa University.[3]

The Colt expedition revealed approximately 20 private houses and three large churches, of which the Northern Church has been dated to the 6th-7th century AD.[4] The findings from the Colt excavation include a group of items that were used for lighting,[5] most of them from the Northern Church and the rooms around the atrium.[6] These included a little tube made from copper alloy (2 cm length, 0.6 cm diameter), which was identified as a lamp wickholder which still contains the remains of a wick (Figure 2).

The wickholder would have been used in a floating-wick oil lamp which probably belonged to one of the glass lamp types that are typical of the Byzantine and Early Islamic periods, including bowl or beaker shaped oil lamps with occasionally

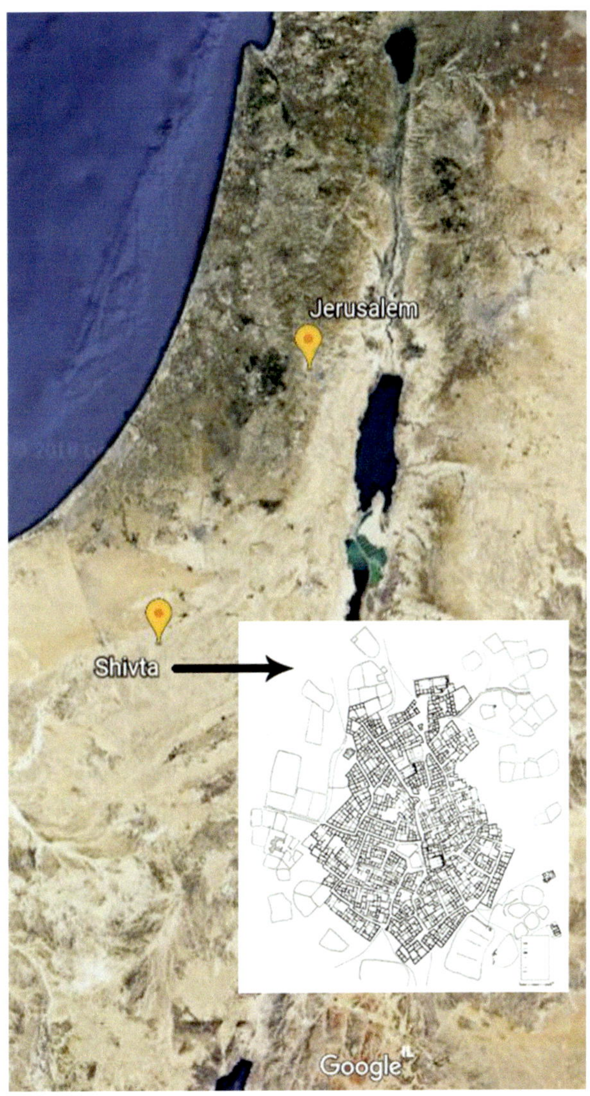

Figure 1. Shivta map (after Hirschfeld 2003).

[1] Hirschfeld 2003; Tepper *et al.* 2018.
[2] See Tepper 2019.
[3] This study was conducted with the frame of the research of the social collapse of the Negev settlements in the Byzantine period at Haifa University. The project has received funding from the European Research Council under the EU's Horizon 2020 research and innovation programme. Extensive excavations were conducted at Shivta during 2016-2019 by Yotam Tepper and Guy Bar-Oz on behalf of the Zinman Institute of Archaeology at the University of Haifa, supported by the Israel Nature and Parks Authority.
[4] Margalit 1987; Maayan-Fanar *et al.* 2018.
[5] Tatcher, Tepper and Ayalon 2019.
[6] Tepper 2019.

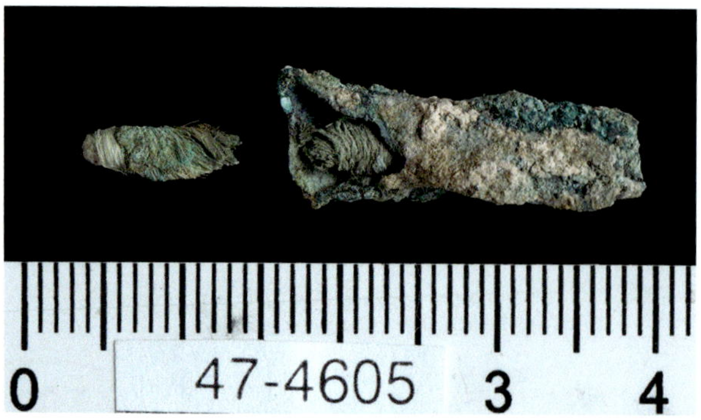

Figure 2. Lamp wickholder from Shivta which still contains the remains of a wick (Photo by Clara Amit, Courtesy of the Israel Antiquities Authority).

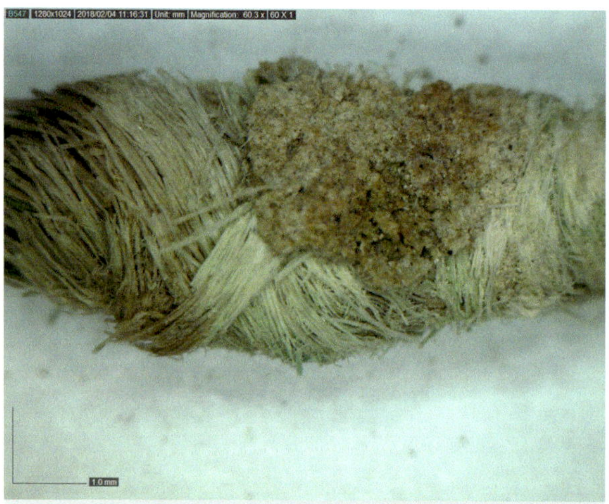

Figure 3. Shivta wick (Photo by Naama Sukenik with Dino lite).

a wick tube at the center of the floor.[7] These lamps fit neatly into the holes of polykandelon that were typically used in churches or public buildings. Similar lamps were found in other sites in the Negev, including: Nessana,[8] Horvat Karkur `Illit[9] and Rehovot (in the Negev).[10] One part of the wick is still in the tube and the other part of the wick was preserved separately. The two parts of the wick are no more than two centimeters in length, and were made of organic fibers that are rarely found in archaeological contexts, since under normal deposition conditions, natural fibers readily perish. This wick was preserved as a result of the dry climate in the region and because of the association of the wick with the metal tube, which prevented the destruction of the fiber by micro-organisms.[11] Only a few parallels of wicks are known in the archaeological record. Our paper will focus on this unique and essential object which was used to shed light in ancient times.

Fiber Description and Identification

The identification of the fiber is based on the morphology of the internal layer of the fiber as observed by light microscopy under different magnifications.[12] In the course of the research, a sample from the archaeological fiber from Shivta was examined (Figure 4) and compared to known fibers including wool, silk, cotton, hemp, and linen, which were commonly used for textile production in the Byzantine period in Palestine.[13]

The microscope analysis of the fibers' morphology eliminated the possibility of animal fiber, including silk fiber, due to the clear differences in diameter. It also eliminated the possibility of the fiber being wool which has a unique cellular appearance of overlapping structure as scales (Figure 5). Additionally, the possibility of cotton fiber was ruled out because the archaeological fiber did not have the twisted ribbon appearance characteristic of cotton (Figure 6).

According to the morphological comparison, the possibility of hemp fiber (*Cannabis sativa* L.) was also ruled out, although they are similar fibers. The internal structure of linen fiber consists of multiple segments of equal length, which are separated from each other by deep linear grooves (Figure 7) and may be contrasted with the internal structure of hemp which is generally not segmented (Figure 8).[14] This character of the cross section was seen in the archaeological fiber (Figure 3), making it possible to determine that the wick is made of linen fibers.

The preserved length of the wick stands at 0.6 cm (some of the wick is invisible in the tube) and 0.4 cm. in width. The separate fragment has the same width of 0.4 cm. and is 1 cm in length. One edge of the separate fragment is complete without any sign of burning or use and in the other edge, tear marks are visible. We can assume that this fragment was torn from the original wick. The two fragments have green stains from the corrosion of the copper. However, the absence of any signs of burning indicates that the wick was never used. According to microscope analysis (by Dino-Lite X 50-200) the linen fiber in the wick was not spun but was rolled into its form in different directions to produce a strong wick. The fiber is relatively coarse (Figure 3) and its diameter is 30-40 micron.[15] A comparison with linen fibers preserved in archaeological textiles, for example from Masada where the

7 Gorin-Rosen and Winter 2010, 172-175, fig. 5:1-2; for reconstruction suggestions, see: Motsianos 2011, 109. Similar but not identical tubes were found at Samaritan cemeteries in Tel Barukh and were identified as amulets. The tubes from Tel Barukh included threads, not fibers and were made of thin folded metal sheets (4.2 cm in length and 1.6 cm in diameter) with a few rolled threads inside. See Tal and Taxel 2015, 120. We cannot exclude the possibility that the objects which were identified as amulets, see Tal and Taxel 2015, 121, were not used, like the one from Shivta, as wicks.

8 Harden 1962, pl. XX: 51-54.

9 Katsnelson 2004, figs 59-60.

10 Patrich 1988, pls. XII: 135, XIII: 137.

11 Andersson Strand *et al.* 2010, 151-152.

12 X 100 – 400; Kvavadze *et al.* 2009; Kvavadze *et al.* 2010.

13 For example see Shamir 2001. The differentiation between animal fiber and plant fiber is simple and can be done at low magnification using a Dino-Lite such as was used in this research. However, the differentiation between flax and hemp is much more difficult and requires high magnification (X 100-400).

14 See also: Kvavadze *et al.* 2009; Kvavadze *et al.* 2010.

15 1 micron = .001 mm.

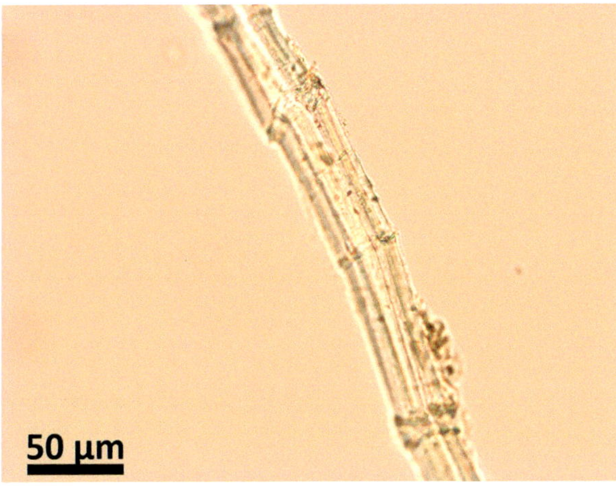

Figure 4. Microscopic view of wick fiber excavated from Shivta (Photo by Naama Sukenik).

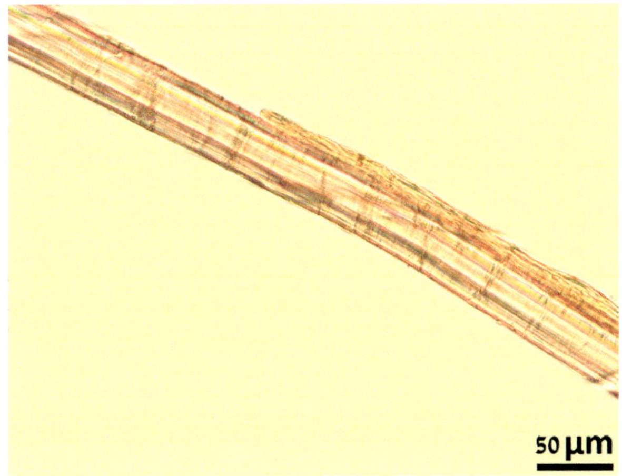

Figure 7. Modern linen fiber under microscope (Photo by Naama Sukenik).

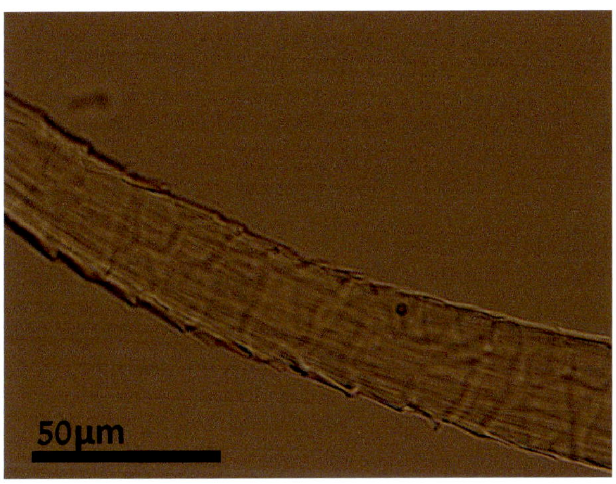

Figure 5. Modern wool fiber under microscope (Photo by Naama Sukenik).

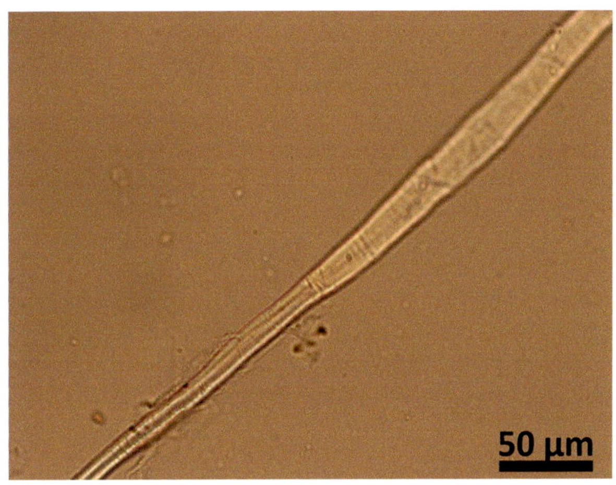

Figure 8. Modern hemp fiber under microscope (Photo by Naama Sukenik).

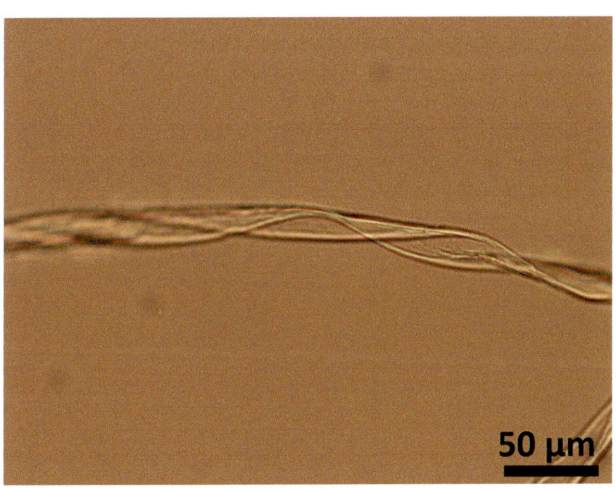

Figure 6. Modern cotton fiber under microscope (Photo by Naama Sukenik).

fiber is 7.5-20 micron,[16] indicates that the wick was not made with high quality fiber.

Discussion and Conclusions

Linen is a plant fiber derived from the stalks of the flax plant (*Linum usitatissimum L.*), characterized by having a strong, flexible fiber reaching over 35 cm in length. Linen had been used since the Neolithic period and became popular as a raw material in the textile industry throughout the world, including the Land of Israel.[17] The first evidence of linen fiber in the archaeological record were fibers discovered attached to a comb in the Wadi Murabba'at caves dated to 10,000 years ago.[18] Linen is mentioned in the Bible many times, as

[16] Shefffer and Granger-Taylor 1994,160.
[17] Forbes 1964, 27; Zohary Hopf, Weiss 2013, 105-106.
[18] Schick 1995, 199-202.

a raw material used for weaving (Deut. 22:11; Hos 2:7) and as clothing (Lev. 13:47).[19] In a few places linen is mentioned as wicks for oil lamps (Ezek. 40: 3; Isa 42: 3). Sources from the 2nd century AD discuss flax often as agriculturally important and as having industrial significance.[20] Flax is mentioned a number of times in the classical literature: for example Pliny the Elder (Plinius, Naturalis Historia: XIX, 2); Strabo (Strabo, Geography: XI, 2, 17) and Josephus, who mention that the population near Jericho and Ein Gedi wore linen clothing because of the hot climate (War, IV, 473). Linen is mentioned many times in Jewish literature which suggests that linen was widely used (See e.g., Mishnah Bava Kamma 10.9; Mishnah, Kil'ayim 9.9; Mishnah Shabbat 6.1).[21] Flax, which requires large quantities of water to grow, was cited in the Jewish literature as a characteristic crop of the Galilee (in Northern Israel) while wool is more characteristic of Judaea (Talmod Bavli, Bava Kama 119a).[22] Flax was grown intensively in Israel in the Jordan Valley at sites such as En-Gedi, Jericho,[23] and in the Bet She'an Valley, which was a major center of production of fine linen.[24] Others center of linen production in Palestine were found in Tiberias.[25] The price edict of Diocletian describes the flax of Scythoplis (in Bet She'an Valley) as the most expensive in all the world.[26] Archaeological findings indicate also that linen was an important raw material for textiles, in addition to wool.[27] Although the organic findings from the Byzantine period are limited, we can assume that the linen industry continued until the Arab conquest and until the use of cotton (*Gossypium sp.*) replaced linen during the Islamic period.[28] Linen textiles dated to the Byzantine period were uncovered in Israel in Rogem Zafir, Tel Sheva and Rehovot (in the Negev)[29] and in Horvat Karkur `Illit.[30]

The most significant uses for linen textiles were functional and not necessarily for decorated garments, probably because wool was the best raw material for dyeing, due to the protein composition of the fiber that allows better absorption of the dye than does linen.[31] According to the archaeological record, linen textiles were used as undergarments, loincloth, socks,[32] sacks,[33] shrouds,[34] fishing nets,[35] and ropes.[36] However, linen wicks are rarely found.

Flax fibers require a mechanical and chemical treatment that required a great deal of knowledge and experience. The process of converting the flax plant to linen cloth is complicated and involves several stages including harvesting the plant before the maturation of the seed; drying the stems in the sun; soaking in running water for 10-14 days in order to ease the separation of the long, fine fibers from the woody outer core and inner cellular tissue in a process called retting; drying the flax stalk in the sun; the physical breaking and combing of the fibers from its woody stem; followed by spinning the fibers into yarn; and weaving the yarn into cloth[37] as depicted in the Beni Hassan tomb painting.[38] In any event, the tow residue that was created during the breaking process, such as some of the waste which contains short fibers and parts of stalks, can also be processed and spun, but the yarn will be coarse and not suitable for clothing.[39]

The finds of linen fibers in archaeological textiles display a wide range of quality and fiber diameters, which depended on the process of the combing and separation of the fibers of the wood. As mentioned before, according to our examination, the wick from Shivta is made by low quality fibers compared to fibers from linen textile. This makes sense because in an object of this type, no high quality fiber is required. Furthermore, it seems that choosing linen as a raw material for wick production rather than other fibers is reasonable since it was common in the local industry in Israel and because the quality of its burning is good.

Use of linen wicks for illumination was mentioned in the Mishana -a collection of the Jewish oral traditions known as the 'Oral Torah' dated to the 2nd century AD. In one passage (Shabbat 2:1) which concerns the fuels and wicks which may be used for a Shabbat lamp,[40] linen fiber is mentioned as an appropriate raw material due to its long combustion and beautiful flame. Other substances that were prohibited in the Mishna are materials that burn unevenly, for short periods of time, or produce unpleasant smoke or odors, including: Ferula (*Ferula communis* L.) which is native to the Mediterranean region, has a long stem and was known to have been used in the wick industry and for other uses; [41] the fiber from the Sodom's Apple (*Calotropis procera*);[42] or *Hosen* -probably some kind of linen waste. Experimental archaeology in those wicks proved that the traditional text was correct and indicated that the linen wick burned for more than four hours in 30 cc of oil and burned evenly.[43] Although cotton fiber is much more effective in burning than linen,[44] in the Byzantine period, cotton was still a rare material that was only imported to Palestine in small quantities.[45]

Although the use of wicks for illumination was vital for life in antiquity, only a few parallel objects from archaeological contexts have been found that can now be identified as wicks. Five wicks were uncovered in their bronze oil lamps from Nijmegen (Netherlands) and from Xanten (Germany) dated to the Roman period. All the wicks made by bast fiber probably

[19] See also Amar 2012, 109-110.
[20] Safrai 2003, 88-91.
[21] See also Safrai 2003, 88-91.
[22] Babylonian Talmud is the central text of Rabbinic Judaism and the primary source of Jewish religious law and Jewish theology (3rd to 5th centuries).
[23] Shamir and Rosen 2015, 327.
[24] Boertien 2013, 37; Shamir 2015, 13.
[25] Forbes 1964, 32; Safrai 2010, 255-256; Kislev *et al.* 2009, 760-770.
[26] Graser 1959, 386.
[27] Shamir and Sukenik 2011, 213-216.
[28] Safrai 2003, 89; Amar 1998, 39.
[29] Shamir 2001.
[30] Figueras *et al.* 2004.
[31] Robinson 1969, 9.
[32] Yadin 1963, 254; Sheffer and Granger-Taylor 1994, 221-222.
[33] Yadin 1963, 259-261.
[34] Shamir 2015, 21.
[35] Yadin 1963, 267-269; Shamir 2005, 108.
[36] Murph *et al.* 2011.
[37] Forbes 1964, 27-32; Strand 2013.
[38] Rooijakkers 2005, 2-13.
[39] Strand 2013, 27.
[40] Shabbat candles lit on Friday evening before sunset to usher in the Jewish Sabbath.
[41] Amar and Shwiky 2003, 43-46; González and Amich 2015, 8.
[42] Friedman and Amar 1997; Amar and Shwiky 2003, 51-58. The silky tuft fibers of the Sodom's Apple seeds give the plant its Hebrew name, *petillat hamidbar* -a wick from the desert. Other identifications of this plant are *Verbascum sinuatum* or *Phlomis brachyodon* which see: Amar and Shwiky 2003, 43.
[43] Amar and Shwiky 2003, 42.
[44] Moullou, Doulos and Topalis 2015, 122.
[45] Samir 2016, 235.

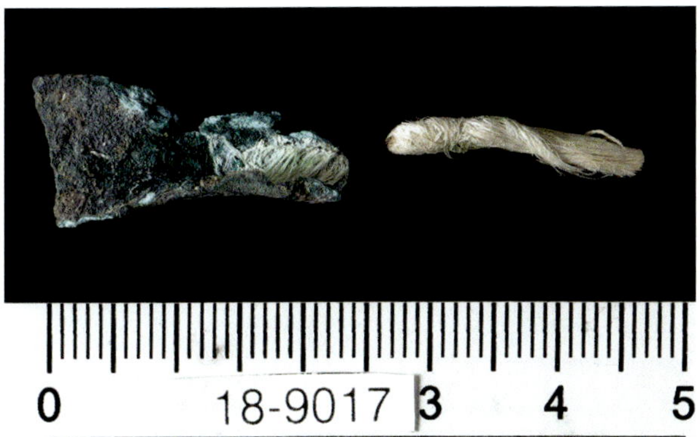

Figure 9. Lamp wickholder from Horvat Karkur `Illit which still contains the remains of a wick (Photo: Clara Amit, Courtesy of the Israel Antiquities Authority).

linen fiber.[46] A charred wick was found in a terracotta lamp in Algeria from the 2nd century.[47] Linen wicks were also found in the new excavation in Israel in Wadi Amram with signs of burning.[48] In Horvat Karkur `Illit a wickholder identical to the object found in Shivta was discovered (Figure 9), with a wick that belonged to a polykandelon dated to the Byzantine period. However, in that case the fiber was made of hemp.[49] Hemp is derived from the stems of *cannabis sativa* L., which is closely related to flax and became a well-known fiber crop.[50] Hemp is taller than flax and its fibers are generally coarser. It has been assumed, therefore, that hemp was not continually used for clothing, but rather for sails, ropes and nets.[51] The use of hemp in the fiber industry was well known in the Byzantine period.[52] Although hemp was mentioned in Jewish literature and was found in archaeological contexts in Isreal,[53] the hemp wick from Horvat Karkur `Illit is unexpected because it was an inferior agricultural plant compared to flax. Furthermore, the quality of the illumination from a hemp wick is less than from a linen wick[54]. The width of the wick from Horvat Karkur `Illit and Shivta is similar, 0.4 mm, indicating similar types of lamp. Other wicks in the archaeological record appeared to be re-used textiles, such as the wicks from Armana dated to the 18th Dynasty,[55] or the objects from Cave 38 in the Judean Desert that were identified as wicks and dated to between the early 9th to late 13th centuries AD.[56] At another site in el-Kom el-Ahmar, situated on the eastern bank of the Nile in Middle Egypt, an object made of flax tow and palm fibers was found.[57] It must be emphasized that all those wicks were not found in lamps and were without any signs of burning and therefore, the identification is uncertain. Hence, to date the wicks from Shivta and Horvat Karkur `Illit are the only wicks that were identified with certainty that belong to the glass lamp type.

Wicks, like the oil, were an important element in the quality of illumination in ancient times[58] but as a vital object they usually biodegraded quickly. Making wicks did not require prior knowledge and they could be made by hand, without any tool. We can assume that the rare wick from Shivta was probably made by hand on site. As we have described, linen fiber was a suitable candidate for making wicks and we can assume that the inhabitants of Shivta chose to light their public buildings with linen wicks. Linen was a common raw material in the textile industry and was readily available. Although flax is not a plant that can grow in Shivta, the flax probably came from further north through commerce. The wick was not made of the best quality raw linen, which was more suitable for the clothing industry, but of poor quality fibers. We can assume that fine fibers were reserved for the garment industry and low-quality fibers were purchased for wicks, probably for a relatively low price. Despite the tiny size of the wick from Shivta –only a few centimeters long– it sheds light on one of the most essential and common objects of antiquity.

Acknowledgments

We wish to thank Prof. Guy Bar-Oz, head of the Laboratory of Archaeozoology at the Zinman Institute of Archaeology at the University of Haifa, who conducted the Shivta excavation (together with Yotam Tepper; 2016-2019) and Ayelet Tacher who identified the object as being a lamp wickholder; and Dr Yael Gorin-Rosen for her help in identifying the lamp and for providing us with her important remarks. The authors gratefully acknowledge Prof. Kosta Mumcuoglu and Dr Anat Weinberger Chon –for their help with the microscope photos. This project was supported by research grants from the European Research Council under the EU's Horizon 2020 research and innovation program (Grant 648427).

Bibliography

Amar, Z. 1998. The Revolution in Textiles in Eretz Israel and Syria in the Middle Ages. *Cathedra* 87, 37-60 (Hebrew).

Amar, Z. 2012. *Flora of the Bible*. Jerusalem: Rubin Mass Press.

Amar, Z. and Shwiky, A. 2003. *Bamme Madlikin*, Tel-Aviv: Machon Eretz Hefetz (Hebrew).

Andersson Strand, E., Frei, K. M., Gleba, M., Mannering, U., and Skals, I. 2010. Old Textiles–New Possibilities. *European Journal of Archaeology* 13/2, 149-173.

Ben-Yehuda, N. 2013. Cannabis - Chanvre - Hemp in Rashi`s Commentary to the Talmud, in J. Banck-Burgess and C. Nübold (eds) *The North European Symposium for Archaeological Textiles* 11, 237-240.

Ben-Yehuda, N. and Murphy, T. 2010. Rope from the Christmas Cave: In search of the Talmudic Hemp. *Archaeological Textiles Newsletter* 50, 14-19.

Figueras, P., Areal Guerra, R., Sheffer, A. and Webber, M. 2004. Tissue Remains, in P. Figueras (ed.) *Horvat Karkur `Illit: A Byzantine Cemetery Church in the Northern Negev* (Final Report of the Excavations 1989-1995), 223-231. Be'er-Sheva': Ben-Gurion University of the Negev Press.

[46] Mitschke *et al.* 2012.

[47] Bussière 1973.

[48] The excavation in Wadi Amram was conducted by Uzi Avner. The artifacts are still in research by the author. The fibers were not found in their lamps or in wickholders. However, they have been identified by Sukenik as wicks due to the burning signs thereon.

[49] Figueras *et al.* 2004, 223-224; Nikolsky *et al.* 2004, 241, 245.

[50] Zohary, Hopf and Weiss 2013, 106-107.

[51] Strand 2013, 24.

[52] Ben-Yehoda 2013.

[53] Ben-Yehuda and Murphy 2010; Murphy *et al.* 2011.

[54] Moullou, Doulos and Topalis 2015, 122.

[55] Kemp and Vogelsang-Eastwood 2001, 160, 223, 263.

[56] Shamir and Baginski 2012.

[57] Huber 2007, 54, 63.

[58] Moullou *et al.* 2012, 239

Forbes, R. J. 1964. *Studies in Ancient Technology* 4, Leiden: E. J. Brill.

Friedman, T. and Amar, Z. 1997. Milkweed: Fibres to Fabric. *Archaeological Textiles Newsletter* 30, 4-7.

González, J. A. and Amich, F. 2015. Plants Traditionally Used for Industrial and Artisanal Purposes in the Arribes del Duero (Spain). *Anales del Jardín Botánico de Madrid* 72 (2), e025.

Gorin-Rosen, Y. and Winter, T. 2010. Selected Insights into Byzantine Glass in the Holy Land, in J. Drauschke and D. Keller (eds) *Glass in Byzantium: Production, Usage, Analyses* (International Workshop Organized by the Byzantine Archaeology Mainz, 17th–18th of January 2008), 165-181. Mainz: Römisch-Germanisches Zentralmuseum.

Graser, E. R. 1959. The Edict of Diocletian on Maximum Prices, in T. Frank (ed.) *Rome and Italy of the Empire*, vol. 5, 305-421. Paterson: Johns Hopkins University Press.

Hirschfeld, Y. 2003. Social Aspects of the Late-Antique Village of Shivta. *Journal of Roman Archaeology* 16, 395–408.

Harden, D. B. 1962. Glass, in D. Colt (ed.) *Excavations at Nessana*, vol. 1, 76-91. Princeton, New Jersey: Princeton University Press.

Huber, B. 2007. The Textiles of an Early Christian Burial from el-Kom el-Ahmar Šaruna, in A. De Moor and C. Fluck (eds) *Methods of Dating Ancient Textiles of the 1st Millennium AD from Egypt and Neighbouring Countries Proceedings of the 4th Conference of the Study Group 'Textiles from the Nile Valley'*, 36-69. Tielt, Lannoo.

Katsnelson, N. 2004. Glass Objects, in P.Figueras (ed.) *Horvat Karkur 'Illit: A Byzantine Cemetery Church in the Northern Negev (Final report of the Excavations 1989-1995)* (Beer-Sheva XVI), 265-291. Be'er Sheva': Ben-Gurion University of the Negev Press.

Kemp, B. J., Vogelsang-Eastwood, G. 2001. *The Ancient Textile Industry at Amarna*. London: Egypt Exploration Society.

Kislev, M. E., Simchoni, O., Melamed, Y. and Maroz, L. 2009. Botanical Remains: Food and Industrial Crop, in N. Paint-Cohen and A. Mazar (eds) *Excavation at Tel Beth-Shean 1989-1996*, vol. III: *The 13th-11th Century BCE Strata in Areas N and A*, 764-771. Jerusalem: Israel Exploration Society.

Kvavadze, E., Bar-Yosef, O., Belfer-Cohen, A., Boaretto, E., Jakeli, N., Matskevich, Z., Meshveliani, T. 2009. 30,000-year-old wild flax fibers. *Science* 325, 1359.

Kvavadze E., Bar-Yosef, O., Belfer-Cohen, A., Boaretto, E., Jakeli, N., Matskevich, Z., Meshveliani, T. 2010. Response to comment on '30,000-year-old wild flax fibers'. *Science* 328, 1634.

Maayan-Fanar, E., Linn, R., Tepper, Y. and Bar-Oz, G. 2018. Christ's face revealed at Shivta: An Early Byzantine wall painting in the desert of the Holy Land. *Antiquity* 92 (364), E8. doi: 10.15184/aqy.2018.150.

Margalit, S. 1987. The North Church of Shivta: The discovery of the first church. *Palestine Exploration Quarterly* 119, 106–121.

Mitschke, S., Leih, S., Becker, P., Meijers, R. and Paetz gen Schieck, A. 2012. Ohne Textilien kein Licht! Untersuchungen zu Römischen Lampendochten, in L. Chrzanovski (ed.) *Le Luminaire antique Lychnological Acts* 3, 251-262. Montagnac: Éditions Monique Mergoil.

Motsianos, I. 2011. Artificial Lighting during Byzantine and Post-Byzantine Period, in I. Motsianos and E. Bintsi (eds) *Light on light, an illuminating story*, Exhibition Catalogue, 90-125. Thessaloniki: Folklife & Ethnological Museum of Macedonia-Thrace.

Moullou, D., Madias, E.-D., Doulos, L. and Bouroussis, C. 2012. Lighting in Antiquity, in M. K. Kostic (ed.) *Balkan Light* (The Fifth Conference), 237-244. Belgrade, Serbia.

Moullou, D., Doulos, L. and Topalis, F. 2015. Artificial Light Sources in Roman, Byzantine, and Post- Byzantine Eras: An Evaluation of their Performance. *Chronos*, 32: 119-132.

Murphy, T. M., Ben-Yehuda, N., Taylor, R. E. and Southon J. R. 2011. Hemp in Ancient Rope and Fabric from the Christmas Cave in Israel: Talmudic Background and DNA. Sequence Identification. *Journal of Archaeological Science* 38/10, 2579-2588.

Nikolsky, V., Figueras, P., Auladell, J. and Guerra, R. A. 2004. Metal objects, in P. Figueras (ed.) *Horvat Karkur 'Illit: A Byzantine Cemetery Church in the Northern Negev* (Final Report of the Excavations 1989-1995), 237-264. Be'er Sheva': Ben-Gurion University of the Negev Press.

Patrich, J. 1988. The Glass Vessels, in Y. Tsafrir (ed.) *Excavation at Rehovot in the Negev.* Volume I: *The Northern Church* (Qedem 25), 134-151. Jerusalem: Institute of Archaeology, Hebrew University.

Robinson, S. 1969. *A History of Dyed Textiles*. Cambridge, Massachusetts: M.I.T. Press.

Rooijakkers, T. 2005. Unravelling Beni Hasan: Textiles production in the Beni Hasan Tomb Paintings. *Archaeological Textiles Newsletter* 41, 2-13.

Safrai, Z. 2003. *The Economy of Roman Palestine*. London: Routledge.

Safrai, Z. 2010. Agriculture and Farming, in C. Hezser (ed.) *The Oxford Handbook of Jewish Daily Life in Roman Palestine*, 246-263. Oxford: Oxford University Press.

Schick, T. 1995. A 10000 Year Old Comb from Wadi Murabba`at in the Desert. *Atiqot* 27, 199-202.

Shamir, O. 2001. Byzantine and Early Islamic Textiles Excavated in Israel. *Textile History*, 32/1, 93-105.

Shamir, O. 2005. Textiles, Basketry, Cordage and Whorls From Mo'a (Moje Awad). *Atikot* 50, 99-152.

Shamir, O. and Baginski, A. 2012. Textiles' Treasure from Jericho Cave 38 in the Qarantal Cliff Compared to other Early Medieval Sites in Israel. Textile Society of America Symposium Proceedings 742, viewed 15 December 2018, <http://digitalcommons. unl.edu/tsaconf/742>.

Shamir, O. 2015. Textiles from the Chalcolithic Period, Early and Middle Bronze Age in the Southern Levant. *Archaeological Textiles Review* 57, 12-25.

Shamir, O. 2016. Textile Trade to Palestine in the Roman Period, in K. Droß-krüpe and M. L. Nosch (eds) 231-245. Münster: Ugarit-Verlag.

Shamir, O. and Sukenik, N. 2011. Qumran Textiles and the Garments of Qumran`s Inhabitants. *Dead Sea Discoveries* 18, 206-225.

Sheffer, A., Granger-Taylor, H. 1994. Textiles from Masada-A Preliminary Selection, in J. Aviram, G. Foerster and E. Netzer (eds) *Masada* IV, 153-256. Jerusalem: Israel Exploration Society; Hebrew University of Jerusalem.

Strand, E. A. 2013. The textile Chaîne Opératoire: Using a Multidisciplinary Approach to Textile Archaeology with a Focus on the ancient Near East. *Paleorient* 38 (1-2), 21-40.

Tal, O., Taxel, I. 2015. *Samaritan Cemeteries and Tombs in the Central Coastal Plain: Archaeology and History of the Samaritan Settlement outside Samaria (ca. 300-700 CE)*. Münster: Ugarit-Verlag.

Tatcher, A., Tepper, T. and Ayalon, E. 2019. Metal, Stone, Ceramic and Glass Objects from Colt`s Excavations at Shivta. *Michmanim*, in press (Hebrew and English abstract).

Tepper, Y. 2019. The archaeological findings from the 'Forgotten Suitcase' in Context: In light of the Colt excavations at Shivta and their contribution to the research of the site. *Michmanim*, in press (Hebrew and English abstract).

Tepper, Y., Erickson-Gini, T., Farhi, Y. and Bar-Oz, G. 2018. Probing the Byzantine/ Early Islamic transition in the Negev: The renewed Shivta excavations, 2015-2016. *Tel Aviv* 45, 120-152.

Yadin, Y. 1963. *The Finds from the Bar Kokhba Period in the Cave of the Letters.* Jerusalem: Israel Exploration Society.

Zohary, D., Hopf, M. and Weiss, E. 2013. *Domestication of Plants in the Old World.* Oxford: Oxford University Press.

Some Thoughts About the Use and the Making of Beeswax Candles in Byzantium

Ioannis Motsianos
Museum of Byzantine Culture, Thessaloniki

Abstract: This paper presents evidence for the adoption and consolidation of the use of beeswax candles in Byzantium and to investigate the infrastructure required for the production of beeswax candles: the tools, the utensils, and methods used for this process during the Byzantine and Post Byzantine periods. We also take into account information from *Encyclopedia, or a Systematic Dictionary of Sciences, Arts and Crafts,* as well as a study of a traditional candle-making workshop of the Monastery of Megisti Lavra on Mount Athos which was in use until the beginning of the 21st century and a candle-making workshop in Paphos, Cyprus.

Keywords: beeswax, beeswax candles, candle-making workshop, Monastery of Megisti Lavra, Paphos, *κηρουλάριος*, *candela*.

Introduction

The Byzantine era saw a consolidation of the use of beeswax candles in both secular and sacred environments. The aim of this paper is to investigate the infrastructure required for the production of beeswax candles: the tools, the utensils, and to identify the methods used for this process during the Byzantine and Post Byzantine periods.[1]

Since candle technology does not yet have direct information from Byzantine written sources, our approach will be based on the traces of technical and technological choices that can be derived not only from candle depictions but also from the characteristics of candle stands. Also, indirect information about candle production can be deduced through the study of candle-making workshops in the monasteries of Mount Athos and in other modern candle-making workshops,[2] assuming that these workshops continued to apply traditional methods. Additional relevant information is available in the *Encyclopedia, or a Systematic Dictionary of Sciences, Arts and Crafts,*[3] which was a general encyclopedia published in France between 1751 and 1772.

In this paper we will focus on the research available about two workshops. The first is a study of a traditional candle-making workshop of the Monastery of Megisti Lavra on Mount Athos[4] which was in use until the beginning of the 21st century. The second is from a poster presentation at the ILA Round Table of Thessaloniki about the mode of operation of a 20th century candle-making workshop in Paphos, Cyprus.[5] The latter candle-making workshop was located in Ktima in the Paphos district of Western Cyprus. Ktima in the 1950s was just a small town of 6000 inhabitants. The workshop was active until about 1957 using, almost exclusively, manual labor. The operation, equipment and technologies of the Paphos candle-making workshop from 1950-1980 are presented through recollections, drawings, and photographs. There is also an attempt to record changes and 'transformations' which occurred in this Cypriot workshop during the stated period.[6]

The information obtained from these studies will be used to investigate whether the production of beeswax candles in the Byzantine world could have had a mass production character or a manual one.

Indications of the wide production and use of beeswax candles in the Byzantine world

The use of beeswax candles in Christian churches seems to have been established in Late Antiquity. Indeed, care was taken that beeswax, along with olive oil, were the only fuels that would be used in Christian churches not only to illuminate them but also to ensure that the worshippers would avoid smoke and ugly odor. Christians, wanting to incorporate the *candelae* (torches, tallow candles, tapers) used widely in pagan rituals into their worship mainly inside their churches, adopted the use of beeswax for the production of *candelae* (it seems that the English word candle arose from this term).[7] Thus they could ensure for their Christian rituals a portable lighting medium with a pleasant smell coming from the beeswax and without a lot of smoke. In their efforts to justify this choice the Christians recruited even older myths and symbolism.[8] It seems that from the 4th century AD and later the use of these fuels, initial selected for solid practical reasons, is also loaded with symbolic value.[9] It was thought that, as beeswax is produced by the bees who are virgins so also the Christ, the Light the True, came by the Virgin Mary.[10] In other words the Christians deployed praise for the honey bee (*huius lampadis apis mater*) who can produce the beeswax without the need for a pregnancy (*nec sexum masculi violant*).[11]

[1] For this issue in detail see Μότσιανος 2011, 69-119.
[2] Μουσιώνη 2000.
[3] Diderot and D'Alembert 1762.
[4] Μαμαλούκος 2000.
[5] Papanikola-Bakirtzi 2011. I thank particularly Dr Papanikola-Bakirtzi who let me use the information of her poster and moreover she contributed to this article transmitting to me her accumulated knowledge about candle making.
[6] Papanikola-Bakirtzi 2011
[7] Lemon 1783,
[8] Fevrier and Poinssot 1959, 154-156.
[9] Vincent 2004, 66.
[10] Μωυσής Μοναχός 2000, 279. Κοτσόκολος and Σωτηρόπουλος 2000, 155-156.
[11] Fevrier and Poinssot 1959, 154-156. Fonseca 1989, 78-79.

At the end of Late Antiquity, professionals working in the production of beeswax candles for lighting seem to have consolidated their place in the economic life of Byzantium.[12] The widespread use of beeswax candles is reflected in the wealth of people who were selling them.

The best confirmation of the extensive use of beeswax is the organization of its trade and the establishment of rules regarding it, as we can conclude from the regulations provided in *The Book of the Prefect or Eparch*, a Byzantine commercial manual addressed to the eparch of Constantinople. Chapter XI: Περὶ τῶν κηρουλαρίων[13] (wax-chandlers and taper-makers) details the rules of conduct for the wax-chandlers and taper-makers during the Middle Byzantine period.[14]

It seems likely that since the Middle Byzantine period Christian churches had their own chandleries or at least particular connections was created between them and the adjoining chandleries. An indication of the widespread use of wax candles during that period is also supported by the fact that the chandleries were typically concentrated around the churches.[15]

Moreover, it seems that, in some cases there were organized candle-making workshops inside the Christian monasteries. Eloquently, the typikon of Empress Irene Doukaina Komnene for the Convent of the Mother of God Kecharitomene mentions that 'The same official (sacristan) must with the help of women inside the convent make the prescribed quantity of candles for the year according to the weight laid down by my majesty for all the lighting of the church, and hand them over to the ecclesiarchissa, who will carry out the lighting of the church in the way that will be described in what follows'[16] and also that 'receiving from the sacristan the made-up candles as prescribed in the rule, both those prescribed to be burnt for daily use and those for each feast'.[17]

The need for beeswax candles during the ecclesiastical and monastic daily life and in particular on the major feast days was a great reason for the development of beekeeping and waxcraft in the 10th-11th century, especially if we also take into account the monastic typika rules established concerning the use of candles. Indeed, these monastic typika seem to have been from the 11th century and then established and continued to be refined until the 20th century. So it does not seem strange that even until the end of the 20th century AD inside the twenty monasteries, twelve sketes and about three hundred cells of Mount Athos, the monks used to consume about eight tons of beeswax candles a year.[18]

Information about beewax candle production based on their depictions

As beewax candles were usually burned completely, and their remnants were likely to be recycled we can only indirectly deduce information about their characteristics, such as their size and shape.

Such indirect information on how to make beeswax candles in Byzantium is provided in the detail of a fresco dated to the 14th century AD[19] where the candles that are distributed appear to have the wick at the top in the shape of a loop. This loop could be used both in the process of dipping and for their suspension in order to dry them. It seems very likely that the same loop served also the hanging of the candles at their point of sale, as depicted in a miniature of the 15th century AD,[20] and the distribution of candles during a memorial service, as shown in a miniature manuscript of the 11th century AD.[21] From the above data it appears that the use of the wick loop for the candles was useful, indeed necessary. The creation of a loop at the top of the cord used for the wick was very convenient for all phases of the production, use and sale of the candles.

In Byzantine art candles are usually represented as tapered, as for example on the *ambo* of Basilica C at Philippi, from the 6th century and now on exhibit in Thessaloniki's Museum of Byzantine Culture[22] and in a mosaic of the 6th century from Syria, today at the Danish National Museum in Copenhagen.[23] This shape is clearly a result of the way they were made, either by immersion or pouring. It seems that because of the gravity the wax tended to roll down until it thickened, such that the amount of wax remaining at the top of the candle is less than that collected at its base.

Procedure of making candles

The basic materials for the production of beeswax candles were the beeswax and cotton for wicks.[24]

An example of the storage of ingots of beeswax in a Mount Athos monastery (Figure 1) recalls to us the analogous way the raw materials for glassmaking were moved into Byzantine glassworks.[25] There were separate processes needed during glassmaking and the glassworking process[26] and these allow us to suppose that something similar happened with the treatment of beeswax. The primary preparation of the raw beeswax, beeswax making in other words, and the secondary beeswax production, namely the working of the produced ingots of beewax, seem to have been two separate processes.

In addition to beeswax, the production of candles initially required the preparation of cord, usually made of cotton, to be used as a wick. This preparation was initially done only manually with the use of a hand tool and later at the time of industrialization with a mechanical tool.

[12] Sorochan 2002, 115.
[13] Koder 1991, 112-116; Dujcev *et al.* 1970, 252.
[14] Νεράντζη-Βαρμάζη1989, 359.
[15] Dagron 2002, 461.
[16] Jordan 2000, 680.
[17] Jordan 2000, 681.
[18] Μωυσής Μοναχός 2000, 281.
[19] Patterson 1983, 315, fig. 36.13.
[20] Chrzanovski, Kaiser 2007, pls 11, 12.
[21] Stornajolo 1910, 23.
[22] Motsianos 2011b, 112, fig. 31.
[23] Biddle 1999, 25, fig. 23.
[24] See Jordan 2000, 696, 697 where there is reference for candles with cotton wicks.
[25] Antonaras 2012, fig. 2.
[26] Antonaras 2012, 6-11 and fig. 2; Antonaras 2017, 5-12.

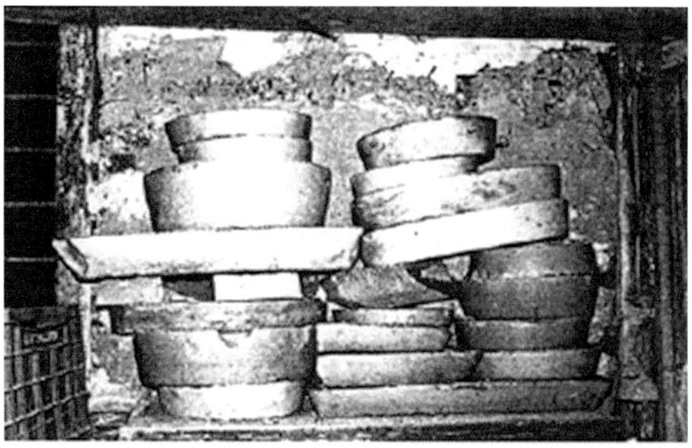

Figure 1. Storage of ingots of beewax in a Mount Athos monastery (Μαμαλούκος 2000, fig. 7).

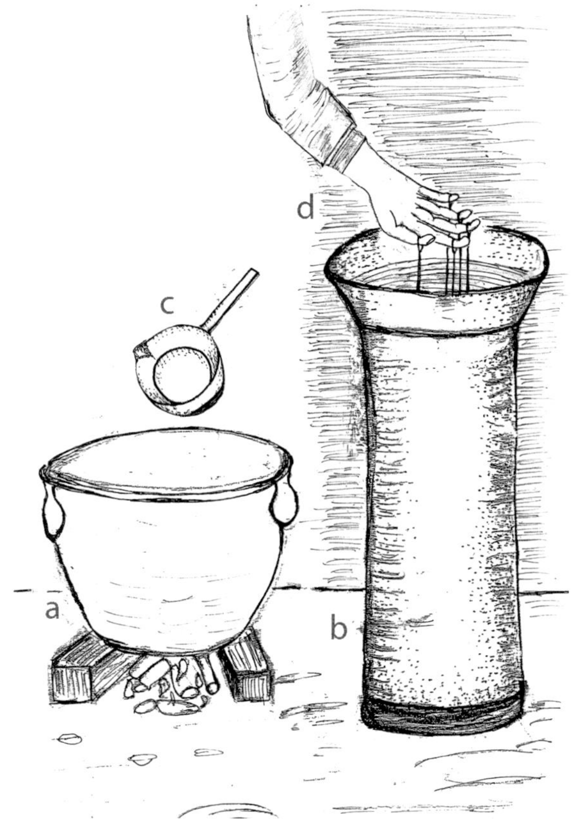

Figure 2. A candle-making workshop in Paphos, Cyprus.
a. The bronze cauldron for melting the wax over a wood fire.
b. The cylindrical tin vat for dipping the candle sticks.
c. The ladle used to move the hot wax from the melting cauldron to thc vat.
d. Process of the dipping wicks hanging by wick loops from the candle maker's fingers.

Equipment of a candle-making workshop

The most basic elements of the equipment of a traditional candle-making workshop, on the basis of which we can assume also the equipment needed by a Byzantine candle-making workshop, were the fireplace, where with wood fire the wax would be melted in special utensils and the tap that supplied the water necessary for the production of candles.

Portable equipment of a candle-making workshop:

a) The copper pot used to melt the beeswax (Figure 2a).[27]

b) The copper ladle with which the molten beeswax was transported (Figure 2c).[28]

c) Metallic or wooden hoops with hooks[29] called in Greek *στεφάνια* (crowns) or *ρόδες* (wheels) for the making of large candles.[30] Small wooden or reed sticks were joined to these hoops. These had nails on each side, from which the loops formed by the folded wicks could be suspended.[31]

d) A copper alloy vessel described by the term *καπέλο* (hat) because of its form.[32] A wick sufficient in size for a large candle was hung above this vessel, and with a ladle the chandler started to pour melted wax over the wick, as many times as needed to get the candle the right thickness. The *καπέλο* vessel was needed to collect the poured wax which did not adhere to the wick. Optionally, a fire beneath the *καπέλο* vessel would keep the wax in a molten state.[33]

e) Special cylindrical containers with a small tap on the bottom where large candles were dipped (Figures 2b, 3).[34]

f) Both the spokeshave and the hand plane are referred to as objects used, at least in recent years, for the finishing of candles.[35]

g) Special constructions to facilitate the preparation of cotton wicks in various lengths and diameters (Figures 4, 5).[36]

Candle making process

Taking into account the evidence derived from the operation of modern traditional candle-making workshops combined with ethnological information, we could restore the candle making process as follows: the chandler put wax in any vessel tall enough to accommodate the length of the candles he wanted to produce; he placed the pot in a stockpot half full of water set over a low heat, to prevent the wax from flaming. In other words, the melting of the wax was done using two copper vessels in a wood-burning fireplace. The fire was low so the wax did not ignite. The molten wax was transported

[27] Σπέης 1995, 47.
[28] Μαμαλούκος 2000, fig. 8.
[29] Μαμαλούκος 2000, figs 9,10.
[30] Μουσιώνη 2000, 236; Vincent 2004, fig. 23; 395. Σπέης 1995, 146; Μουσιώνη 2000, 236 and note 4.
[31] Μουσιώνη 2000, 236 and note 5; Patterson 1983, 315, fig. 36.13.
[32] Σπέης 1995, no. 578.
[33] Σπέης 1995, 47.
[34] Μαμαλούκος 2000, fig.12.
[35] Diderot and D'Alembert 1762, pl. III; Σπέης 1995, 146.
[36] Diderot and D'Alembert 1762, pl. III; Σπέης 1995, 146.

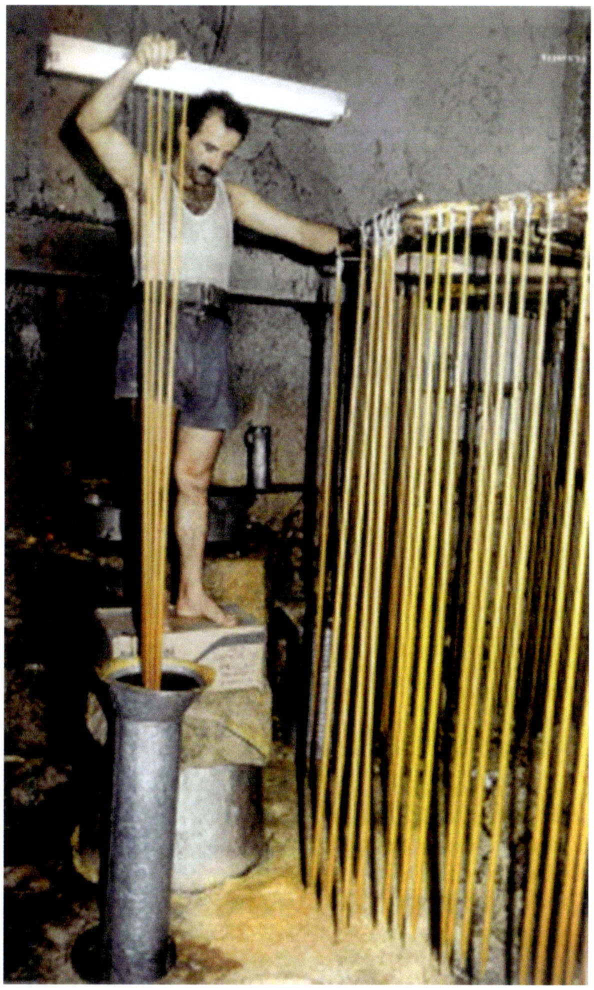

Figure 3. Construction of large candles in a modern Greek traditional candle-making workshop in the 1980s (Greek magazine *ΦΩΤΟγράφος* no. 38, April 1995).

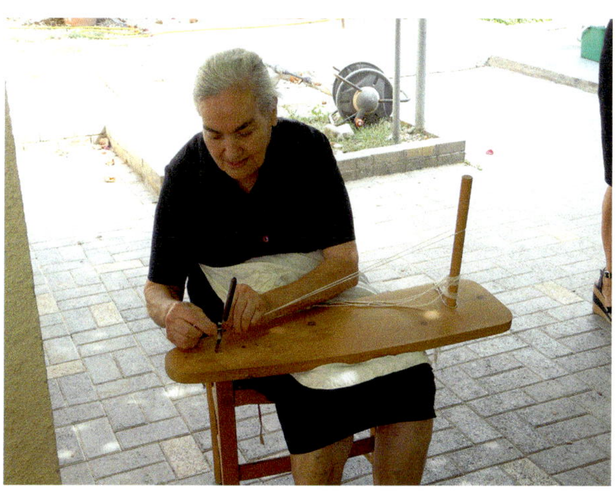

Figure 4. A candle-making workshop in Paphos, Cyprus. The wooden plank is used for measuring and cutting the candle wicks in various sizes and cutting and twisting the cotton wicks.

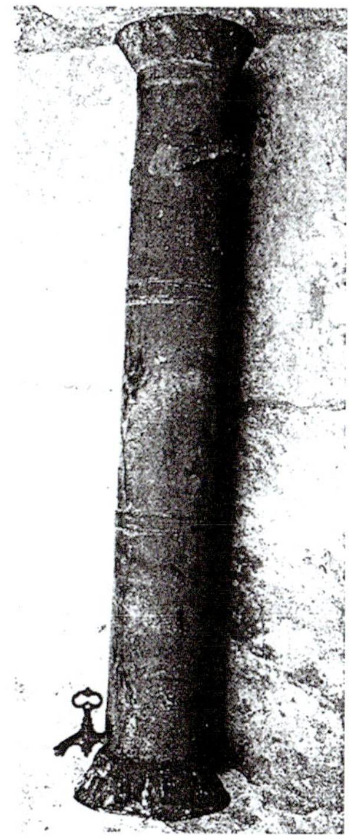

Figure 5. Special cylindrical container with a small tap on the bottom where the large candles were dipped. Monastery of Megisti Lavra, Mount Athos (Μαμαλούκος 2000, figure 12).

from the vessels to the various candle making points with special ladles (Figure 2). [37]

Methods for making candles:

• By immersion[38]

The simplest method seems to reflect the improvised process that is given to us by ethnological information. Previously beekeepers used to make candles in their homes using the following procedure: they cut cotton wicks, dipped them in a vessel containing melted wax and finally they rubbed them to 'straighten' them.[39]

During this process the wicks were dipped into the melted wax while they were hung on the fingers[40] (Figure 2d), or on a special construction from which more than one wick was supported giving an image similar to one depicted on a photograph taken in a modern traditional candle-making workshop (Figure 3). Both these options were also suitable for mass-production. Then the candles were cleansed of any lumps and any surplus wick removed.[41] The Modern Greek term for candles made by immersion is *βουτηχτά*[42] (immersed).

[37] Μαμαλούκος 2000, 254.
[38] Vincent 2004, fig.15; Μουσιώνη 2000, figs 3, 4; Peclet 1827, 76-77.
[39] Ριζοπούλου-Ηγουμενίδου 2000, 405.
[40] Μωυσής Μοναχός 2000, 280-281.
[41] Μωυσής Μοναχός 2000, 280-281.
[42] Μουσιώνη 2000, 244.

Larger candles were dipped into special cylindrical containers which had a small tap on the bottom (Figure 5) from where water or surplus wax could be evacuated.[43] Such a process is depicted in a photograph of a modern traditional candle-making workshop (Figure 3) in which a vessel similar to this and to the one used in the candle making workshop of Paphos (Figure 2b) is depicted and this process is mentioned in a religious book dated 1772.[44] This information leads us to suppose that candles of medium and large size were made by immersion of the wicks supported from wooden or metallic hoops around which small hooks were joined.[45]

• By pouring the melted wax[46]

In some cases the chandler could pour melted wax over wicks hung either from wooden dowels, reeds, metallic hooks or even from his fingers using the wax ladle as many times as needed to get the candle to the desired thickness.[47] Surplus wax was collected in a vessel which had the form of an inverted top hat.[48]

• Using molds

We do not have information or any indirect testimonies for producing candles with molds in Byzantium. However, even in Western Europe, where it appears that the art of candle-making in molds developed in the 14th century AD,[49] there was always the difficulty that the beeswax tended to stick to the walls of the molds and thus it was difficult to remove the candles from the molds without destroying them.[50] Maybe this method was more suitable for candles made by animal fats.

Candle drying

This was the final stage of candle making. The candles appear to have been hung while still dripping to dry and harden, and the presence of loops also supports this theory (Figure 6).[51] Then there followed the finishing using a spokeshave or hand plane and the finishing of the bottom edge with a hot knife.[52]

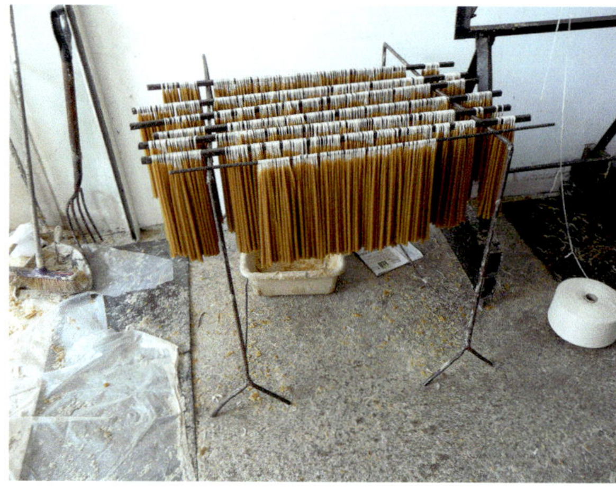

Figure 6. A candle-making workshop in Paphos, Cyprus. Candle drying.

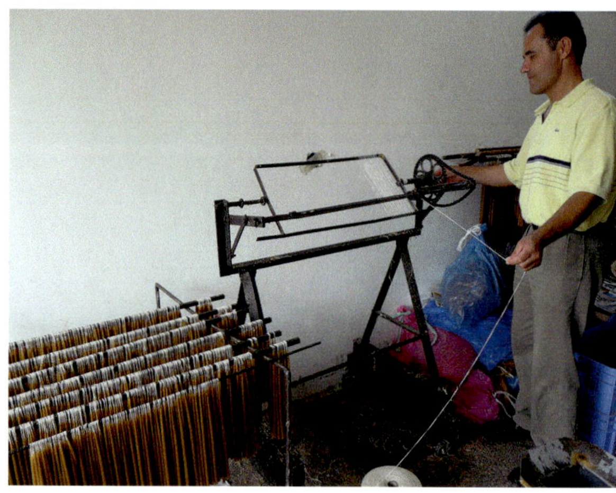

Figure 7. A Candle-making workshop in Paphos, Cyprus. Wrapping the cotton wick around iron frames.

Innovations in candle-making workshop operations

The following data are indicative of the innovations adopted at different times depending on the region and the economic conditions prevailing in it.

At the beginning of the 20th century the candle-making workshops in some Russian monastic foundations in Mount Athos, such as that of Skete of Saint Andrew (near Karyes), take a different form. These workshops are characterized by their large size, sophisticated fixed and mobile equipment and, above all, by order and organization. These candle-making workshops are another example of the industrialization of production even in Mount Athos, a process that was postponed by the decline of the Russian influences after 1917 and resultant changes in its civil society and economy after the incorporation of Mount Athos into the Greek state.[53]

Sometime around 1957/58 major improvements took place at the candle-making workshop, when the aging owner, reluctant to any change, accepted the need to adopt the 'modern technologies' in the form of manually powered machinery to wrap the cotton string around iron frames (Figures 7-9). Appropriately, elongated rectangular containers suitable for these frames were also introduced.[54] Later, the use of natural gas instead of wood was used to heat the wax. These improvements in the candle production line directly affected the number of staff needed in the workshop. Several manual laborers were fired. The family could now rely solely on the

[43] Μαμαλούκος 2000, 255.
[44] Μακαρίου Κυδωνέως 1772, 7.
[45] Μαμαλούκος 2000, 254; Μουσιώνη 2000, 244.
[46] Vincent 2004, fig.15; Μουσιώνη 2000, figs 3, 4; Peclet 1827, 76-77.
[47] Peclet 1827, 77; Μαμαλούκος 2000, 255.
[48] Σπέης 1995, 146.
[49] Chrzanovski, Kaiser 2007, 306, fig. 222.H,I,K.
[50] Forbes 1966, 138-139.
[51] Peclet 1827, 77; Patterson-1983, 315, fig. 36.13.
[52]

[53] Μαμαλούκος 2000, 256.
[54] For a similar and likely contemporary innovation in Mount Athos, see Μαμαλούκος 2000,255.

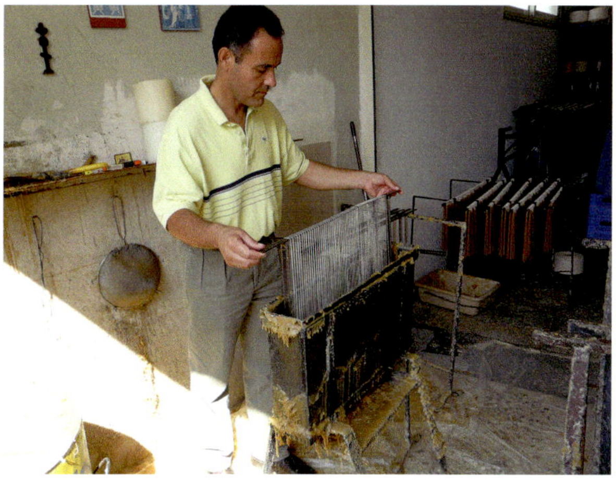

Figure 8. A Candle-making workshop in Paphos, Cyprus. The first dipping of the frames in a rectangular tin vat containing hot melting wax.

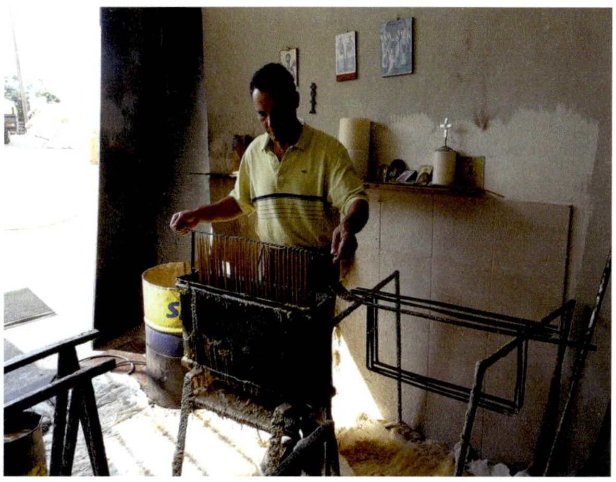

Figure 9. A candle-making workshop in Paphos, Cyprus. The last dipping of the wrapped frames. In the background notice the barrel used for melting the wax with the use of gas.

work of its members with the occasional hiring of seasonal workers during periods of high demand such as Christmas and especially Easter.

The information about the mode of operation of the candle-making workshop in Paphos before 1957/58 is important for us as we can realize that the *Encyclopedia, or a Systematic Dictionary of Sciences, Arts and Crafts*[55] describes a way of candle-making which reflected production techniques related to the industrial revolution but which took place in some parts of Greece and Cyprus much later.

Conclusions

The organization of traditional candle-making workshops such as the workshop of the Monastery of Megisti Lavra on Mount Athos and in Paphos both of which were in use until recent times let us realize that the need for mass candle production probably led the candle makers to adopt innovations and new techniques. However, the fact that the mass candle production in recent times was achieved thanks to the use of tools and machines invented during the period of the industrial revolution, initially presented in the *Encyclopedia, or a Systematic Dictionary of Sciences, Arts and Crafts*,[56] shows that in Byzantine period the candle makers used simpler processes, when they wanted to increase production they could simply make use of small wooden or reed sticks in order to suspend more wicks.

Bibliography

Antonaras, A. 2012. *Fire and sand: ancient glass in the Princeton University Art Museum*. Princeton, N.J: Princeton University Art Museum.

Antonaras, A. Ch. 2017. *Glassware and glassworking in Thessaloniki: 1st century BC-6th century AD*. Oxford: Archaeopress.

Biddle, M. 1999. *The Tomb of Christ*. Gloucestershire, Stroud (Glos): Sutton Publishing.

Chrzanovski, L. and P. Kaiser (eds) 2007. *Dark ages? Licht im Mittelalter/L'eclairage au moyen age*, Exhibition Catalog. Historiches Museum Olten. Milano: Edizioni Et.

Dagron, G. 2002. The Urban Economy, Seventh–Twelfth Centuries, in Laiou, A. E. (ed.) *The Economic History of Byzantium: From the Seventh through the Fifteenth Century*, 393-461. Washington: Dumbarton Oaks Trustees for Harvard University.

Diderot, D. and D'Alembert, J. L. R. 1762. *Recueil de planches, sur les sciences, les arts liberaux, et les arts mechaniques, avec leur explication*. Paris: Briasson.

Dujcev, I., Nicole, J. and Freshfield, E. H. 1970. *To eparchikon biblion: The book of the Eparch; Le Livre du préfet*. London, Variorum reprints.

Fevrier, P. A. and Poinssot, C. 1959. Les cierges et l'abeille. Note sur l'iconographie du baptistere decouvert dans la region de Kelivia (Tunisie). *Cahiers Archéologiques* 10, 149-156.

Fonseca, C. D. 1989. Lavoro agricolo e tempo liturgico, in G. Musca (ed.) *Uomo e ambiente nel Mezzogiorno normanno-svevo, Atti delle ottave Giornate normanno-sveve*, Bari, 20-23 October 1987, 67-87. Bari: Edizioni Dedalo,

Forbes, R. J. 1966. *Studies In Ancient Technology VI: Heat and Heating, Refrigeration, Light*. Leiden: Brill.

Jordan, R. 2000. Kecharitomene: Typikon of Empress Irene Doukaina Komnene for the Convent of the Mother of God Kecharitomene in Constantinople, Date: 1110–1116', in J. P. Thomas, A. Constantinides-Hero and C. Giles (eds) *Byzantine monastic foundation documents: a complete translation of the surviving founders' typika and testaments*, 649-724. Washington, D.C.: Dumbarton Oaks Research Library and Collection.

Koder, J. (ed.) 1991. *Das Eparchenbuch Leons des Weisen*, Corpus fontium historiae byzantinae 33. Wien: Osterreichischen Akademie der Wissenschaften.

Κοτσόκολος, Γ. and Σωτηρόπουλος, Δ. 2000. Μελισσοκομικές καταγραφές από την υμνολογία της Εκκλησίας μας, in *Η μέλισσα και τα προϊόντα της, ΣΤ΄ τριήμερο εργασίας, Νικήτη, 12-15 Σεπτεμβρίου 1996*, 155-160. Αθήνα: Πολιτιστικό Ίδρυμα ΕΤΒΑ.

[55] Diderot and D'Alembert 1762.

[56] Diderot and D'Alembert 1762.

Lemon, G. W. 1783. *English etymology, or a Derivative dictionary of the English language*. London: G. Robinson.

Μακαρίου Κυδωνέως τοῦ Τριγώνη 1772. *Προσκυνητάριον τῆς Βασιλικῆς καὶ Σεβασμίας Μονῆς Μεγίστης Ἁγίας Λαύρας τοῦ Ἁγίου Ἀθανασίου τοῦ ἐν τῷ Ἄθῳ*. Venice.

Μαμαλούκος, Σ. Β. 2000. Το κηροπλαστείο της Μονής Μεγίστης Λαύρας του Αγίου Όρους, in *Η μέλισσα και τα προϊόντα της, ΣΤ΄ τριήμερο εργασίας, Νικήτη, 12-15 Σεπτεμβρίου 1996*, 246-258. Αθήνα: Πολιτιστικό Ίδρυμα ΕΤΒΑ.

Motsianos, I. 2011. Φως ιλαρόν: Ο Τεχνητός Φωτισμός Στο Βυζάντιο (Joyful Light: The Artificial Lighting in Byzantium). Unpublished PhD dissertation, Πανεπιστήμιο Θεσσαλίας. <http://thesis.ekt.gr/thesisBookReader/id/31739#page/1/mode/2up>, viewed 15 December 2018.

Motsianos, I. 2011b. Artificial Lighting during Byzantine and Post-Byzantine Period, in I. Motsianos and E. Bintsi (eds) *Light on light, an illuminating story, Exhibition Catalogue*, 90-125. Thessaloniki: Folklife & Ethnological Museum of Macedonia-Thrace.

Μουσιώνη, Λ. 2000. Παραδοσιακό κηροπλαστείο στη Λάρισα, 'Ο Μαρκογιαννόπουλος', in *Η μέλισσα και τα προϊόντα της, ΣΤ΄ τριήμερο εργασίας, Νικήτη, 12-15 Σεπτεμβρίου 1996*, 235-245. Αθήνα: Πολιτιστικό Ίδρυμα ΕΤΒΑ.

Μωυσής Μοναχός 2000. Περί των αγιορειτικών κηροποιείων, in *Η μέλισσα και τα προϊόντα της, ΣΤ΄ τριήμερο εργασίας, Νικήτη, 12-15 Σεπτεμβρίου 1996*, 279-283. Αθήνα: Πολιτιστικό Ίδρυμα ΕΤΒΑ.

Νεράντζη-Βαρμάζη, Β. 1989, Συντεχνίες εμπόρων στην Κωνσταντινούπολη τον 10ο αι., in *Η καθημερινή ζωή, in Βυζάντιο*, Πρακτικά Α' Διεθνούς Συμποσίου, 352-361. Αθήνα: Κέντρο Βυζαντινών Ερευνών.

Papanikola-Bakirtzi, D. 2011, *A Candle-making Workshop in Paphos, Cyprus*, 4th Round Table of International Lychnological Association, Thessaloniki, 11-14 October 2011. Poster presentation.

Patterson-Ševčenko, N. 1983. *The Life of Saint Nicholas in Byzantine Art*, Monografie, Centro Studi Bizantini, Bari 1. Turin: Bottega d' Erasmo.

Peclet, E. 1827. *Traite de l'eclairage*. Paris: De Malher et Cie.

Ριζοπούλου-Ηγουμενίδου, Ε. 2000. Η παραδοσιακή μελισσοκομία στην Κύπρο και τα προϊόντα της (μέλι, κερί) κατά τους νεότερους χρόνους, in *Η μέλισσα και τα προϊόντα της, ΣΤ΄ τριήμερο εργασίας, Νικήτη, 12-15 Σεπτεμβρίου 1996*, 390-408. Αθήνα: Πολιτιστικό Ίδρυμα ΕΤΒΑ.

Σπέης, Τ. 1995. *Παραδοσιακές τέχνες και επαγγέλματα*. Αθήνα: Υπουργείο Πολιτισμού-Διεύθυνση Λαϊκού Πολιτισμού; Συλλογή Εταιρείας Λαογραφικών Μελετών.

Sorochan, S. 2002. Light for life and death in early Byzantine Empire, in D. Zhuravlev (ed.) *Fire, Light and Light Equipment in the Graeco-Roman World*, 111-119. Oxford: British Archaeological Reports, International Series 1019.

Stornajolo, C. 1910, *Miniature delle Omilie di Giacomo Monaco (Cod. Vatic. gr. 1162) e dell'Evangeliario greco Urbinate (Cod. Vatic. Urbin. gr. 2)*. Rome: Danesi.

Thomas, J. P., A. Constantinides-Hero and G. Constable (eds) 2000. *Byzantine monastic foundation documents: a complete translation of the surviving founders' typika and testaments*. Washington, D.C.: Dumbarton Oaks Research Library and Collection.

Vincent, C. 2004. *Fiat Lux. Lumière et luminaires dans la vie religieuse du XIIIe siècle au début du XVIe siècle*. Paris : Éditions du Cerf.

The Virtual Lighting of the Rotunda's Mosaics

Pelli Mastora
Archaeologist, Ephoreia of Antiquities of the City of Thessaloniki

Abstract: This article examines the types of lighting that operate, or are depicted as operating, in the Rotunda's mosaics. Three kinds of light: the virtual 'natural' light, the noetic light, and the virtual 'artificial' light- coexist in the mosaics. Their coexistence achieves an illusionistic rendering of the constructed space and permits an uninterrupted view of a world which, although imaginary, changes in response to variations in the light, almost like the real world. The lighting of the mosaics is the result of scenographic ingenuity, which is comparable to the ritualistic representations that Constantine commissioned in order to implement his religious policy.

Keywords: wall mosaics, tesserae, lighting, lamp, candlestick, hanging gems, theatrical presentation, scenography, Constantine I.

The Rotunda, located north of Thessaloniki's late Roman palace complex, is the city's most important and imposing monument. From an architectural point of view, it constitutes an example of a Constantinian-era central-plan building, which in this case, according to S. Ćurčić, was built as a mausoleum for Constantine I (306-337).[1] According to the recent work of Ch. Bakirtzis and P. Mastora, the mosaic decoration in the dome was original to the building's construction as a mausoleum for Constantine, rather than as a later addition from the time of the building's conversion into a Christian church, as was previously thought.[2] This article examines the types of lighting that operate, or are depicted as operating, in the hemispherical dome's mosaic decoration.

The mosaic decoration is divided into three zones and is comprised of: 1) in the lower zone, eight panels which depict opulent buildings behind praying men. This zone is dominated by a golden background in constant interplay with the light; 2) in the middle zone, a representation of a group of white-clad men with strong movements in an open-air space; and 3) in the upper zone, the standing figure of Constantine with a halo in front of a silver medallion borne by four angels or winged goddesses of Victory (*Nikes*). The middle of these three zones, as well as the figure of Constantine and the bodies of the angels/*Nikes*, is almost completely destroyed.

As is well known, the art of wall mosaic is connected with the properties of light because the glass tesserae of gold, silver, and vivid light colors reflect light, thus making the light part of the representation. That is why the golden background of Byzantine mosaics is usually interpreted as a symbolic depiction of the otherworldly, transcendent, and divine light.[3] In addition, however, to the golden light of the mosaic background, which is a reflection of the natural light coming into the building, the Rotunda's mosaics also feature other types of light originating from different sources.

In this paper, I will first examine the way in which the natural light coming into the building illuminates the mosaics, and how it was taken into account for keeping the overall design of the decorative program. The natural light's effect on the mosaics can be observed primarily in the design's lower zone, both because it is the best preserved and because the high percentage of gold tesserae enhances their reflective property. In bright natural light, the gold tesserae shine and diffuse light. In low lighting, they lose their brilliance and are transformed into a dull and dark color. Both instances highlight the different types of lighting in the mosaic representations, which, on the one hand, complement each other, and, on the other hand, effectively utilize, in different ways, natural light's presence, absence, or changes in intensity. The other types of lighting besides natural light can generally be divided into three categories: 1) virtual 'natural' light, which is a reflection of the natural light in the mosaic; 2) noetic light, which is emitted by the mosaic's representation of the central medallion; 3) virtual 'artificial' light, which emanates from the lamp and candlestick images present in the buildings depicted on the mosaic decoration.

Natural light

Natural light enters through the eight large arched windows and the, originally, eight lunettes which were opened in the building's circular wall.[4] The windows are situated above the recesses of the ground level and their location coincides with the vertical axis of the eight panels in the mosaic's lower zone. The smaller lunettes are located at the base of the dome along the vertical axis of the bands dividing the panels in the design's first zone, and are part of the dome's mosaic design (Figure 1).

Originally, the lunettes inclined toward the interior of the building floor, which reflected the incoming light toward the

[1] Ćurčić 2000, 11-14.
[2] Bakirtzis and Mastora 2011, 33-45; Bakirtzis 2012, 51-117, which also contains the relevant bibliography; Bakirtzis and Mastora 2016, 26-34.
[3] For the interpretation of the golden background of the Rotunda's mosaics as part of the iconography of the Second Coming and the Heavenly Jerusalem, see Nasrallah 2005, 493-4, which also includes the relevant bibliography.

[4] The dazzling light entering through the building's large windows is one of the key features which distinguish the Christian mausoleums of Constantine's reign from the mausoleums of the Tetrarchy; see Johnson 2009, 74-82. Johnson, however, does not include the Rotunda among the Constantinian mausoleums. Ćurčić, on the other hand, refers to the importance of the large windows as one of the common features between the Rotunda and Constantinian Christian shrines of Rome, see Ćurčić 2000, 12.

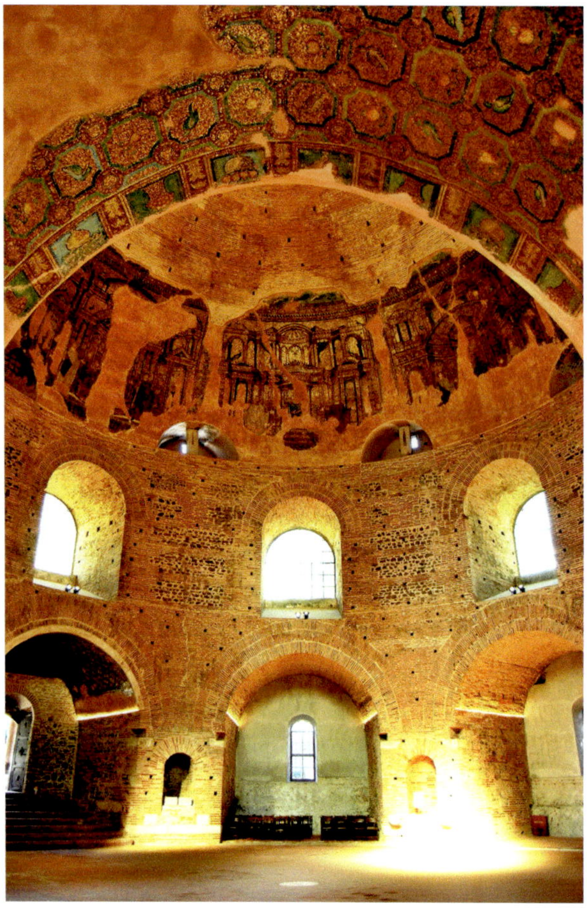

Figure 1. Thessaloniki, Rotunda's mosaics. Natural light enters through the large arched windows and the lunettes.

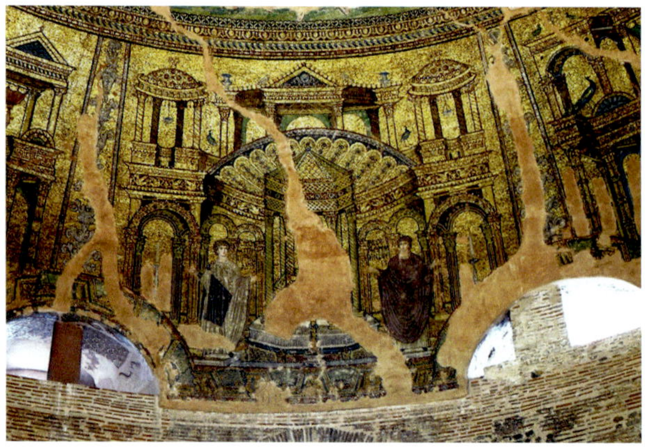

Figure 2. South panel: The incoming light from the window is directed at the center of the panel.

dome, as I. Iliadis has demonstrated.[5] Today, these openings differ from the originals both in number and form.[6] Another important change, related to the reflection of the incoming light, is the loss of the marble slabs which once covered the building's interior surfaces.[7] One can, nevertheless, still make out the way in which the mosaic program utilizes the windows and lunettes as sources of illumination.

The mosaic program's division into eight panels in the lower zone is directly related to the identical number of both windows and lunettes. The light that enters through them harmonizes with the organization of the representations in the panels. In each panel, the representation is organized symmetrically around a central subject which lies on the vertical axis that passes through the middle of the panel and the large window. The incoming light from the window is directed at the center of the diametrically opposed panel, illuminating the panel's central subject and emphasizing its importance (Figure 2). Between the windows, but at a higher level, are the lunettes, which admit a smaller amount of light, inasmuch as they are themselves smaller than the windows. The incoming light from the lunettes complements and balances the panels' central light, which comes from the large windows. We could thus say that the windows and the lunettes function as spotlights focused on the panels both directly and obliquely. Measurements of the incoming light have further shown that the light's intensity varies in relation to the daily movement of the sun, and that specific areas of the mosaic program are illuminated more or less in accordance with the sun's position throughout the year.[8]

Moreover, the central-plan layout of the large windows high in the building permits the passage of abundant diffused light into the interior, as well as providing direct light to the mosaic at the key of the dome, where the surface becomes almost flat as the curvature of the hemisphere decreases (Figure 3). The incoming natural light reflects off the gold, silver, and light-colored glass tesserae and the dome's apex becomes a kind of mirror which gathers and reproduces natural light. In this respect, it is no coincidence that the representation at the key of the dome draws together a large number of subjects related to light (see below). These brilliant, shining objects make the dome's apex the center of light, which, as we shall see shortly, then emits light onto the representations in the two lower levels of the mosaic program.

Virtual 'natural' lighting

As natural light falls upon the mosaics it induces their golden background which in turn is diffused and fills the place with its golden light, within which stand opulent buildings. The same light shines through the openings in the buildings, as though coming from a source behind them. In this respect, the natural light reflected in the golden background appears as an inherent part of the composition, becoming a light immanent in the image, a virtual 'natural' light. The virtual 'natural' light is diffused throughout the space and 'illuminates' the buildings as if it were an external light source, revealing their dimensions and creating a sense of depth.

The coloring of the buildings in this virtual natural light is no different from the golden background of the representations. The buildings are decorated with semi-precious stones rendered with silver and shiny glass tesserae in brilliant shades of red, green, and blue, materials, in other words,

[5] Iliadis 2005, 467-478.

[6] On the alterations to the windows, see Theocharidou 1991-1992, 62.

[7] Hébrard 1920, 34; Theocharidou 1991-1992, 67-68.

[8] Iliadis 2005, 468-471.

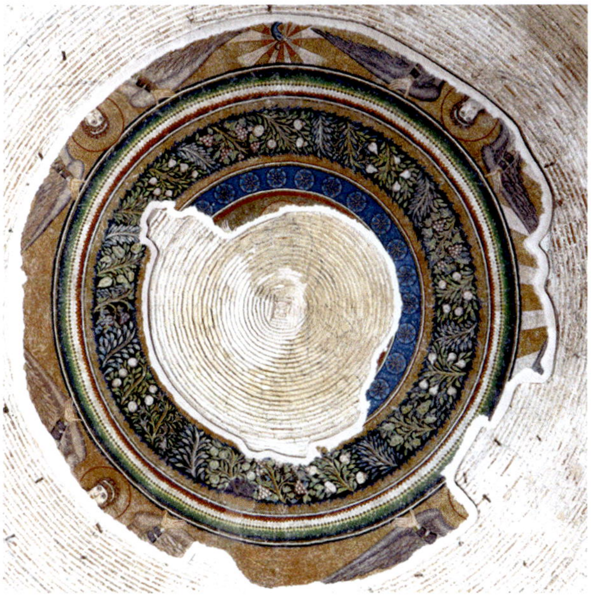

Figure 3. The glittering medallion at the key of the dome.

which reflect the light and are consequently also subject to the variations in the virtual 'natural' light. One can observe an indicative example of these variations in the light on the buildings in the N and S panels, in the middle of each of which is depicted a quarter-sphere recess with two tetrastyle wings (Figure 2). The virtual 'natural' light is concentrated in the middle of the quarter-sphere recess and on the facades of the tetrastyle wings. At the edges of the quarter-sphere recess, the golden virtual light recedes and is replaced by beams of yellow light until it is finally lost in the shadows of the entablature (Figure 4). The shadows at the edges of the recess are immutable, regardless of the time of day, as are the shaded back columns and the adjacent sides of the buildings. This is because their shading owes to the virtual 'natural' light, which, in comparison with the changing natural light, is fixed, as I will explain below.

The impression that the buildings are lit by an external light source is also created by the intensification of light on either side of a bright central point. The light is concentrated in the middle of the buildings, on the surfaces facing the viewer, and gradually spreads and intensifies to the left and right of the illuminated central point (Figure 5). Thus there is a virtual 'natural' light emanating from a source in front of the mosaic, which coincides with the frontal and receding oblique-frontal natural light coming into the building from its windows and lunettes. And as the natural light moves, illuminating individual subjects on the panels' representations in accordance with the movement of the sun, it enters into the world of the mosaic representation. In this way, the windows and lunettes act like theater spotlights, creating steady scenic lighting so that the buildings are illuminated in the same way theater stage buildings are spotlighted for the viewer who watches the events taking place in front of them.

Noetic light

Noetic light is the second source of the mosaic's lighting. It becomes perceptible during the least bright hours of the day, when the light coming into the building wanes, the gold's glow

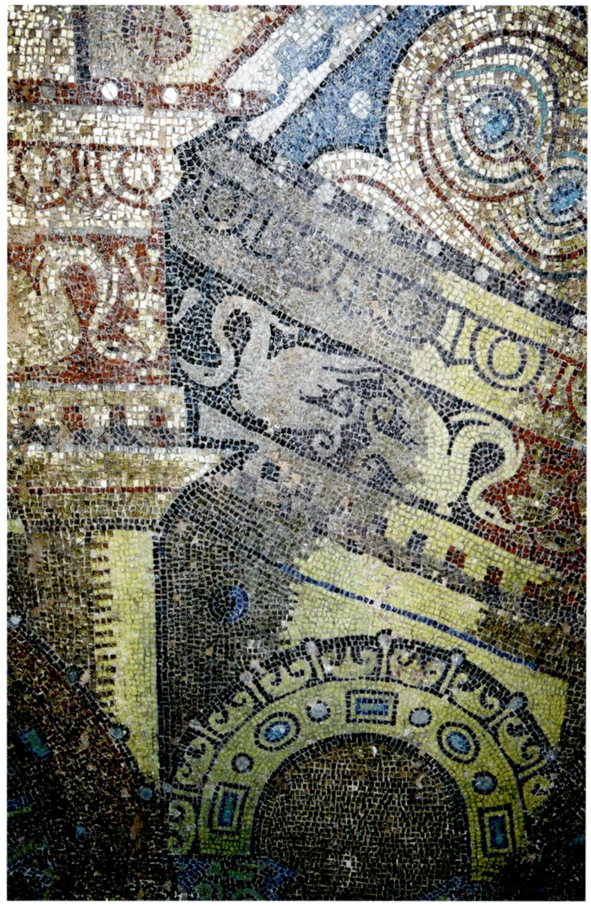

Figure 4. South panel, detail. The light recedes at the edges of the quarter-sphere recess of the central building.

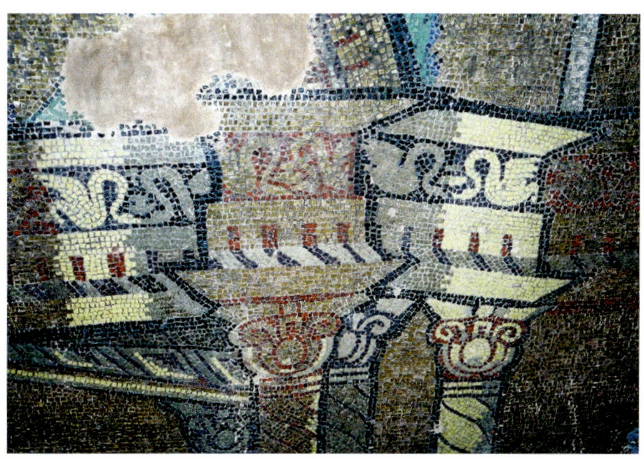

Figure 5. West panel, detail. The light is concentrated in the middle of the building, and gradually spreads and intensifies to the left and right.

fades and the gold turns from light into a dull, dark color.[9] As the gold darkens, the light-colored glass tesserae which, under strong natural light, are seen as simple yellow and light

[9] On the golden background's change to a dull, dark color under low-light conditions, see Frances 2003, 14, 21.

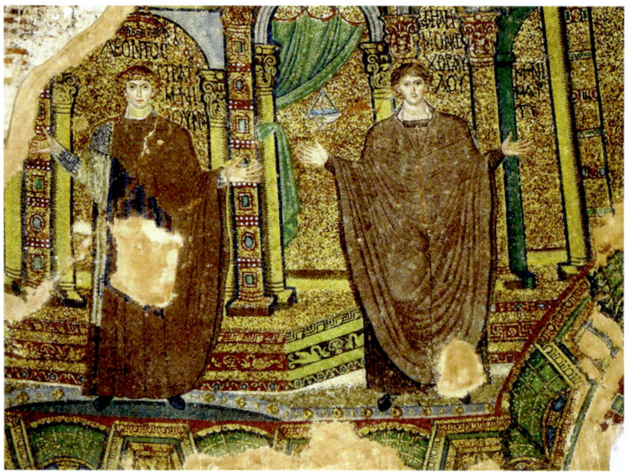

Figure 6. Southeast panel, detail. Bright beams appear on the floor.

green beams, become brighter. The light is thus rendered in a painterly way, with the tonal gradation of the hues of glass tesserae which, because of their vivid, brilliant hue and the luster of the glass, preserve their reflective properties in low-light conditions better than the gold and silver tesserae.

This light comes from high above, from a source above and beyond the architectural members of the lower zone's panels. This light falls vertically and obliquely on the buildings, generating a vivid glow primarily in the apices of the domes, such as, for example, in the golden dome of the ciboria in the NE and SE panels. Slightly divergent beams of yellow light spread to the receding or side surfaces of the buildings and to their curved niches, such as, for example, in the parapet separating the upper buildings from the lower ones in the N and S panels. Bright beams also appear on the floors as if the light were passing through the spaces between the buildings and reflecting, diffused, on the glossy surfaces of the marble slabs (Figure 6).

The destroyed male figures of the mosaic program's middle zone moved under the same vertical and slightly oblique light, as the shadows cast by their bodies indicate. The slightly inclining vertical shadows on the green grass indicate the materiality of the figures, which are depicted with vivid choreographic movements in a highly illuminated natural landscape.

The source of this light, which originates from a single brilliant center and spreads in all directions, is that depicted in the dome's central medallion (Figure 2). As I mentioned previously, concentrated at this point are all the brilliant, shining objects, which emit an otherworldly, unwaning light: a haloed figure on the medallion's silver background, a ring of silver stars, and a golden crown of fruits and a rainbow. The glittering medallion is held up by haloed angels/*Nikes* flying over the golden background. Between them appear a phoenix to the east and another brilliant object to the north, now destroyed,[10] which emit blinding white light. This light,

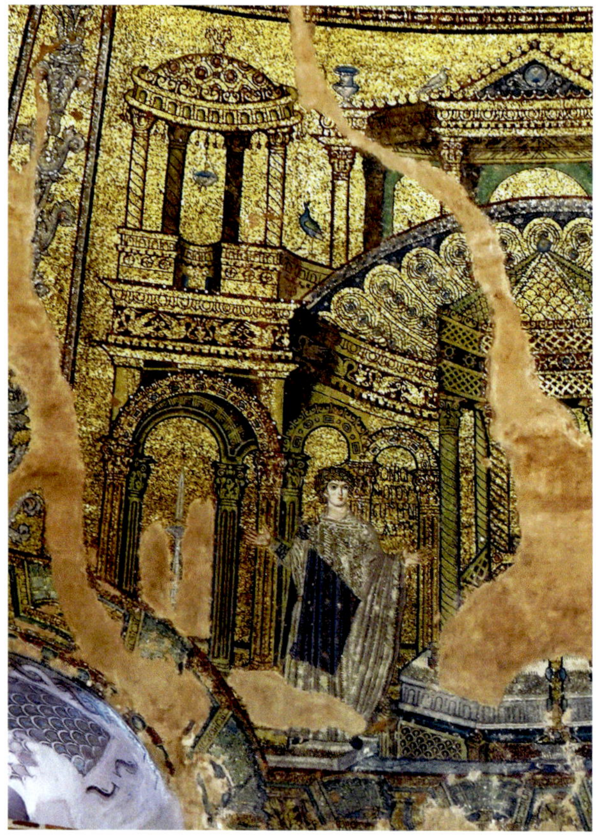

Figure 7. South panel, detail. Candlestick and hemispheric lamp.

which I call noetic, penetrates the two lower zones of the mosaic composition, forming a link between them.

Virtual 'artificial' light

The third light source comes from the burning lamps and standing candlesticks found inside the buildings depicted on the mosaic's lower level. These are different types of silver objects, which show structural similarities to lamps and candlesticks dating chiefly to the late Roman period. The lamps, together with the candlesticks, craters, censers, and jeweled crowns that adorn the Rotunda's buildings, comprise a rich set of objects which are mentioned in the *Liber Pontificalis* as imperial gifts to churches and mausoleums, and as treasures that are also associated with gifts, as well as funerary possessions in the tombs of Roman officials.[11]

The light from the lamps and candlesticks is not only reflected on the building's architectural members, but it also illuminates the interior spaces (Figure 7). This becomes perceptible in low-light conditions, when the golden background darkens and the buildings' golden facades recede, losing their brilliance. The interior spaces then appear to be illuminated by the lamps and candlesticks. Lamps or candlesticks, or both means of lighting, are carefully distributed throughout various areas of the buildings in all the panels. In the buildings of the N and S panels, there are candlesticks below tetrastyle ciboria and

[10] Torp conjectured that it was a cross, see Torp 1963, 36. Korres assumed this view in his proposed reconstruction of the mosaic; see Sotiriou 1970-1972, 195.

[11] Mastora and Motsianos 2011, 129-138.

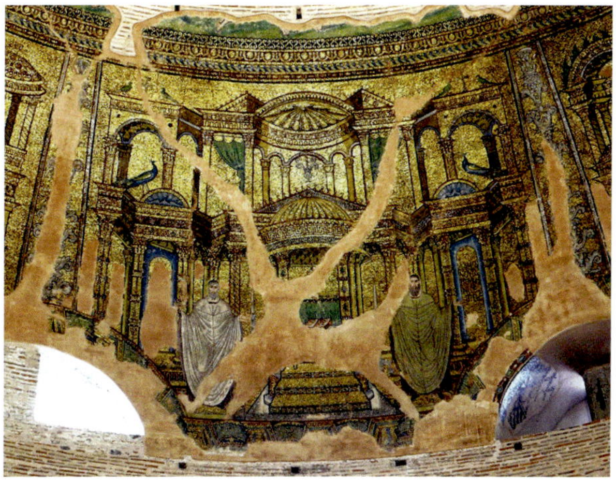

Figure 8. Northwest panel. Candlesticks are located near the central ciborium.

hemispheric lamps hung with chains from the middle of the buildings' domes. The candlesticks' lit candles produce light that diffuses throughout the interior of the ciboria and is reflected on the arches and columns. The lamps in the upper buildings are lit, but their light is directed upward due to their hemispheric shape, concentrated in the domes and not diffusing like the light of the candles. It also does not produce reflections on the interior surfaces of the columns, which are in shadow in the background.

In the central ground floor building in both the NE and SE panels, there are lamps with a conical foot and attached suspension fittings. The drawn curtains of the arched openings reveal lamps, which suggest that they illuminate the interior spaces. In this respect, it is no coincidence that the lamps appear to be both hanging and tabletop, judging by the shape of their bases. Unlike the hemispheric lamps of the tetrastyle ciboria in the N and S panels, these are hung for safer use. In the buildings of the NW and SW panels, the candlesticks are located near the central ciborium, such that the flames from the lit candles illumine not only the building's interior but also the ciborium's columns (Figure 8).

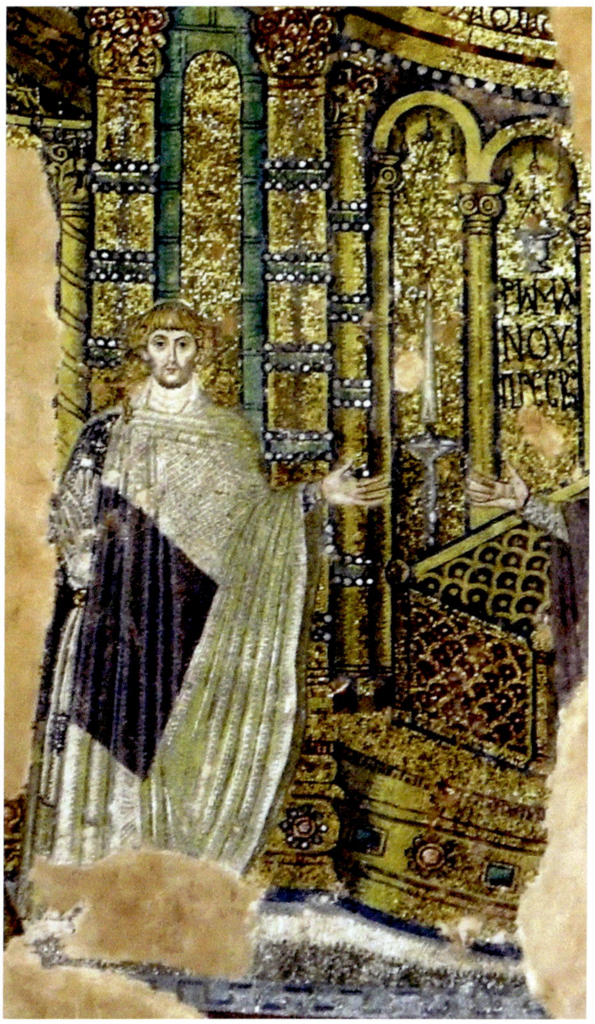

Figure 9. West panel, detail. Lamps and candlesticks are found in very close proximity.

In the recess of the W panel, lamps and candlesticks are found in very close proximity (Figure 9). The light is so strong that the columns of the recess gleam and the candle's upwardly directed light, supported by the diffused light from the candlesticks, forms arched beams of light on the golden background. The rendering of the light's color with gradated yellow and green tesserae, which penetrate into the golden background, i.e., the gradation of the light into the light, is an original stylistic invention, which indicates the active role of the light that was produced by the lighting devices depicted

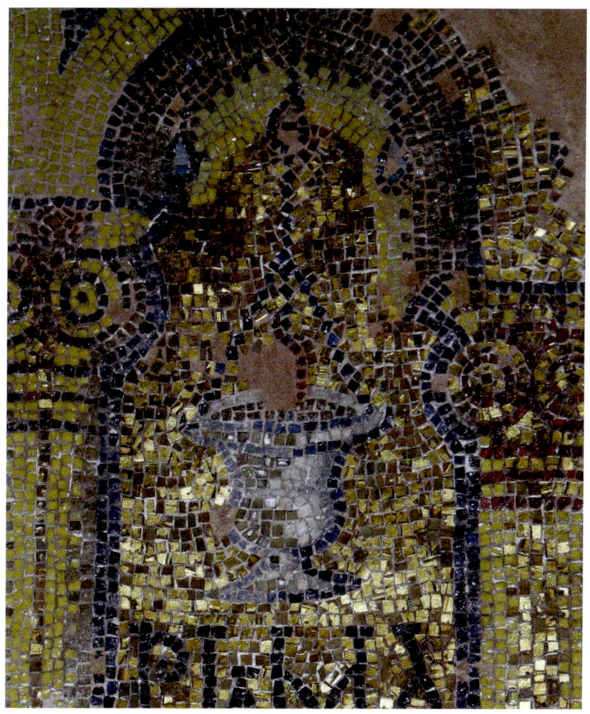

Figure 10. West panel, detail. Rendering of the light's color with gradated yellow and green tesserae.

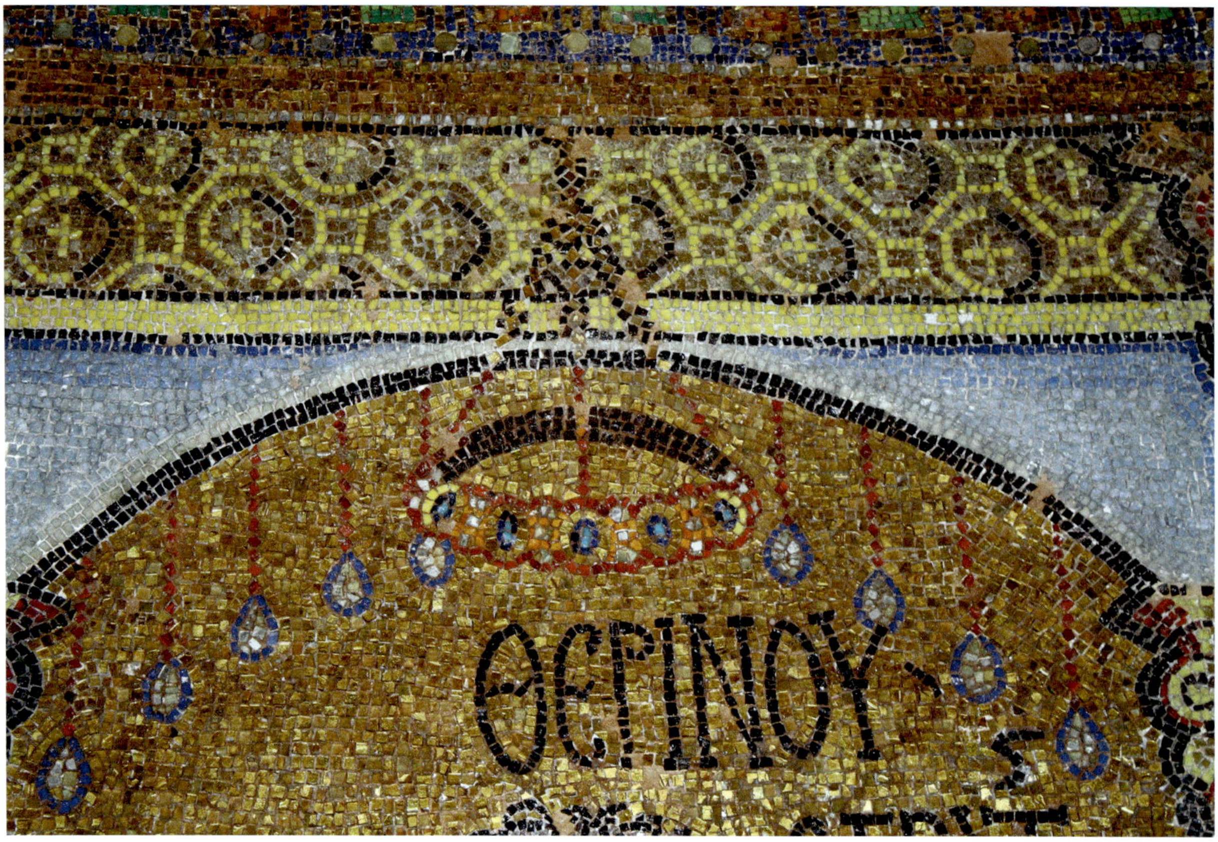

Figure 11. Northeast panel, detail. Hanging semi-precious stones and jeweled crown in the entrance of a building.

in the representation (Figure 10). Indeed, the intensity of the 'artificially' produced light is such that it brings out the brilliance of the blue and red semi-precious gemstones hanging from the numerous metal chains attached to the interiors of the arches of the semicircular recess.

Hanging semi-precious stones, which can be seen in other buildings of the Rotunda's mosaic program, likely indicate pearls or other precious stones and function as small mirrors reflecting the light in accordance with the way in which each building is illuminated (Figures 10, 11). As objects, they belong to the category of offerings (*oscilla*) with symbolic and apotropaic properties, which, in antiquity, were customarily hung on buildings' architraves and arches.[12] Hanging precious or semi-precious gemstones from chains has been associated with the public display of spoils during the celebration of victories and with rites of imperial worship.[13] The pearl jewelry is selectively hung at particular points in the buildings where there is either imperial regalia (a relief bust of the emperor with Nikes in the NE/SE panels, a orb between the dolphins in the W panel) or other objects which, as noted above, are associated with imperial donations (jeweled crowns in the NE/SE panels (Figure 11), monogrammatic crosses in the SW/NW panels, lamps in the N/S and W panels). This differentiation makes it clear that they are offerings which add not only to the magnificence of the buildings but also to the triumphant celebration taking place in front of the buildings depicted in the Rotunda's dome. Moreover, it is worth noting that the suspension of precious or semi-precious stones, primarily in buildings dedicated to worship, is associated with the stone's ability to reflect natural light and to glow in the dark, thus creating a constant source of illumination and connecting it with various symbolisms.[14]

The Rotunda's light devices are depicted as parts of the real world which functions within the virtual world, i.e., inside the world of the image, assisting in its continual illumination. When the golden light of the representations' background, the virtual 'natural' light, recedes during the least bright times of day, we can see that the flames of the lamps and candlesticks illuminate all the more brightly the buildings' interiors, just as real buildings are illuminated at night. This verisimilitude is heightened by the realistic way in which the mosaics render the lighting devices and other objects, which not only appear in the representations but actually function in that virtual world: the flames of the lamps sway from the force of the air circulating through the arched openings; smoke billows out from the censers in all directions; the water flows through the fountains, sparkling in the light.

[12] Taylor 2005, 83-105.

[13] On the correlation between hanging jeweled chains and spoils, see Knauer 1993, 13-46, esp. 41-42, no. 74.

[14] Although the depiction of hanging stones is generally rare, it is not unprecedented, such as we see in the representations on coins of the Temple of Capitoline Jupiter and the Temple of Divus Augustus in Rome (see Fishwick 1992, 238), and the mosaic decoration of the building at the Antonine Baths in Carthage (see Yacoub 1995, 358-9, fig. 178a-b).

Conclusions:

Based on our analysis of the kinds and sources of light for the mosaic in the dome of the Rotunda, we can draw the following conclusions:

1. The organization of the design in the mosaic's lower zone took into account the number and location of the building's windows and lunettes, making the most of the way they illuminate the mosaic representations.
2. The incoming natural light illuminates the mosaics and is transformed by the reflective properties of the gold, silver, and glass light-colored tesserae into virtual 'natural' light, which virtually illuminates the buildings in the panels, participating in the depiction and style of the mosaic and suggesting its interpretation. When the natural light wanes, the virtual natural light continues to 'shade' parts of the buildings.
3. The noetic light that emanates from the representation of the central medallion and penetrates into the design's middle and lower zones indicates that both of these two different worlds depicted therein exist under this light. In the mundane world of the shining architectural members, the noetic light is interwoven with the virtual 'natural' light to render the buildings' depth and shape. When the natural light disappears, i.e., when the day grows dark, the noetic light intensifies since it is an unwaning, otherworldly light emanating from the celestial bodies.
4. The depiction of lighting devices inside the buildings creates a light that becomes perceptible in conditions of low natural light, and thus enhances and complements the noetic light. This, however, is an 'artificial' light, the product of human ingenuity, and its use inside the buildings accents the material dimension and the opulence of the objects depicted.
5. The three kinds of light: the virtual 'natural' light, the noetic light and the virtual 'artificial' light, coexist in the mosaic's lower zone in a state of constant interplay with the building's natural light. Their coexistence achieves an illusionistic rendering of the constructed space. Their emergence, in accordance with the intensity of the natural light, and their distinction as different types of light, permits an uninterrupted view of a world which, although imaginary, changes in response to variations in the light, almost like the real world. In this respect, it is no coincidence that the blending of the properties of the three types of light and their interrelationships is fully felt in the depiction of the opulent buildings. It thus heightens the theatrical presentation of the event in which the praying men, who are piously gathered together in front of the buildings, are participating.
6. The three types of light create an illusory impression of light which give verisimilitude to the depicted subjects, while the world they inhabit, a world of virtual reality, is vivified.

The lighting of the Rotunda's mosaics is the result of scenographic ingenuity, which is comparable to the ritualistic representations that Constantine commissioned in order to implement his religious policy. According to Eusebius of Caesarea, the emperor, participating as a hierophant in the celebration of Easter, made the all-night resurrection vigil as bright as day by placing towering wax candles throughout the city; these lamps/torches illumined by their flames the whole area so that the night of the holy vigil would become more brilliant than daytime.[15] Thus also is the carefully illumined world of the Rotunda's mosaic representations, which depict the emperor himself in the dome. Eusebius describes Constantine's thrice-blessed soul in the heavenly realms together with God himself, freed from every mortal and earthly encumbrance, gleaming in a resplendent garment of light.[16]

Bibliography

Bakirtzis, Ch. and Mastora, P. 2011. Are the mosaics in the Rotunda in Thessaloniki linked to its conversion into a Christian church?, in *IX International Symposium 'Niš and Byzantium', Niš, 3-5 June 2010*, 33-45. Niš: NKC.

Bakirtzis, Ch. (ed.) 2012. *Mosaics of Thessaloniki, 4th-14th century*. Athens: Kapon.

Bakirtzis, Ch. and Mastora, P. 2016. Σχετίζονται τα ψηφιδωτά της Ροτόντας με τη μετατροπή της σε χριστιανικό ναό. *Προθήκη-δελτίον βελτιώσεως* 1, 26-34.

Bakirtzis, Ch. and Mastora, P. *Το μαυσωλείο του Μεγάλου Κωνσταντίνου στη Θεσσαλονίκη* (The mausoleum of Constantine the Great in Thessaloniki) (forthcoming).

Ćurčić, S. 2000. *Some Observations and Questions Regarding Early Christian Architecture in Thessaloniki*. Thessaloniki: Ephoreia of Byzantine Antiquities of Thessaloniki.

Eusebius of Caesarea, *Life of the Emperor Constantine*.

Fishwick, D. 1992. On the Temple of Divus 'Augustus'. *Phoenix* 46, 232–55.

Frances, R. 2003. When all that is Gold does not Glitter: On the strange history of looking at Byzantine Art, in A. Eastmond and L. James (eds) *Icon and World, The Power of Images in Byzantium, Studies presented to R. Cormack*, 13-23. Aldershot; Burlington: Ashgate.

Hébrard, E. 1920. Les travaux du Service Archéologique de l'Armée d'Orient à l'Arc de Triomphe 'de Galère' et à l'église Saint-Georges de Salonique. *Bulletin de Correspondance Hellenique* 44, 5-40.

Iliadis, I. 2001. The natural lighting of the mosaics in the Rotunda at Thessaloniki. *Lighting Research and Technology* 33.1, 13-24.

Iliadis, I. 2005. Το φυσικό φως και τα ψηφιδωτά του τρούλου της Ροτόντας στη Θεσσαλονίκη, in Ch. Bakirtzis (ed.) *Proceedings of the VIIIth Conference of the International Committee for the Conservation of Mosaics (ICCM), Wall and Floor Mosaics: Conservation, Maintenance, Preservation), Thessaloniki, 29 October - 3 November 2002*, 467-478. Thessaloniki: European Centre of Byzantine and Post-Byzantine Studies, 9th Ephoreia of Byzantine Antiquities.

Johnson, M. J. 2009. *The Roman Imperial Mausoleum in Late Antiquity*. New York: Cambridge University Press.

Knauer, E. R. 1993. Roman Wall Paintings from Boscotrecase: Three Studies in the Relationship between Writing and Painting. *Metropolitan Museum Journal* 28, 13-46.

Mastora, P. and Motsianos, I. 2011. Concerning the material, shape and functioning of lighting devices, prompted by depictions on monuments from Late Antiquity, in I. Motsianos and E. Bintsi (eds) *Light on light, an illuminating*

[15] Eusebius of Caesarea, *Life of the Emperor Constantine* IV, 22.

[16] Eusebius of Caesarea, *Life of the Emperor Constantine* I, 2.

Story, 129-138. Thessaloniki: Folklife & Ethnological Museum of Macedonia-Thrace.

Nasrallah, L. 2005. Empire and Apocalypse in Thessaloniki: Interpreting the Early Christian Rotunda, *Journal of Early Christian Studies* 13.4, 361-396.

Sotiriou, M. 1970-1972. Προβλήματα της εικονογραφίας του τρούλλου του ναού Αγ. Γεωργίου Θεσσαλονίκης. *Δελτίον της Χριστιανικής Αρχαιολογικής Εταιρείας* 5-6, 191-203.

Taylor, R. 2005. Roman Oscilla: An Assessment. *RES: Anthropology and Aesthetics* 48, 83-105.

Theocharidou, K. 1991-1992. Η Ροτόντα της Θεσσαλονίκης. Νέα στοιχεία και αποσαφηνίσεις με αφορμή τις αναστηλωτικές εργασίες. *Δελτίον της Χριστιανικής Αρχαιολογικής Εταιρείας* 16, 57-75.

Torp, H. 1963. *Mosaikkene I St. Georg-Rotunden I Thessaloniki. Et hovedverk i tidlig-bysantinsk kunst.* Oslo: Gyldendal.

Yacoub, M. 1995. *Splendeurs des Mosaïques de Tunisie.* [Tunis]: Editions de l'agence nationale du patrimoine.

Light and Lighting Devices in Wall Paintings of Byzantine Churches in Thessaloniki

Ioannis G. Iliades
Electrical and Lighting Engineer, Kavala, Greece

Abstract: In the wall paintings of Byzantine churches lighting devices such as candles, tapers, candelabra, candlestands, torches, oil lamps, etc. are depicted. This paper will try to gather information, as far as it is possible, about the types of lighting sources and the rendering of light and will also examine the devices from the artistic point of view to evaluate whether their virtual forms have any similarities with those which are found in churches nowadays. Focusing on the murals of Byzantine churches in Thessaloniki of *c.* 13th-14th centuries, which became known as the Palaeologan Renaissance, we will also attempt to make, in some cases, comparative lighting device analyses of the light and devices depicted in paintings of the same or later period. The examination will include the lighting devices which are depicted not only in the wall paintings in the Byzantine churches at Thessaloniki but also those in the Holy Monastery of Timios Prodromos near Serres.

Keywords: Greece, Thessaloniki, Serres, Byzantine churches, wall paintings, lighting, lighting devices.

Thessaloniki[1]

The Church of the Holy Apostles

The church dates to the beginning of the 14th century, a period considered to be the 'golden era' of Thessaloniki.[2] The monastery was founded by Patriarch Niphon I and was dedicated to the Theotokos.

Above the south entrance of the ambulatory to the naos is portrayed the Tabernacle[3] (Figure 1). The artist painted the Tabernacle as described in chapter 4:2 of the Book of Zechariah. In the text there is a reference to seven oil lamps.[4]

This wall painting depicts a candlestand with seven tapers, instead of seven oil lamps. Though the bases of the candles are circular, it is depicted in a line. The central candle is larger, it looks like a taper and it is depicted in yellow color. The rest of them are white colored and thinner.

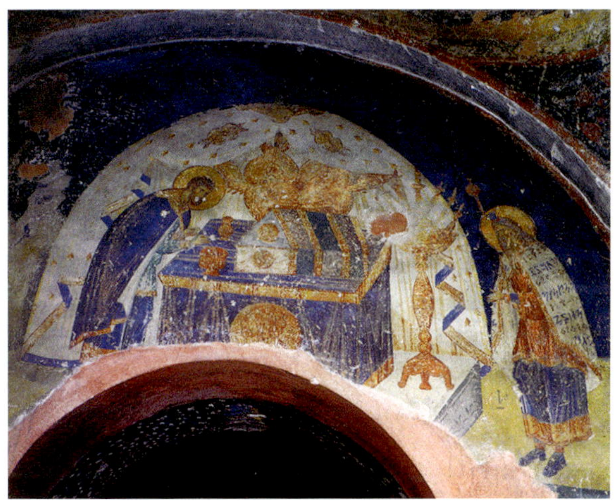

Figure 1. The Church of the Holy Apostles in Thessaloniki. Depiction of a candlestick with seven tapers in the Tabernacle.

As concerns the rendering of light:

1. It seems that the light source is located outside, opposite and above the Tabernacle.
2. The scene is very bright and white light fills the entire space.
3. There is a sense of brightness around the candlestand.

The artist makes innovations, i.e., he uses candles instead of oil lamps.

The Church of Christ the Savior

The small Byzantine Church of Christ the Savior, dates to the 14th century,[5] and stands on modern Egnatia Street in Thessaloniki, very close to the Arch of Galerius and Rotunda.

The church is an aisled, tetraconch, cross–in–square structure with a tall dome and eight arched windows (Figure 2). We will focus on those scenes which have depictions of lighting devices, such as the scenes of the Little Entrance and Reading the Gospel Passage.[6]

* This paper was initially published in http://www.archaeology.wiki/blog/2015/10/26/light-lighting-devices-wall-paintings-byzantine-churches-thessaloniki-2/.

[1] I am most grateful to the Ephorate of Antiquities of Thessaloniki City (ex 9th Ephorate of Byzantine Antiquities of Thessaloniki) for granting me permission to carry out the research in the interior of the churches in Thesssaloniki. I would also like to thank the Archimandrite Nectario Triantafyllidi and the Priest Eythimio Mertzani for their help of my research in the church of Hagia Aikaterini.

[2] Nikonanos 1998, 10; Stepfan 1986, 4; Xyngopoulos 1953, 5-6.

[3] Stephan 1986, 132-135.

[4] He asked me, 'What do you see?' I answered, 'I see a solid gold lampstand with a bowl at the top and seven lamps on it, with seven channels to the lamps (New International Version).

[5] Kourkoutidou-Nikolaidou 2008, 36-37; Xyngopoulos 1952, 67-75.

[6] Kourkoutidou-Nikolaidou 2008, 90-98; Iliades 2013, 32-45.

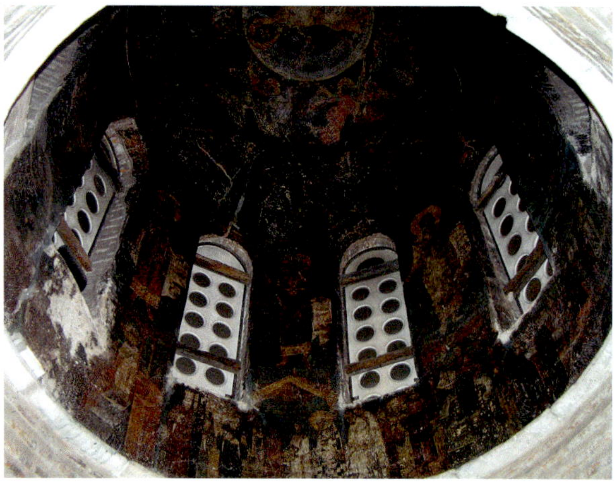

Figure 2. The Church of Christ the Savior. The dome looking west.

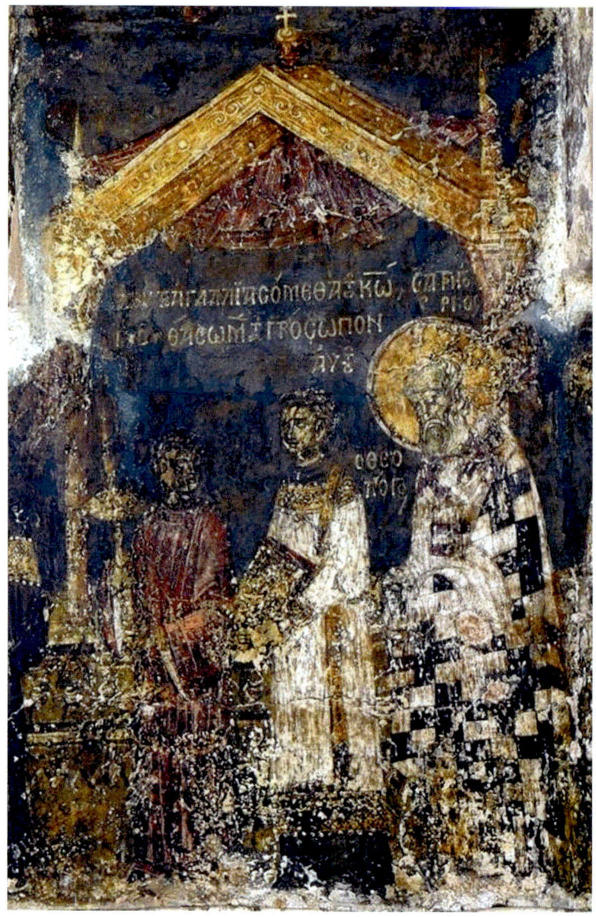

Figure 3. The scene of the Little Entrance.

1. The scene of the Little Entrance

In front of the Royal Door stands the deacon with a Gospel book, a youth holding a candlestick, and a hierarch (Figure 3) who, according to the inscription, is Saint Gregory the Theologian.[7]

The candlestick has a golden yellow color, the color of bronze, and the flame of the candle is reddish. This shape of candlestick is still found in churches to this day (Figure 4). The painter depicted the faces and objects in the scene with different intensities of illumination. He rendered more severely the face of the Saint, the Gospel with a golden color and their old vestments with a golden yellow light. The propylon of the sanctuary scene is illuminated with a golden yellow light. The same color is found in the halo of the hierarch.

According to our study of the natural light the illumination source is opposite and outside the scene. Probably the painter depicted the morning sunlight that enters through the window of the dome which is opposite the scene.

2. The scene of Reading the Gospel Passage

On the ambo stands a deacon reading the Gospel and holding a candlestick with a light blue flame (Figures 5, 6).

In front of him two youths, one wears a dark blue garment and holds a candlestick, the flame of which is light red-orange and the other youth holds an open Gospel book in one hand and a candlestick in the other, with the same color flame. The base of the candlestick has three small supporting points and probably it is an altar candlestick.[8] Their faces are illuminated by candlelight. The Gospel and the vestments of the deacon are illuminated by whitish light (Figure 6).

The display of candlesticks in the Divine Liturgy and the efficiency of light contain realistic elements (Figure 7). The

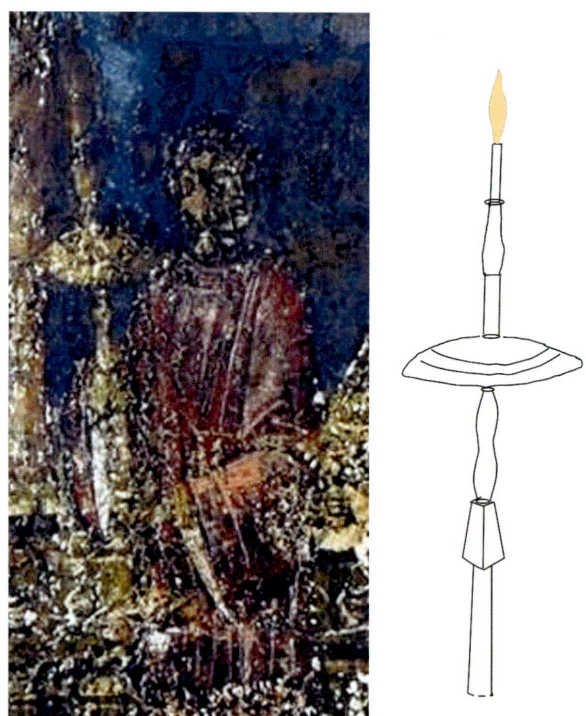

Figure 4. The scene of the Little Entrance. Detail of a youth holding a candlestick and a drawing of the candlestick.

[7] Kourkoutidou-Nikolaidou 2008, 90, fig. 91.
[8] Augustin 1907, entry 'Altar candlesticks'.

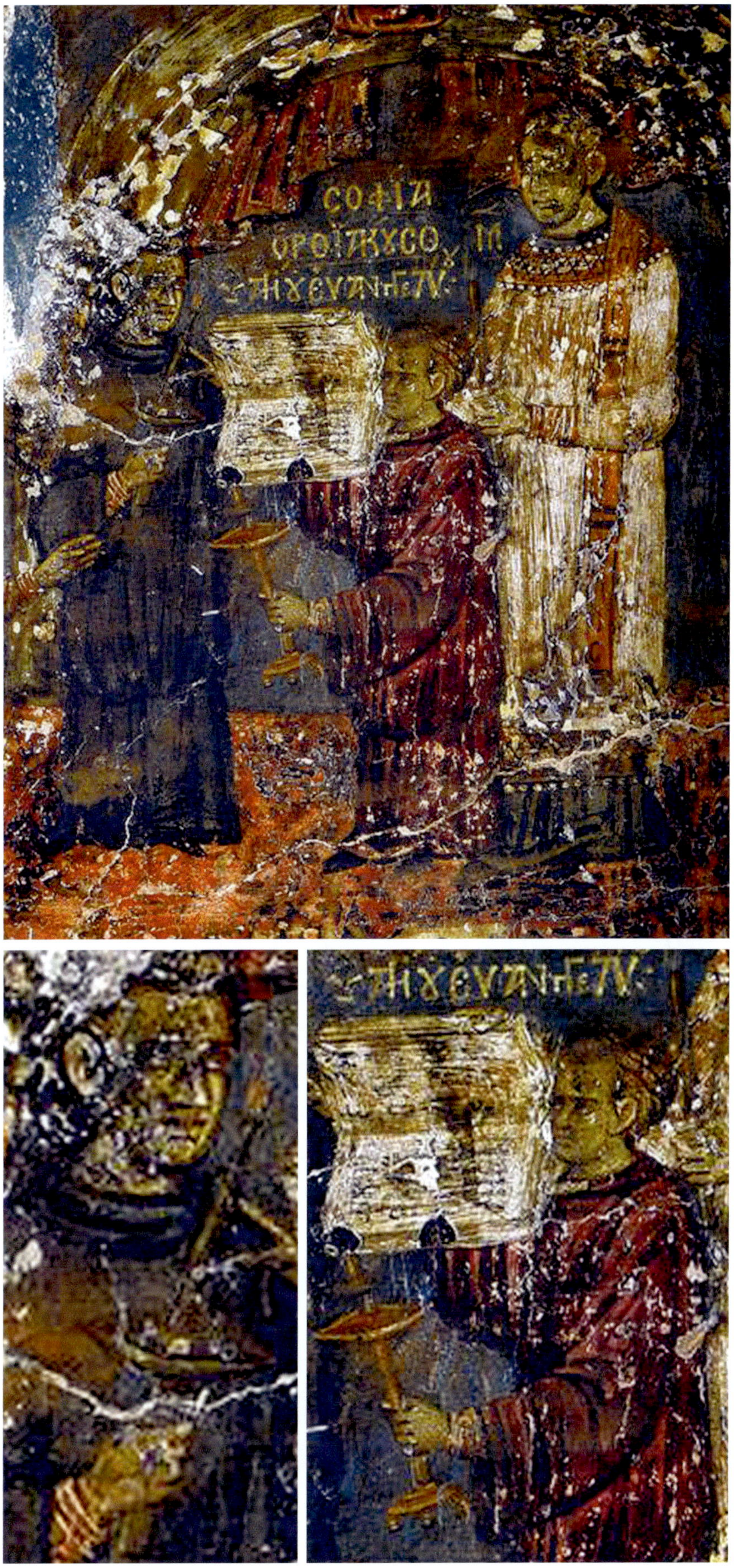

Figure 5. The scene of Reading the Gospel Passage, and details.

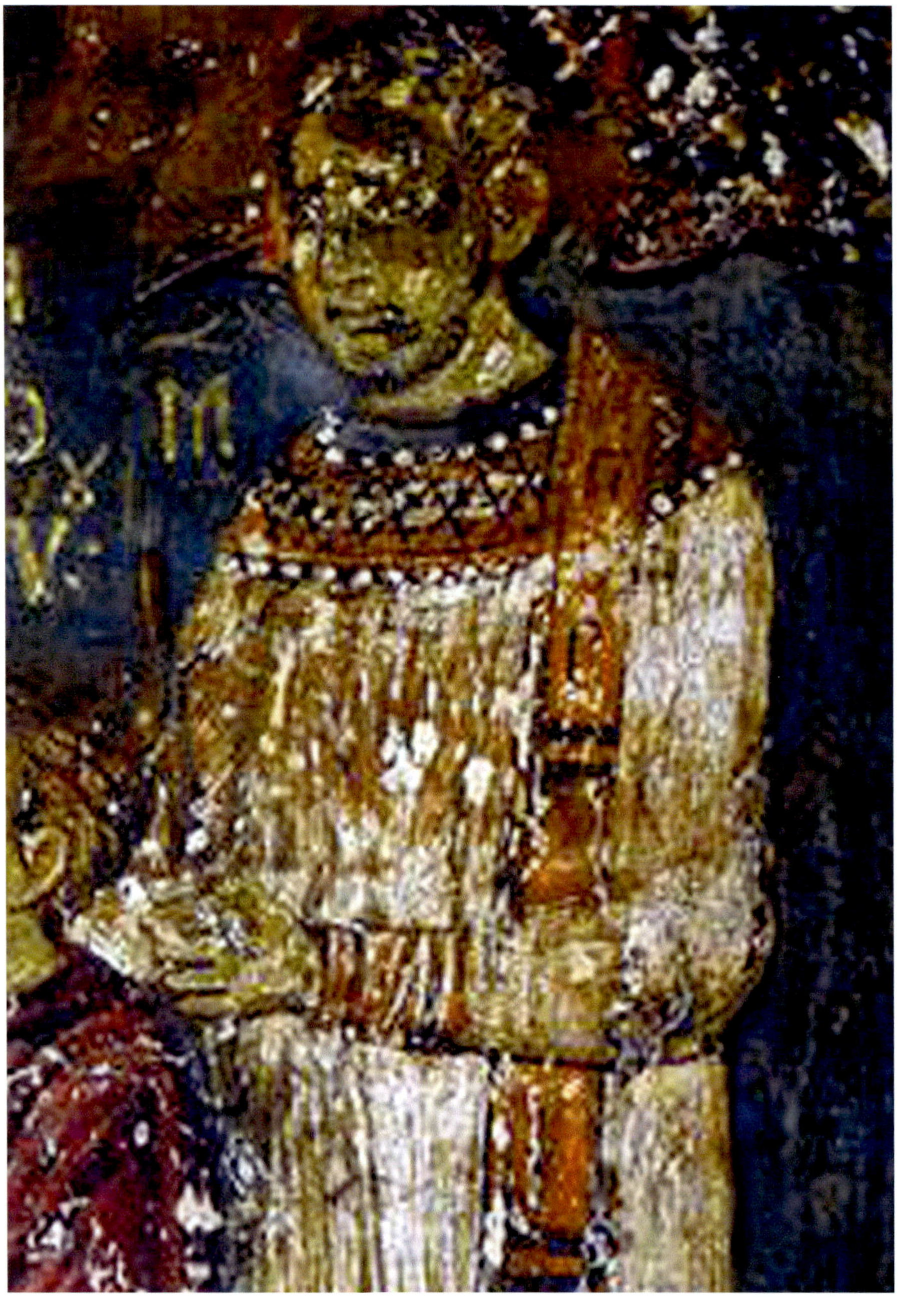

Figure 6. Details from the scene of Reading the Gospel Passage. Two youths and deacon holding candlesticks.

painter depicted lighting devices which were apparently in use at that era.

The painter used the light blue color for the flame of the candles, probably because he wanted to render the Holy Light and the spirituality of the scene.[9] Panselinou in *The Byzantine Art* contributes significantly to the understanding of this subject.[10]

The final scene of the Divine Liturgy and beneath it the prophet Elijah are depicted on a high vaulted architectural structure.[11] The light in the vaulted architectural structure can be observed on three orientations as to the incoming viewer/visitor: upper left, frontal, upper right.

[9] The PhD dissertation of L. James (James 1996, 4-17) opened new horizons on the subject of manipulation of light in Byzantine Art. N. Panselinou in fig. 134 published the miniature of the Transfiguration of Christ according to the Theological Treatises of Emperor John VI Kantacuzenos dating between the years 1347 to 1355 (In Codex Paris gr. 1242, fol. 92b). In this scene the light is rendered by different intensities of light blue color.

[10] Panselinou 2010, 267-279.

[11] Kourkoutidou-Nikolaidou 2008, 97-99, fig. 96.

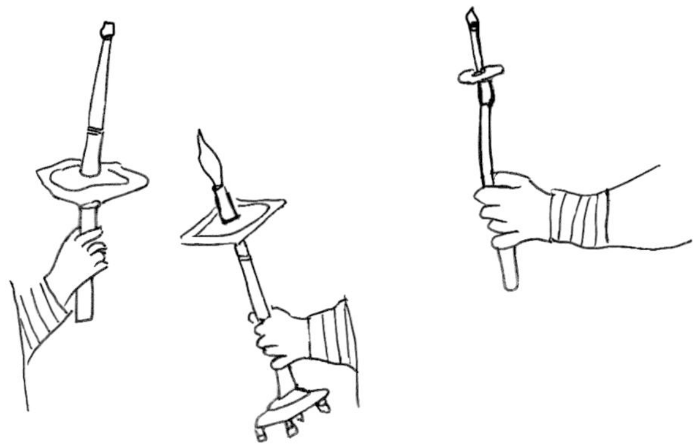

Figure 7. Drawing of candlesticks from the scene of Reading the Gospel Passage.

The Church of Hagios Nikolaos Orphanos

The small church of Hagios Nikolaos Orphanos was the catholicon of a monastery, which was founded in an uninhabited area very close to the eastern walls.[12] The church has been dated to the second decade of the 14th century and is a single aisle basilica with U-shaped ambulatory along the three sides and a triconch sanctuary at the east. It is actually a metochion (dependence) of the Vlatades Monastery and it functioned as a church even during the Ottoman era.

In the interior there are remarkable murals reflecting the spirit of the Renaissance in arts and letters in Byzantium during the Palaeologan era in Thessaloniki. These have has been associated with the workshop of the renowned Thessalonian artist Kalliergis.[13]

The scene of the Betrayal by Judas

The scene of the Betrayal by Judas provides information about the lighting devices which are depicted in it.[14] A lantern is shown with a candle inside it. At the back, left to the viewer, there are two torches with an intense red light and on the right, a guard holds two torches together and another guard holds a torch with a red and golden yellow flame.

The scene of the Betrayal is depicted in several wall paintings in Byzantine churches. For example:

a. In the Monastery of Saint Nikolaos Anapafsas in Meterora,[15] are depicted two torches and a third torch-lantern with a conical roof, which has two small doors in front of it. The format looks like the torchlight of the scene from Hagios Nikolaos Orphanos. The flame of torches, intense red, has a different color from the third torch (golden yellow).

b. In the Monastery of Philanthropinon on the island of Ioannina[16] and in the scene of Betrayal (1531/32) on the wall painting in the katholikon two torches are depicted with intense red colored flames and a third depiction shows a red flame in a form of a lantern.

c. The Monastery of Stavronikita. The Katholicon is decorated with frescoes by the famous painter Theophanes the Cretan and his son Symeon. In the Betrayal fresco (16th century) are depicted three torches with intense red flame and a lantern with two small doors on its front side. In the background of the painting of Jesus Christ's Trial before Annas and Caiaphas, a lantern is depicted hung from the arm of a small iron mast. The mast is supported on the surrounding building wall (enclosure). It seems that it is not shedding light from its lateral openings. If we closely examine the background around the lantern, we may observe that it looks brighter. This is caused either by the light of the lantern or the technique of *sfumato*[17] the artist used.

The Church of Hagia Aikaterini

The northwest part of Thessaloniki near the Byzantine wall is adorned by the elegant Church of Hagia Aikatereni.[18] We do not know its original name, nor when it was founded. Features of the architecture suggest the late 13th or early 14th century. The church is a complex tetrastyle cross-in-square with five domes and an ambulatory terminating in two chapels at the east end, a type which was very common in Palaeologan Thessaloniki.

The fragmentarily preserved frescoes were discovered during restoration work in 1947-1951.[19] Co-officiating hierarchs and the Communion of the Apostles survive in the sanctuary, scenes from the life of Christ in the naos, and angels and prophets adorn the dome. The choice of subjects suggests that the church belonged to a monastery and was dedicated to Christ.

On the west wall of the nave there is a mural painting depicting the Healing of the Paralytic, a scene that provides us with a source of information about light, the portrayal of light and lighting conditions. The miracle takes place at Bethesda[20] in front of a colonnaded building with a five-domed roof. On the front of the colonnade and above the capitals, the artist has painted five lanterns (Figure 8). These are obviously fixtures designed for exterior lighting and more precisely as lights for illuminating streets.[21]

[12] Kirchhainer 2001; Mavropoulou-Tsioumi 1992, 137-144; Tsitouridou 1986, 33-45; Xyngopoulos 1964/65, 90-98; Kourkoutidou-Nikolaidou and Tourta 1997,73-85.

[13] Bakirtzis *et al.* 2003, 71-73.

[14] In the scene of the Betrayal by Judas the natural landscape is conveyed with color variations, giving the impression of shadow and light. In the dim background the metal of weapons gleams in the moonlight and in the light of the lighting devices.

[15] Sofianos and Tsigaridas 2003, 205.

[16] Acheimastou-Potamianou 1994, 284, fig, 174.

[17] Sfumato is a painting technique in which the colors blend softly into each other, rather than objects or shapes having sharp outlines or hard edges; Gombrich 1998, 303;

[18] Milet 1916, 182-183; Diehl *et al.* 1918, 181-183; Pelekanidis 1951, 1-88, 24; Vocotopoulos 1982, 98-99; Hadzitryphonos 1991, 424-425.

[19] Hadzitryphonos 2004, 270-274.

[20] Kazamia-Tsernou 1992, 199.

[21] Iliades 2011, 139-141, figs 3-5. The depiction of lanterns in the murals is a rare or perhaps unique event. The detailed depiction of the lanterns lead us to the conclusion that at the time of the program creating the murals in the church, there were similar lights in the public areas. Most attractive is the view that probably the same shape of lanterns were to be found on the streets of Thessaloniki and particularly in the neighborhood of Hagia Aikaterini.

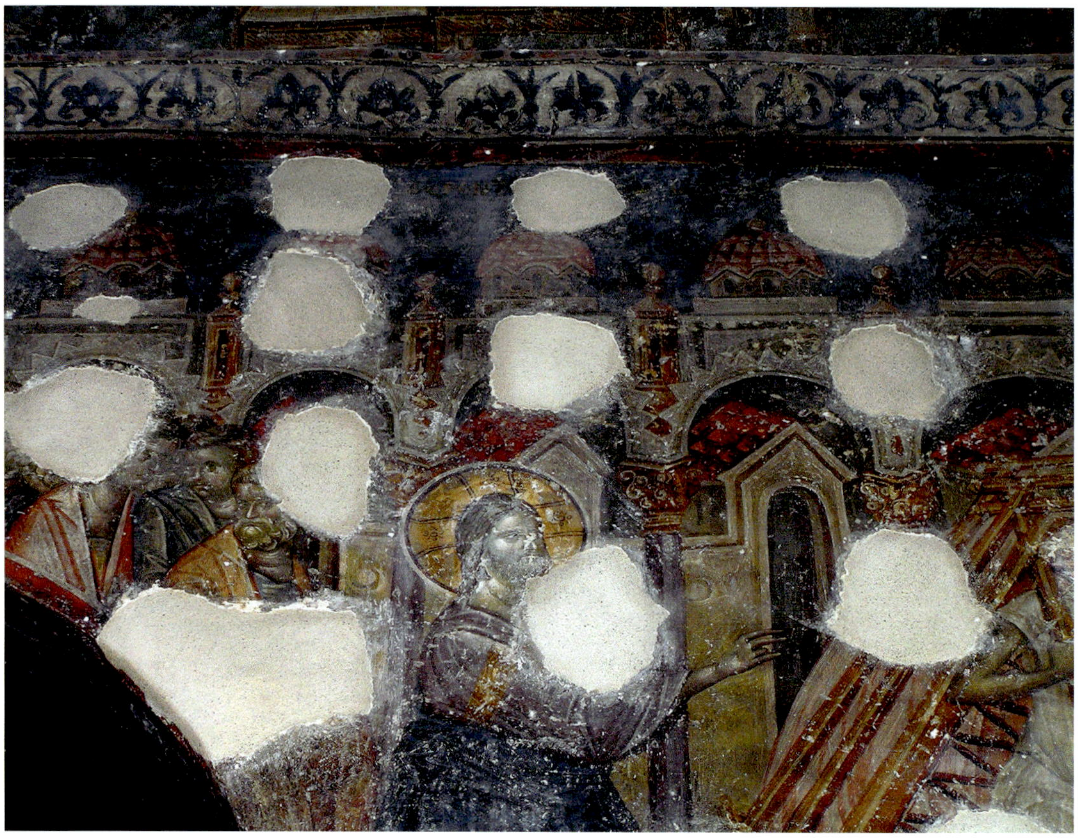

Figure 8. The church of Hagia Aikaterini in Thessaloniki. The Healing of the Paralytic.

Looking closely at the first lantern, on the viewer's left, it is possible to distinguish the candle, which is supported on a base of sheet metal in a zigzag shape (Figure 9). The base of the lantern may be made of sheet copper. The candle appears to be burning, but it is not shedding light.

This detail enables us on the one hand to assess the conditions in which the miracle is taking place and, on the other, to ascertain certain features of these lights, such as their form, shape, size, kind of light source, etc.

We make two proposals concerning the iconographical form of the lanterns,[22] because the mural is not in good condition.

The first proposal:

The main body of the lantern is cuboid, with an opening in the middle of the front side and another one on the side of the wall of the colonnade (Figure 10).

The second proposal:

According to the notes of the painter and mosaic artist Kamaraki the shape of the third lantern seems to be hexagonal (Figure 11).

The upper part of the lantern is shielded with a tapering lid that rises in the center to a ring supporting a globe. At the base of the ring there may be holes to vent the smoke and heat.

The second lamp is depicted in the same way, while the third is not lit. The lamp base is fixed on the wall by an angled bracket, which keeps the lantern away from and above the point of attachment.

It's quite interesting how the painter portrayed the base of each lantern on the wall. Each of the bases has a different angle. In this way the artist wanted to express the change of position of viewer. This is, then, a simple, realistic, narrative rendering of a historic event. The painter is concerned with a historic detail, but not at a specified time. As regards the lighting conditions, it is not possible to extract any further information from the scene. What conditions were depicted in the scene of the miracle? Is it in the light we perceive or the light apprehended by the artist in expressing a transcendent world?

Serres

The Monastery of Timios Prodromos near Serres

The Monastery of Timios Prodromos was founded by the monk Ioannikio in about 1270 in a deep ravine of Mountain Menoikion and in distance 12 km northeast of the city of

[22] Iliades 2011, 139-141.

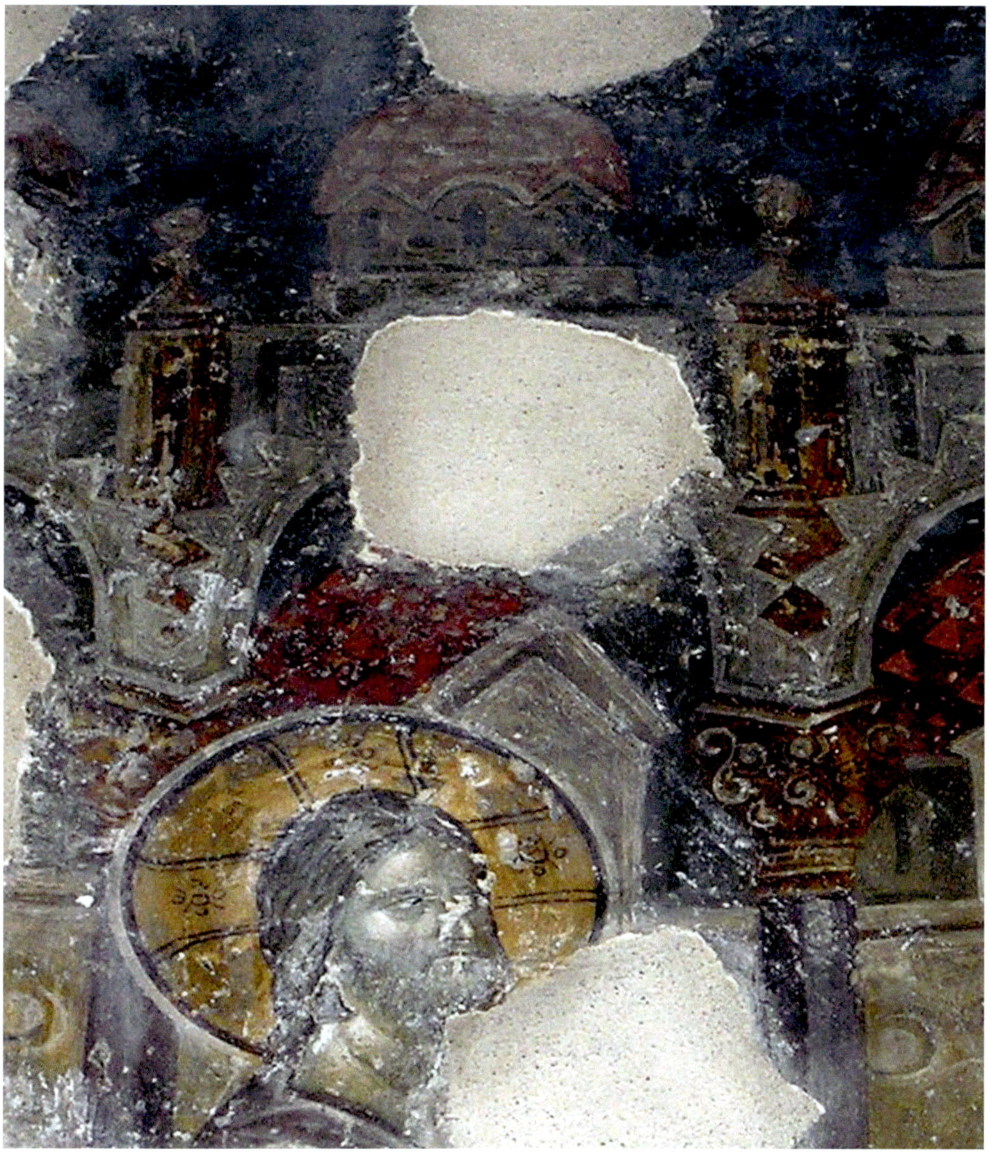

Figure 9. The Healing of the Paralytic. The detail of the lanterns.

Serres.[23] The foundation of the church dates to the beginning of the 14th century. In its present form it consists of an exonarthex, the Enate, the Mesonyktiko, the main church and the Holy Altar. Along the west side of the katholikon extends the exonarthex, a structure dated to 1849.[24] The wall paintings were done in 1852 by the painter Stergios Nevrokopites. On the west of the exonarthex candlestands are depicted in two scenes, Chapter 1 of the Apocalypse and Chapter 4 of the Apocalypse (Figure 12).

The painter depicted the flame of the candles on the candlestands with light blue color. The forms of lighting devices are similar to those found in the katholikon, where on their base is engraved the date 1870 (Figure 13).

[23] Strati 1989, 12-23; Xyngopoulos 1973, 1- 8.

[24] Strati 2007, 145-159.

Conclusions

Concluding I will focus on two issues:

As their references, the painters had standard lighting devices that were either depicted in earlier murals or used by priests in churches during the period of the iconography. Even today we see in churches, lighting devices similar to those of the murals.

It is quite interesting how the painter utilized color for the flames of the lighting devices. Of course, this has to do with the artistic movement that each artist belonged to. Depicting the flame of the candles with an intense red, the artist wants to create more tension on the issue. However, in some cases, the color of the flame is the same as that which is actually emitted by the lighting devices. But when the artist depicts the flame with a blue or cyan color, he wants to declare truth,

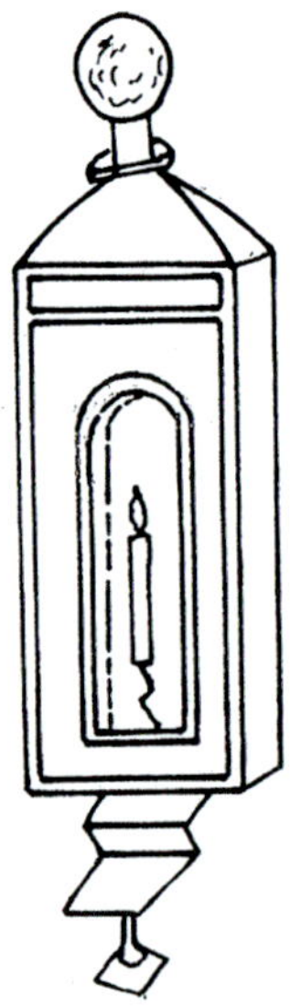

Figure 10. The lantern in the mural. The first proposal.

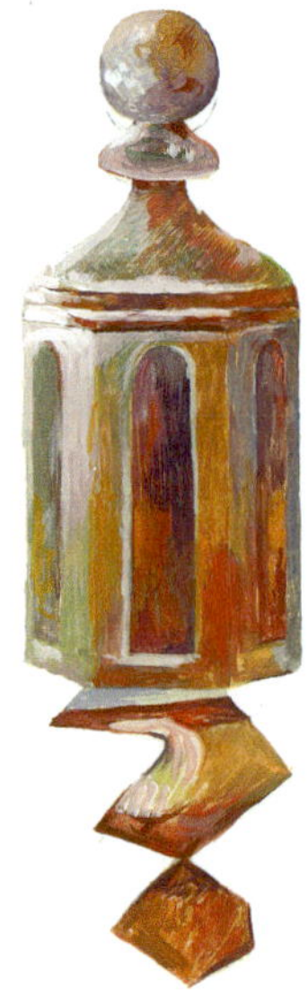

Figure 11. The lantern in the mural. The second proposal (Drawing by Dimitra Kamaraki).

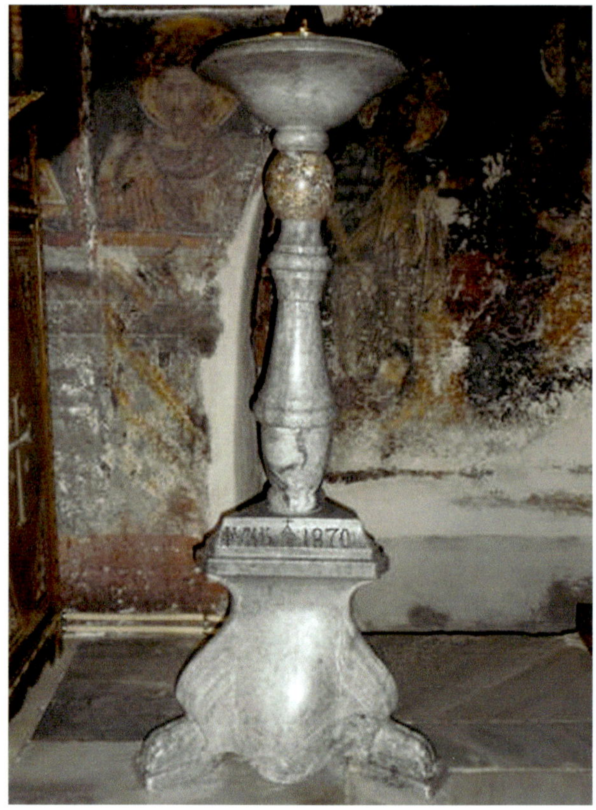

Figure 13. The Monastery of Timios Prodromos. Enate. The candelabrum.

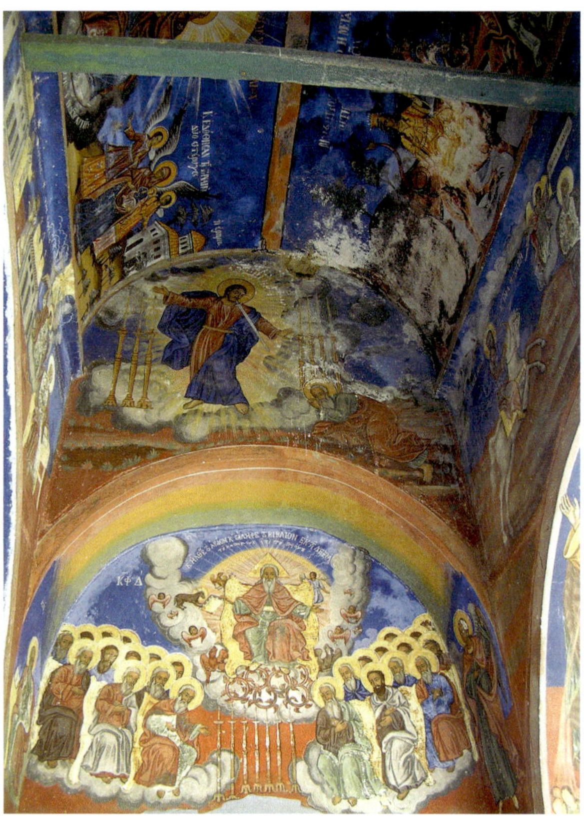

Figure 12. The Monastery of Timios Prodromos near Serres. Exonarthex. The depiction of the candelabra.

harmony, tranquility, celestial consciousness and spirituality. Perhaps it symbolizes the 'Uncreated Light', the light that was revealed on Mount Tabor at the Transfiguration of Jesus.

Bibliography

Acheimastou-Potamianou, M. 1994. *Byzantine wall paintings*, Athens, Ekdotike Athenon.

Acheimastou-Potamianou, M. 1994. The *Murals of the Philanthropinon Monastery in the Island of Ioannina*. Athens: Adam-Pergamos.

Augustin, J. 1907. The Catholic Encyclopedia, vol. 1, New York.

Bakirzis, Ch. 1996. *Byzantine monasteries in Eastern Macedonia and Thrace (Synaxis, Mt. Papikion, St. John Prodromos Monastery), Mount Athos and Byzantine Monasticism*. Birmingham: A. Bryer and M. Cunningham.

Bakirtzis, N. 2011. The architecture of the Synodikon Hall. The Monastic use of an eighteenth-century secular reception space, in *The Synodikon Hall at the Monastery of Timios Prodromos near Serres*, 13-17. Leukosia/Nicosia: Foundation Anastasios G. Leventis.

Barasch, M. 1987. *Cambridge studies in the History of Art, Giotto and the language of gesture*. New York: University of Cambridge.

Chatzidakis, M. 1985. Βυζαντινής Θεσσαλονίκης εγκώμιον. *Νέα Εστία* 118, issue 1403, 514-522.

Chatzidakis, M. 1986. *The Cretan Painter Theophanis. The Wall-Paintings of the Holy Monastery of Stavronikita*. Mount Athos: Holy Monastery of Stavronikita.

Diehl, C., Le Tourneau, M. and Saladin, H. 1918. *Les monuments chrétiens de Salonique*. Paris: E. Leroux.

Hadzitryphonos, E. 1991. Konstruktion und Baumaterial von Hagia Ekaterini in Thessaloniki, in *Resumes des Communications XVIIIe Congres International des etudes Byzantines, 8-15/8/1991*, 424-425. Moscou.

Hadzitryphonos, E. 2004. *Το περίστωο στην υστεροβυζαντινή εκκλησιαστική αρχιτεκτονική. Σχεδιασμός και Λειτουργία*. Thessaloniki: European Centre for Byzantine and Post-Byzantine Monuments.

Iliades, I. G. 2003. Light and Climatic Conditions, in Ch. Bakirtzis (ed.) *Ayios Nikolaos Orphanos-The Wall Paintings*, 134-139. Athens: Akritas.

Iliades, I. G. 2011. Exterior lighting devices in the wall painting of Hagia Aikaterini, in I. Motsianos and E. Bintsi, *Light on light, an illuminating story, Exhibition Catalogue*, 139-143. Thessaloniki, Folklife & Ethnological Museum of Macedonia-Thrace.

Iliadis, I. G. 2013. The depiction of light icons at the church of the Christ the Savior. *Iconostasion, Bulletin of Iconology Center of Chrysorroyiatissa Holy Monastery* 4, 35-45.

James, L. 1996. *Light and Colour in Byzantine Art*. Oxford: Clarendon Press.

Kazamia-Tsernou, M. 1992. The healing of the paralytic the paleochristian and byzantine iconography. Unpublished PhD dissertation, Aristotle University of Thessaloniki.

Kirchhainer, K. 2001. *Die Bildausstattung der Nikolauskirche in Thessaloniki: Untersuchungen zu Struktur und Programm der Malereien*, Marburger Studien zu Kunst und Kulturgeschichte 3. Weimar: Weimar VDG.

Kourkoutidou-Nikolaidou, E. 2008. *The church of the Christ the Saviour*. Athens: Kapon.

Kourkoutidou-Nikolaidou, E. and Tourta, A. 1997. *Wandering in Byzantine Thessaloniki*. Athens: Kapon.

Nikonanos, N. 1998. *Άγιοι Απόστολοι Θεσσαλονίκης*. Thessaloniki: Institute for Balkan Studies.

Mavropoulou-Tsioumi, Chr. 1992. *Ο Άγιος Νικόλαος Ορφανός*. Thessaloniki.

Milet, G. 1916. *L'École grécque dans l'architecture byzantine*. Paris: E. Leroux.

Panselinou, N. 2010. *Βυζαντινή ζωγραφική: η βυζαντινή κοινωνία και οι εικόνες της*. Αθήνα : Καστανιώτης.

Rautman, M. L. 1984. The Church of the Holy Apostles in Thessaloniki: a Study in Early Palaeologan Architecture. Unpublished PhD dissertation, Indiana University.

Stephan, Chr. 1986. *Die Mosaiken und Fresken der Apostelkirche zu Thessaloniki*. Baden-Baden: Wernersche Verlagsgesellschaft mbH, Worms.

Strati, A. 1989. *The Monastery of Timios Prodromos at Serres*. Athens: Achaeological Receipts Fund.

Strati, A. 2007. *Η ζωγραφική στην Ιερά Μονή Τιμίου Προδρόμου Σερρών (14ος-19ος αι.)*. Thessaloniki: Mygdonia.

Sofianos, Z. D. and Tsigaridas, N. E. 2003. *Ιερά Μονή Αγίου Νικολάου Αναπαυσά. Ιστορία-Τέχνη / The Holy Monastery of St. Nikolaos Anapafsas, History-Art*. Trikala: Genesis.

Tambaki, S. 1998. *Η Θεσσαλονίκη στις περιγραφές των περιηγητών 12ος-19ος αι. μ.Χ.* Θεσσαλονίκη: Κέντρο Βυζαντινών Ερευνών.

Tsigaridas, N. E. 2008. *Οι τοιχογραφίες του παρεκκλησίου του Αγίου Ευθυμίου (1302/3) στον ναό του Αγίου Δημητρίου*. Thessaloniki: Pournaras.

Tsigaridas, N. E. 2008. *Manuel Panselinos from the Holy church of the Protaton*. Thessaloniki: Ageioritiki Estia.

Tsitouridou, A. 1986. *Ο ζωγραφικός Διάκοσμος του Αγίου Νικολάου Ορφανού στη Θεσσαλονίκη: Συμβολή στη μελέτη της Παλαιολόγειας ζωγραφικής κατά τον πρώιμο 14ο αιώνα*. Θεσσαλονίκη: Κέντρο Βυζαντινών Ερευνών.

Velenis, G. 1981. Οι Άγιοι Απόστολοι της Θεσσαλονίκης και η Σχολή της Κωνσταντινούπολης. *Jahrbuch der Österreichischen Byzantinistik* 32/4, 457-467.

A Beacon from the Castle of Kitros, Pieria, Greece

Efterpi Marki
Archaeologist, Honorary Director of Antiquities

Abstract: This paper describes a special construction found at the medieval castle of Kitros which could be interpreted as a *φρυκτωρία* (beacon), i.e., a system of direct communication based on the use of fire which allowed for the rapid transfer of messages over long distances. The beacon in the castle of Kitros was created by the Franks in the early 13th century after they had captured Kitros in 1204 following a siege.

Keywords: Beacon, *φρυκτωρία*, Kitros, Franks, light signals.

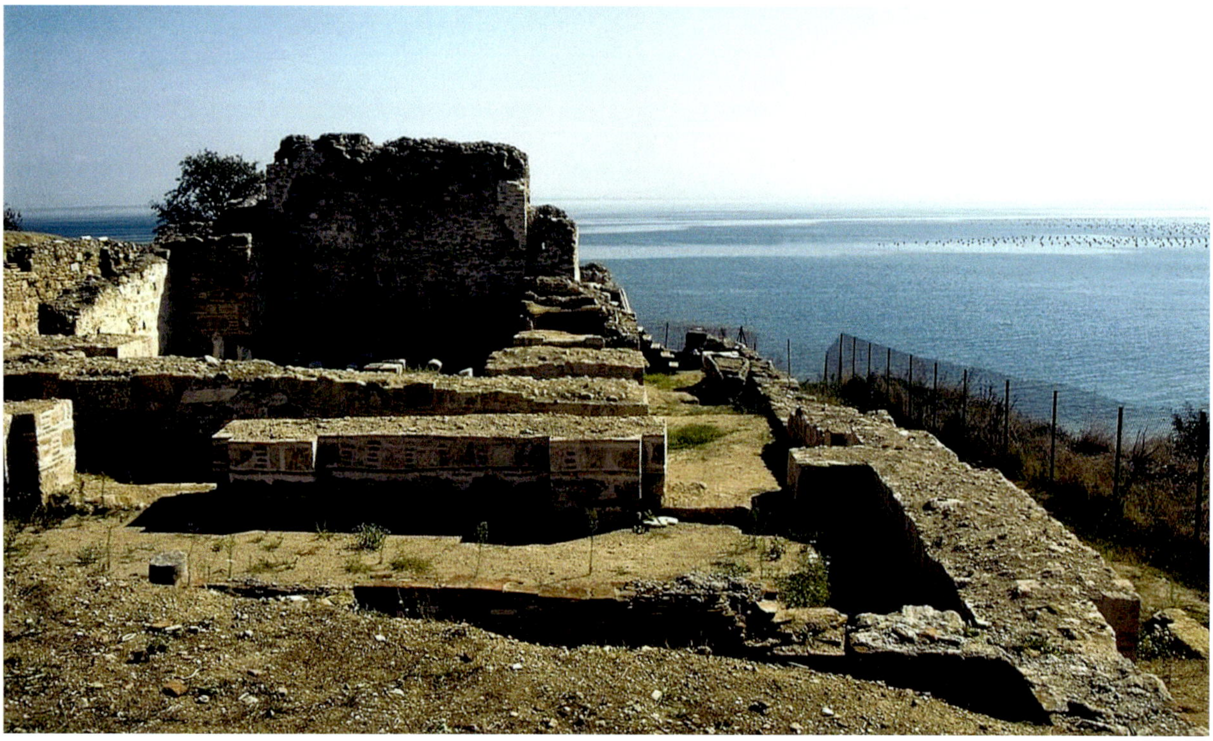

Figure 1. Kitros: View of the cathedral converted into a camp.

The word *φρυκτωρία* (beacon),[1] denotes the signals by beacons or alarm-fires and is related to a so-far-unique in Macedonia and continental Greece *φρυκτώριον* (a beacon-tower) of Kitros. It was excavated in 1983 on the ruins of the a Byzantine castle (Figure 1),[2] founded in the 6th century on the west coast of the Thermaikos Gulf, at the site of the Acropolis of ancient Pydna, one kilometer south of the village of Makrygialos in the Prefecture of Pieria. Kitros, the principal Byzantine town in Pieria and an important harbor, enjoyed an economic heyday in the 12th century as seat of the homonymous *katepanikion* (Catepanate, i.e., an administrative division of the empire). Its prosperity was brought to an end by the *golden bull* or *chrysobull* of Alexios Angelos in 1199 AD, granting the Venetians the right to establish a trading station at Kitros.

Beacons were a system of direct communication based on the use of fire and allowed for the rapid transfer of messages over long distances. A message would be transmitted by lighting a torch inside a specially made construction that was situated upon a hill or mountaintop to gain greater visibility.[3]

It seems that the beacon in the castle of Kitros was created by the Franks in the early 13th century after they had captured Kitros in 1204, following a siege of the town.[4] The Franks built the beacon by re-using second-hand materials. It seems that they used this beacon construction for a special fire that could send light signals for the prompt warning to the surrounding forts and to Thessaloniki in the case of any raid. Also, they could warn the ships for safe passage.

[1] Liddell and Scott 1938, vol. 4, 582.
[2] Marki 2001.
[3] For the warning system in Byzantium, see Pattenden 1983.
[4] Marki 1988.

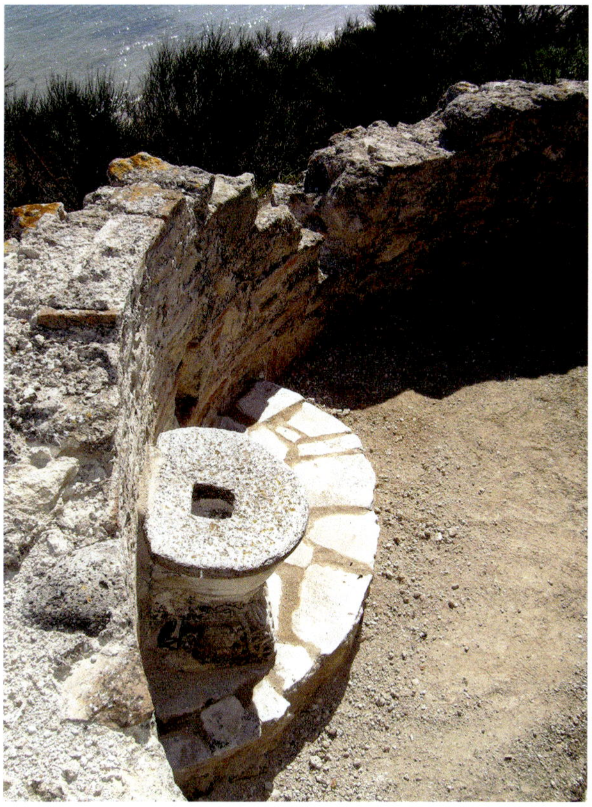

Figure 2. The beacon at the castle of Kitros.

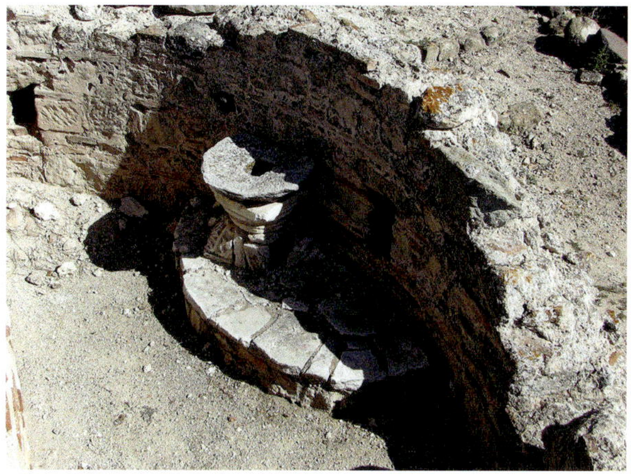

Figure 3. The beacon at the castle of Kitros.

The historical events associated with the construction of this particular beacon are the Frankish siege and subsequent fall of the Byzantine castle of Kitros in 1204.[5] Because of the resistance of the local people, the Franks, after the castle fell, burnt all the buildings inside and outside the castle, as well as the harbor facilities, and then turned the burned cathedral into a camp.[6] The Franks systematically fortified the ruined episcopal residence and converted it into a fortress-cum-citadel, which included an open walled courtyard to the west, passageways, a central assembly hall with a timber balcony, a tall, three-story keep, a kitchen, a cistern, an underground escape door and a beacon tower.[7] The camp was created by rebuilding over the ruins of the Byzantine cathedral and included a double gate that led, through a walled courtyard, to corridors housed under pointed arches. The corridors, which were the transformed church's corridor, led to the dining room (the central area of the church) and the knights' dormitories, and finally to a high donjon rising above the sanctuary's hall.[8] The three-story tower, where the prince of Kitros, Virich von Daum, lived, served as a donjon. It contained a storage room with a well on the ground floor, a private space with a fireplace on the first floor and a balcony based on a machicolation on the upper floor.

A perimeter in contact with the north side of the church-camp, a water tank, an underground escape gate and a kitchen can be added to the interventions made by the Franks.

The insecurity of the knights and the sense of being in hostile territory led them to the creation of the phrycotry, which was used to monitor the passage of enemy ships, to illuminate the high coast at night, and to inform neighboring fortresses and Thessaloniki about a possible raid. The dimensions of this structure, rectangular in plan view, are 45 cm x 54 cm and its height is 51 cm. It used as a cover the ruined arch of the burnt cathedral at the eastern end of which it was constructed. It rests on a pedestal having a height of 12 cm, on which were placed welded parts of two capitals and on top of them, two column bases (Figure 3), all perforated by aligned rectangular holes having dimensions 11 cm x 12 cm, in which thin sticks of wood, tinder or torchwood would be burnt. The parts of the columns used in the lower part of the beacon belong to a type of impost capital.[9] One of them, dating back to the third quarter of the 6th century, is adorned with acanthus leaves and a vase from which a sprout emerges. A tile found nearby would have covered the upper surface hole and probably was the medium used to control the fire and the smoke used to generate the light signals.

Bibliography

Kazhdan, A. P. (ed.) 1991. *The Oxford Dictionary of Byzantium*, Prepared at Dumbarton Oaks, Vols. 1-3. New York-Oxford: Oxford University Press.

Liddell, H. and Scott, R. 1938. *Μέγα Λεξικόν της Ελληνικής Γλώσσης*, Αθήναι 1938, vol. Δ΄.

Marki, E. 1988. Ανασκαφές βυζαντινής Πύδνας. *Το Αρχαιολογικό Έργο στη Μακεδονία και Θράκη* 2, 195-197 (in Greek with an English abstract).

Marki, E. 1990. Παρατηρήσεις στον οικισμό της αρχαίας Πύδνας, in *Μνήμη Δημήτρη Λαζαρίδη, Πρακτικά Συνεδρίου*, Καβάλα 1986, 45-50. Θεσσαλονίκη: Αρχαιολογικό Μουσείο Καβάλας.

Marki, E. 2001. *Κίτρος: Μια πόλη κάστρο της βυζαντινής περιφέρειας*. Θεσσαλονίκη: Τα τραμάκια.

Marki, E. 2005. Pieria 2004. *Το Αρχαιολογικό Έργο στη Μακεδονία και Θράκη* 18, 407-411 (in Greek with an English abstract).

Pattenden, P. 1983. The Byzantine Early Warning System. *Byzantion* 53, 258-299.

[5] Marki 1988.

[6] Marki 1990, 49.

[7] Marki 2004.

[8] Marki 2005.

[9] See Kazhdan 1991.

Lessons of Modernity from the Past: Some Amazing Parallels Drawn from Antique and Medieval Lychnological Economics

Laurent Chrzanovski
Doctoral School of History and Humanities, Lucian Blaga State University, Sibiu
Founder and Secretary-General, International Lychnological Association

Abstract: Rich civilizations without good fuel for lighting, long distance imports, fuel taxes and bureaucracy, fuel poors, monopolies, monopsonies as well as children working in industries. All these topics, most of them being considered as a part of our society and of nowadays hottest challenges in global macro-economics can be found in different ancient and medieval civilizations, in the Classical world and then in Medieval Europe but also in other continents. This article desires to draw some parallels, teaching us that, alas, we learn very little and our systems do often reproduce mistakes made up to two thousand years ago...

Keywords: Artificial light, ethno-archaeology, macro-economy, Ancient-Modern parallelism.

For this volume I decided not to write a 'classical' lamp paper, as in other researches prepared for publication (on the lamps found by the University of Warsaw at Novae (Svishtov, Bulgaria) and on the lamps of the Bouvier corpus, the second largest collection in Europe after Benaki's on Egyptian and Near Eastern lamps), which are full of lighting devices reflecting the transition from the Roman world to the Medieval and Byzantine periods.[1] Instead I have chosen to follow some ideas born out of visits to both the exhibitions mentioned in my introduction to this volume, in a sort of wink that the history of Ancient and Medieval lighting makes to some modern colossal mistakes of governance ...

As a matter of fact, the 'golden age' of the Roman Empire, with its successful globalization, extremely low taxation, smooth bureaucracy, existence of a real 'middle class' and complete integration of Romanized elites from all over the conquered provinces –which is exemplified by having a 'foreigner,' born in *Lugdunum* (Lyon), crowned as the third Emperor Claudius– has been but a short parenthesis in world history.

From the end of Augustus' reign, i.e., the first decade of the 1st century AD, to the death of Marcus Aurelius in AD 180, when the Empire reached its maximal extension, everything went smoothly: two languages, one law code, one measuring system, one single value, a 2.5% tax on inter-provincial trade, and a 30% tax on goods coming from outside the Empire. Within this framework, it is easy to understand how lamps made in the enormous factories located near the best clay banks (Tunisia, Asia Minor, Central Italy) could be sold in markets as far away as Gaul or the German provinces at the same price as locally-produced lamps.

This period finds, in several points, similarities with what we lived through in Europe at the beginning of the 2000's. Then, a decade after the end of the Cold War, a real perspective of building something named Europe, an existent middle class and the respect of an *ad minima* of trade fair-play as cast by the World Trade Organization, allowed a real and competitive globalization to develop, with its strengths and weaknesses, but which made people believe again that 'the dream machine' was on, similar to the era after the second World War.

What we have witnessed in less than 18 years –but whose roots were already well established before– exactly parallels what happened after the death of Marcus Aurelius.

A cohort of incompetent or corrupt rulers with a plethoric and out-of-control state administrations have plunged some of our societies into a constant yet not easy to see pauperization with an almost complete death of the middle class and with generations condemned to earn less than their parents for the first time in a century. Our small little western world is run, exactly like the Late Roman Empire, by less by 5% of the people and the population suffers a yearly increase of 'Byzantine' bureaucracy, constant increases in taxes and contributions, disrespect of international trade agreements, and sectorial kingdoms of monopolies and monopsonies ruling the economic world.

No doubt that if we turn our clocks back to antiquity, we could find some astonishing similitudes.

1. The disappearance of oil lamps in continental Europe as a consequence of fuel poverty

In cities like *Aventicum* (Avenches, Switzerland), as soon as the mid 2nd century, we see classical Roman disk lamps disappear to give place, in less than fifty years, to a panorama of wheel-made open lamps which would allow any kind of animal fat to be used, a direct result of the ever-rising costs of the liquid fuels (vegetal oils, available and used mostly in the Mediterranean area but also in the East).

This phenomenon can be seen in almost all the Transalpine area, with some exceptions where mold-made oil lamps were in use for at least two centuries longer. For example, their use continued in the cities of the Rhine Valley, which had a massive military presence, and implies there was a perfectly functional mechanism of constantly-supervised transport

[1] For the Bouvier collection, see Chrzanovski 2012.

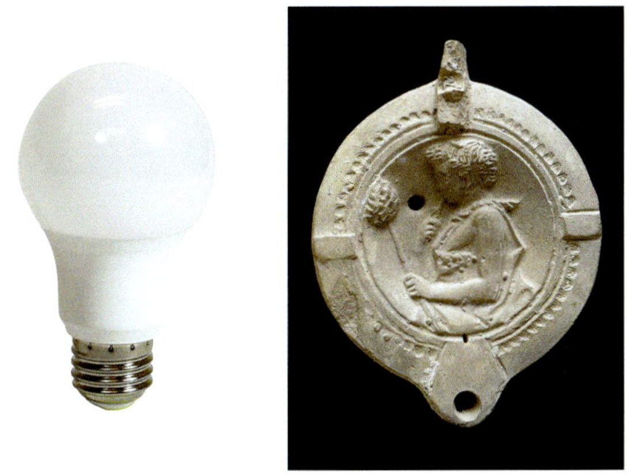

Figure 1. Today and Yesterday: a good lamp costs the equivalent of a one hour wage of the worse paid employees in western EU states. (Left: 75W LED bulb, versus right: a third-century lamp from Aegina, roughly contemporary with the Edict of Diocletian, ©Kelsey Museum).[1]

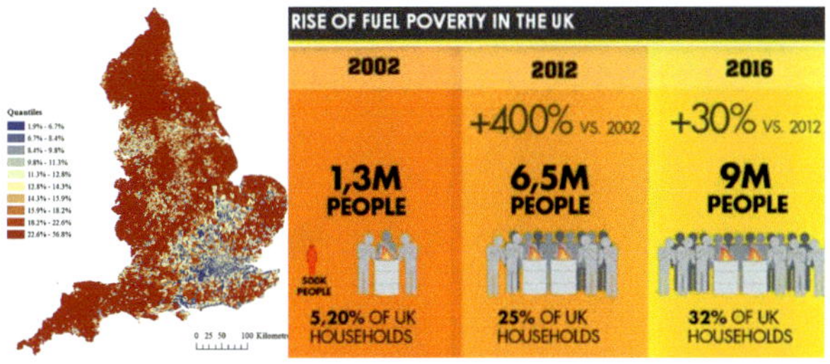

Figure 2. England, 2018: Relative percentage of fuel poor households using 10% indicator (© Robinson, Bouzarovsk, Lindley, 2018, fig. 3) and a poster showing the dramatic increase in the amount of poor in the last decades.

logistics in order to receive from the Mediterranean not only military supplies but also all that a 'standard' Roman life implied, including sufficient oil for lamps. [2]

As nowadays for gas or petrol, in some huge regions the total dependence on vital products which needed to be imported from far away generated poverty, especially if we keep in mind that in antiquity, most of the best fuels for lamps could also be considered as food stuffs (olive oil, pork or beef tallow, etc.).

Then as well as now, the lamp price itself was never a problem: according to Diocletian's edict on maximal prices, the worse paid of all employees, the *per diem* field laborer, could buy one lamp for each hour of work, while a scribe could buy a lamp for each four lines written. Today, it is exactly the same: as a French employee paid with the compulsory minimum wage (SMIC)[3] (7.83€ net/hour for 2018) can buy with a hour's wage a state-of-the-art 75W LED bulb produced by a well-known manufacturer.

Energy consumption costs are a huge problem, and today the issue appears even more dramatic than 'yesterday.' It seems as if more and more Europeans are living in 'third-century Roman Transalpine' conditions. The number of *fuel poor*, i.e., households spending more than 10% of their incomes to heat (and also to light[4]) their homes reached 32% of UK families (EU's most impacted country for this topic). Some households have to sacrifice up to 56.8% (!) of their incomes to afford to buy the needed energy.[5]

In many rural regions of Central Europe, we have observed exactly what happened in Continental Europe, starting with the 3rd century AD. As energy prices increased yearly, families gathered around the hearth which was used for heating and cooking as well as for lighting. Cheap and dangerous kerosene lanterns also reappeared massively. Lenin stated once that 'Communism is Socialism plus electrification' hence the electric grid is even more capillary in these countries than in Western Europe. Although electricity is available almost everywhere in Central Europe, in a number of houses it is now used only for two tools: the fridge and the smartphone plug.

In fact, if we look back in history, the very separation of the three uses of energy (heating, cooking, and lighting) is a precious indicator of the social wealth of the golden age of the Roman Empire, especially when contrasted with the pre-Roman period. This functional separation only slowly returned to Continental Europe during the 17th century, with the increased use of heating stoves, first present in the wealthiest houses, then progressively in other, mainly urban, ones.

In contrast in the Mediterranean area, blessed by nature as far as lighting supplies are concerned, oil lamps continued to be used, thanks to the abundant presence of non-combustible but well-burning oils such as olive oil, castor oil, linen oil, sesame oil, palm oil and many others.

There, only the progressive disappearance of molded lamps, to be replaced by 'good old' wheel-made ones (as in the Phoenician period!), is the main problematic. It corresponds almost certainly to a general diminution of the whole European population after the fall of the Western Roman Empire as well as to the sea and land insecurities in several regions of the Eastern Roman Empire. Manufacturers who maintained a staff of qualified personal needed to produce mold-made lamps and master the big oven's temperature had no more reason to exist as their activity was entirely dependent on the mid- and large-scale distribution of their products.

Without these innovating factories mastering the whole process and inventing new archetypes, even provincial factories, which would make molds by over-molding an existing imported lamp instead of using an archetype, quickly ceased to exist.

Slowly but surely, the Eastern Roman regions and then the whole Byzantine empire converted to what was then the most common 'Arabic' lamp, the 'double-saucer' type, where one

[2] 'Aegina; Lamp; Clay.', viewed 26 July 2018, <http://quod.lib.umich.edu/k/kelsey/ x-0000.00.0782/0000_00_0782. University of Michigan Library Digital Collections>.

[3] https://www.service-public.fr/particuliers/vosdroits/F2300

[4] Cf. Simcock *et al.* 2016.

[5] Robinson *et al.* 2018.

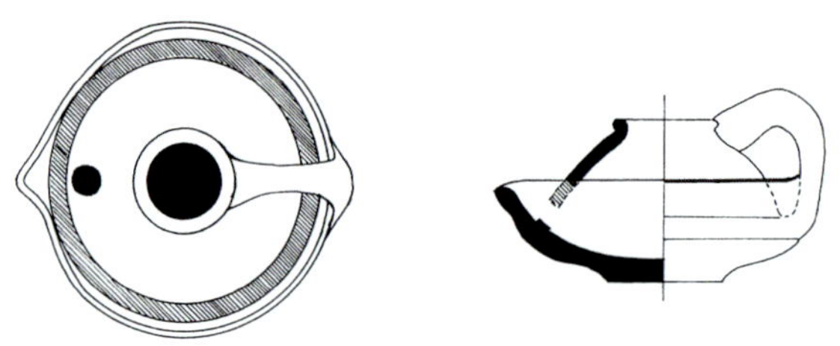

Figure 3. Kubiak type E in the Fustat lamps classification; drawing: Kubiak 1970, 9.

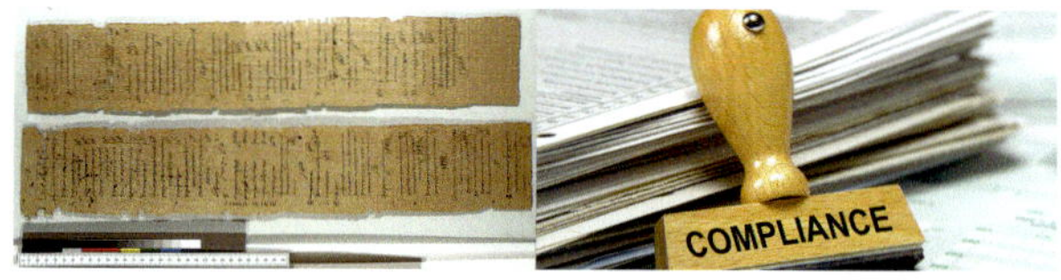

Figure 4. Signs of bureaucracy.

saucer acts as the base of the reservoir and the wick-spot, and the second saucer -large at first and then, through the centuries shrinking in size- acts to close the reservoir, holds the wick and allows the addition of a long handle. This type, born in Egypt and the Near East in the 9th century was soon produced in Asia Minor, in Cyprus -where the type was nicknamed 'crusader's lamp'- and in other regions around the Ionian and the Aegean seas.

The wealth of the Muslim world, then in full expansion, is a further key to understanding this return of the wheel in lamp-making. Multifunctional workshops with their own, relatively small, ovens, could immediately adapt their diversified production (from lamps to bowls, from jugs to cups, etc.) to the needs of the local market, as it was their own -and single-market, thus reflecting the new 'economic model.' Ceramics, with few exceptions, were no longer the objects of medium or long-distance trade, in contrast to primary goods and luxury artifacts. Yet Southern and Eastern Mediterranean potters were not insensitive to fashion, as we can see all through the Islamic world; other regional lamps, were produced with an identical shape, from the Eastern borders of the Persian world to the Caliphate of Córdoba.

In addition, it is this same period when, both in Byzantium as well as in the Arabic world, we witness a huge movement we could call a form of 'iconophobia,' which affected all functional domestic objects. Molds and all the richness of motif they could add to a lamp were really no longer fashionable. Even in the Christian world, during the last centuries when molded lamps were produced in Asia Minor, Constantinople and the Black Sea, motifs were already restricted, showing only geometric or vegetal patterns if decorated. The added value of using a mold was to be able to shape the handle in form of a cross, a human or animal head, or a leaf.

2. The roots of fuel poverty

2.a. Bureaucracy and monopolies

Visiting recently the headquarters of the Romanian branch of one of the world's major telco companies, we viewed with no astonishment the departmental floor list posted besides the elevator. Out of 30 floors 28 had office spaces allocated according to the usual 'pyramid of power' so typical in today's world: open-office zones for all the lower pay employees occupied the first 15 floors, then progressively each floor housed fewer employees dealing with increasing more 'important' departments. The whole culminated with the 28th floor, which house only the 'office of the Chief Executive Officer'. Up to that, everything normal. Yet what was -and is now in many companies- above the Chief Executive Officer, as a Damocles' sword? The two top floors, 29th and 30th, with a beautiful view on all Bucharest, was home to the almighty 'compliance department'.

This phenomenon is a typical example of a bureaucracy required by state law but whose administration has run mad and is totally out of control. State rules, norms and their cohort of compulsory forms to submit on almost all sectors of a business have generated a dramatic increase in employees, not so much within the State administration itself, but within the private sector. Of course, the final price of every single service or product reflects the wages of all the paper-pushing people paid by the business sector but needed to interact with the State.

In our lychnological research, we have already found this kind of absurdity. The scene takes place in overly bureaucratic Ptolemaic Egypt. At that time, oil provided a considerable source of income for the Egyptian State. Indeed, as was established by the laws governing the incomes promulgated by Ptolemy Philadelphus, oil was a complete royal monopoly and strictly controlled. The State was the owner of presses, fixed the purchase prices of the cereals, the sale price of the various oils and was the single entity who could grant licenses for the trade of these commodities.

Only the big landowners could obtain an exemption, which consisted only in being able to produce the oil in a quantity necessary to operate their farms for one year and giving it to their whole staff for working purposes. Furthermore, only the quantity of seeds necessary to plant the same surfaces cultivated during the preceding year could be kept, and any surplus had to be sold to the State at a compulsory price lower than the one obtained by small independent farmers. In return, no State oil press could be established in villages controlled by a large landowner. The main sanctuaries also benefited from preferential treatment, but only for the sesame oil which was reserved for worship services.

A mine of information on the daily consumption of oil has been preserved for us thanks to the study[6] of an exceptional document, the Cornell Papyrus II.1, preserved since 1923 at Cornell University (USA). This invaluable testimony is said to have been discovered at the beginning of the 20th century at Gerza, the ancient Philadelphia, at the heart of the most prosperous agricultural region of the country, the Fayum. Drafted in Greek, it is dated to the reign of Ptolemy Philadelphus, and describes with great rigor the expenses in oil for lighting made in the enormous land property of Apollonius, treasurer-general of the monarch. The document,

[6] Cf. Westerman 1924; Samuel 1963-1964 and Brent Sandy, 1989.

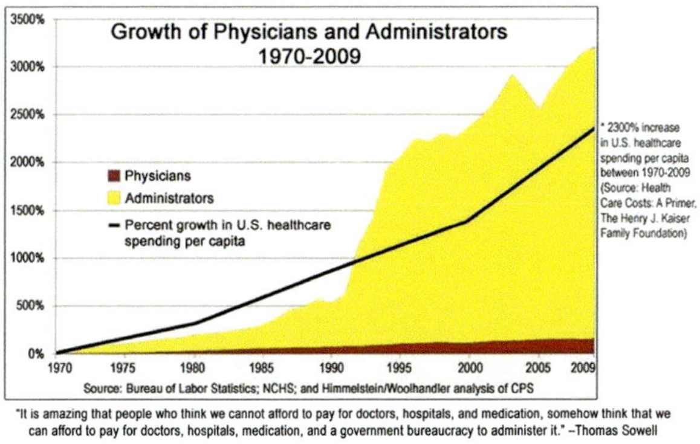

Figure 5. The growth of administrators vs physicians in the USA.

drafted by one of the bookkeepers of Apollonius, a man named Zenon, is a complete administrative register of the company, recording day after day the quantities of oil granted to each of the different employees of the farm. All the expenses of the liquid are noted, as well as the name of the person receiving the oil and, if needed, the use that this employee had to make with the oil.

Precisely dated 258-257 BC, it covers two whole months of the Ptolemaic calendar, the 1st of *Apellaeus* to the 30th of *Audnaeus*, corresponding to our current November 9th to January 6th. The daily measures expressed there are *kotyles*, which are approximately the equivalent of a quarter of our liter, more exactly 24.65 centiliters.

Straightaway we can easily recognize the bureaucracy generated not only by record keeping of the oil consumed, but also by its capillary management: the recorded measures comprise amounts of oil as small as an eighth of a *kotyle* (three centiliters!). Additionally, the fuel which was not consumed during a given night was deducted from the quantity given to that same employee for the next day.

We have to think about the fact that this extremely long document consisting entirely of all numbers, names and measures, was certainly made in two or three copies, which were then sent to and processed by the State administration. We can observe that in 258 BC as well as today, useless 'compliance' –i.e., when it goes way too far, as in our case for three centiliters (!) of oil– needed a huge number of people both in the private sector and in the state administration to compile, submit, read, approve and archive all those forms.

Ptolemaic Egypt, in contrast to the one-stop-shop offered by Roman fiscal laws, had then generated parallels to what we see now not only in the EU, but also in more liberal countries, as the tragic scheme of the growth of healthcare administrators per physician working in the USA since 1970 shows better than any possible comment. If the number of doctors doubled, the number of administrators grew by more than 3000%, resulting in a final cost for the patient which grew following the same pace. This is the hidden price of 'compliance.' Every one of us is paying today, no matter what we buy!

Did some Ptolemy survive in some EU technocrats' brains when they decided to privatize the production of energy goods and services, including the state-owned and managed energy grids? Well, Egypt did the same for sesame and castor oil, the two most common fuels for lamps. Each year, state bids allowed a single individual or a company owned by stakeholders to have a monopoly of oil sales within a nome (province).

But there was a problem, as 'the harvest was stored in government granaries, after being purchased from its growers by government agents at fixed prices, and distributed for chopping and pressing to government-operated factories. The workers were paid by the government at fixed rates for each unit of production. Extracting machinery was rigidly guarded. The oil produced was wholesaled and retailed at fixed prices. No oilseeds, oil, or oil-producing tools were permitted to exist outside of the above arrangement'.[7]

Hence what was the incentive of someone, as for example today's gas or electricity operators, to win the bid? In his diachronic paper, J. Wickersham deduced well that, in fact, as nowadays, the benefit ratio of the oilmen resided in their mere skills to be... compliant! They had to hire as few as possible good administrators, had to pay them as little as possible and had to provide to the State all necessary documents justifying their activity. Wickersham concludes brilliantly that, simplifying the question, for a small *nome*, if the bid was one talent, the winner had to hire at least 40 administrators and after having paid all wages and taxes, he could remain with one talent, meaning a 100% benefit. All that, of course, was reflected in the total poverty of rural Egypt, and to a price of oil fixed by the State which included all bureaucratic work needed by both parties. As a result, as nowadays, this fuel was overpriced in comparison with other commodities where monopolies did not exist. Workers were underpaid to allow bureaucrats, both from the State and from the private sector, to receive good wages for... *paper-pushing*!

2.b. Monopsonies

To complete the EU current energy framework, in contrast to Ptolemaic Egypt, even the production is no longer State-owned. A sole distributor can buy fuel from different producers (i.e., a monopsony); the best example is to be found in medieval Japan.[8] There, at the end of the 9th century AD, the oil merchants of Ôyamazaki near Kyoto, started a prosperous journey that would last at least 600 years. First, they became the only provider for the prestigious sanctuary of Iwashimizu Hachiman, for which, in exchange of 'sponsoring' the principal celebrations of the holy place, they received the right to buy, tax-free, all the seeds of *egoma* cultivated on the enormous piece of land owned by the sanctuary. This was the start of the impressive growth of their corporation. Its leaders, with outstanding trading and diplomatic skills, succeeded in winning the monopoly to provide the Imperial court with this excellent illumination oil and then, progressively, convinced a lot of landlords (shoguns) to have a custom-free pass when transiting or selling on their territory. From simple resellers of oil of a tiny area, the Ôyamazaki oil merchants became a real monopsony, most of the producers of egoma in Japan feeling comfortable to have a consistent buyer. At the same time, the Ôyamazaki 'corporation' also became a transport business

[7] Wickersham 1970.

[8] Gay 2009.

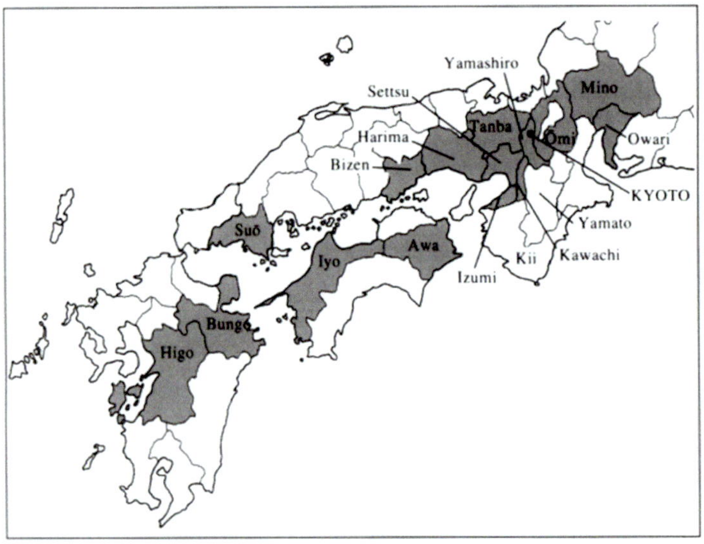

Figure 6. The territories where the Ôyamazaki oil merchants had trade privileges (© Gay 2009, 18).

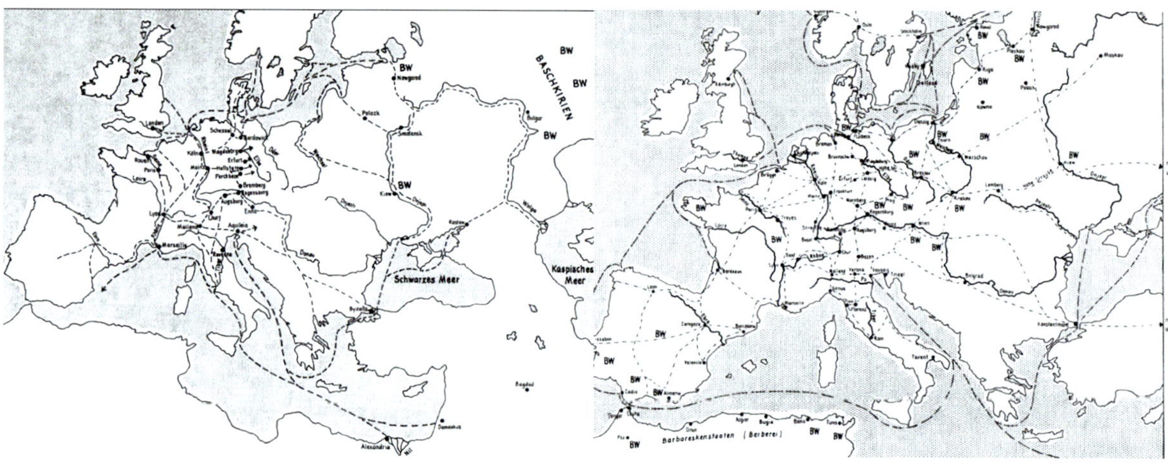

Figure 7. The situation faced by Charlemagne and Barbarossa (left, wax trade from 5th to 10th centuries AD) and the consequences of their wise politics circa 1400 (right). The acronym BW (Bienenwachs) highlights the most important regions where beekeeping was practiced. Both figures belong to Warnke's fundamental study.[11]

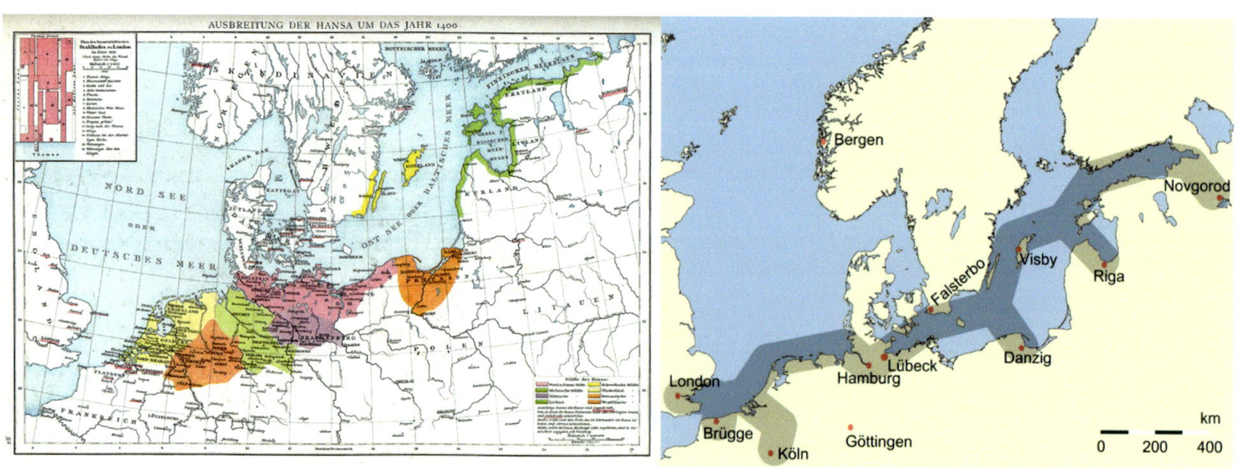

Figure 8. Maximum expansion of the territories controlled by the Hanseatic league in 1400 (G. Droysen, *Allgemeiner Historischer Handatlas*, 1886, pl. 28) and the main trade routes followed by the Hanseatic commercial fleet (©Wikipedia).

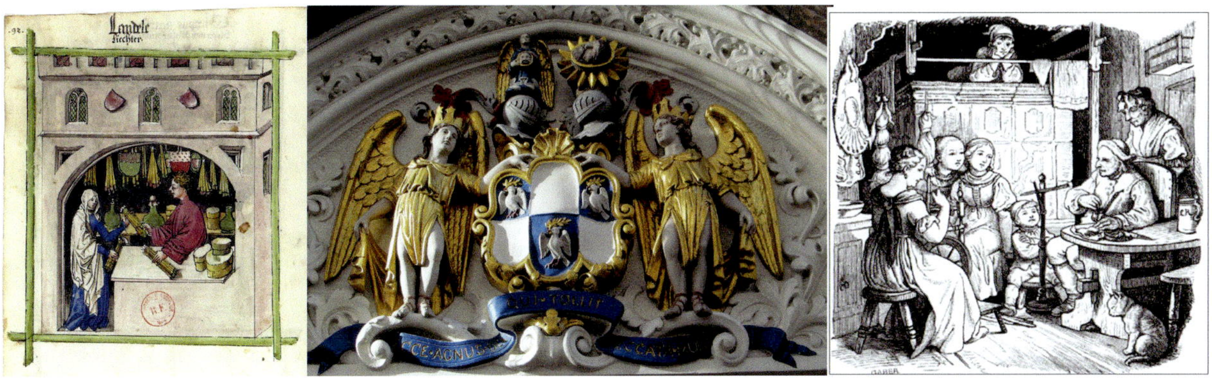

Figure 9. Left: A wealthy candle merchant in an image from Îbn Butlan's 'taccuinum sanitatis' on a Rhine-Valley 15th century re-transcription and bilingual translation (Latin-German) © Bibliothèque Nationale de France, Latin Manuscript 9333, fol. 92v. Middle: the magnificent coat of arms at the entrance of the Tallow Chandlers' Hall at Dowgate Hill, London. Right: the tragic consequences of taxing even tallow –a 'middle-class' English family at night. While the household economic wealth is witnessed by the nice ceramic heating stove, the quality of the garments, the lighting they could afford is... the prehistoric vegetal rush-light, set in the center of the room.

as well as a bank, proposing loans to landlords. By the 15th century, almost a half of Japan was their trade monopoly, so similar to nowadays where some French, German or Italian energy holdings have won the monopoly of distribution of gas or electricity in regions or entire countries of the EU.

In continental Europe, the medieval guilds had this part of the pie. Since the Early Byzantine Empire candles, being the most sacred light source, were crafted only by wax-makers (*keroularioi*), licensed by the imperial court. The sentence in the case of mixing wax with tallow was ... death.

One of the most important guilds of Constantinople was thus born. Its tasks, obligations and duties being perfectly described in the *Book of the Prefect* enacted by Emperor Leon VI (866-912), in its Chapter 11, where the rules applicable to wax and taper merchants are established. Per force it was one of the 20 guilds which had privileges to collect taxes in the name of the State.[9]

In continental Europe, this phenomenon took an unexpected amplitude, and wax imports (exactly as the case with contemporary gas and petroleum imports) generated, alone: new trade routes and new international geo-strategic decisions which would shape the continental history until – and even after– the modern era.

As a matter of fact, wax was indispensable to both European courts and churches: wax seals were the only way to prove that a document was legal, wax candles were the symbol of church power and one of the principal causes of its enormous yearly benefits.[10] As in antiquity, wax was necessary to make the best bronze or metal molded objects, from bells to canons. All were made with the 'lost wax' casting technique.

Wax collection became a State priority and consequently a monopoly for a guild, a church or ruler in each region or city. An important aspect of the enormous challenge faced by a continent where apiculture was totally insufficient to satisfy its needs is witnessed by the decisions and efforts made by some of history's most famous rulers: the King of the Franks (since 768), the King of the Lombards (since 774) and the Holy Roman Emperor Charlemagne (from 800 to his death in 814), Holy Roman Emperor Frederic I 'Barbarossa' (who ruled from 1155 to his death in 1190), and his vassal, Henry the Lion (Duke of Saxony and Bavaria from 1142 for Saxony respectively 1146 for Bavaria to 1180). [11]

Charlemagne decided to boost apiculture in all the kingdoms he progressively ruled, mainly in the Frankish kingdom, by granting to peasants starting this activity a total tax exemption and a very generous remuneration for the wax produced. Knowing that Charlemagne's wise decisions had to be improved but would never cover the needs of the Holy Roman Empire, Frederic I made an extremely significant decision in medieval 'international' history, i.e., to establish free trade markets open to traders belonging to the great cities of Russia and Ukraine (Novgorod, Smolensk, Kiev) and to the vassals of the Empire. This decision was expanded by the creation of the Hanseatic League, establishing a sort of German monopoly of trade navigation in the Baltic Sea, and from which later consequences would be the foundations/conquests of the German free cities such as Sczecin (Stettin), Kaliningrad (Königsberg), and so on.

Those choices, which deeply influenced the course of Central European history, were taken because beeswax was massively available for acquisition in two areas: present day Russia and Ukraine, accessible to both the Holy Roman Empire and the Islamic World, including the huge markets of Baghdad and Damascus commonly frequented by Byzantine traders. Slavs had been beekeeping since ages for honey and were not, until the Late Middle Ages, used to lighting their households with wax candles, thereby creating enormous surpluses of the raw wax material. Islam as well as Judaism forbid the use of animal products for religious and also domestic lighting, while being huge fans of pastries and products based on honey. No wonder coins of medieval Baghdad are discovered each year

[9] Nicole 1894, cf. Chapitre XI. Des cérulaires ou fabricants de cierges, 52-54.

[10] Cf. Vincent 2004.

[11] Warnke 1987, 545-569.

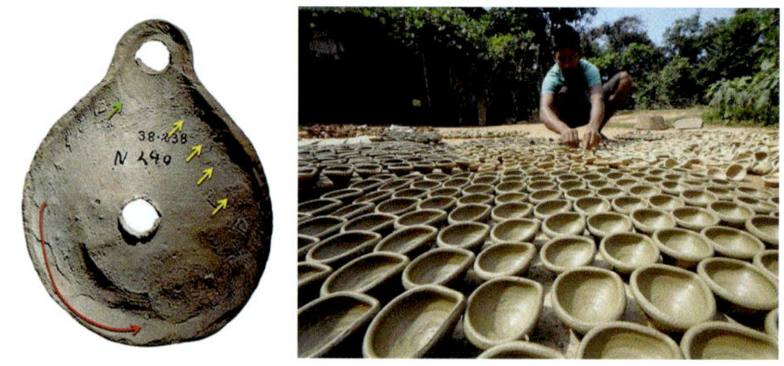

Figure 10. Left: Finger actions needed to ensure a good repartition of the clay in the mold. On the upper part of a lamp found at Beit Nattif (©Lichtenberger, Moran 2018, 5, fig. 5,). Right: Clay lamp-making in India today for the festival of light (Diwali): the child of a potter helping his family.

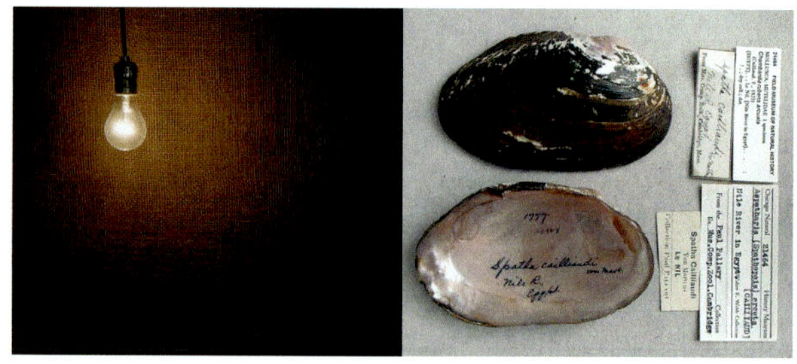

Figure 11. Left: A bulb simply hanging from the ceiling. Right: A Nile mussel found in Egypt (Field Museum, © Mussel Project).

as far away as Poland, Lithuania, the Scandinavian countries and other parts of Central Europe.

Upon arrival at its final destination, beeswax would be kept under very strict surveillance and was, in most parts of the European continent, the monopoly of very rich guilds. Some powerful states went even further, over-taxing even tallow candles, as was the case in late Medieval England, where the Worshipful Company of Tallow Chandlers was founded by Royal Decree in 1462, with catastrophic effects on the capacity of households to light their homes after dawn.

If we look at the price a household pays now in the EU (average, data from Eurostat), some 0.20€ per kwh, this amount is obtained by adding the price given by the Producer, some 0.05€, the price to pay to the private companies running the transmission and distribution grids, some 0.02€ and different taxes paid by both the producers and distributors, some 0.03€, a total of *c.* 0.10€ per kwh. So why do we pay double? Taxes, bureaucracy, low-cost energy for the industry, and profits for all levels of the process, from the producer until the last reseller in contact with the final customer. We combine the Ptolemaic cases with the medieval Japanese and the English case, taking the worse of each, and here we are!

3. The European moral hypocrisy on children at work: a lighting example from antiquity

Recently, a new French law[12] (called 'Sapin II' as it was initiated by former Finance Minister Michel Sapin) outraged all the Chief Executive Officers of the country. With a lot of absolutely unenforceable measures, it points to the Chief Executive Officer as the exclusive scapegoat of all National and International business, moral and ethical problems. The most ridiculous part of the law is certainly the one regarding the possibility to fine or even to condemn to prison a Chief Executive Officer who did not verify if any of the subsidiary suppliers of the company, wherever in the world it is established, abuses worker's rights as they are set into French laws, i.e., underpaying or exploiting employees and, in the worse scenario, employing children, a criminal case under French law.

This totally hypocritical new law is written as if the whole world were France and does not take into account any third-party national labor laws or rules. As most imports for any component of a EU-produced item may be coming from Asia, it will be very interesting to see the first judgements if any prosecutor does decide to enforce this chapter of the law. Furthermore, besides not taking into account the different national laws ruling the minimal age allowed to enter the labor market in Asian countries, it hides a profound ignorance of poverty within the EU.

Let's take the example of Romania, where the rural populations are the first victims of the EU's megalomaniac agricultural politics which focus on helping huge landowners and breeders. Cultivating small surfaces or growing just a few animals is a consequence of a geo-topography where industrial-scale agriculture is absolutely impossible, unless going back to communist-style *kolkhozes*. All these rural folk are excluded from EU financial aids, supermarket supply chains and, now more and more, from local markets where EU rules are enforced. For instance, there are now requirements for costly certifications and labels of compliance, even the possession/loan of a compulsory refrigerator if selling dairy products, and so on.

As a consequence, in one of the most fertile and uncontaminated areas of the EU, which has continuously dealt with ancestral techniques of micro-agriculture, children live as their parents in total poverty, way under the 'poverty indicators' set by bureaucrats. Non-governmental organizations may draw the alarm signal, but note that in 2018, one of every two Romanian children is living under the poverty line and must work after school to help his family.

Such unacceptable pockets of poverty, affecting more than 14 countries of the richest 'entity' on the globe, the EU, have their exact parallels in antiquity and also exist nowadays in third-world countries, where the under-proletariat parts of societies are huge. Hence children have no other option than to contribute to their family's incomes by working.

Lamps are, again, a precious indicator that an identical phenomenon existed in parts of the Ancient world where the populations were so poor that even slavery was not needed, local inhabitants being the new real slaves. On the single careful attempt to study *dermatoglyphs* (fingerprints) on lamps made in Alexandria, and found during the Polish excavations

[12] Loi no 2016-1691 du 9 décembre 2016 relative à la transparence, à la lutte contre la corruption et à la modernisation de la vie économique.

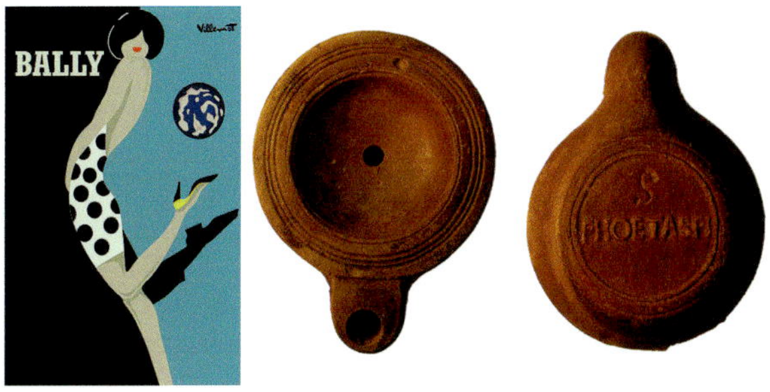

Figure 12. Left: Bally, a world-known Swiss-made shoe brand since 1851, now a luxury multinational company with not a single production site in Switzerland; only the Headquarters remain there. Right: The first types of lamps produced by Phoetaspus at Alexandria (Collection Bouvier).

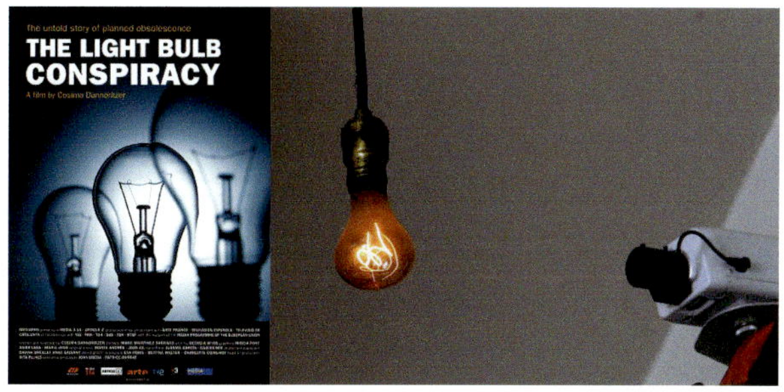

Figure 13. Left: The poster of the documentary movie on the 'Phoebus cartel' and what the producers wanted to avoid at all costs. Right: The Livermore bulb, lighting every day for 117 years!

at Kôm el-Dikka,[13] the researchers found out that the creators of all those elegant Roman imperial lamps where no one other than ... ten to twelve year-old children, who performed one of the key manual actions needed for the manufacture of mold-made lamps: equalizing and smoothing the clay in the mold, as it has been recently shown by a pioneer study led on Beit-Natif lamps in Israel.[14]

4. Fuel-rich poor

This case is extremely rare in history, but we can find examples both in antiquity and in our contemporary world. As a matter of fact, with the exception of cultures where lamps are traditionally obtained directly from rough materials (stone or wood, such as the lamps of the Arctic tribes), or where no lamps other than natural containers were used by the whole society (sea shells, as in ancient Mesopotamia or in the Orkney Islands until modern times), there are some examples where even the poor were able to access a good energy source but could not afford to have a man-made lamp for using it. This is frequently the case, nowadays, in rural areas of countries where electricity is very cheap, generally because it is a granted by the state as part of its model of development, as in China.[15] There, light bulbs are simply hung, screwed into a socket fixed at the end of the electric cable.

In Byzantine times, we find a breathtaking analogy at Baouit, a site renowned for its famous monastery, founded at the end of the 4th century and which was visited by hundreds of pilgrims from all over the Christian world, such as the monastery of Abu Mena, from the 7th to the 9th century. Within the monastery there were extraordinary bronze lamps, *polykandela*, as well as clay lamps[16] with Christian epigraphy.[17] Yet, probably because of the strict discipline of poverty of the monks, dictated by the founder of the monastery, Saint Apollo, the daily lighting devices, used with the best available oils (radish oil and rapeseed oil) were nothing less than ... the common shells one could find simply by walking on the shores of the Nile (*Chambaria rubens*).[18] The shells were big enough to be used as lamps for the short time they were needed.[19]

5. Relocation of production sites

Delocalization and relocations may not always be a result of the high cost of local skilled employees, but may also be provoked by excessive taxation, social contributions or instabilities or obsolete bureaucracies. This can have an enhanced effect on the local economy by impoverishing the local customer base. Recently in the EU, first, the industries flew away. For more than two decades, the brains are doing the same. Is that so new? Absolutely not. It has been the case of Northern and – even if in a minor extension– Central Italy since the beginning of the 2nd century AD. Local archaeological evidence reflects a real impoverishment of almost all the previously rich cities such as *Mediolanum* (Milan), *Augusta Taurinorum* (Torino), *Laus Pompeia* (Lodi Vecchio) and so on. No more monumental buildings are constructed, poor materials are found in necropolises and production areas are abandoned.[20]

Northern Italy, mainly the regions of Veneto and Emilia, was the homeland of the inventors of the *Firmalampe*, a lamp with quite a rough aspect but which had the ability to provide the illumination equivalent of two lamps if the wick was pulled up onto the nozzle-channel. These lamps saw a 'boom' in the recently-conquered Transalpine and Danubian Provinces. Almost systematically bearing the name of their factory, we can follow their export and their consecutive imitations and counterfeits.

At a certain point, two of the main producers of the Firmelampen, Strobilius and Phoetaspus, opened branches in Southern Gaul, Alexandria and in the East. Some researchers even suggest that these two lampmakers actually left Italy for Egypt, since the first Nilotic productions are more or less contemporary with the last original Cisalpine ones, where these makers had had successful businesses at least for three generations.

[13] Dzierzykray-Rogalski and Grzeszyk 1991.
[14] Lichtenberger and Moran 2018.
[15] Cf. Xiujie 2009.
[16] The most prestigious of them are now on exhibit in the Louvre in a special room dedicated to the monastery, the metallic ones having been published in Bénazeth 1992.
[17] Lyon-Caen 2010.
[18] Romanus *et al.* 2008.
[19] For a complete list of images of such shells found in Egypt and other African countries, see <http://mussel-project.uwsp.edu/db/db.php?p=div&l=spp&n=839>, viewed 27 July 2018.
[20] Chrzanovski 2006.

6. Planned obsolescence, a lamp-producers' invention

The progressive disappearance of all kinds of local workshops whose specialty was to repair any kind of household machine leaves us with a world where we just buy, use and trash. This can be viewed as the consequence of a decision taken between the two World Wars by all major producers of... lamp bulbs!

In the mid-20th century electric lighting relied on bulbs made with glass and the famous Edison filament and which were easy to replace. Moreover, early bulbs was so strong and well-made that they could last years and years, as it is shown by an amazing example which has recently made the headlines of many newspapers: a light bulb bought in 1901 and still working non-stop for 117 years in a Firehouse in Livermore, California![21] This bulb is now so famous that a webpage is available only for it, where a webcam takes a picture of it each 30 seconds and broadcasts it live around the globe.[22]

For lamp bulb production companies, this was exactly what was not tolerable for the growth of their businesses: their products were so good that despite their success and the yearly growth of the number of consumers, sales were stable if not decreasing. Hence, in 1924, all major electric bulb companies from all over the world gathered in Geneva and determined that the maximum working time for a bulb has to be, in a compulsory way, one thousand hours. Their engineers found solutions by weakening the filaments, the bulb socket was no longer removable, and the job was done. All companies respecting this rule were being admitted into the '*Phoebus cartel*'. Planned obsolescence was born, and is now ruling most of our domestic tools: printers, fridges, washing machines, smartphone, and so on, as is masterly shown in Cosima Dannoritzer's award winning documentary movie '*The light bulb conspiracy*' (2010).

Bibliography

Bénazeth, D. 1992. *L'Art du métal au début de l'ère chrétienne.* Paris: Éditions de la Réunion des musées nationaux.

Brent Sandy, D. 1989. *The production and use of vegetable oils in Ptolemaic Egypt* (Bulletin of the American Society of Papyrologists, Supplement 6). Atlanta: Scholars Press.

Chrzanovski, L. 2006. L'urbanisme des villes romaines de Transpadane. Montagnac: Éditions Monique Mergoil.

Chrzanovski, L. 2012. La transition lychnologique byzantino-arabe en Egypte et au Proche-Orient: lampes coptes et musulmanes de la collection Bouvier. *Peuce SERIE NOUĂ XI*, 179-230.

Dzierzykray-Rogalski, T. and Grzeszyk, C. 1991. Les dermatoglyphes (empreintes des lignes papillaires) relevés sur des lampes alexandrines de Kôm el-Dikka (Alexandrie). *Cahiers de la Céramique* éyptienne 2, 125-128.

Gay, S. 2009. The Lamp-Oil Merchants of Iwashimizu Shrine: Transregional Commerce in Medieval Japan. *Monumenta Nipponica* 64:1, 1-52.

Lichtenberger, A. and Moran, K. S. 2018. Ancient fingerprints from Beit Nattif: studying Late Roman clay impressions on oil lamps and figurines. *Antiquity* 92:361, 1-6 (online at:<https://www.cambridge.org/core/services/aop-cambridge-core/content/view/242426A0AC1A51E6A44ACD238B84C7D4/S0003598X18000029a.pdf/acient_fingerprints_from_beit_nattif_studying_late_roman_clay_impressions_on_oil_lamps_and_figurines.pdf>, viewed 27 July 2018).

Lyon-Caen, C. 2010. Les lampes du monastère de Baouit, in A. Boud'hors and C. Louis (eds) *Études coptes XI. Treizième journée d'études (Marseille, 7-9 juin 2007)*, Cahiers de la bibliothèque copte 17, 31-44. Paris: De Boccard.

Nicole, J. (ed.) 1894. *Le livre du Préfet ou l'édit de l'Empereur Léon le Sage sur les corporations de Constantinople*, Genève: Georg. (book available online at <https://archive.org/details/lelivreduprfet00nico>, viewed 27 July 2018).

Robinson, C., Bouzarovski, S. and Lindley, S. 2018. 'Getting the measure of fuel poverty': The geography of fuel poverty indicators in England. *Energy Research & Social Science* 36, 79-93.

Romanus, K., Van Neer, W., Marinova, E., Verbeke, K., Luypaerts, A., Accardo, S., Hermans, I., Jacobs, P., De Vos, D. and Waelkens, M. 2008. Brassicaceae seed oil identified as illuminant in Nilotic shells from a first millennium AD Coptic church in Bawit, Egypt. *Analytical and Bioanalytical Chemistry* 390, 783-793.

Samuel, A. E. 1963-1964. Illumination by castor oil - P. Cornell I. *Bulletin of the American Society of Papyrologists* I, 32-38.

Simcock, N., Walker, G. and Day, R. 2016. Fuel poverty in the UK: Beyond heating? *People Place and Policy Online*, 10/1, 25-41.

Vincent, C. 2004. *Fiat lux: lumière et luminaires dans la vie religieuse du XIIIe au XVIe siècle* (Histoire religieuse de la France, 24). Paris: Le Cerf.

Warnke, C. 1987. Der Handel mit Wachs zwischen Ost- und Westeuropa im frühen und hohen Mittelalter. in A.A.V.V., *Untersuchungen zu Handel und Verkehr in früh mittelalterlichen Mittel- und Nordeuropa* IV, 545-569.

Westerman, W. L. 1924. Account of lamp oil from the estate of Apollonius. *Classical Philology* 19, 229-260.

Wickersham, J. M. 1970. The Financial Prospects of Ptolemaic Oilmen. *Bulletin of the American Society of Papyrologists* 7:2-3, 45-51.

Xiujie, W. 2009. *Ein Jahrhundert Licht. Eine technikethnologische Studie zur Beleuchtung im chinesischen ländlichen Alltag.* Wiesbaden : Otto Harrassowitz.

[21] <https://www.nouvelobs.com/sciences/20160607.OBS2 02 2/la-belle-histoire-de-l-ampoule-qui-brille-depuis-115-ans.html>, viewed 27 July 2018.

[22] <http://www.centennialbulb.org/cam.htm>, viewed 27 July 2018.

Short Biographies

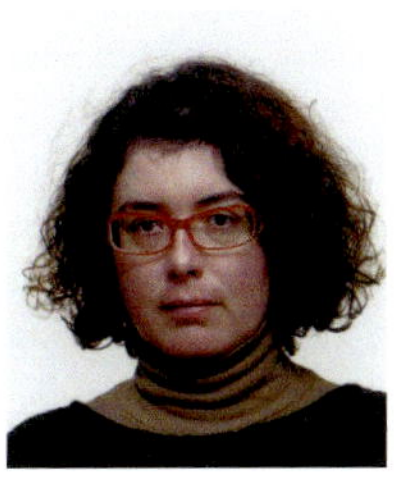

AKRIVOPOULOU SOPHIA is an archaeologist for the Hellenic Ministry of Culture and Sports. Educated in the Aristotle University of Thessaloniki, both as an undergraduate and doctoral student, she has participated in many archaeological projects in Central and Western Macedonia until her placement with the Ephorate of Byzantine Antiquities of Thessaloniki in 2006. In March 2018 she was appointed as Director of the Department of Byzantine and Post Byzantine Antiquities in the Ephorate of Antiquities of Rhodope.
Email: akrivopoulou.s@gmail.com

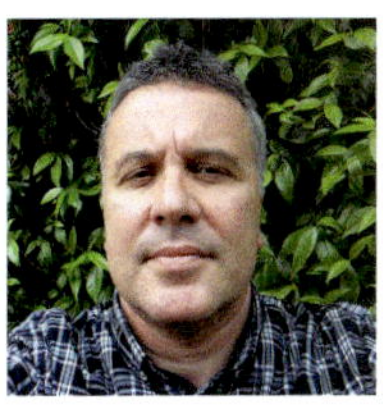

ANDROUDIS PASCHALIS is Assistant Professor in Byzantine Art and Archaeology for the Department of History and Archaeology of the Aristotle University of Thessaloniki. Educated at the Aristotle University of Thessaloniki, he earned his Diploma in Architecture in 1988. He earned his DEA (Diplôme d'Etudes Approfondies) and Doctoral Degree (Doctorat) in Byzantine Archaeology from the University Paris I- Panthéon Sorbonne in 1990 and 1995 (Professor J.-P. Sodini) and his Master of Arts in Conservation Studies from the Institute of Advanced Architectural Studies at the University of York, UK (1995). He has published six books and over 120 articles on Byzantine, Ottoman and Islamic art and architecture. He has participated in over 150 International and National Congresses on Byzantine, Ottoman and Islamic art and architecture and has conducted over 100 projects of restoration of historic monuments in various regions of Greece, Mt Athos included.
Email: archaio22@gmail.com

ANGELKOU EVANGELIA was born in Thessaloniki. She is a graduate of the History and Archaeology Department of the Aristotle University of Thessaloniki and holds a MA degree in the Protection, Conservation and Restoration of Monuments from the Faculty of Engineering of the Aristotle University of Thessaloniki, and a PhD in Archaeology from the History and Archaeology Department of the National and Kapodistrian University of Athens. Her PhD dissertation treated the tools and the professional and commercial activities of the Episcopal complex at Louloudies Kitros (5th-9th century). From 1998 to 2005 she worked as an archaeologist of the Hellenic Ministry of Culture, working in Thessaloniki and Kavala and since 2007 at the Museum of Byzantine Culture, where she is currently Deputy Director and Head of the Icons and Art Collections Department. She has written studies and articles on minor objects, lamps, tools, jewelry, coins and pottery, as well as on Thessaloniki's topography through postcards.
Email: evagelkou@hotmail.com

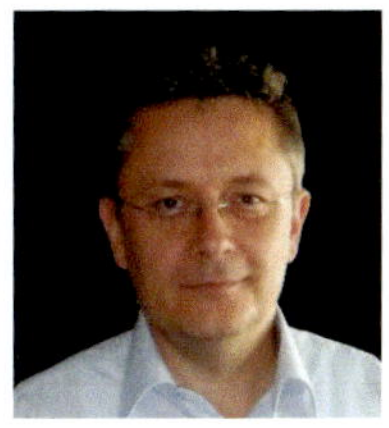

ANTONARAS ANASTASSIOS C. a specialist in the history of glass, jewelry and textiles, is an archaeologist head of the Exhibition Department at the Museum of Byzantine Culture, Thessaloniki. His books include: *Glassworking, Ancient and Medieval: Terminology, Technology and Typology; A Greek-English-English-Greek Dictionary; Roman and Early Christian Glassworking: Vessels from Thessaloniki and its Region; Fire and Sand: Ancient Glass in the Princeton University Art Museum; Artisanal Production in Ancient and Byzantine Thessaloniki: Archaeological, Literary and Epigraphic Evidence;* and *Glassware and Glassworking in Thessaloniki: 1st Century BC- 6th Century AD.* Antonaras has organized numerous exhibitions and symposia, and has published numerous articles on objects from Thessaloniki. He is member of the editorial board of scientific journals in both Europe and the USA. He is the president of the International Association for the History of Glass.
Email: aantonaras@culture.gr

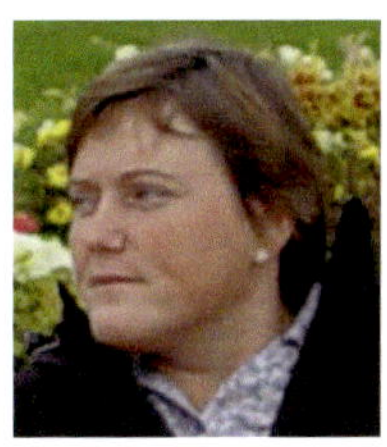

AVDUSINA SVETLANA, a specialist in the Viking Age in the territory of Old Rus', studied at the Moscow State University in Department of Archaeology. In 2000 she joined The State Historical Museum where she participated in the preparation of an exhibition 'Holy Russia' in The Louvre museum (2010) and in the series of exhibitions devoted to the Vikings in the National Museum of Denmark (2013), the British Museum (2014) and in Berlin's state museums (2014-2015). Every year she takes part in the excavations of the medieval site of Gnezdovo near Smolensk. She is the curator of a collection of medieval Russian hoards.
Email: avdusina@gmail.com

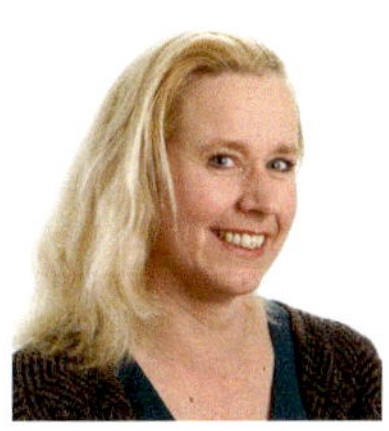

BERG NILSSON LENA, is an archaeologist specialized in medieval mining. A field she has been working within more than 20 years, largely within the project Atlas över Sveriges bergslag (Atlas of the mining districts of Sweden). The last ten years she has been working as a consultant in archaeology, cultural history and cultural environment within the consultant group Lands-kapsarkeologerna. She is a part-time PhD-student in archaeology at Stockholm University in Sweden, where she also earned her MA degree and she participate in several historical research committees at Jernkontoret (The Swedish steel producers' association). Currently she is also working within the project 'Medieval mining districts in Sweden – the essential outlands' where she and six other archaeologists and historians is writing a book on medieval metal and mining industry and the mining districts effect on landscape, economy and social transformation.

BITTERLI-WALDVOGEL THOMAS, was born in Basel and studied medieval history, modern history, human and physical geography and medieval archaeology in Basel and Hamburg. Having obtained a Master of Arts degree in Basel in 1980, he was involved in the compilation of the *Burgenkarte der Schweiz* [Map of Swiss Castles] (sheet no. 4, 1985) and from 1985 to 1990, he worked for the *Inventar der Historischen Verkehrswege der Schweiz (IVS)* [Swiss Inventory of Historical Transportation Routes]. He was subsequently involved in several major and minor projects in the area of architectural research, cataloguing (Map of South Tyrolean Castles, Second-World-War bunkers and pillboxes, fortified buildings as part of a Swiss inventory of individual structures) and archaeological excavation (deserted Alpine settlements, castle excavations). He has published various papers on topics relating to castle research (e.g., timber construction, writing implements, lighting, water supply and components of fortification). Since 1997, he has worked in a part-time role managing the office of the *Schweizerischer Burgenverein* [Swiss Castle Association] and from 2005 to 2007 he was involved in updating the *Burgenkarte der Schweiz* [Map of Swiss Castles] (Wabern 2007).
Email: thomas.bitterli@hispeed.ch

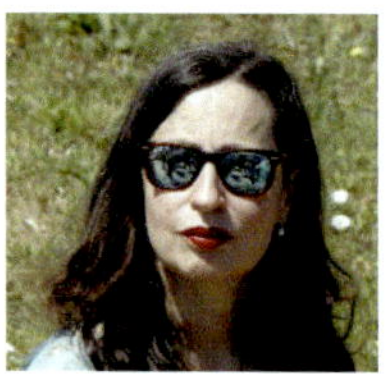

CHEIMONOPOULOU MARIA was born in Thessaloniki. She is a graduate of the Department of History and Archaeology of the Aristotle University of Thessaloniki and holds a postgraduate diploma in the interdisciplinary postgraduate program entitled 'Protection, conservation and restoration of cultural monuments' of the Faculty of Engineering of the same university. From 1995 to 2005 she worked as an archaeologist at the 9th Ephorate of Byzantine Antiquities, focusing on the recording and studying of minor objects and coins from the excavation at Louloudies of Kitros in Pieria. In 2006 she was appointed permanent archaeologist at the Ephorate of Antiquities of Imathia, where she now serves. She received her PhD from the National and Kapodistrian University of Athens with a thesis on early Byzantine jewelry from Louloudies Kitrous. She has published articles on issues of rescue excavations, pottery, miniature and monumental Byzantine art.
Email: cheimaria@yahoo.gr

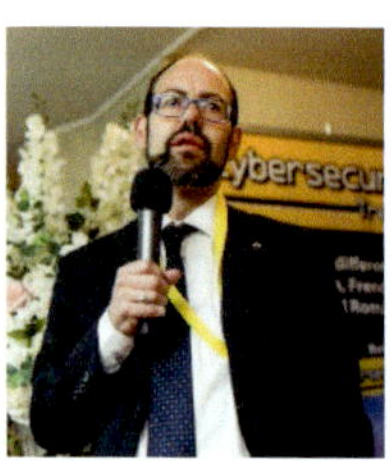

CHRZANOVSKI LAURENT, Prof. Univ. HDR Dr, Postdoc. Founder and Secretary-General of the International Lychnological Association (www.lychnology.org). With a PhD in Roman Archaeology, a Postdoctoral Research Degree in History and Sociology and an EU Habilitation to direct PhDs in History and related sciences, Laurent is Professor at the doctoral School of the '*Lucian Blaga*' State University of Sibiu and invited Professor at the Universities of Warsaw and Lyon II 'Lumière'. He regularly holds postdoctoral courses within several major EU Universities. He is the author/editor of 28 books, of more than a hundred scientific articles and of as many general-public articles. Specializing mainly in transdisciplinary approaches in archaeology, his studies focus on lighting devices and their socio-economic as well as religious meaning through time. He is also a winner of the ICOM prize for the best itinerant exhibition of South-Eastern EU, as well as a proud holder of the Gold Medal of the SEP (aka 'French Nobel'). Besides archaeology, Laurent works since 2010 in the domain of cyber security, where his main areas of research are focused on the relationship between human behavior and the digital world. As such, he is member of the ITU (UN) roster of experts. He founded and manages 4 major yearly networking dialogue events and 4 PPP quarterly journals in as many languages (*www.cybersecurity-dialogues.org*). Both the dialogues and the quarterly are awarded and promoted since 2015 as '*Best practice examples for the European Continent*' by the UN. You can reach Laurent on Linkedin, Academia as well as at l.chrzanovski@icloud.com

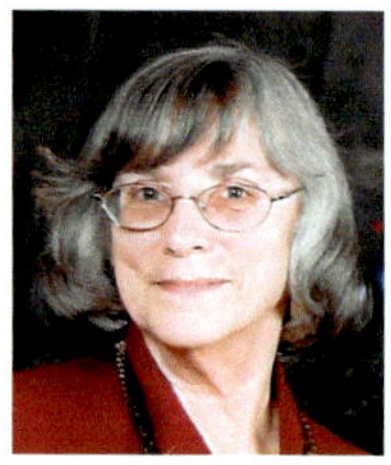

GARNETT KAREN S. is a long time resident of California. Born in Texas, and raised in Pennsylvania and California, she received her BA from Ohio State University, Columbus, Ohio and MA from the University of Texas, Austin, Texas, USA. Her archaeological research is focused on the Late Roman terracotta lamps from the Gymnasium and Fountain of the Lamps excavations in Ancient Corinth, Greece. In addition to her archaeological research, her various careers have included Motherhood, Artist, Art Historian, Lecturer, CAD Designer, Software Quality Assurance Manager and Technical Writer. Currently she manages the writing of technical documentation for the Intellectual Property Division of VeriSilicon, a fabless semiconductor company.
Email: karen.garnett@gmail.com

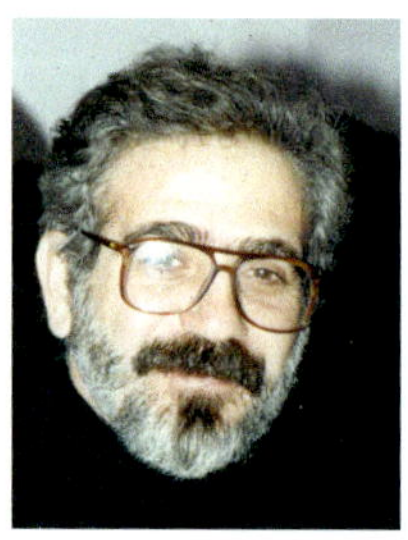

ILIADES IOANNIS G. is an Electrical and Lighting Engineer. He received his PhD from the University of Patras, Department of Electrical and Computer Engineering. Between 1979 and 2012 he worked for the Hellenic Ministry of Culture. He was a visiting Prof. at the Democritus University of Thrace. He designed and implemented numerous lighting studies for museums, such as the Archaeological museum of Amphipolis, the museum of Byzantine Culture, the museum in the castle of Chlemousti and the Byzantine Museum in Argos. In addition, he designed and implemented night lighting studies for monuments and Byzantine churches in both Greece and Cyprus. His papers and his concept studies about the illumination of museums' exhibition rooms, the lighting of monuments as well as the natural lighting of Byzantine churches, have been published in national and international scientific journals and reviews.
Email: ioheliades@yahoo.com

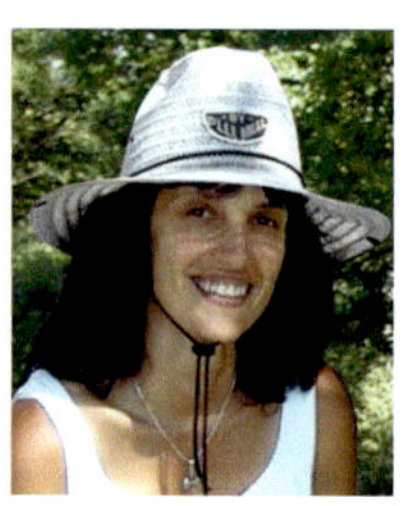

INKOVA MARIELA is Head of the Department of Archaeology at the National Museum of History in Sofia, Bulgaria. Her doctoral dissertation is on the 'Old Bulgarian Belts, end of the 7th - beginning of the 11th centuries'. She has specializations in Museology under the Sejour cultural program of the French Ministry of Culture and the Maison de la Culture Institute in Paris and Strasbourg; in archaeology at the Maison des Sciences de l'Homme under the guidance of Professor Jean-Pierre Sodini and at the

Roman-German Museum in Mainz, Germany at the invitation of Professor F. Daim. Her researches are mostly in the field of Bulgarian medieval culture and metal-plastics, and her main contribution, the conceptual thesis on one of the attributes of Old Bulgarian clothing, the belts, as a symbol of status in the social hierarchy of the Bulgarians. She was the scientific head of the archaeological research in the first Bulgarian capital Pliska and at the Early Byzantine and Medieval sites in southern Bulgaria. Since 2017 she is a member of the international research team for the archaeometric, technological and morphological research on the Preslav gold treasure of the Roman-German Central Museum in Mainz, the National Archaeological Institute with the Museum in Sofia and the Veliki Preslav Archaeological Museum. She participated in international conferences on archaeology in Bulgaria, France (1999, 2018), Greece (2011) and the 22nd International Congress on Byzantine Studies, Sofia, 2011.
Email: inkovam@mail.bg

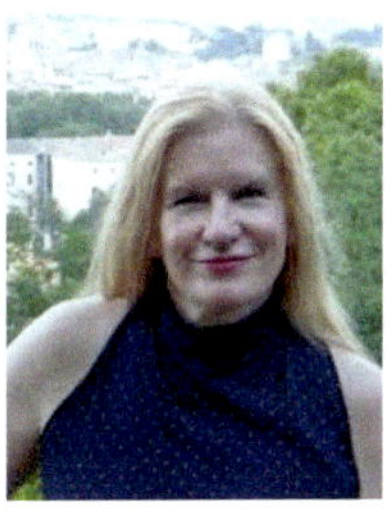

KARIVIERI ARJA, a specialist in Roman and Early Christian art and architecture, pottery, oil lamps, and cultural change in Late Antiquity, joined Stockholm University in 2003. Educated at Turku University and Åbo Akademi, she earned her licenciate degree in history of art, PhD in archaeology and docent degree in Classical archaeology. She was the director of the Finnish excavations of the Early Christian church in Arethousa, Greece, in 1999-2004 and the field director of the documentation of the House of Caecilius Iucundus for the Swedish Pompeii Project in 2000-2010. She was the president of the ILA from 2003 to 2015. She is the Viktor Rydberg professor of Classical Archaeology and Ancient History at the Department of Archaeology and Classical Studies in Stockholm University. Currently (2017-2020) she is on leave from Stockholm to serve as director of the Institutum Romanum Finlandiae. She is presently engaged in directing a project on ancient Ostia, financed by the Academy of Finland and Tampere University.
Email: arja.karivieri@antiken.su.se

KOUTOUSSAKI LAMBRINI, neé à Kavala (Grèce); 1988-1989 Apprentissage de la langue allemande à Saint-Gall, avec diplôme; 1989-1990 Cours de préparation aux études universitaires à Fribourg (maturité suisse) et examens réussis; 1990-1996 Études bilingues à la Faculté des Lettres de l'Université de Fribourg en Archéologie; Classique, Histoire ancienne et Grec ancient; Décembre 1996 Licence, branche principale Archéologie classique (Université de Fribourg); 2001 Formation de bibliothécaire à la Bibliothèque Centrale de Zurich (formation spéciale pour assistants de l'Université de Zurich, responsables des Bibliothèques universitaires); 2006 Formation d'enseignante langues (FEL). Cours d'introduction. Expérience professionnelle: 1991-1994 Stagiaire/responsable de secteur dans les fouilles suivantes en Grèce : Naxos (Université d'Athènes), Philippi (Université de Thessalonique), Béotie (American School of Athens), stagiaire au Musée d'Argos (École Française d'Athènes); 1991-2000 Stagiaire puis assistante scientifique au Service archéologique du canton de Fribourg; 1993-2000 Stagiaire puis collaboratrice scientifique aux fouilles grecques sur l'île d'Amorgos (dir. Prof. Marangou); 1995-1999 Stagiaire puis collaboratrice scientifique à la rédaction centrale du LIMC (Lexicon Iconographicum Mythologiae Classicae) à Bâle; 2000-2001 Recherches, étude et travail de rédaction dans le cadre du projet FRIMA à Fribourg. Activités actuelles: Depuis juin 2001 Assistante à l'Institut d'Archéologie Classique de l'Université de Zürich, avec charge d'enseignement et responsabilité de la bibliothèque de l'Institut; Dès avril 2005 Enseignante de grec moderne à l'École-Club Migros, Fribourg.
Email: artemis@mysunrise.ch

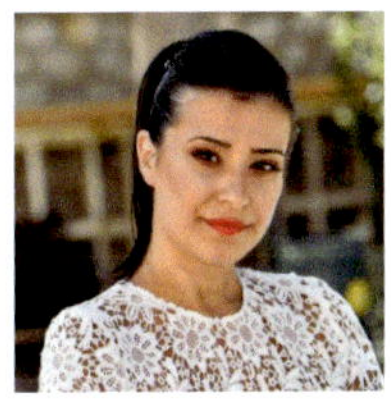

LEVENTELI PARASKEVI is a PhD student in Byzantine Archaeology in the Aristotle University of Thessaloniki, from where she also earned her undergraduate and Master degrees, specialized in Byzantine art and painting. She worked as field archaeologist in many projects in Northern Greece, including the excavation of Thessaloniki's Metropolitan Railway and the digital documentation of the Art Collections of the Hellenic Ministry of Culture and Sports. Moreover, she participates in archeological conferences and published papers in academic journals.
Email: pkleventeli@yahoo.gr

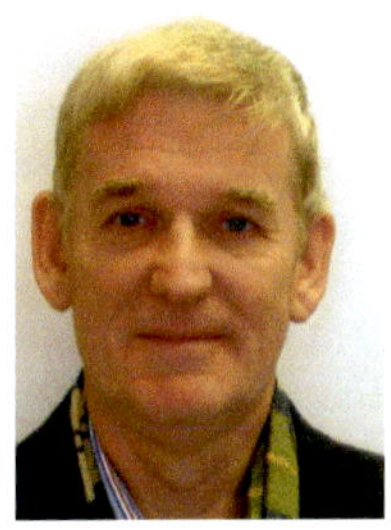

LIGHTFOOT CHRISTOPHER S., a specialist in ancient glass and Roman art, joined The Metropolitan Museum of Art in 1999. He contributed to the reinstallation of the permanent Greek and Roman Galleries that opened in April 2007. He curated the Lod Mosaic exhibition in 2010–2011 and Ennion: Master of Roman Glass in 2014–2015. Educated at the University of Oxford, he earned his undergraduate and doctoral degrees (PhD) in Literae Humaniores. He was formerly a Research Assistant in the Department of Greek and Roman Antiquities, The British Museum (1982–1986), and the Director of the Amorium Excavations Project, Turkey (1993–2014). He is presently working on the publication of the Metropolitan Museum's collection of terracotta oil lamps from the Cesnola Collection. After twenty years as the curator of the Roman collections in the Department of Greek and Roman Art, he recently retired and now works as a freelance archaeologist.
Email: amorclight@gmail.com

MARKI EFTERPI was born in Thessaloniki. She graduated from the History and Archaeology Department of the Aristotle University of Thessaloniki, from which she also obtained her PhD. She was Director of the 9th Ephorate of Byzantine Antiquities having as her major research responsibility the implementation of rescue excavations in Thessaloniki and Pieria. She has researched the Early Christian necropolis of Thessaloniki, which was the subject of her doctoral dissertation and has published more than 100 studies on the results of the rescue excavations she has supervised in Pieria, and more specifically in the Middle Byzantine Episcopate of Kitros and the Episcopal complex at Louloudies Kitros.
Email: eumarki@yahoo.gr

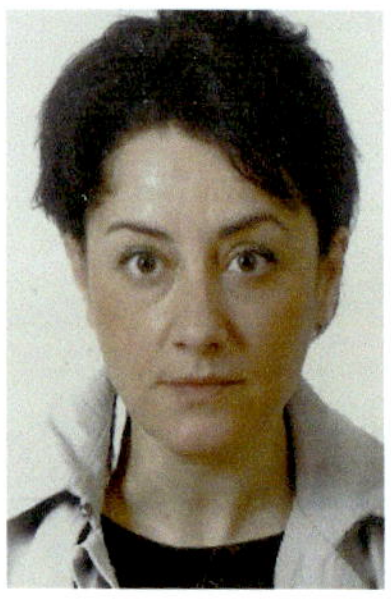

MASTORA PELLI has been an archaeologist at the Ephorate of Antiquities of Thessaloniki City since 1997. She earned her BA in History and Archaeology from the School of History and Archaeology, Faculty of Philosophy, Aristotle University of Thessaloniki her MA in Byzantine Archaeology (as above), and her PhD in Greek Civilization from the Open University of Cyprus. She is a member of the Greek Archaeological Mission in Agios Georgios in Peyeia, Cyprus. She has contributed to the publication of the *Bulletin de l'Association International pour l'Etude de la Mosaïque Antique* as a correspondent-editor of entries relating to Greek publications on mosaics, and has been a regular contributor to the magazine 'Iconostasion'. Her main fields of interest include wall mosaics and the history of archaeology in Thessaloniki.
Email: pellinmstr@gmail.com

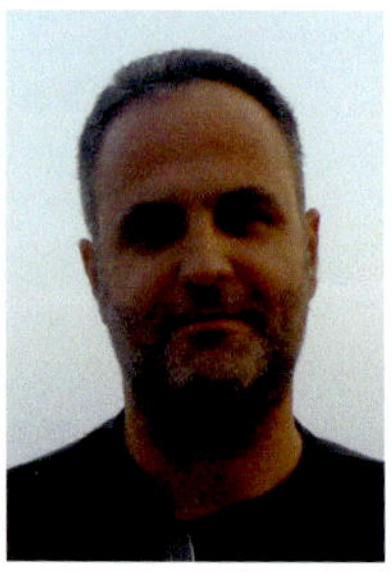

MOTSIANOS IOANNIS born in Thessaloniki and since 1995 has been an archaeologist working at the Museum of Byzantine Culture, Thessaloniki. He has studied Archaeology (BA, Aristotle University of Thessaloniki, 1990), Museum Studies (Postgraduate Interdisciplinary Course in Museology, Aristotle University, 2004). His PhD dissertation at the University of Thessaly, 2011, treats the evolution of artificial lighting during the Byzantine and Post-Byzantine periods. He is the author of more than 10 papers in scientific journals. He organized the 'Lighting in Byzantium', 4th International ILA Round-Table, 11-14 October 2011, in Thessaloniki and the Exhibition 'Light on Light: an Illuminating Story', Thessaloniki, Folklife & Ethnological Museum of Macedonia-Thrace, 31 October 2011-11 June 2012. He is also the co-editor of the exhibition catalogue *Light on light: an illuminating story*, Thessaloniki 2011. From 2003 he has been an active member of the 'International Lychnological Association' and from 2009 a member of its governing Committee. His research interests include artificial lighting in the Byzantine and Post Byzantine Periods.
Email: motsianos@gmail.com

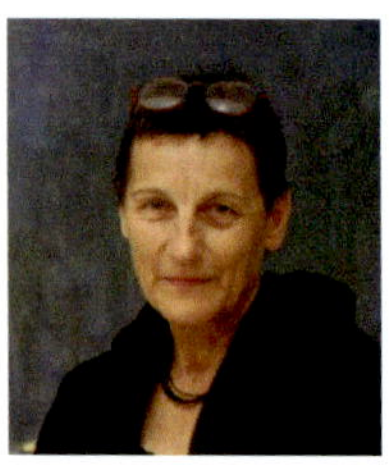

PERKO VERENA, Assistant Professor, PhD, is a writer and poet, and a specialist in Roman ceramic and heritage studies. She was formerly a Research Assistant in The City Museum of Ljubljana and at University of Ljubljana (1988-1997). She joined The Regional Museum in Kranj (from 1997) as a senior curator and Assistant Professor at the University of Ljubljana. At the University of Ljubljana, she earned her doctoral degree in the field of Roman Archaeology. She also holds a doctoral degree in museology from the University of Zagreb. Her publication Museology, Public Archaeology, Museum of Region Kras is a fundamental theoretical work in the Slovenian language for heritage studies. She has curated many exhibitions and participated in archaeological excavations. She serves as a member and heritage interpreter for many heritage projects. She is the initiator and founder of heritage doctoral studies at the University of Ljubljana. She is an active member of many international professional associations including ILA, ICOM, ICOMOS, INTERPRET EUROPE and Rei Cretariae Romanae Fautores.
Email: verena.vidrih-perko@gorenjski-muzej.si

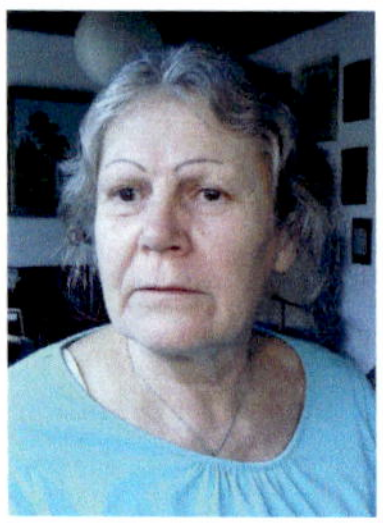

ROSENTHAL-HEGINBOTTOM RENATE is an independent scholar who earned her PhD from Hebrew University, Jerusalem in 1975, where her thesis was 'The North Church at Sobota and the Tri-apsidal Churches in the Near East'. (published as *Die Kirchen von Sobota und die Dreiapsidenkirchen des Nahen Ostens*, Wiesbaden, 1982. She has participated in numerous excavations including: 1965–1967 Mampsis, director Avraham Negev; 1968 Binyanei Ha'uma, director Michael Avi-Yonah; 1977 Tel Yoke'am, director Amnon Ben-Tor; 1980–2000 Tel Dor Expedition, director Ephraim Stern. Her research *specialties and foci* includes the material culture of the Hellenistic, Roman and Byzantine periods in the Levant, in particular ceramics, jewelry, burial customs and visual arts with the emphasis on ethnic and religious affiliation. Her publications include the finds from the Mampsis excavations, the Tel Dor Expeditions, the Jewish Quarter Excavations (1969–1982), the Roman kiln site at the Jerusalem International Convention Center and the Western Wall Plaza excavations.
Email: renate34@gmx.de

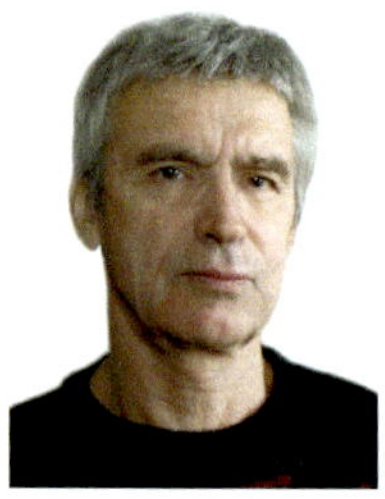

RYZHOV STANISLAV G. specializes in both the ancient and medieval archaeology of Chersonesos. After graduating from Kharkov University, he has worked in the Chersonesos Museum since 1963. Annually he conducts the excavation and restoration of the ancient and Byzantine residential quarters and temples of ancient Chersonesos. In 1972-1976 he curated the restoration of unique polychrome floor mosaics of the early Byzantine churches. He has authored numerous publications on the history and archaeology of Chersonesos and served as curator of the international project 'The early-byzantine Church architecture of Chersonesus Taurica' (2004). He has co-authored the catalogues *The Legacy of Byzantine Cherson* (2011) and *Antique painted ceramics of Tauric Chersonesos* (2018). Currently he is the head of the architectural and archaeological department of the State Museum-Preserve 'Tauric Chersonesos', and is overseeing the creation of an open-air exposition, and preparing to publish collections reflecting various aspects of the two-thousand-year history of Chersonesos.

Email: staspolis@list.ru

SEIDEL YVONNE received her PhD in 2005 from Friedrich-Schiller Universität Jena with her thesis subject 'Künstliches Licht im individuellen, familiären und öffentlichen Lebensbereich'. Her 2000 MA theses preented 'Öllampen der Sammlung Antiker Kleinkunst der Friedrich-Schiller Universität'. Since 2016 she has been working on the FWF

project (P 29 452-G25) Cult Continuity at the Summit of Hemmaberg. Previously from 2010-2012 she worked on finds of the amphitheater from Carnuntum. Since 2005 she has provided graphic work for various ephesian projects and in 2007-2008 she worked on the scientific structure of the picture database UNIDAM at the University of Vienna, Department of Classical Archaeology. In 2002 she provided both conception and design of the exhibition: Oil lamps–lights of antiquity. The History of Lighting, in the City Museum Jena. She has participated in various other excavations and exhibitions.
Email: yvseidel@gmx.net

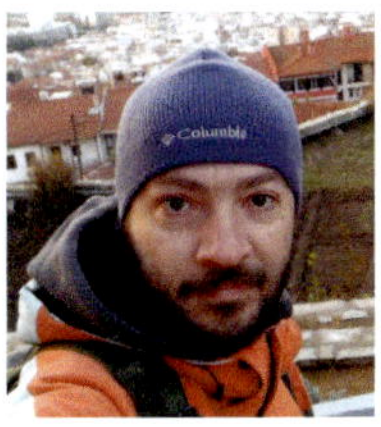

SLAMPEAS PERIKLIS is an archaeologist specialized in Late Roman, Byzantine and Ottoman pottery. He earned his undergraduate degree from Aristotle University of Thessaloniki and his Masters degree from the International Hellenic University. From 2004 to 2006 he took part in many archaeological projects, studying and recording pottery from the Roman Agora of Thessaloniki, Dion, Corinth and Crete. From 2008 to 2013 he was the in charge of the pottery recordings for the excavation of Thessaloniki's Metropolitan Railway Station. Presently he is working on his PhD in Early Ottoman Pottery from Thessaloniki.
Email: slambeas@yahoo.com

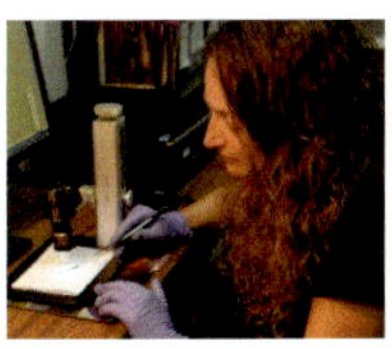

SUKENIK NAAMA is curator of organic material at the Israel Antiquities Authority (IAA) and is a specialist on organic materials including archaeological textiles, basketry, leather and dyes. She holds a PhD in Archaeology and wrote her thesis on: 'Dyes in textiles from the early Roman Period: chemical, historical and archaeological aspects.' Her MA thesis was on 'The Textile Dyes from Mo`a.' Her special area of study is the archaeology of textile production, including the investigation of textile dyes with a focus on identifying dyes with HPLC analysis. Her articles can be seen at: https://antiquities.academia.edu/naamasukenik
Email: naamasuk@gmail.com

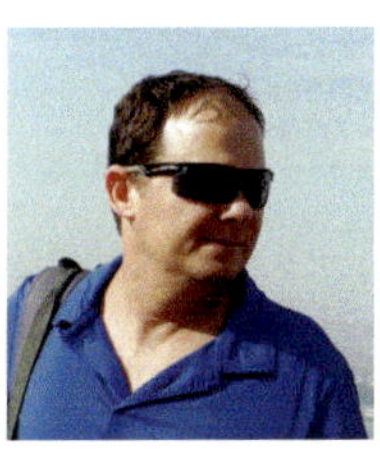

TEPPER YOTAM, a Field Archaeologist and specialist of the Hellenistic, Roman and Byzantine periods, received his Ph.D. from Tel Aviv University (Legio in the Roman period) and published books on various topics (Roman Roads; the Historical sites of Beth Shearim and Legio-Kefar 'Othnay). As a researcher and archaeologist with the Israel Antiquities Authority (IAA), he has directed numerous excavations and surveys in Israel for the last three decades. One of his most interesting excavations discovered an Early Christian Prayer Hall from the end of the Roman period in the region of a Roman legionary base. A survey and excavation conducted at the Megiddo prison compound and its environs, uncovered the daily life of civilians and soldiers. His Post-Doc at the University of Haifa focused on the causes of the collapse of the agricultural Byzantine society in the Negev, using agro-archaeology methods. These studies underlined the ancient agricultural installations and the relict field systems, which dotted the hinterland of Byzantine settlements.
Email: yotamtepper@gmail.com

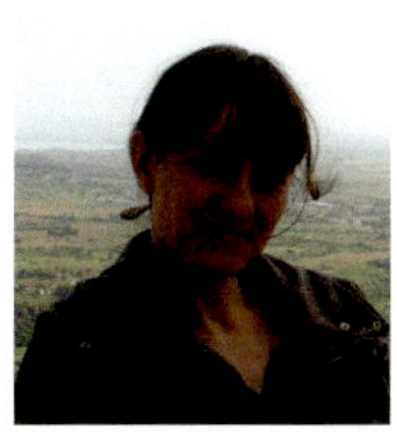

TRATNIK VESNA is presently engaged as a Research Assistant in The Archaeological Department, National Museum of Slovenia and is working on a PhD thesis focusing on Late Antiquity in Slovenia. Educated at the University of Ljubljana, she earned her undergraduate and master degrees in Roman Archaeology. She gained early field work experience as a student, working a few seasons with the Italian Archaeological Mission in Luxor, Egypt. She previously worked for the Center for Preventive Archaeology, The Institute for the Protection of Cultural Heritage of Slovenia from 2008-2014 and is the author of several articles and monographs related to Roman Archaeology and the Migration Period in Slovenia.
Email: tratnik.vesna@gmail.com

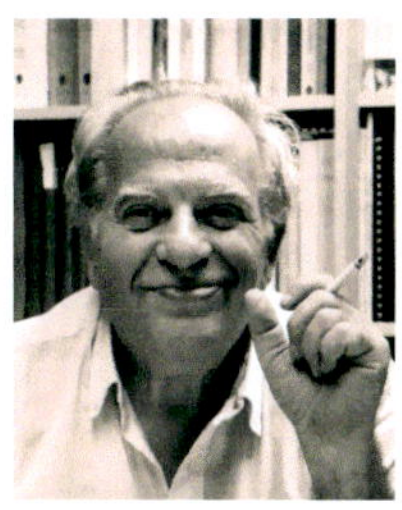

VELENIS GEORGE, emeritus professor of Byzantine Archaeology at the Aristotle University of Thessaloniki, is a specialist in Byzantine Architecture and Epigraphy. Educated at the same university that he served for more than twenty years, in the faculties of both Architecture and History and Archaeology, he earned his PhD in 1984 and a few years later he joined the department of History and Archaeology. He oversaw the restoration projects of numerous monuments and historical sites in Greece, Turkey, Cyprus and the Balkans, including the Roman forum of Thessaloniki and the Theater of Dodona. He has published numerous books, papers and studies on a variety of fields, including Byzantine painting, sculpture and even pottery, receiving national and international awards for his contributions.
Email: velenis@hist.auth.gr

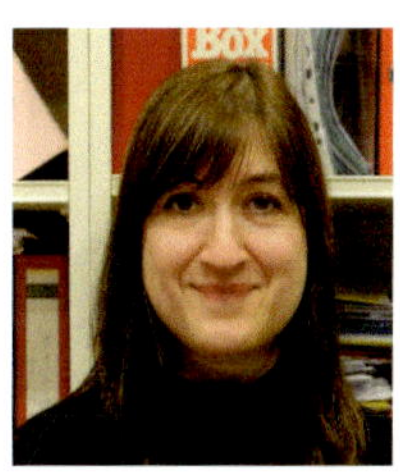

VERYKOKOU MARA joined the Byzantine and Post Byzantine collection of the Benaki Museum, Athens in 2003 as an assistant curator. She obtained her MA in History of Art from the University of Aberdeen and her MA in Museum Studies from UCL. She has been involved as a member of the research team in various exhibition projects, the most recent being *'Heaven and Earth. Art of Byzantium from Greek Collections'* (National Gallery of Washington 2013-14, J. Paul Getty Museum, Los Angeles 2014, Art Institute of Chicago 2014-15) and participated in the coordination of *'Russian Religious Art from Russia to Greece'* (Benaki Museum, Athens, 2017-18). Her current research focuses on the collection of late antique and medieval medical instruments of the Benaki Museum. She is a PhD candidate at the National and Kapodistrian University of Athens, School of History and Archaeology, on the subject: *'Exhibition policy and communication in the Benaki Museum from the 1970s to the present (1970-2015) in the context of periodic exhibitions'*. She is responsible

for the Byzantine and Post Byzantine collection of the Benaki Museum, Athens.
Email: verykokou@benaki.gr

VINCENT CATHERINE, est une historienne médiéviste française. Elle a fait ses études à Normale Sup Sèvres. Professeur des Universités elle enseigne l'Histoire du Moyen-âge à l'Université Paris-Nanterre. Elle est membre ‹ senior › de l'Institut Universitaire de France, et appartient au Centre de recherche d'histoire sociale et culturelle de l'Occident (CHISCO). Elle est l'un des principaux spécialistes des confréries religieuses et des pèlerinages chrétiens. Thèmes de recherche : Histoire religieuse et sociale des XIIIe - XVe siècles en Occident; Formes de la pratique religieuse (vie des paroisses, culte des saints, confréries, initiatives caritatives, expressions de la dévotion par les prières et les gestes); Production hagiographique en langue vernaculaire à la fin du Moyen Âge; Sanctuaires et pèlerinages au Moyen Âge. Elle a écrit: Des ouvrages de recherche: *Des charités bien ordonnées: les confréries normandes de la fin du XIIIe siècle au début du XVIe siècle*, Paris, École Normale Supérieure de Jeunes Filles, 1988. *Les confréries dans le royaume de France, XIIIe -XVe siècle*, Paris, Albin Michel, 1994. *Fiat Lux :* lumière *et luminaires dans la vie religieuse du XIIIe au XVIe siècle*, Paris, Le Cerf, 2004, ' Histoire religieuse de la France ' no. 24. Et participé à des ouvrages à caractère pédagogique: *Introduction à l'histoire de l'Occident médiéval*, Paris, Le Livre de Poche, 1995, collection 'Références', 223 p. Église *et société en Occident, XIIIe - XVe siècle*, Paris, Armand Colin, 2009, ' Collection U '. Activités de recherche : Membre du CHISCO ; Séminaire mensuel à l'ENS-Ulm : programme consultable sur la page http://www.histoire.ens.fr/Pratiques-et-cultures-religieuses,277.html ; Créatrice et directrice de l'*Inventaire des sanctuaires et lieux de pèlerinage chrétiens en France*. site: http://sanctuaires.coldev.org/
Email: catherine.vincent14@wanadoo.fr

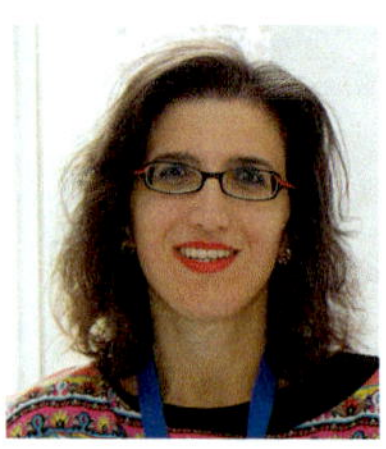

YANGAKI ANASTASIA G. is senior researcher at the Institute of Historical Research of the National Hellenic Research Foundation (IHR/NHRF). She studied archaeology (BA) at the University of Crete, Byzantine archaeology (MA and PhD) at the University Paris I – Panthéon-Sorbonne and received a special diploma on Museology (Diplôme Spécial de Muséologie) from the École du Louvre. She has also been teaching at various Universities as well as at the Hellenic Open University. She is the author of *La céramique des IVe-VIIIe siècles ap. J.-C. d'Eleutherna; Glazed Pottery from the "Hagioi Theodoroi" site in Akronauplia (11th-17th c.)*, co-editor of the books: *University of Crete, Educational Collection of Medieval and Modern Pottery. Catalogue*; *Byzantium without glamour. The 'humble' objects and their use in the everyday life of the Byzantines* and author of numerous publications. Her main research interests concern the archaeology of towns and the material culture (pottery and minor arts) of Byzantium. She also investigates the cultural contacts of Byzantium with the West. She has participated in various European research programs, excavations, surveys, NHRF research programs and museum collaborations.
Email: yangaki@eie.gr

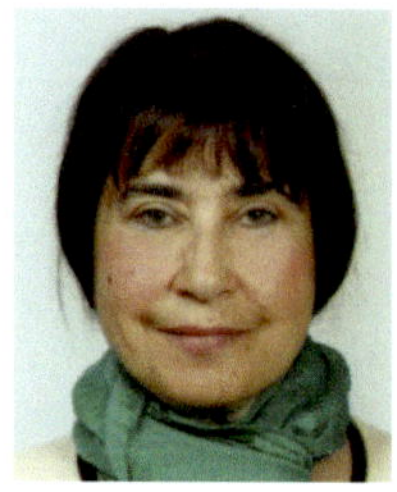

YASHAEVA TATYANA YU. specializes in Christian archaeology and Byzantine church applied art. After graduating from the Simferopol University in 1979, she works in the Museum-Preserve 'Tauric Chersonesos'. In 1979-2014 she carried out archaeological studies of the country and cave chapels, Christian churches and monasteries of Byzantine Cherson. She is the author of several publications on the Christian history of the Byzantine Cherson. She was a participant of the international project 'The early-byzantine Church architecture of Chersonesus Taurica' (2004) and scientific editor and one of the authors of the catalogue *The Legacy of Byzantine Cherson* (2011). Currently, she heads the Byzantine department of the State Museum-Preserve 'Tauric Chersonesos,' is publishing a collection of church utensils and encolpion crosses from Cherson excavations and is preparing a re-exposure of the permanent Byzantine exposition. She holds international scientific conferences on Christian archaeology and publishes scientific collections based on the conference materials.
Email: tatjus2@mail.ru

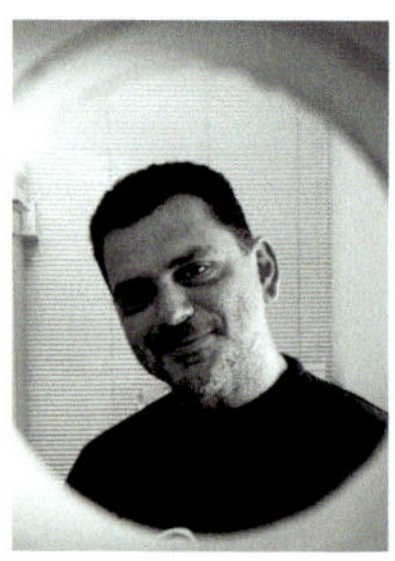

ZACHARIADIS STAVROS specializes in Late Roman pottery. He had his undergraduate and master's degrees from the Aristotle University of Thessaloniki, where he is currently a PhD candidate. He has been an associate of the Aristotle University's excavation in Philippi and worked under Professor G. Vellenis on the Lamps from the Roman Forum of Thessaloniki. Currently he works as a pottery expert for the Metropolitan Railway of Thessaloniki excavation. He is also a partner at the Aristotle University's excavation in Amyntaion, Florina and collaborates with an international team on publishing the Theatre of Philippi excavation.
Email: stzachariadis@gmail.com